ADVANCES IN CHEMICAL ENGINEERING
Volume 7

CONTRIBUTORS TO THIS VOLUME

Ralph Anderson
Robert S. Brown
Benjamin Gal-Or
Robert V. Macbeth
Knud Østergaard
J. M. Prausnitz
William Resnick
Larry J. Shannon

ADVANCES IN
CHEMICAL ENGINEERING

Edited by

THOMAS B. DREW

*Department of Chemical Engineering
Massachusetts Institute of Technology
Cambridge, Massachusetts*

GILES R. COKELET

*Division of Chemistry and Chemical Engineering
California Institute of Technology
Pasadena, California*

JOHN W. HOOPES, JR.

*Atlas Chemical Industries, Inc.
Wilmington, Delaware*

THEODORE VERMEULEN

*Department of Chemical Engineering
University of California
Berkeley, California*

Volume 7

Academic Press · New York · London 1968

Copyright © 1968, by Academic Press Inc.
ALL RIGHTS RESERVED.
NO PART OF THIS BOOK MAY BE REPRODUCED IN ANY FORM,
BY PHOTOSTAT, MICROFILM, OR ANY OTHER MEANS, WITHOUT
WRITTEN PERMISSION FROM THE PUBLISHERS.

ACADEMIC PRESS INC.
111 Fifth Avenue, New York, New York 10003

United Kingdom Edition published by
ACADEMIC PRESS INC. (LONDON) LTD.
Berkeley Square House, London, W.1

Library of Congress Catalog Card Number: 56–6600

PRINTED IN THE UNITED STATES OF AMERICA

CONTRIBUTORS TO VOLUME 7

Numbers in parentheses indicate the pages on which the authors' contributions begin.

RALPH ANDERSON,* *United Technology Center, United Aircraft Corporation, Sunnyvale, California* (1)

ROBERT S. BROWN, *United Technology Center, United Aircraft Corporation, Sunnyvale, California* (1)

BENJAMIN GAL-OR, *Department of Aeronautical Engineering, Technion-Israel Institute of Technology, Haifa, Israel; also at Department of Chemical Engineering, University of Pittsburgh, Pittsburgh, Pennsylvania* (295)

ROBERT V. MACBETH, *United Kingdom Atomic Energy Authority, Atomic Energy Establishment, Winfrith, England* (207)

KNUD ØSTERGAARD, *Department of Chemical Engineering, The Technical University of Denmark, Copenhagen, Denmark* (71)

J. M. PRAUSNITZ, *Department of Chemical Engineering, University of California, Berkeley, California* (139)

WILLIAM RESNICK, *Department of Chemical Engineering, Technion-Israel Institute of Technology, Haifa, Israel* (295)

LARRY J. SHANNON, *United Technology Center, United Aircraft Corporation, Sunnyvale, California* (1)

---

* *Present address:* CETEC Corporation, Mountain View, California.

CONTENTS

LIST OF CONTRIBUTORS v
CONTENTS OF PREVIOUS VOLUMES ix

Ignition and Combustion of Solid Rocket Propellants
ROBERT S. BROWN, RALPH ANDERSON, AND LARRY J. SHANNON

I. Introduction. 1
II. Ignition. 6
III. Steady-State Combustion 29
IV. Combustion Instability 52
V. Combustion Termination 57
    Nomenclature 65
    References 66

Gas-Liquid-Particle Operations in Chemical Reaction Engineering
KNUD ØSTERGAARD

I. Introduction. 71
II. Gas-Liquid-Particle Processes 73
III. Gas-Liquid-Particle Operations 79
IV. Theoretical Models of Gas-Liquid-Particle Operations. 81
V. Transport Phenomena in Gas-Liquid-Particle Operations 90
VI. Summary and Conclusions 130
    Nomenclature 131
    References 133

Thermodynamics of Fluid-Phase Equilibria at High Pressures
J. M. PRAUSNITZ

I. Introduction. 140
II. Fugacities in Gas Mixtures: Fugacity Coefficients 144
III. Fugacities in Liquid Mixtures: Activity Coefficients. 154
IV. Effect of Pressure on Activity Coefficients: Partial Molar Volumes. . . 160
V. Dilute Solutions of Gases in Liquids at High Pressures 166
VI. Concentrated Solutions: High-Pressure Vapor-Liquid Equilibria . . . 170
VII. Liquid-Liquid Equilibria in Binary Systems 184
VIII. Gas-Gas Equilibria in Binary Systems 190
IX. Liquid-Liquid Equilibria in Ternary Systems Containing One Supercritical Component. 194
    References 203

The Burn-Out Phenomenon in Forced-Convection Boiling

Robert V. Macbeth

I.	Introduction.	208
II.	The Meaning of Burn-Out	210
III.	The Important System-Describing Parameters	225
IV.	The Linear Relation between Burn-Out Flux and Inlet Subcooling	235
V.	Possible Contractions among the System-Describing Parameters.	238
VI.	The Local-Conditions Concept of Burn-Out	241
VII.	The Low-Velocity Burn-Out Regime	246
VIII.	The Correlation of Water Data for Uniformly Heated Channels	249
IX.	The Equivalent-Diameter Hypothesis	273
X.	The Effect on Burn-Out of Nonuniform Heating	274
XI.	The Technique of Modeling Forced-Convection Burn-Out.	280
	Nomenclature	286
	References	287

Gas-Liquid Dispersions

William Resnick and Benjamin Gal-Or

I.	Introduction.	296
II.	Experimental Studies	300
III.	Effect of Surface-Active Agents	327
IV.	Mathematical Models.	333
	Nomenclature	388
	References	390

Author Index	397
Subject Index	409

CONTENTS OF PREVIOUS VOLUMES

Volume 1

Boiling of Liquids
  J. W. Westwater

Non-Newtonian Technology: Fluid Mechanics, Mixing, and Heat Transfer
  A. B. Metzner

Theory of Diffusion
  R. Byron Bird

Turbulence in Thermal and Material Transport
  J. B. Opfell and B. H. Sage

Mechanically Aided Liquid Extraction
  Robert E. Treybal

The Automatic Computer in the Control and Planning of Manufacturing Operations
  Robert W. Schrage

Ionizing Radiation Applied to Chemical Processes and to Food and Drug Processing
  Ernest J. Henley and Nathaniel F. Barr

AUTHOR INDEX—SUBJECT INDEX

Volume 2

Boiling of Liquids
  J. W. Westwater

Automatic Process Control
  Ernest F. Johnson

Treatment and Disposal of Wastes in Nuclear Chemical Technology
  Bernard Manowitz

High Vacuum Technology
  George A. Sofer and Harold C. Weingartner

Separation by Adsorption Methods
Theodore Vermeulen

Mixing of Solids
Sherman S. Weidenbaum

AUTHOR INDEX—SUBJECT INDEX

Volume 3

Crystallization from Solution
C. S. Grove, Jr., Robert V. Jelinek, and Herbert M. Schoen

High Temperature Technology
F. Alan Ferguson and Russell C. Phillips

Mixing and Agitation
Daniel Hyman

Design of Packed Catalytic Reactors
John Beek

Optimization Methods
Douglass J. Wilde

AUTHOR INDEX—SUBJECT INDEX

Volume 4

Mass-Transfer and Interfacial Phenomena
J. T. Davies

Drop Phenomena Affecting Liquid Extraction
R. C. Kintner

Patterns of Flow in Chemical Process Vessels
Octave Levenspiel and Kenneth B. Bischoff

Properties of Cocurrent Gas-Liquid Flow
Donald S. Scott

A General Program for Computing Multistage Vapor-Liquid Processes
D. N. Hanson and G. F. Somerville

AUTHOR INDEX—SUBJECT INDEX

Volume 5

Flame Processes—Theoretical and Experimental
  J. F. Wehner

Bifunctional Catalysts
  J. H. Sinfelt

Heat Conduction or Diffusion with Change of Phase
  S. G. Bankoff

The Flow of Liquids in Thin Films
  George D. Fulford

Segregation in Liquid-Liquid Dispersions and Its Effect on Chemical Reactions
  K. Rietema

AUTHOR INDEX—SUBJECT INDEX

Volume 6

Diffusion-Controlled Bubble Growth
  S. G. Bankoff

Evaporative Convection
  John C. Berg, Andreas Acrivos, and Michel Boudart

Dynamics of Microbial Cell Populations
  H. M. Tsuchiya, A. G. Fredrickson, and R. Aris

Direct Contact Heat Transfer between Immiscible Liquids
  Samuel Sideman

Hydrodynamic Resistance of Particles at Small Reynolds Numbers
  Howard Brenner

AUTHOR INDEX—SUBJECT INDEX

ADVANCES IN CHEMICAL ENGINEERING
Volume 7

IGNITION AND COMBUSTION OF SOLID ROCKET PROPELLANTS

Robert S. Brown, Ralph Anderson,* and Larry J. Shannon

United Technology Center, United Aircraft Corporation
Sunnyvale, California

I. Introduction	1
A. Types of Solid Propellants	2
B. Solid-Propellant Rocket Motors	3
II. Ignition	6
A. Propellant-Ignition Phenomena	8
B. Igniter Design	21
C. Flame Propagation	24
D. Chamber Filling	29
E. Ignition-Pressure Transients	29
III. Steady-State Combustion	29
A. Combustion of Double-Base Propellants	31
B. Combustion of Composite Propellants	35
C. Erosive Burning	50
IV. Combustion Instability	52
A. Pressure-Coupled Instability	52
B. Velocity-Coupled Instability	55
C. Bulk-Coupled Instability	56
V. Combustion Termination	57
A. Rapid-Depressurization Termination	58
B. L^* Termination	62
C. Fluid-Injection Termination	63
Nomenclature	65
References	66

I. Introduction

The utilization of combustible solid materials in propulsion devices is by no means a modern development. There is evidence that the first rockets were used in India around 2000 B.C., and the Chinese are known to have experimented with pelletized gun powder in rockets during the Tang dynasty

*Present address: CETEC Corporation, Mountain View, California.

(618 to 907 A.D.). Solid-propellant rockets were used in warfare around 1232 A.D., and since then have received considerable attention for this application. The British, for example, used artillery rockets in the early 19th century with notable success. Modern interest in rockets has developed from the need for intercontinental defense systems, improved tactical weapons, and man's ever-present desire to explore space.

For chemical engineers, the application of solid propellants to rocket motor applications represents one specific example of using for practical gain a change in the chemical state of matter taking place under controlled conditions. Other examples include the combustion of coal and wood for a wide variety of applications, the ablation of material for heat shields, the use of packed catalytic reactors and of fluidized-bed reactors, and the explosive combustion of dust particles. On a fundamental basis, all these processes can be characterized by the equations describing the simultaneous transport of mass and energy in a reactive environment. The coefficients and boundary conditions determine the particular configuration and account for the wide variations in characteristic temperatures, pressures, and velocities of the various processes.

The one unique characteristic of solid rocket propellants which distinguishes them from all other solid fuels is the incorporation of sufficient reducing and oxidizing components within the solid phase to effect the transformation to gaseous products. Although the combustion process still can be analyzed within the broad framework of simultaneous mass and energy transport, this feature considerably complicates the analysis. Hence, the study of propellant combustion has practically been a separate field of research, and one in which chemical engineers, though qualified by background and training, have not been particularly active.

A. Types of Solid Propellants

Solid propellants may, depending upon their composition and physical structure, be divided into three general classes—double-base, composite, and composite modified-double-base propellants. In double-base propellants, the fuel and oxidizing components are parts of the same molecule and have as their principle components nitrocellulose and explosive plasticizer, usually nitroglycerin. Other materials may be added in smaller proportion to serve as stabilizers, nonexplosive plasticizers, coolants, lubricants, opacifiers, and burning-rate modifiers, or to confer desirable properties on the product. Double-base rocket-propellant compositions derive largely from ballistites and cordites, which have been used as gun propellants for many years. The principal innovations in adapting them for rocket applications have been in the techniques of fabricating the large grains needed for this purpose.

Composite propellants are made by embedding a finely divided solid oxidizing agent in a plastic, resinous, or elastomeric matrix. The matrix material usually provides the fuel for the combustion reaction, although solid reducing agents are sometimes included. Composite-type propellants have been made in a great variety of compositions. Oxidizing agents which have been used include ammonium nitrate, sodium nitrate, potassium nitrate, ammonium perchlorate, and potassium perchlorate. Asphalt, natural and synthetic rubbers, aldehyde, urea and phenolic resins, vinyl polymers, polyesters, and nitrocellulose are among the matrix materials which have been employed as binders. The common characteristic of these varied compositions is a markedly heterogeneous structure with adjacent regions of oxygen-rich and oxygen-deficient materials. In this respect, they may be considered to be related to black powder, although their physical form derives from modern plastics technology.

A third type of propellant, the composite modified-double-base propellant, represents a combination of the other two types. These propellants are made from mixtures of nitroglycerine and nitrocellulose or similar materials, but with crystalline oxidizers such as ammonium perchlorate also included in the matrix.

B. Solid-Propellant Rocket Motors

In a solid-propellant rocket motor, the propellant is contained within the wall of the combustion chamber, as shown in Fig. 1. This contrasts with liquid systems, where both the fuel and oxidizing components are stored in tanks external to the combustion chamber and are pumped or pressure-fed to the combustor. In hybrid systems, one component, usually the fuel, is contained in the combustion chamber, while the other component is fed to the chamber from a separate storage tank, as in liquid systems. The solid-propellant motor also has an ignition system located at one end to initiate operation of the rocket. The supersonic nozzle affects the conversion of

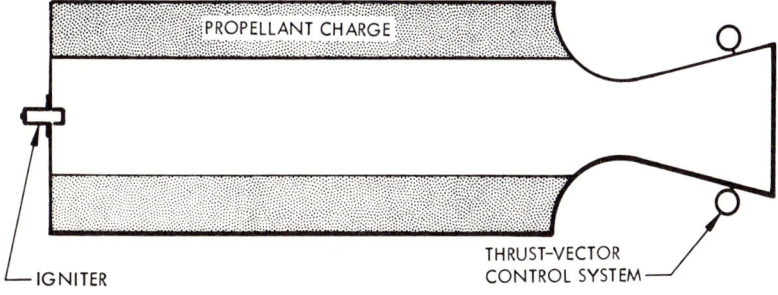

Fig. 1. Typical solid-propellant rocket motor configuration.

thermal energy developed by the propellant combustion process to kinetic energy. In addition to these components, the rocket may contain a nozzle closure to maintain specified pressure prior to ignition, aerodynamic or thrust-vector control devices, and termination units for neutralizing or reversing the thrust of the rocket.

By accelerating the gaseous combustion products through the exhaust nozzle, a thrust is imparted to the nozzle and motor case. This thrust is determined by the time rate-of-change of the total momentum of the bounded fluid, as indicated by the expression

$$F = \frac{\lambda \dot{m}}{g_c} V_e + (P_e - P_a) A_t \tag{1}$$

where F is the thrust, A_t the area of the nozzle throat, g_c the gravity constant, $\dot{m}$ the mass flow rate, P_a the absolute pressure of the atmosphere, P_e the absolute pressure at the nozzle exit plane, V_e the exhaust velocity, and λ a factor to account for the expansion angle of the nozzle.

Assuming isentropic expansion of the combustion gases through the nozzle and $P_e = P_a$, the exhaust velocity can be determined from the equation

$$\frac{F}{\dot{m}} = \frac{V_e}{g_c} = \left[\left(\frac{2\gamma}{\gamma - 1} \right) \frac{RT_f}{m_w} \left(1 - \frac{T_e}{T_f} \right) \right]^{1/2} = I_{sp} \tag{2}$$

where I_{sp} is the specific impulse of the propellant, T_e the absolute temperature at the nozzle exit plane, T_f the absolute adiabatic flame temperature of the propellant, γ the ratio of specific heats, m_w the molecular weight, and R the gas constant. Thus,

$$F = \dot{m} I_{sp} \tag{3}$$

The ratio of rocket thrust to propellant mass flow, commonly called the specific impulse (I_{sp}) of the propellant, represents a measure of the force developed per unit mass flow of propellant. From Eq. (2), it is apparent that high propellant-flame temperatures and low molecular-weight combustion products are required to produce high I_{sp}.

Another consideration in the design of a rocket motor is the boost velocity, or velocity which the vehicle will attain when all the propellant has been consumed. Neglecting drag losses, this velocity becomes

$$V_b = I_{sp} g_c \ln \left(\frac{1}{1 - f_m} \right) \tag{4}$$

where f_m is the ratio of the mass of the propellant to the total mass of the motor, and V_b is the boost velocity.

Equations (3) and (4) provide a basis for comparing solid propellants to liquid- and hybrid-propellants. In comparison to specific impulses of 290 sec^{-1} and greater for liquids and hybrids, solids have a typical range from 240 to 260 sec^{-1}. However, Eq. (4) shows that the mass fraction of the motor (ratio of propellant weight to total motor weight) also has a significant influence on motor performance. Because of their inherent simplicity, typical solid-propellant rocket motors have mass fractions around 0.90, compared to 0.80 for liquids and hybrids. Thus, solid-propellant motors can generally produce a higher boost velocity than comparable liquid or hybrid systems. In addition, solids have a significantly higher density than liquids (110 lb/ft^3 compared to 75 lb/ft^3), and this permits the use of more propellant in a given volume, an important consideration in systems which operate from a fixed-volume launch complex, such as silos (Minuteman) or submarine launch tubes (Polaris). Solids are also particularly attractive where simplicity, instantaneous operation capability, and reliability are extremely important, and where the safety considerations of tactical weapons are important.

In addition to the energy requirements of solid propellants, Eq. (3) shows that consideration must be given to the mass-flow rate of the combustion products through the nozzle. Because all solids burn on the exposed surface, the mass flow of propellant combustion products is given by the equation

$$\dot{m} = \rho_p \dot{r} A_B \quad (5)$$

where A_B is the area of burning propellant surface, $\dot{r}$ the linear propellant burning rate, and ρ_p the propellant density. Because the flow through the nozzle is always supersonic, the flow capacity of the nozzle can be written

$$\dot{m} = \frac{\Gamma P_c A_t g_c}{(RT_f/m_w)^{1/2}} = P_c A_t g_c / c^* \quad (5a)$$

where c^* is the characteristic velocity, P_c the absolute pressure in the combustion chamber, and Γ a function of the specific heat ratio.

The burning-rate studies on many propellants show that most burning-rate data can be correlated by the empirical relation

$$\dot{r} = a P_c^n \quad (5b)$$

where a is the burning-rate factor, and n is the pressure exponent of the propellant burning rate.

These relations can be combined to provide a means of calculating the equilibrium chamber pressure:

$$P_c^{1-n} = \frac{\rho_p a c^* A_B}{g_c A_t} \quad (5c)$$

Another consideration is the time duration of the thrust, or "burn" time of the motor. The burn time required is determined by the particular mission, and when combined with the linear burning rate, determines the thickness of the propellant charge in the direction of burning. The grain or web thickness, the nozzle diameter, the required burning rate, and vehicle-drag considerations all combine to determine the particular motor diameter and length.

The designer must also be concerned with methods for initiating the combustion process within the motor. The design of the igniter represents an extremely important factor in the reliability of the rocket motor, a consideration which has presented some difficulties over the years. The operational characteristics of the igniter also determine the rate, or "ignition shock," at which the thrust achieves the steady-state design thrust. For manned flights, this rate cannot exceed certain tolerances. On the other hand, anti-missile systems require that full thrust be achieved as rapidly as possible.

In addition, a number of other propellant combustion characteristics must be considered. For example, when the flow of combustion gases parallel to the propellant surface becomes excessive, the linear propellant burning-rate becomes dependent upon the mass velocity through the propellant port. Erosive burning, as this phenomenon is called, can change the mass flow through the nozzle, the thrust, and the ballistics of the vehicle's flight path. The erosive-burning characteristics of propellants set the minimum diameter for the initial propellant port and the resulting motor diameter.

From these many considerations, the propellant formulator is often faced with preparing a propellant with rather precise I_{sp} and burning-rate characteristics. As a result, there is considerable interest in understanding the basic propellant combustion and ignition processes and in developing the capability to prepare from theory propellants with the desired burning-rate and I_{sp} characteristics.

Over the past 15 years, many studies have been conducted to determine the basic physical and chemical mechanisms of combustion and ignition. Although the studies have served to illustrate many of the overall characteristics of these processes, a quantitative characterization of ignition and of combustion has yet to be accomplished. The purpose of this article is to review the studies conducted to date, to indicate the significant accomplishments, and to show the areas which are not understood and therefore need further study. The discussion of ignition is presented first, since it is the initial stage of the combustion process.

II. Ignition

The most important consideration in the design of an operational ignition system is the time interval from the initiation of the igniter to the generation of the design thrust level in the motor. As indicated previously, the mission requirements often set the limits within which the motor-ignition cycle must operate.

The time necessary to achieve complete ignition of a solid-propellant grain and bring the motor up to operating pressure can be divided into three phases, as shown in Fig. 2 (B14). During the first phase of the ignition process, the propellant is subjected to the action of either a rocket-exhaust, a pyrotechnic, or a hypergolic igniter alone. When critical propellant-ignition conditions are reached (i.e., autoignition temperature), the propellant surface at that point ignites, initiating combustion. When this occurs, the first phase of the motor-ignition process is complete. The time delay of this phase is a function of the ignition requirements of the propellant and of the operational characteristics of the igniter system.

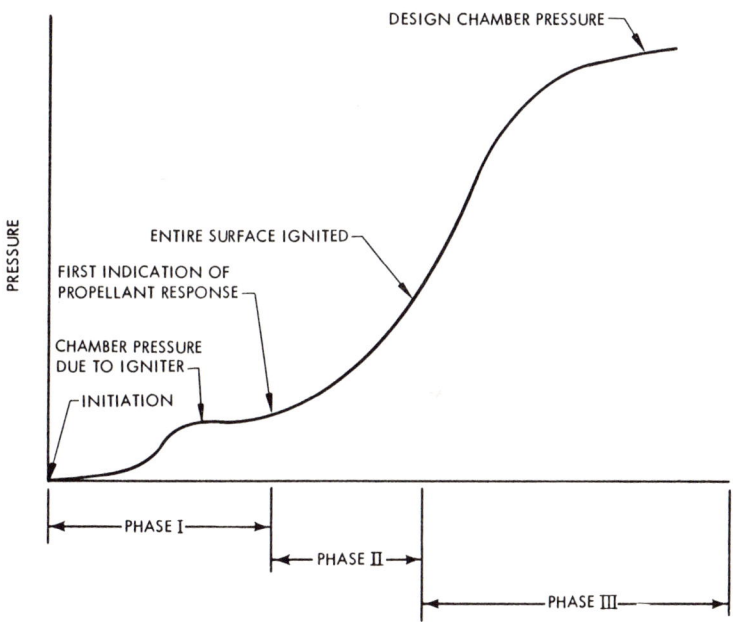

FIG. 2. Diagram of chamber-pressure transient during ignition (B14).

In the second phase, the heating of the unignited portion of the propellant surface continues, promoting ignition of the remaining propellant surface. This period is characterized by the propagation of the flame front across the unignited areas of the propellant surface. The time required to propagate the flame to the entire grain surface is determined by the heat-transfer rate from the mixture of igniter exhaust products and hot propellant-combustion products flowing across the unignited propellant surface.

During this period, the mass discharge rate through the motor nozzle increases as the flame spreads over the propellant, causing an increase in the chamber pressure (as shown in Fig. 2) which is described by

$$\frac{dP_c}{dt} = \frac{\Gamma^2 c^{*2}}{g_c V_c}\left[\rho_p a P_c^n A_B(t) + \dot{m}_i - \frac{A_t g_c P_c}{c^*}\right] \quad (6)$$

where $\dot{m}_i$ is the mass-flow rate of igniter,

$$\Gamma = \gamma^{1/2}\left(\frac{2}{\gamma+1}\right)^{[(\gamma+1)/2(\gamma-1)]}$$

and t is the time. Initiation of combustion at all points on the propellant surface completes the second phase of the ignition process.

During the third phase, the motor chamber continues to fill with propellant combustion products until the steady-state pressure has been reached. The chamber-pressure transient during this phase is described by

$$\frac{dP_c}{dt} = \frac{\Gamma^2 c^{*2}}{g_c V_c}\left[\rho_p a P_c^n A_B + \dot{m}_i - \frac{A_t g_c P_c}{c^*}\right] \quad (7)$$

Equations (6) and (7), coupled with the analysis of the processes occurring in phase I, provide insight into the parameters which control the overall motor-ignition process. In phase I, the most important consideration is the time required from initiation of the igniter to the attainment of propellant combustion at some point on the propellant surface, i.e., the propellant ignition-delay. During the second phase, the most important consideration is the burning area as a function of time, i.e., $A_B(t)$, which is dependent upon the igniter mass-flow, the ballistic design parameters of the motor, and the propellant burning-rate. These same parameters also characterize the third phase of the motor-ignition process. The sections that follow present a discussion of some of important processes which govern the selection of values for these parameters.

A. Propellant-Ignition Phenomena

The most important consideration in each of these phases, and hence in the design of an ignition system for a solid-propellant rocket motor, is to determine the conditions the igniter must generate in the combustion chamber to achieve ignition of the grain. In other words, the critical conditions required by the propellant to sustain the combustion process must be defined. Then, the means of generating these conditions can be determined and the optimum igniter designed.

Experimental studies have shown that the response of propellants, and hence their ignitability characteristics, depend upon the type of propellant, the physical nature of the propellant surface, the heat flux from the igniter, the pressure that the propellant "sees" during the ignition interval, and the initial propellant temperature. Complete understanding of these experimental facts has not yet been achieved. However, some insight in them has been developed, and the following sections discuss the main theoretical models that have been proposed to describe solid-propellant ignition processes.

1. Thermodynamic Approach to Propellant Ignition

Historically, early investigators speculated that once the solid propellant received a critical quantity of energy from the igniter, combustion reactions would be initiated and ignition thereby achieved. This led to a consideration of the total energy release by the igniter, and resulted in an igniter design based on the weight of igniter material. Motor-test experiments (A2) suggested that there was some value in this approach, but scaling on the basis of igniter weight did not always prove reliable.

2. Kinetic Approach to Propellant Ignition

In the search for a better approach, investigators realized that the ignition of a combustible material requires the initiation of exothermic chemical reactions such that the rate of heat generation exceeds the rate of energy loss from the ignition reaction zone. Once this condition is achieved, the reaction rates will continue to accelerate because of the exponential dependence of reaction rate on temperature. The basic problem is then one of critical reaction rates which are determined by local reactant concentrations and local temperatures. This approach is essentially an outgrowth of the bulk thermal-explosion theory reported by Frank-Kamenetskii (F2).

a. Thermal-Ignition Theory. The first attempts to apply the kinetic approach to propellant ignition were directed towards the behavior of double-base formulations. The principal ingredients in these formulations were known to undergo exothermic decomposition in the solid phase. Hence, it was assumed that ignition occurred when the solid temperature was raised to the level where these reactions reached runaway heating conditions. According to this approach, the critical rate-controlling mechanism leading to ignition is the initiation of rapid exothermic solid-phase chemical reactions at or near the solid-propellant surface.

Hicks (H6) and Frazer and Hicks (F3) considered the ignition model in which exothermic, exponentially temperature-dependent reactions occur within the solid phase. Assuming a uniformly mixed solid phase, the one-dimensional unsteady heat-flow equation relates the propellant temperature, depth from the surface, and time by the nonlinear equation:

$$\frac{\partial T}{\partial t} = \frac{\alpha\,\partial^2 T}{\partial x^2} + BQ \exp(-E/RT) \tag{8}$$

where B is the preexponential factor, E the activation energy, Q the heat of reaction, R the gas constant, T the propellant temperature, x the distance from the propellant surface, and α the solid thermal diffusivity.

Hicks showed that the time required to achieve ignition depends on the manner in which the external heat flux is applied. If the propellant surface is heated continuously, the surface temperature will continue to rise until runaway reaction conditions are reached [curve (a) of Fig. 3]. If the heat flux is terminated just before runaway reaction conditions are achieved, then a sudden drop in surface temperature can occur, followed by a long time-delay before the surface temperature again begins to rise [as shown in curve (b)]. If the flux is removed too soon, the temperature will drop continuously and ignition will not be achieved [as shown in curve (c)].

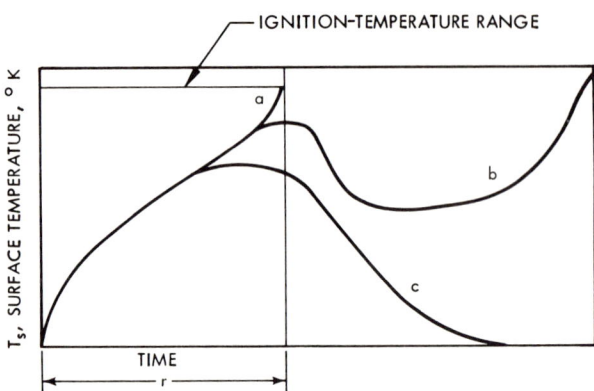

FIG. 3. Surface-temperature history predicted by thermal-ignition theory (P8).

Under these conditions, calculations showed that the time interval between the initiation of heating and the attainment of runaway reaction conditions can be roughly divided into two intervals. During the first interval, the propellant does not generate any "self-heating" from exothermic chemical reactions, the only source of energy being the external source. In the second interval, propellant reactions contribute significantly to the total heating rate of the propellant. The time required to generate runaway reactions [i.e., τ for curve (a) in Fig. 3] thus depends on the rate of external heating and the rate of propellant self-heating.

The calculations indicated that no criterion except runaway heat generation applies universally. However, under a wide range of conditions, one useful approximation can be developed from the results. When the time interval for external heating is long compared to the time during which propellant reactions contribute, then runaway reactions occur at a unique value of the surface temperature regardless of the heat-transfer rate to the solid surface. This approximation represents a very useful simplification. If ignition can be defined in terms of a critical surface temperature, the time required to achieve ignition can be calculated from the analysis of transient heat conduction in an inert medium. For example, if the heating rate is constant over the entire heating period, the surface temperature can be calculated from the equation

$$(T_s - T_0) = \frac{2\dot{Q}}{k}\left(\frac{\alpha t}{\pi}\right)^{1/2} \tag{8a}$$

where k is the thermal conductivity of the propellant, T_s the propellant surface temperature, $\dot{Q}$ is the heating rate, and T_0 the initial propellant temperature. The time required to achieve ignition can therefore be calculated by a simple rearrangement:

$$\tau^{1/2} = \frac{k}{2\dot{Q}}(T_{ai} - T_0)\left(\frac{\pi}{\alpha}\right)^{1/2} \tag{8b}$$

where T_{ai} is the propellant-surface ignition temperature and τ is the ignition-delay time. The form of Eq. (8b) suggests that a plot of $\log(\tau)^{1/2}$ versus $\log \dot{Q}$ should give a slope of -1 if the ignition-temperature concept is valid. In addition, several forms of Eq. (8b) can be used to correlate experimental data. A logarithmic plot in the form $\tau^{1/2}$ versus $\dot{Q}$ tends to minimize experimental scatter in the ignition delay. A plot of τ versus $\dot{Q}^2$ on log coordinates provides a better picture of the experimental scatter in the measured ignition times, even though it magnifies the errors in the flux measurements. A third form, resulting from a plot of $\tau\dot{Q}$ versus $\dot{Q}$, presents a clearer representation of the total energy requirements, although the times involved are not readily apparent.

Baer and Ryan (B1) have extended the calculations of Hicks and Frazer by analyzing the effect of the activation energy on the time required to achieve runaway reaction conditions. Their calculations demonstrated that the slope on a plot of $\log(\tau)^{1/2}$ versus $\log \dot{Q}$ is related to the activation energy of the propellant-heating reaction by the expression

$$S = 4.2(RT_0/E) - 1 \tag{9}$$

The studies of Hicks and Baer were based on a constant heat flux to the propellant surface. Price (P8) considers the more important complex cases which have been analyzed, and the reader is referred to his work for the details.

Altman and Nichols (A4) were the first to test the constant ignition-temperature approach experimentally. These investigators ignited samples of double-base propellants with electrically heated wires located on the propellant surface. The ignition delay was measured from the time of application of the heat to the emission of the first visible light. The heat flux was varied by changing the voltage. Their experimental results, which are shown in Fig. 4, agree with the predicted ignition delays using a constant autoignition temperature.

Price (P9) has also investigated the ignition characteristics of JPN, a double-base propellant, in an arc-imaging furnace. Price's data show agreement with the predictions of Eq. (8b) for heat fluxes below 1.5 cal/cm^2-sec. Above this flux level, the data deviated from the theoretical predictions.

Although the thermal-ignition theory was developed for double-base propellants, several investigators have attempted to correlate the ignition characteristics of composite propellants using this approach. Baer and Ryan (B1) have correlated ignition data for a polysulfide–ammonium perchlorate

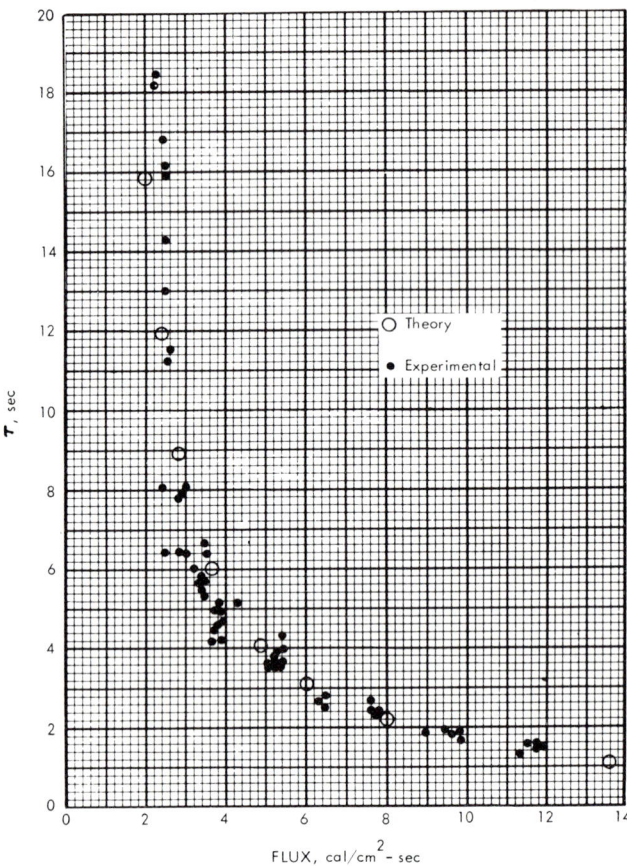

FIG. 4. Comparison of theory with experiment for Hot-wire ignition of composite solid propellants (A4).

propellant at fluxes below 10 cal/cm^2-sec, and find the activation energy of the solid-phase reaction from Eq. (9) to be 28,000 cal/gm-mole-°K.

Beyer (B7, B8) has reported data for several composite-propellant formulations, and finds similar agreement with the thermal-ignition theory at low fluxes. At high heating fluxes, however, he observed a significant decrease in ignition delay with increasing pressure of the gas environment surrounding the propellant test sample. Baer (B2) has also measured ignition-delay times under convective heating conditions; he was able to correlate the data using the thermal-ignition theory, but also observed a significant effect of environmental pressure upon ignition delay.

b. Gas-Phase Ignition Theory. Although the thermal-ignition theory appears to correlate some of the observed ignition phenomena, it is apparent under detailed experimental examination that several serious deficiencies exist. First, there is no provision in the thermal-ignition model to account for the effect of external pressure on ignition delay. Secondly, the thermal-ignition theory is based on exothermic reactions in the solid phase. This type of reactive mechanism is reasonable in double-base propellants, but early investigators could not justify similar reactions in composite propellants. A third objection to the thermal theory, when applied to composite propellants, is that there is no provision to account for the heterogenous structure of the propellant, and thus no way to account for the mixing of the reactants that must occur.

These objections lead McAlevy and Summerfield (M3, M4) to postulate that the chemical reactions between propellant constituents which result in runaway heating conditions occur in the gas phase at some small but finite distance from the solid surface. They reasoned that gas-phase reactions would respond to changes in pressure, and hence could account for the observed pressure dependence of the ignition delay. According to this mechanism, the igniter raises the propellant surface temperature to the point where the solid fuel and oxidizer start to decompose. The resulting pyrolysis products then diffuse away from the solid surface, intermix, and react by a second-order gas-phase reaction, with the heat liberated being transferred back to the solid surface. Ignition is assumed to occur when the rate of the gas-phase reaction reaches runaway conditions. At this point, the heat transfer back to the solid surface is assumed to be sufficient to sustain the pyrolysis processes.

To test this hypothesis, McAlevy and Summerfield reasoned that the propellant-ignition reactions should be strongly dependent on the chemical composition of the gaseous environment in contact with the solid surface. If this environment contained oxidizing species, such as oxygen, then the ignition delay should be significantly shorter than the ignition delay in an inert environment.

McAlevy (M3, M4) developed an approximate mathematical analysis for the ignition of a pyrolyzing fuel in an oxidizing environment as shown in Fig. 5. Hermance (H4, H5) expanded the analysis by incorporating the contributions which were previously omitted. However, the effects of natural

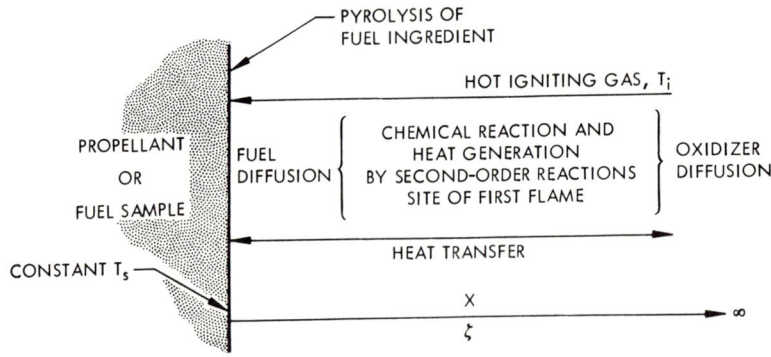

FIG. 5. Gas-phase ignition of propellant samples in shock tube (P8).

convection resulting from the large temperature gradients in the gas phase have been omitted in both these studies. The mathematical model can then be written

Mass:

$$(\partial C_f/\partial t) = D(\partial^2 C_f/\partial x^2) - C_f C_o Z \exp(-E/RT_g) \quad (10)$$

$$(\partial C_o/\partial t) = D(\partial^2 C_o/\partial x^2) - n' C_f C_o Z \exp(-E/RT_g) \quad (11)$$

Energy:

$$(\partial T_g/\partial t) = (\partial^2 T_g/\partial x^2) + (Q/\rho c) C_f C_o Z \exp(-E/RT_g) \quad (12)$$

with the boundary and initial conditions

For $t \leq 0$: $T_g = T_i$; $C_f = 0$; $C_o = C_{oi}$ at $x = 0$

For $t > 0$: $T_g = T_s$; $f(C_f) = \text{const}$; $D(\partial C_o/\partial x) = 0$ at $x = 0$

$T_g = T_i$; $C_f = 0$; $C_o = C_{oi}$ as $x \to \infty$ (13)

where n' in Eq. (12) is the stoichiometric ratio in the gas-phase chemical reaction as represented by the reaction

$$\text{Fuel} + n'(\text{oxidizer}) = \text{products} \quad (14)$$

and C_f is the concentration of fuel species, C_o the concentration of oxidizing species, D the diffusivity, C_{oi} the initial concentration of oxidizing species, C_o' the concentration of oxidizing species at the propellant surface, T_g the gas temperature, T_i the initial gas temperature, Z the Arrhenius preexponential factor, c the specific heat, and ρ the gas density.

Two limiting cases for gasification at the fuel surface were considered. In case 1, the fuel concentration was assumed constant and independent of time, i.e., $f(C_f) = C_f$; and in case 2, it was assumed that the fuel mass flux was constant and independent of time or pressure, i.e., $f(C_f) = - D \, \partial C_f / \partial x = \dot{m}$. Case 1 was identified with a condensed phase behaving as a boiling liquid or subliming solid, and case 2 with a polymer undergoing irreversible decomposition at constant temperature.

Hermance performed a series of numerical calculations to show the effects of the important independent parameters on the predicted ignition-delay time when various ignition criteria are used. According to his results, the effect of oxidizing-gas concentration on ignition delay depends on the ignition criterion as well as on the reaction kinetics and the transport parameters. Hence, comparison between theory and experiment cannot be made until the criterion for determining ignition theoretically can be established.

Although these studies represent a significant improvement over the original model proposed by McAlevy, Price (P8) has emphasized that two further refinements are required before this model can be compared with experimental observations. First, the present analysis does not incorporate the correct energy balance in the gas phase. Both Hermance and McAlevy assumed a constant gas-phase density even though their calculations show that large temperature gradients exist. In other words, the effect of natural convection on the temperature and concentration profiles is not considered.

The second and more serious limitation concerns the extension to include the concurrent decomposition of the fuel and oxidizer within the propellant. The studies of Hermance and McAlevy are limited to the configuration shown in Fig. 5, i.e., to the consideration of only the counter-diffusion of fuel and oxidizing species. In composite propellants, however, both the solid-fuel and solid-oxidizing species undergo initial decomposition reactions prior to ignition. As a result, both fuel and oxidizing species must diffuse away from the solid surface concurrently, in addition to the possible counter-diffusion of external oxidizing gases.

c. Hypergolic-Ignition Theory. Hypergolic ignition occurs when reactive oxidizers, such as fluorine and chlorine trifluoride, contact the surface of a solid propellant. This contact results in spontaneous exothermic heterogeneous reactions between the reactive external oxidizer and the solid-propellant surface; such reactions release heat at the reaction interface, and very rapidly raise the surface temperature of the propellant to a level at which stable combustion can proceed spontaneously without additional external heat.

To describe hypergolic heating, Anderson and Brown (A10) proposed a theoretical model based upon spontaneous exothermic heterogeneous reactions between the reactive oxidizer and a condensed phase at the gas–solid interface. In these studies, the least complex case was considered, i.e., the one in which the solid phase is instantaneously exposed to a stagnant (nonflowing) gaseous oxidizer environment. This situation can be achieved experimentally provided the sample to be tested is suddenly injected into the desired environment in a manner designed to minimize gas flow.

In the theoretical treatment, the heat- and mass-transfer processes shown in Fig. 6 were considered. Simultaneous solution of the equations describing the behavior of the unsteady-state reaction system permits the temperature history of the propellant surface to be calculated from the instant of oxidizer propellant contact to the runaway reaction stage.

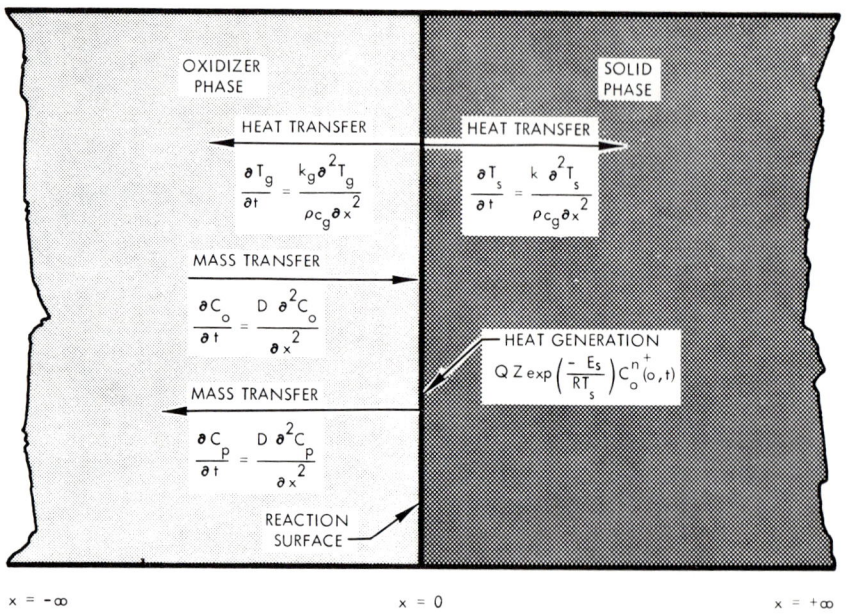

FIG. 6. Essential features of hypergolic-ignition theory (A10).

Inspection of the numerical solutions of the equations shows that, with the exception of $E_s = 0$ kcal/mole, the rate of surface temperature increase with time is very large once the surface temperature reaches approximately 420°K—on the order of 10^8 °K/sec. Because typical autoignition temperatures are of the order of 625°K for composite propellants, the particular value of the ignition temperature does not affect the computed numerical value of the ignition-delay time.

Parametric studies showed that mass diffusion in the gas phase could be neglected under most conditions. The calculations also show that the selection of the hypergolic combination (i.e., the gaseous oxidizer and the propellant system) fixes all of the parameters except the initial temperature and the oxidizer concentration. A general solution of the model shows that the ignition-delay time is approximately rated to the gaseous oxidizer concentration by the relation

$$\tau \propto \frac{1}{C^{2n^+}} \tag{15}$$

where C is the oxidizer concentration and n^+ is the order of surface reaction.

A comparison between data and theory is shown in Fig. 7 for a very reactive oxidizer. These results show reasonable agreement between the theoretically predicted and experimentally observed effect of oxidizer concentration on the ignition delay, assuming a first-order heterogeneous reaction. The validity of this assumption is discussed in detail in reference (A10). It should be noted that the concentration of reactive oxidizer has been changed both by changes in the total pressure of a pure oxidizing environment and by changes in the mole fraction of oxidizer in a constant-pressure environment. These results agree with the predictions of Eq. (15), and thereby provide some indication of the validity of the analytical approach.

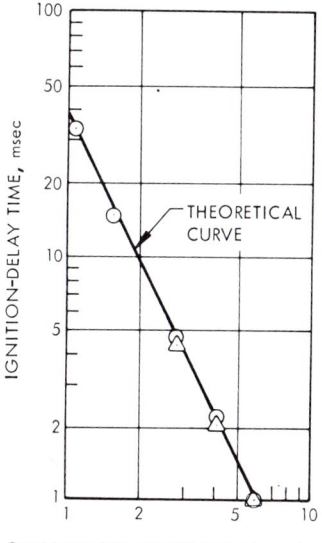

Fig. 7. Best-fit agreement between experimental and calculated ignition data for PBAA/AN composite propellant (P8). Key: ○, pure-oxidizer environment; △, oxidizer-nitrogen environment at 55 psia total pressure.

Anderson and Brown then suggested that the effects observed by McAlevy (M3, M4) of oxygen on the ignition characteristics of solid fuels might be the result of exothermic heterogeneous reactions, since the experimental observations of McAlevy as well as those of Shannon and Anderson (S3), can be correlated by Eq. (15). Shannon has also extended the original treatment of Anderson and Brown to include the effects of adsorption and desorption on the predicted results.

d. Heterogeneous-Ignition Theory. The hypergolic model describes the behavior of externally applied oxidizing gases, but says nothing about what occurs within a composite propellant to achieve ignition in inert environments. In this sense, the oxidizing gases are simply another energy source, like radiative energy from an arc-imaging furnace.

However, the fact that these reactions occur at temperatures below the characteristic propellant "ignition temperatures" calculated from Eq. (8b) does suggest possible propellant reactions which must be initiated to ignite composite propellants in inert environments. To elucidate the possible role of gas–solid reactions in composite-propellant ignition in nonreactive environments, a qualitative ignition-response model has been proposed (A6, A7, A8, A9). It is postulated that the primary preignition reactions occur at the propellant surface or in the subsurface layers, and that these reactions are generally between gaseous molecules and the solid organic polymer matrix. For propellants with oxidizers that have high decomposition temperatures, solid–solid or gas–liquid reactions might be important.

The reaction mechanism can be visualized by considering, as an example, ammonium perchlorate-based composite propellants. The propellant surface is pictured as a mosaic composed of oxidizer cemented together by the polymer matrix. The external ignition stimulus acts to increase the temperature of the propellant; as the temperature increases, the oxidizer undergoes decomposition, while the polymer experiences only limited thermal degradation. As the ammonium perchlorate decomposes [a process that can be detected at 85° to 110°C (B9)] oxygen as well as the gases Cl_2, NO, $HClO_4$, and HCl are released. These gases then diffuse to the oxidizer–polymer interface and into the adjoining gas phase. If present in adequate concentrations, all of these oxidizing gases are sufficiently reactive to attack and degrade the polymer surface, leading to the liberation of heat and gaseous fuel species. These fuel species are then available to undergo further exothermic reaction in the adjoining gas phase. Energy released by the surface and gas-phase exothermic processes augment the external energy sources and raise the propellant surface temperature to the critical level, i.e., to the autoignition temperature.

A major limitation of the heterogeneous-ignition theory is the lack of a complete analytical treatment. Though difficult, it is being pursued at the present time. In differential equation form, the energy balance at the propellant surface is expressed as

$$-k\frac{\partial T}{\partial x} = Q + Z\, C_o(P)\exp(-E_s/RT) \qquad (16)$$

Equation (16) suggests that for a given propellant the ignition delay is a unique function of flux and pressure.

Beyer (B8) has recently reported the analysis of arc-imaging-furnace data obtained on a variety of composite-propellant formulations. He showed the data could be correlated by the expression

$$\tau = \left[\frac{d^3}{\dot{Q}^6} + \frac{b^3}{P_c^{6.15}}\right]^{1/3} \qquad (17)$$

The form of correlation, which is shown in Fig. 8, is suggested by the solution of Eq. (8) subject to Eq. (16) as a boundary condition.

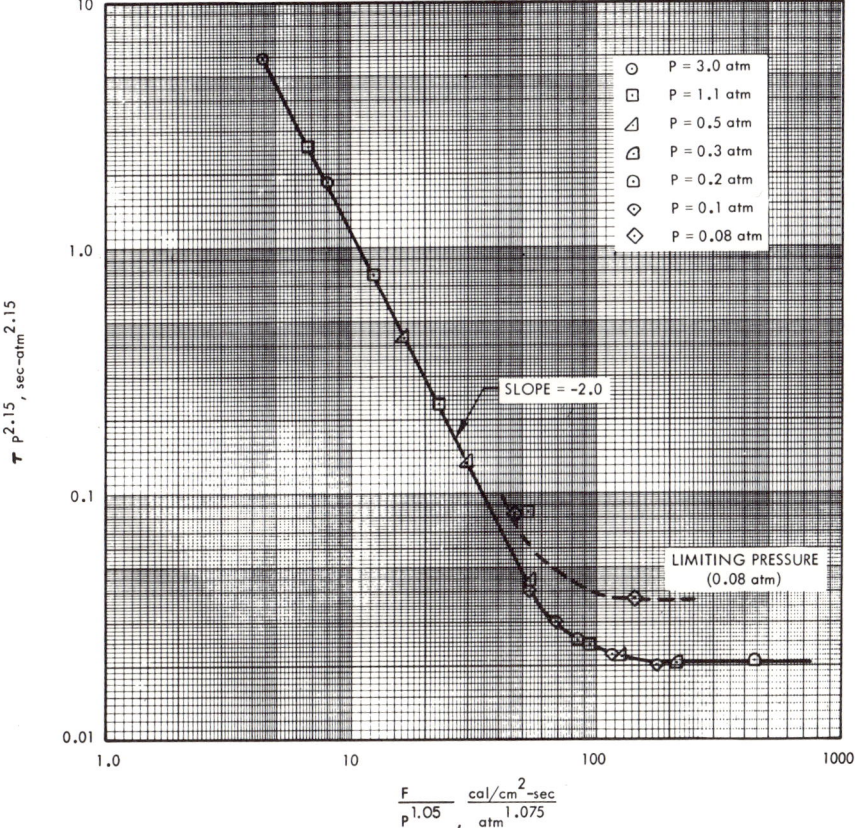

FIG. 8. Master ignition curve (B8).

Two observations on the correlations can be made. First, these results tend to invalidate one of the major objections to the application of the thermal-ignition theory to composite propellants, namely that heterogeneous interfacial reactions within the solid phase are not possible. Secondly, the effect of pressure on propellant ignitability can be qualitatively explained.

The pressure dependence of the ignition-delay time has been the subject of considerable controversy. Altman (A4) reported that the ignition-delay time was insensitive to pressure, whereas Baer, Ryan, and Summerfield indicated a small pressure-dependence. Arc-imaging-furnace data, on the other hand, show a very strong dependence on pressure, as shown by Beyer and Fishman (F1). Atlman employed very low fluxes in his study; as shown in Fig. 9, the ignition delay is independent of pressure in this regime. The flux levels used by Baer and Summerfield were somewhat higher, and a mild pressure-dependence is observed in this flux region. At still higher fluxes and low pressure levels, as are found in arc-imaging studies, the ignition delay becomes quite sensitive to pressure. The apparent inconsistency can be qualitatively explained in terms of the heterogeneous-reaction ignition model by including the influence of pressure on the site and magnitude of the surface reaction (A8).

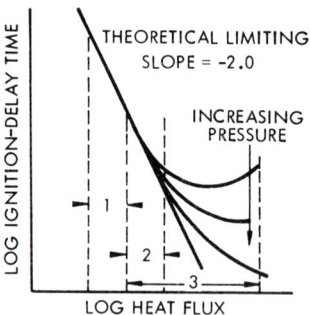

Fig. 9. Ranges of heat fluxes studied by various investigators. (1) Altman—P_c^0; (2) Ryan and Summerfield—P_c^0-$P_c^{0.2}$; (3) arc-imaging-furnace studies (A8).

The three approaches to the analysis of basic ignition mechanisms have recently been reviewed in detail by Price (P8). As he indicated, none of these approaches can be considered as verified, although each offers potential insight into the basic mechanisms. Proponents of each of these approaches are actively studying the problem; it is hoped that definitive conclusions will be forthcoming in the near future.

B. Igniter Design

Though the detailed mechanisms controlling propellant ignition are still the subject of study and further quantitative characterization, these basic research studies have shown that the kinetic approach to ignition appears to be basically sound. This means the ignition device should be designed on the basis of the heat-transfer characteristics of the igniter exhaust products, and not on the basis of the total energy in the igniter charge. Therefore, the heat-transfer characteristics of the igniter must be known, in addition to propellant ignitability data, in order to design an adequate ignition system.

There are a variety of igniter designs which are currently employed in solid-propellant rockets. These types include rocket-exhaust (pyrogen), pyrotechnic, and hypergolic igniters, each of which can be located in the head-end closure of the motor or in the exhaust nozzle at the aft-end of the motor. The heat-transfer information appropriate to each of these possible combinations is discussed in the following sections.

1. *Rocket-Exhaust Igniters*

A rocket-exhaust igniter is essentially a small rocket motor which is initiated by a squib (a small charge of pyrotechnic material and an exploding bridge wire); the combustion products from the igniter then flow across the surface of the main motor propellant, and, as a result of this flow, heat is transferred to the motor propellant, raising the propellant temperature to the point where ignition occurs. The energy transfer in this type of igniter is predominately by convection and radiation. If the igniter exhaust products contain significant quantities of solid particles, these particles can also contact the propellant surface and contribute significantly to the overall heat-transfer rate (heat flux).

a. Heat-Transfer Characteristics of Head-End Rocket-Exhaust Igniters. The typical design of a head-end rocket-exhaust igniter is such that the igniter exhaust jet expands from a relatively small igniter nozzle into a flow chamber of relatively large diameter. Typical diameter ratios vary from 5:1 up to 15:1. Because of these large expansions, igniters may be designed with several exhaust jets directed toward the propellant surface, rather than with axially directed exhaust jets. In either case, the flow configuration and resulting boundary layer structure are characteristic of developing flows and must be defined by experiment.

Mullis (M10), Bastress (B4), and Carlson and Seader (C1) have conducted experimental studies to determine the heat-transfer characteristics of typical rocket-exhaust igniters. In these studies, the total rate of heat transfer to the propellant or simulated propellant surface was measured as a function of mass flow rate, geometry, and impingement angle between the igniter exhaust jet and the solid propellant. Although these studies were directed toward developing generalized correlations to be used in the design of igniters, they were not wholly successful, basically because of the limited information available on the flow characteristics of ducted supersonic jets.

Mullis made measurements of the heat flux for axially directed jets and for jets directed 30° to the motor axis. For the axially directed jets, Mullis found that the flux 4 to 10 diameters downstream of the contact point could be correlated by the boundary layer relations

$$\frac{N_{Nu}y}{N_{Nu}} = \left[0.273 + \frac{0.546 D_p}{y^+}\right] \frac{\sum_{n=1}^{\infty} G_n \exp(-\lambda_n^2 y^+)}{\sum_{n=1}^{\infty} G_n/\lambda_n^2 \exp(-\lambda_n^2 y^+)} \quad (18)$$

where $y^+ = (y'/D_p)N_{Re} \cdot N_{Pr})^{-1}$, y' is the distance from the impingement point, N_{Nu} is the Nusselt number, the G_n are constants in the equation for the entry heat-transfer coefficient (K1), and λ_n is the eigenvalue (K1). The value of N_{Nu} is given by the Dittus–Boelter equation

$$N_{Nu} = \frac{hD_p}{k} = 0.023 N_{Re}^{0.8} N_{Pr}^{0.67} \quad (19)$$

where h is the heat-transfer coefficient, D_p the diameter of the motor port, N_{Pr} the Prandtl number, and N_{Re} the Reynolds number.

For the region near the attachment point, Mullis found a strong effect of axial position on flux, but no satisfactory general correlation for this effect. In addition, he found no quantitative relation for the heat-transfer characteristics of jets directed toward the propellant surface. Under most conditions studied by Mullis, the radiation contribution is approximately 10% of the convective flux. The effects of solid-particle impingement were not investigated.

Carlson and Seader (C1) generally confined the correlation of heat-transfer data to the region 4 to 10 diameters downstream of the impingement point. They found the important motor-length parameter to be the distance from the point of attachment of the igniter jet and not the motor length itself. They also investigated the effects of canted nozzles and of slots in the propellant grain, and compared sonic versus supersonic nozzles in the igniter; however, they were unable to account quantitatively for the observed effect of the ratio of motor-port area to motor-nozzle area.

Bastress (B4) investigated the effect of incorporating particles of a solid such as Al_2O_3 in the exhaust of a rocket-exhaust igniter. The general conclusion from these studies was that condensible vapors or condensible particles had little effect on the igniter-produced heat transfer, outside of any radiation contributions.

Beyer (B8) has measured ignition-delay times as function of igniter mass-flow for head-end rocket-exhaust igniters in small test motors. His data suggest that the igniter heat-transfer coefficient is related to the igniter mass-flow rate raised to the 0.6 power. According to Eq. (8b), τ varies as $\dot{Q}^{-2}$. If $\dot{Q}$ varies as $\dot{m}^{0.6}$, then τ is proportional to $\dot{m}^{-1.2}$, which was shown to correlate the data adequately, as shown in Beyer's report (B8). He also showed that the ambient pressure is largely unimportant, provided the igniter mass-flow rate is sufficient to choke the exit nozzle of the motor.

b. Aft-End Igniter. Carlson and Seader (C1) have also determined the heat-transfer characteristics of aft-end rocket-exhaust igniters. Their results correlate on the basis of the relation

$$h \propto \dot{m}^{0.5} D_p^{-1.5} \tag{20}$$

They also considered the depth of penetration of the igniter jet into the motor. They found that the motor L/D_p ratio had little effect on the heat transfer provided L exceeded the depth of penetration of the igniter jet. Their results also showed that high penetration is desirable and can be achieved by the use of high igniter mass-flow rates in conjunction with supersonic igniter-exhaust nozzles.

2. Pyrotechnic Igniters

A pyrotechnic igniter consists of a basket of inorganic combustible materials, such as boron-potassium nitrate, which are usually in the form of pellets. This basket is also filled with a squib which provides a means of initiating the igniter.

Although pyrotechnic igniters were the first design used for solid-propellant motors, little information has been obtained on the motor heat-transfer characteristics for this type of igniter. A notable exception is the work of Bastress (B4), who measured the igniter fluxes much as Mullis and Carlson did for rocket-exhaust igniters. Though generalized correlations were not obtained, Bastress did conclude that condensible vapor or liquid drops in the igniter exhaust reduce the effectiveness of the igniter. In general, his results showed the strong effect of igniter mass-flow rate that was observed by Mullis and Carlson.

Beyer (B8) has measured ignition delays in small test motors as a function of igniter mass-flow rate. He showed that the data correlated with the results obtained using a rocket-exhaust igniter, i.e., on the basis of the mass-flow rate produced by the igniters.

Several evaluation studies have been conducted as part of the aft-end igniter development for large solid-motor demonstration programs. However, these results were specific to the particular motor configuration, and no attempt was made to develop a generalized correlation in these programs.

3. *Hypergolic Ignition*

The hypergolic-ignition technique consists of a system deploying a spontaneously reactive or hypergolic fluid to the surface of a solid-propellant motor grain. The exothermic reaction between the hypergolic fluid and the major components of the solid-propellant grain initiates burning of the propellant.

Beyer (B8) has recently reported experimental data obtained in small test motors under atmospheric and altitude conditions. At atmospheric pressure, his results showed the observed ignition delay to be a function of the delivery rate, as shown in Fig. 10. Additional data obtained in small test motors by Fullman and Nielsen (F6) are shown for comparison. These latter investigators conducted studies on the effects of various injectors, with delivery from both the head end and the aft end. Their results indicate that the hollow-cone injector is the most efficient. This subject has been treated in more detail by Miller (M7).

C. Flame Propagation

A knowledge of the ignition characteristics of the propellant and the heat-transfer characteristics of the igniter permits the igniter designer to determine the propellant ignition-delay for a particular system. The next question is "How fast does the flame spread across the propellant surface?" The answer to this question determines the burning area on the propellant surface as a function of time; this is the function $A_B(t)$ required to solve Eq. (6) for the chamber pressure as a function of time.

Brown (B11, B14) and deSoto (D1) have considered this problem theoretically in two separate studies. The basic approach in these studies was to incorporate the variation of the heat flux produced by the igniter over the entire propellant surface. The basic model considers the propellant as a two-dimensional body with the igniting environment flowing over one surface, as shown in Fig. 11. While heat transfer to the surface may occur by a number of modes, as shown in the figure, heat transfer to the exposed propellant surface is considered, in the model, to occur only by the possible simultaneous processes of convection, radiation, and hypergolic attack. In addition to the above considerations, one-dimensional energy and mass balances were used to describe the variation of igniter gas temperature and gas-phase oxidizer concentration above the surface of the propellant grain.

The following equations form the theoretical model:

(1) The two-dimensional energy equation simulating the thermal behavior of the solid propellant:

$$\frac{\partial T}{\partial t} = \alpha \left[\frac{\partial^2 T}{\partial x^2} + \frac{\partial^2 T}{\partial y^2} \right] \tag{21}$$

where y is the distance along the propellant surface.

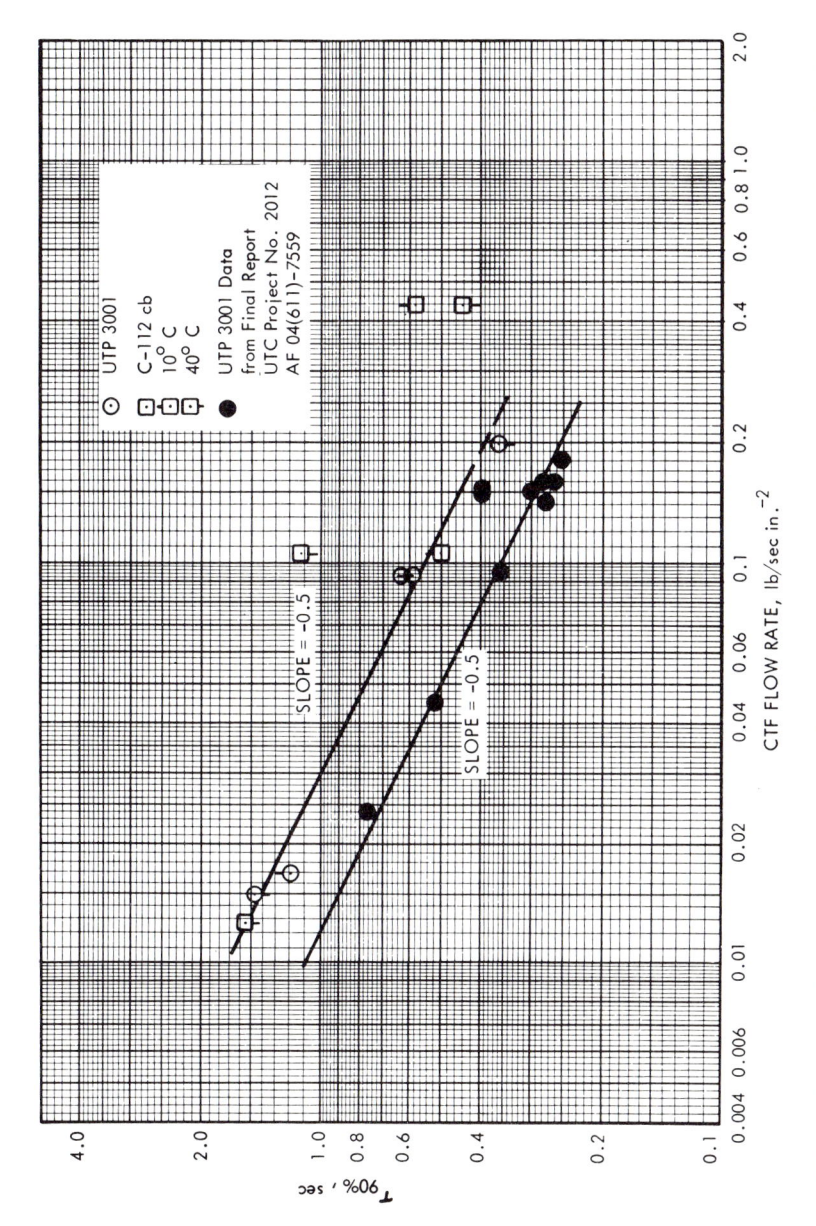

FIG. 10. Ignition delay as a function of mass-flow rate from hypergolic igniter at an initial pressure of 1 atm (B8, F6).

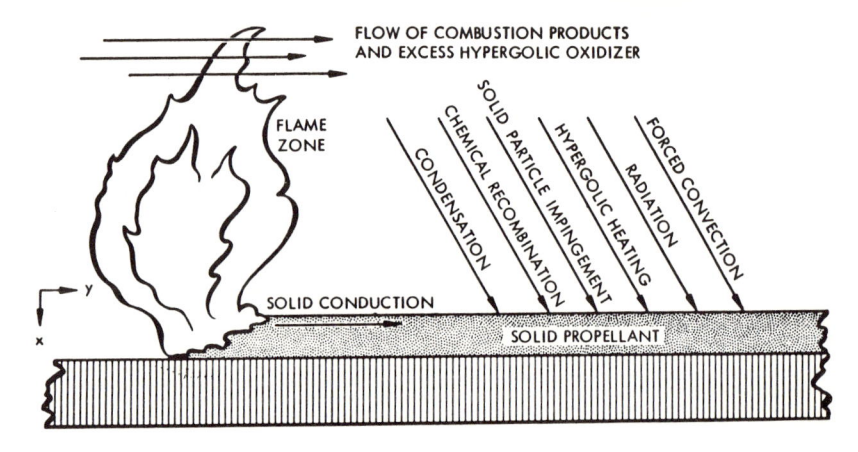

FIG. 11. External modes of energy transfer in solid-propellant ignition (B14).

(2) The equations describing the possible nonreactive and/or reactive forms of energy transfer at the gas–solid interface:

$$\dot{Q}_{\text{convection}} = h(T_g - T_s) \tag{22a}$$

$$\dot{Q}_{\text{radiation}} = \varepsilon\sigma(T_g^4 - T_s^4) \tag{22b}$$

$$\dot{Q}_{\text{hypergolic}} = ZC_o \dot{Q} \exp(-E/RT_s) \tag{22c}$$

where ε is the emissivity and σ the Boltzmann constant.

(3) The one-dimensional equation describing the variation of igniter energy content:

$$-\frac{dT_g}{dy} = \frac{W}{Gc_g A_p} [h(T_g - T_s) + \varepsilon\sigma(T_g^4 - T_s^4)] \tag{23}$$

where c_g is the specific heat of the gas, G the mass velocity, A_p the port area, and W the wetted perimeter of the port.

(4) The one-dimensional equation describing the variation of oxidizer concentration:

$$-\frac{dC}{dy} = \frac{Wk_c(C_o - C'_o)}{A_p V} \tag{24}$$

where V is the linear gas velocity and k_c the mass-transfer coefficient.

Equations (21)–(24) permit the temperature history of the propellant grain during ignition to be calculated. Ignition of any point on the surface is assumed to occur when the propellant autoignition temperature is reached at that point. The propagation rate can then be predicted from the different times at which the different positions on the propellant surface are ignited. The important basic assumption in this approach is the applicability of the autoignition temperature concept. The use of this concept under many practical conditions has been well substantiated by the work discussed in Section II,A,2. However, under conditions of extremely high flux or low ignition pressure, the autoignition temperature is not invariant. Hence, suitable modifications in the analysis must be made.

The examination of this model and results of numerical solutions indicate that ignition propagation is determined by the absolute value and variation of heat transfer along the surface of the grain, both before and after the first instant of ignition on the grain surface. Therefore, the important variables are: (1) igniter flow rate, (2) port diameter, (3) gas temperature, (4) gas composition, and (5) motor pressure.

In nonreactive environments, the energy transfer along the surface of the grain is controlled primarily by convective heat transfer, and hence by the gas temperature. The structure of the gaseous boundary layer over the surface of the propellant is also important. The boundary layer is, in turn, a function of gas mass velocity, port diameter, and igniter orientation. If conditions are such that there is little variation in igniter-produced flux along the surface of the grain, then all points on the surface are raised to the propellant autoignition temperature almost simultaneously, and a rapid rate of ignition propagation results. Such a phenomenon is most likely to be observed with high-velocity flows, large-diameter ports, or in cases when well-developed gas flows are produced by the igniter. However, conditions of developing flow, low velocities, or small port diameters should result in greater spatial variation of convective energy transfer. In these instances, elements of surface area downstream of the point of maximum heat transfer will be at a much lower temperature when ignition is first achieved. Hence, a much longer period of time will be necessary to achieve ignition at the downstream points.

In reactive environments, e.g., in conditions of hypergolic ignition, the igniter mass velocity and port diameter determine the variation of gaseous-oxidizer concentration along the axis of the port. Essentially the same considerations given to convective energy transfer apply to reactive systems. Large mass velocities and large port diameters should result in little axial variation of oxidizer concentration. In such instances, the only difference between the ignition times of propellant areas along the port axis should be the difference in the time of the first contact by the oxidizing species. Hence, the limiting rate of ignition propagation should be the linear velocity of the oxidizer species. Greater variation of igniter gas concentration, which would be expected for lower flow rates and small diameter ports, will result in slower ignition propagation rates.

The above considerations, as applied to nonreactive and/or reactive igniter systems, have not been adjusted for the perturbing effect of the addition of propellant combustion products during the propagation period. This addition of combustion products will modify the heat-transfer characteristics of the igniting environment by disrupting the igniter flow patterns, by increasing the mass flow rate and decreasing the concentration of hypergolic agents, and (if the igniter gas temperature differs greatly from that of the propellant combustion products) by changing the mixed gas temperature.

During flame propagation, a constantly increasing amount of propellant combustion products is added to the igniter products, thereby producing a constantly increasing mass velocity of total gas. As a first approximation to the real situation, the following expression has been used in the theoretical analysis to account for this effect:

$$h = a\dot{m}^g \qquad (25)$$

Jensen (J3) has conducted experimental studies in both laboratory and test motors to determine the value of the exponent g. The results obtained in a 3-in.-diameter test motor show that a value of $g = 0.5$ correlates the data. Using this correlation, experimentally observed propagation rates could be predicted with reasonable accuracy using Eqs. (21)–(24). A typical comparison between theory and experiment is shown in Fig. 12.

These studies have concentrated on the performance of head-end ignition. Because of the increasing application of aft-end igniters, Jensen also investigated the propagation characteristics of the latter. A comparison of the propagation rates obtained from aft-end and head-end igniters shows that flame propagation can be extremely slow with an aft-end igniter. However,

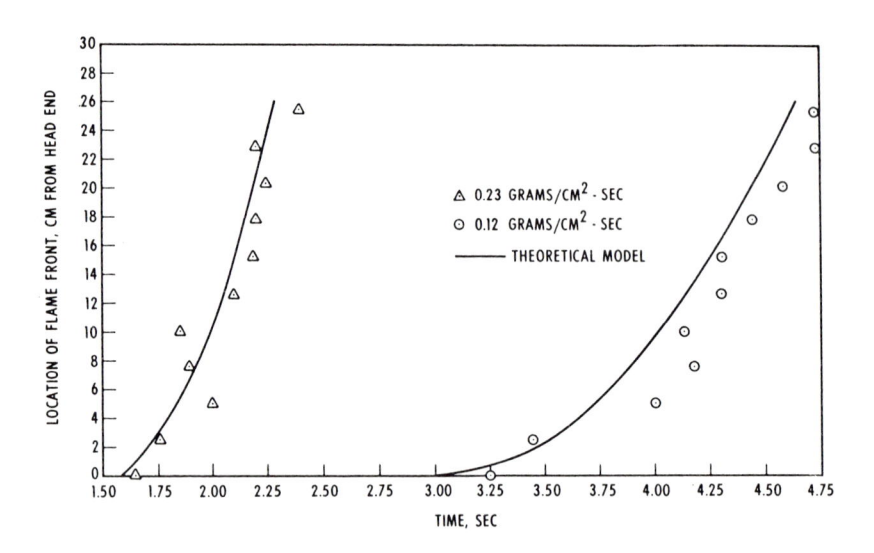

FIG. 12. Ignition propagation in laboratory system (J3).

the results for aft-end igniters agree with theory and the measured heat-transfer data.

Brown (B11) has also reported propagation rates obtained in gaseous hypergolic igniters. According to Eqs. (21)–(24), the propagation rate should approximate the linear velocity of the igniter gases flowing through the motor. Jensen's data show reasonable agreement with this conclusion.

D. Chamber Filling

The results of the studies discussed in Section II,C permit calculations to be made of the time required for the flame to spread to the entire propellant surface. Once this phase of the motor-ignition process has been completed, the time required to fill the combustion chamber and establish the steady-state operating conditions must be computed. This can be done by the formal solution of Eq. (7). Because this equation is a Bernoulli type of nonlinear equation, the formal solution becomes

$$\frac{P_s^{1-n} - P_i^{1-n}}{P_c^{1-n} - P_i^{1-n}} = 1 - \exp\left[\frac{(n-1)^2 c^* t}{L^*}\right] \qquad (26)$$

where $L^* = V_c/A_t$, P_i is the chamber pressure due to the igniter, P_s the pressure during the ignition of the motor, and V_c the volume of the combustion chamber.

E. Ignition-Pressure Transients

To test the validity of combining the results of studies of propellant ignitability, igniter heat-transfer, flame propagation, and chamber filling, Jensen (J2) compared measured and predicted ignition-pressure traces using a head-end rocket-exhaust ignition. He reasoned that the quantitative analysis of igniter action and the ignition-propagation theory described in Section II,D, together with Eqs. (8a) and (26), should result in a complete and quantitative analysis of the motor-ignition transient. Experimentally measured ignition-propagation patterns were substituted into the ballistic analysis to obtain a predicted pressure transient. The calculated pressure transient is shown by the solid line in Fig. 13. The agreement shown with the experimental points not only validates the propagation measurements, but, more importantly, indicates that, if accurate predictions of ignition heat-transfer characteristics are developed, quantitative predictions of ignition-propagation rates and motor ignition-pressure transients can be developed.

III. Steady-State Combustion

The discussion of the important design considerations of solid-propellant motors presented in Section I has shown the importance of the steady-state burning rate of the propellant. The particular mission for a rocket motor to

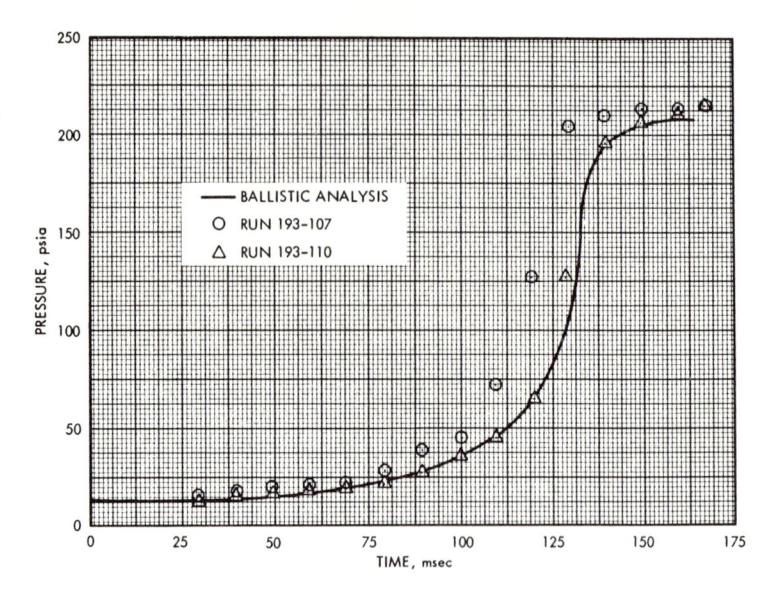

FIG. 13. Comparison of predicted and experimental pressure transients (J2).

which the designer is addressing his attention usually provides a definition of the range of energy and burning-rate requirements for the propellant. In other words, the particular velocities and motor sizes which must be achieved determine the boost velocity and motor diameters which can be used to achieve the desired objective. Hence, the ranges of propellant burning rate and I_{sp} which can be used to achieve the desired result are often fixed within rather narrow limits. Methods for predicting I_{sp} are based on well-established thermodynamic principles, and therefore predictions of the thermodynamic performance of candidate propellants can readily be made.

The prediction of burning-rate characteristics, on the other hand, has not been possible. This has caused rocket designers to adopt a trial-and-error approach to the development of specific propellants to meet specific mission requirements. In an effort to reduce the large development effort required for each new propulsion system, considerable basic research effort has been directed toward the definition and quantitative characterization of propellant combustion mechanisms. The ultimate objective of this effort is to provide methods for predicting the burning-rate characteristics of particular propellant formulations.

Even if this objective could not be attained completely, definition of the combustion mechanism would provide a basis for the development of specific propellants for novel uses, such as stop–start applications. The results of such basic research would also provide guide lines for eliminating those troublesome combustion phenomena, such as instability, detonation, and erosive burning, which can have an adverse effect on the operation of the motor.

With these goals in mind, several investigators have undertaken to set down quantitative expressions which will predict propellant burning rates in terms of the chemical and physical properties of the individual propellant constituents and the characteristics of the ingredient interactions. As in the case of ignition, the basic approach taken in these studies must consider the different types of propellants currently in use and must make allowances for their differences. In the initial combustion studies, the effort was primarily concerned with the development of combustion models for double-base propellants. With the advent of the heterogeneous composite propellants, these studies were redirected to the consideration of the additional mixing effects.

One extremely important point to realize is that different propellant types may have different rate-controlling processes. For example, the true double-base propellants are mixed on a molecular scale, since both fuel and oxidizing species occur on the same molecule. The mixing of ingredients and their decomposition products has already occurred and can therefore be neglected in any analysis. On the other hand, composite and composite modified-double-base propellants are not mixed to this degree, and hence mixing processes may be important in the analysis of their combustion behavior.

A. Combustion of Double-Base Propellants

The approach taken in the development of an analytical model for the combustion of double-base propellants has been based on the decomposition behavior of the two principal propellant ingredients, nitrocellulose and nitroglycerin. The results of several studies reviewed by Huggett (H12) and Adams (A1) show that nitrocellulose undergoes exothermic decomposition between 90° and 175°C. In this temperature range, the rate of decomposition follows the simple first-order expression

$$-dC/dt = 10^{20} Z \exp(-49{,}000/RT) \tag{27}$$

These studies have also shown that the decomposition proceeds through a series of intermediate steps to finally yield N_2, CO, CO_2, and H_2O at pressures typical of solid rocket motors.

Like nitrocellulose, nitroglycerin also undergoes a slow first-order exothermic decomposition at temperatures below 140°C. As the pressure is increased, this decomposition reaction is followed by a sudden explosive reaction. Evidence suggests that the explosive reaction is autocatalyzed by the accumulation of NO_2. The combined results of several studies indicate that the rate of nitroglycerin can be approximated by the expression

$$-dC/dt = 10^{19} Z \exp(-45{,}000/RT) \qquad (28)$$

These results were used by Rice and Ginell (R2) and by Parr and Crawford (P2) to develop a theoretical analysis of the combustion mechanism of double-base propellants. The model of the combustion zone is essentially identical in both studies and is shown schematically in Fig. 14. Because of the decomposition of the solid ingredients, Rice assumed that the initial exothermic decomposition reactions could be analyzed as a single reaction occurring at the propellant surface. The initial products of decomposition vaporize and undergo partial oxidation reactions. This gives rise to a boiling porous reaction zone, commonly called the "fizz reaction" zone, which can be visually observed on the propellant surface. Rice has assumed that the reaction rates in this zone can be characterized by a single second-order Arrhenius expression, even though a complex series of interrelated reactions actually occur. The gaseous products from the fizz reaction zone then flow away from the propellant surface, and are ultimately converted to final combustion products at a significant distance away from the propellant surface.

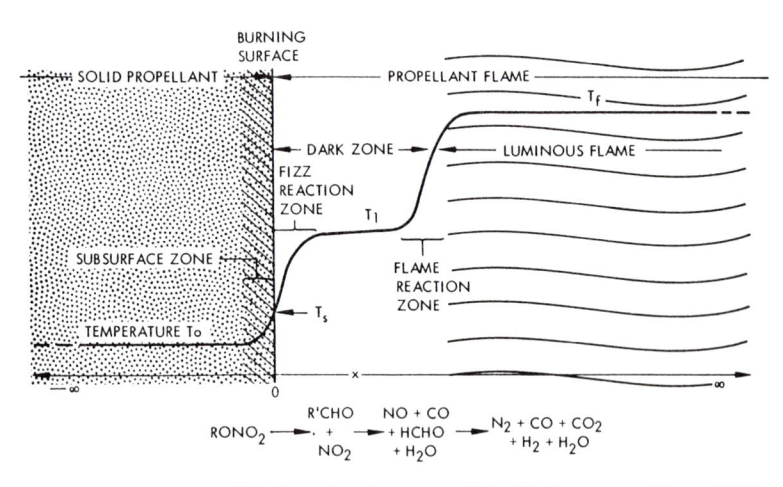

FIG. 14. Schematic model of combustion zone of double-base propellants (H12).

The equations resulting from both these studies are extremely complex, and contain several reaction parameters not readily evaluated from separate experiments. Figure 15 shows a comparison between experimental data and the two theoretical presentations. The theoretical curve was obtained by curve-fitting through adjusting the unknown parameters. This comparison shows that the theoretical expressions have sufficient flexibility to adequately correlate experimental burning rates. However, the value of the activation

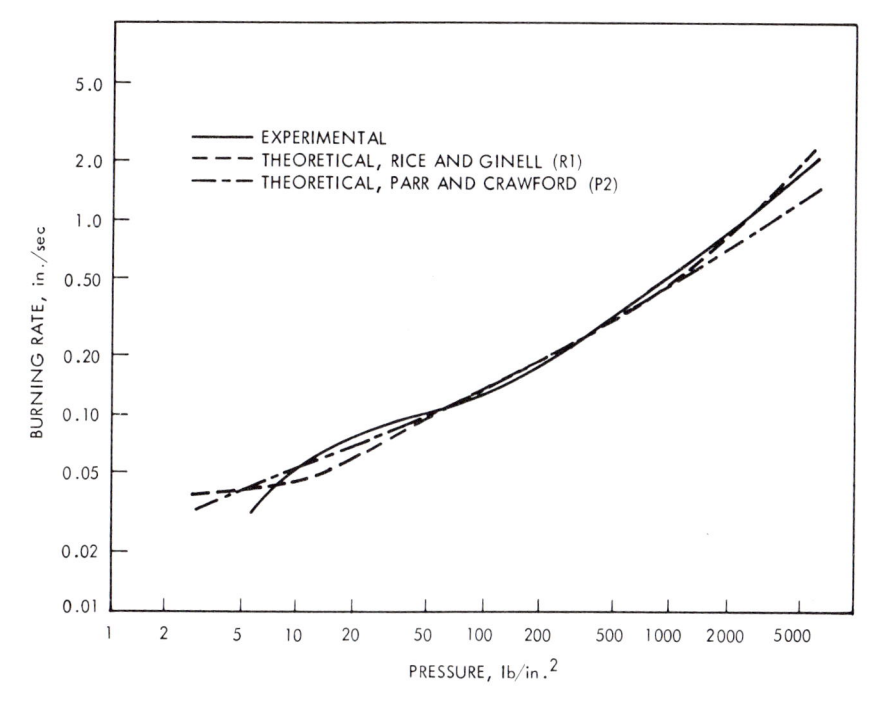

FIG. 15. Comparison of experimental and theoretical burning-rate curves (H12).

energies and frequency factors used to correlate the data do not correspond to the values expected from the decomposition kinetics of the individual propellant constituents, or with values characteristic of the oxidation reactions expected in the gas-phase combustion processes.

More recently, Rosen (R3), Spalding (S5), and Johnson and Nachbar (J4) have considered a simplified approach using the analysis of laminar-flame propagation velocities. According to these investigators, the principal exothermic reactions occur in the gas phase. Some of the heat liberated by these reactions is then transferred back to the solid surface to sustain the endothermic surface-gasification processes. Thus, the temperature profile within the reactive zone is quite similar to that of Rice and Crawford. However, gasification of the solid surface is assumed to be endothermic, while exothermic reactions were considered in the studies discussed previously.

The solutions obtained by Rosen show that at low pressures, the burning rate becomes linear in pressure and the surface pyrolysis characteristics are not important. At high pressures, however, the burning rate becomes independent of pressure and is determined almost entirely by the decomposition reactions at the solid surface. Rosen points out that this simple model can account for "pressure plateaus," i.e., high-pressure regions where the burning rate is observed to be independent of pressure.

Spalding has considered essentially the same problem but has used the centroid-rule approach to obtain an approsimate solution. In addition, Spalding considered two surface conditions. In the first, the surface regression was by the pyrolysis kinetics of the solid phase, and Spalding has found essentailly the same effect of pressure on burning rate as Rosen found. In his second case, Spalding assumed the surface to be controlled by the vaporization of a liquid layer. Under these assumptions, the Clausius–Clapeyron equation relates the pressure and the surface temperature. The solution of the resulting equations shows that the burning rate increases with increasing pressure, then reaches a maximum which depends on the heat of vaporization of the liquid, and finally (as the surface temperature reaches the adiabatic-flame temperature) decreases with further increases in pressure.

It is well known that pressure plateaus and pressure–burning-rate maxima are observed in some double-base propellant formulations. However, the conditions under which they appear are not seen to be consistent with the prediction of Rosen or Spalding. For the case using the Arrhenius-type surface pyrolysis condition, the burning rate becomes independent of pressure when the surface temperature approaches the propellant flame temperature. Because activation energies of 40 kcal/mole characterize pyrolysis reactions, the propellant burning rate would be many orders of magnitude higher than the fastest decomposition rates which have been observed for solid fuels and for the highest burning-rate propellants currently available. In the case of the liquid surface analyzed by Spalding, it seems highly unlikely that an organic material heated to the propellant flame temperature will vaporize without significant pyrolysis; hence, the predicted conditions are probably unrealistic in comparison to observation.

Another difficulty with the adiabatic models of Rosen and Spalding is that the resulting equations do not predict a lower deflagration limit of pressure. It is well established experimentally that double-base propellants will not sustain combustion below a certain critical pressure, which in many cases is as high as 200 psi. To account for this phenomena, Spalding and Johnson and Nachbar have proposed that the combustion process is in reality not truly adiabatic. This, as Spalding has suggested, would be particularly true in laboratory burning-rate studies where a small cigarette-like strand of propellant is burned in a large-volume container having relatively cool walls.

In an effort to account for this effect, Spalding and Johnson have considered the effect of radiation losses from the solid surface. Johnson has also included a provision for a volume rate of energy loss, but he does not state the basis for the term. In the case of surface pyrolysis, the resulting equations predict the relations of burning-rate versus pressure as shown in Fig. 16.

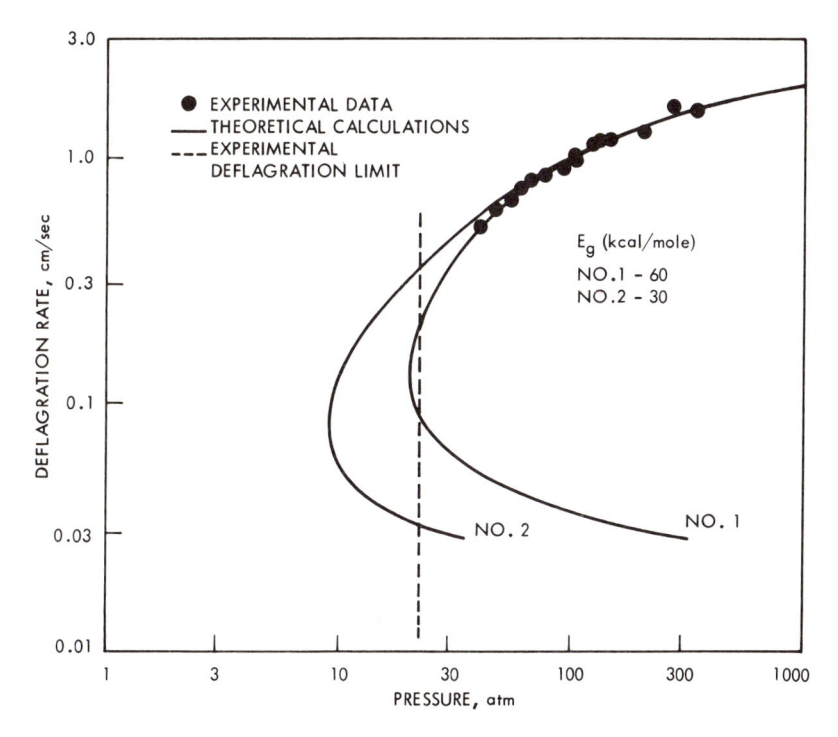

FIG. 16. Variation in deflagration rate of pure ammonium perchlorate with pressure at 25°C (J4).

Thus, within the theoretical framework, provision is made for the lower-pressure deflagration limit observed esperimentally.

Spalding has carried this analysis further by considering the effects of radiation losses when the surface is controlled by the vaporization of a liquid phase. Spalding argues qualitatively that the resulting equations predict an upper-pressure deflagration limit; that is, a pressure above which the propellant will not burn. This phenomenon has been observed in some propellants (H12). Under certain conditions, the surface-vaporization conditions can be used to predict a lower-pressure deflagration limit as well; Spalding did not have sufficient experimental data to develop a comparison between theory and experiment for low-pressure conditions.

B. Combustion of Composite Propellants

Since an understanding of the combustion characteristics of composite propellants requires information on the chemical behavior of the individual oxidizer and fuel, a discussion of these areas is presented as a prelude to the discussion of propellant combustion.

1. *Decomposition Behavior of Ammonium Perchlorate*

A considerable amount of research has been conducted on the decomposition and deflagration of ammonium perchlorate with and without additives. The normal thermal decomposition of pure ammonium perchlorate involves, simultaneously, an endothermic dissociative sublimation of the mosaic crystals to gaseous perchloric acid and ammonia and an exothermic solid-phase decomposition of the intermosaic material. Although not much is presently known about the nature of the solid-phase reactions, investigations at subatmospheric and atmospheric pressures have provided some information on possible mechanisms. When ammonium perchlorate is heated, there are three competing reactions which can be defined: (1) the low-temperature reaction, (2) the high-temperature reaction, and (3) sublimation (B9).

Heath and Majer (H3) have recently used a mass spectrometer to study the decomposition of ammonium perchlorate. Decomposition was detected in the range from 110° to 120°C. At this temperature, there were ions in the mass spectrum caused by NH_3, $HClO_4$, Cl_2, HCl, nitrogen oxides, and O_2. The appearance of the species NO, NO_2, O_2, and Cl_2 in the decomposition products under very low pressure (i.e., in the absence of gas-phase molecular collisions) indicates that the principal decomposition reactions take place in the crystal and not in the gas phase.

Experimental investigations indicate that the mechanism of the high-temperature solid decomposition (>350°C) is apparently quite different from that of the low-temperature process. In the high-temperature reaction, sublimation and solid decomposition are competitive processes. Bircumshaw and Phillips (B10) have made simultaneous measurements of sublimation and decomposition rates in an initially evacuated apparatus. According to their results, 60% of the total weight loss occurred by vaporization at 300°C, while 80% of the weight loss was caused by sublimation at 400°C. This would indicate that sublimation is the fastest reaction process in the decomposition of pure ammonium perchorate in the high-temperature regime, in the absence of retarding inert-gas pressure.

Catalysts which enhance the burning rate of composite propellants are generally believed to accelerate the decomposition of ammonium perchlorate, but the catalytic mechanism is still not very clear. The important observed aspects of this catalysis can be summarized as follows:

(1) Transition-metal oxides are particularly effective decomposition and burning-rate catalysts. The metal elements can demonstrate variable valence or oxidation states.

(2) The introduction of electron donors or acceptors can modify oxidizer decomposition rates.

(3) Varying the semiconducting properties of the catalyst crystal affects the rate of ammonium perchlorate decomposition.

(4) Catalytic activity is a result of physical contact; the reaction ceases when particles no longer present a salt–additive interface.

(5) The effectiveness of all catalysts is related to the sizes of the particles of ammonium perchlorate and of the catalyst, i.e., to the surface-to-surface contact area.

(6) Iron compounds are common catalysts for ammonium perchlorate propellant systems, and burning-rate augmentation is a strong function of catalyst surface area and shape.

(7) A redox cycle involving the ferric–ferrous couple may be the key mechanism in combustion catalysis by iron compounds.

(8) Gaseous-product distribution may be altered by some catalytic agents.

The self-deflagration rate of ammonium perchlorate at typical rocket pressures is of the same order as the burning rate of many ammonium perchlorate-based propellants; thus, the ammonium perchlorate deflagration may well be a controlling factor for propellant burning rates. Evidence of the interrelation between deflagration and propellant burning rate may be deduced from data showing the variation of composite burning rates with pressure, ammonium perchlorate particle size, and catalyst or inhibitor effects.

At normal ambient temperatures, ammonium perchlorate which has been pressed in the form of a strand can be ignited at pressures above 22 atm. The burning rate increases with pressure and decreasing particle size. Adams (A1) found that the deflagration wave progressed through pressed pellets at roughly 1 cm/sec at 1000 psia. The rate increased with the 0.57 power of pressure, and was somewhat dependent on the particle size of the powder from which the pellets were pressed. Friedman *et al.* (F5) investigated the effects of pressure, particle-size distribution, and catalysts on the deflagration rate, and they reported results that agreed rather well with those of Adams.

Hightower and Price (H7) have studied the combustion of large single crystals of ammonium perchlorate. Observation of the surface of the crystal by high-speed motion pictures during burning revealed a very intricate pressure-dependent structure of ridges, troughs, and depressions. The pattern on any one sample remained substantially unchanged as the surface receded, indicating that a stable three-dimensional combustion-zone structure prevailed. In addition, strong evidence for the presence of a melt on the surface of the deflagrating crystal was noted, indicating that the decomposition processes studied in most laboratory experiments at temperatures in the 200°–500°C range remain relatively slow in the solid-phase portion of the deflagration wave. This result raises the possibility that some other reaction path is dominant in determining the burning rate. Their results appear to indicate that most of the decomposition of the condensed phase occurs in a (noncrystalline) melt rather than in or on the crystal. If supported by further study, these findings will have far-reaching effects in the development of steady-state combustion models for composite propellants.

The presence of catalysts markedly changes the deflagration rate. The greatest rate increase is produced by copper chromite, a well-known hydrogenation catalyst. Some additives which catalyze the process at higher pressures may inhibit it strongly at lower pressures. The catalyst effect is related to catalyst particle-size and concentration, but these factors have not been studied extensively.

It has been postulated (F5) that the lower-pressure limit is caused by radiant heat loss to the surroundings. Detailed theoretical examination of this postulate by Johnson and Nachbar (J4) shows that, while it is qualitatively reasonable, it requires the assumption of a larger rate of energy loss from the system than can plausibly be accounted for by radiation. Although radiation is undoubtedly one factor influencing this limit, another would appear to be the pressure dependence of the heat of decomposition. Analysis of the gaseous products indicates that in the region of the lower limit, appreciable amounts of nitric oxide are present in the final products, whereas little is observed at higher pressures. If nitric oxide is an intermediate product of the decomposition and fails to decompose below a certain pressure, as has been observed for nitric ester propellants, its persistence will lower the net energy release and contribute to the limit process.

Combustion may be induced by lower pressures by preheating the strand, by providing a sufficient incident radiant flux, by adding certain catalytic agents, and by adding small amounts of various fuels.

A marked change in burning behavior occurs in properly restricted strands at pressures above 5000 psig, as reported by Irwin *et al.* (I1). Above 5000 psig, the pressure dependence increases sharply and obeys the expression

$$\dot{r} = 2 \times 10^{-7} P^{1.75} \quad \text{in./sec} \tag{29}$$

These investigators postulated that the observed increase in burning rate results from an increased burning surface area, i.e., from surface breakup under the action of the very high pressures existing in the closed bomb. Little information exists to explore the possibility that a change in the general burning-rate law (i.e., a change in the deflagration mechanism) might be partially responsible for the increased burning rate and pressure dependence at these higher pressures.

2. *Degradation of Polymers*

The second major component of a composite solid propellant is the organic polymer material, which serves to bind the oxidizer and other additives into a solid matrix. The response of this polymer surface to intense transient heating under reactive gas exposures would appear to be a pivotally important factor in the formulation of complete theories of composite solid-propellant combustion. From this standpoint, two types of polymer decomposition mechanism are of importance—pure-thermal and the oxidative-thermal degradation mechanisms.

Extensive research has been conducted to determine the thermal-decomposition properties of polymers, the products of their degradation, and the kinetics involved in their reaction during pyrolysis (M1). Complete comprehension of the mechanism involved in thermal degradation requires, among other facts, knowledge of these three fundamental aspects:

(1) The change of molecular weight of the polymer as a function of temperature and extent of degradation.

(2) The qualitative and quantitative composition of the volatile and nonvolatile products of degradation.

(3) The rates and activation energies of the degradation processes.

Thermogravimetric data indicate that the structure of a polymer affects stability in a neutral environment (H1). A polymer such as Teflon, with carbon–carbon bonds which are (by comparison) easily broken, and with strong carbon–fluorine bonds, is quite stable thermally. However, polyethylene, also with carbon–carbon bonds but containing carbon–hydrogen bonds which are broken relatively easily in comparison with the carbon–fluorine bond, is less stable than Teflon. In turn, polyethylene is more stable than polypropylene. This difference in stability is probably caused by tertiary carbon–hydrogen bonds in polypropylene. Polypropylene is more stable than polyisobutylene or polystyrene, which decompose principally by "unzipping" mechanism.

Data accumulated with regard to the nature of the products of thermal degradation indicate that some polymers, such as polytetrafluoroethylene and poly-2-methylstyrene, yield almost 100% monomer on pyrolysis in vacuum at temperatures up to about 500°–600°C. In contrast, polyethylene pyrolyzed under similar conditions yields a spectrum of hydrocarbon fragments varying in molecular weight from 16 (CH_4) to about 1000. Intermediate between these two extremes are polymers that pyrolyze to a mixture of monomers and chain fragments of varying sizes. There are also polymers, such as polyvinyl chloride, polyvinyl fluoride, and polymethylacrylate, which on pyrolysis yield, in addition to the more customary fragments, fragments not related in structure to the parent polymer chains (M1).

In contrast to the extensive work of the pure thermal degradation of polymers, less fundamental chemical information is available on the mechanism of oxidative degradation of polymeric materials. As another point of departure, the oxidation mechanism of simple olefins has been thoroughly investigated, and the hydroperoxidation mechanism for these reactions has been put on a sound quantitative basis. Simple thermal oxidation of hydrocarbon polymers begins almost instantly on exposure to oxygen at unhindered tertiary carbon atoms and proceeds autocatalytically. The structure of a polymer also determines its reactivity with molecular oxygen. Highly branched polymers in which the branches are not excessively bulky are more easily oxidized than linear hydrocarbon polymers. Only in sterically hindered polymeric hydrocarbons, such as poly(vinylcyclohexane), poly(allylcyclohexane), or polystyrene, is the resistance toward thermal oxidation imparted to the polymer by the side chain.

Thermal-oxidative decomposition has been studied by Beachell and Nemphos (B5), Grassie and Weir (G2), Notely (N3), Parker (P1), and Ryan (R4) for a variety of polymer systems. The principal results of these studies are:

(1) Significant surface regression of polymers appears to start at about 340°C.

(2) Endothermic pyrolysis reactions may occur below 300°C which do not result in significant weight loss of the polymers.

(3) Exothermic oxygen–polymer reactions apparently occur before ignition.

(4) Ignition in oxygen appears to start when regression of the polymer begins.

(5) Results of the fast pyrolysis reactions appear to be reasonable extrapolations of the results from conventional tests.

Unfortunately, the majority of the polymer investigations cited have been performed under the classic conditions of slow heating rates or isothermal conditions in vacuum environments. However, the residence time of a polymer element at the solid surface during the normal ignition and combustion of a solid propellant is usually on the order of milliseconds. Thus, a major portion of the heating and reaction sequence will involve transient processes. Furthermore, the pressure level is normally several hundred pounds per square inch in most actual combustion environments. Direct application of low-pressure isothermal decomposition data to propellant combustion is a questionable procedure. In addition, it is likely that the rate of the surface degradation produced by intense surface heating is quite different from that controlling bulk degradation. The need for additional studies of polymer decomposition at rapid heating rates, high pressure, and transient conditions is readily apparent. This is important for the determination of combustion mechanisms, since the rate-determining step may shift with temperature, pressure, heating rate, composition, catalysts, and extent of reaction.

3. Propellant Combustion

Returning to the behavior of propellants, the essential difference between the combustion mechanism for heterogeneous or composite propellants and the mechanisms controlling the combustion of homogeneous propellants was first suggested by Rice (R1) in 1945. He indicated that the diffusional mixing between the decomposition products of the solid fuel and the solid oxidizer is an important consideration in the overall combustion process. Geckler (G1), in an early review, has further emphasized this point but has indicated that possible exceptions could result if there is melting and subsequent mixing in the liquid phase.

The basic approach taken in the analytical studies of composite-propellant combustion represents a modification of the studies of double-base propellants. For composite propellants, it has been assumed that the solid fuel and solid oxidizer decompose at the solid surface to yield gaseous fuel and oxidizing species. These gaseous species then intermix and react in the gas phase to yield the final products of combustion and to establish the flame temperature. Part of the gas-phase heat release is then transferred back to the solid phase to sustain the decomposition processes. The temperature profile is assumed to be similar to the situation associated with double-base combustion, and, in this sense, combustion is identical in the two different types of propellants.

a. Sandwich Model. The principal result of these studies has been the incorporation of the effects of the mixing of ingredients into a theoretical description. Geckler first suggested that the combustion theory of composite propellants be based on the behavior of alternate slabs of solid fuel and oxidizer, as shown in Fig. 17. More recently, Schultz *et al.* (S1), in their review have suggested the use of the Shvab–Zeldovich procedure in the analysis. Nachbar (N1) has extended the mathematical studies to the point where preliminary predictions of the propellant burning rate can be made.

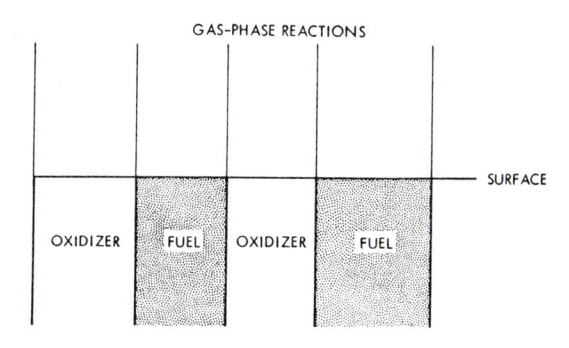

FIG. 17. Sandwich model for combustion zone of composite propellants (N1).

Basic to the sandwich model is the two-temperature concept. According to this concept, the linear regression rate of the fuel and oxidizer sandwiches must be nearly identical; i.e.,

$$\dot{r} = Z_0 \exp(-E_0/RT_0) = Z_x \exp(-E_f/RT_f) \tag{30}$$

Because the activation energy and preexponential factor for the fuel and oxidizer pyrolysis reactions are not identical, the only way for Eq. (30) to be valid is for $T_o \neq T_f$.

In his analysis, Nachbar considered the following mass and energy equations:

Mass:

$$\dot{m}\frac{\partial \tilde{Y}_k}{\partial x} + \dot{m}_y \frac{\partial \tilde{Y}_k}{\partial y} - \frac{\partial}{\partial x}\left(\rho D \frac{\partial \tilde{Y}_k}{dx}\right) + \frac{\partial}{\partial y}\left(\rho D \frac{\partial \tilde{Y}_k}{\partial y}\right) = \dot{\omega} \tag{31}$$

where $\tilde{Y}_k$ is the mole fraction of the Kth species, $K = 1, 2$, and $\dot{\omega}$ is the heat generation rate in the sandwich model.

Energy:

$$\dot{m}\frac{\partial T}{\partial x} + \dot{m}_y \frac{\partial T}{\partial y} - \frac{1}{c_g}\left[\frac{\partial}{\partial x}\left(k_g \frac{\partial T}{\partial x}\right) + \frac{\partial}{\partial y}\left(k_g \frac{\partial T}{\partial y}\right)\right] = \dot{\omega} \tag{32}$$

where k_g is the thermal conductivity of the gas, $\dot{m}_x$ the mass flow rate in the x direction, and $\dot{m}_y$ the mass flow rate in the y direction.

The relative thicknesses of the fuel and the oxidizer slab are determined by the stoichiometry of the particular propellant formulation. At the surface of the oxidizer slab, the solid oxidizer is assumed to vaporize, producing the gaseous oxidizer decomposition products. At the fuel surface, a similar assumption is made.

The principal difficulty with these equations arises from the nonlinear term $\dot{\omega}$. Because of the exponential dependence of $\dot{\omega}$ on temperature, these equations can be solved only by numerical methods. Nachbar has circumvented this difficulty by assuming very fast gas-phase reactions, and has thus obtained preliminary solutions to the mathematical model. He has also examined the implications of the two-temperature approach. Upon careful examination of the equations, he has shown that the model predicts that the slabs having the slowest regression rate will protrude above the material having the faster decomposition rate. The resulting surface then becomes one of alternate hills and valleys. The depth of each valley is then determined by the rate of the fast pyrolysis reaction relative to the slower reaction.

In making a preliminary comparison between predicted and measured burning rates, Nachbar has shown that the present analysis does not predict the observed effect of pressure on the burning rate. In fact, the model predicts the burning rate to be independent of pressure. Clearly, this is an unsatisfactory situation. Barrère (B3) has suggested that inclusion of both the correct form of the reaction-rate term and the pressure effects in ammonium perchlorate decomposition might provide the generalizations required to rectify the pressure-effect limitations of the present analysis.

b. Granular Diffusion-Flame Theory. Summerfield and co-workers (S7) have considered an alternate approach to the analysis of composite-propellant combustion. In this analysis, a steady-state flame is assumed to exist over a pyrolyzing surface in much the same way as in the combustion of double-base propellants. It is further assumed that no chemical reactions of any consequence occur within the solid. At the surface, the reaction can either be endothermic, as in the case of fuel pyrolysis or some oxidizers, or exothermic, as for ammonium perchlorate. The vaporization products of the fuel or the oxidizer are released in the form of pockets which then react in or with the surrounding atmosphere. Summerfield assumes that the average mass content of each pocket is very much smaller than that of the average oxidizer crystal, but that the mass of the pocket is somehow related to the mass of the crystal. Thus, according to Summerfield, the propellant combustion zone contains a number of oxidizer granules in various stages of decomposition and reaction, as shown in Fig. 18.

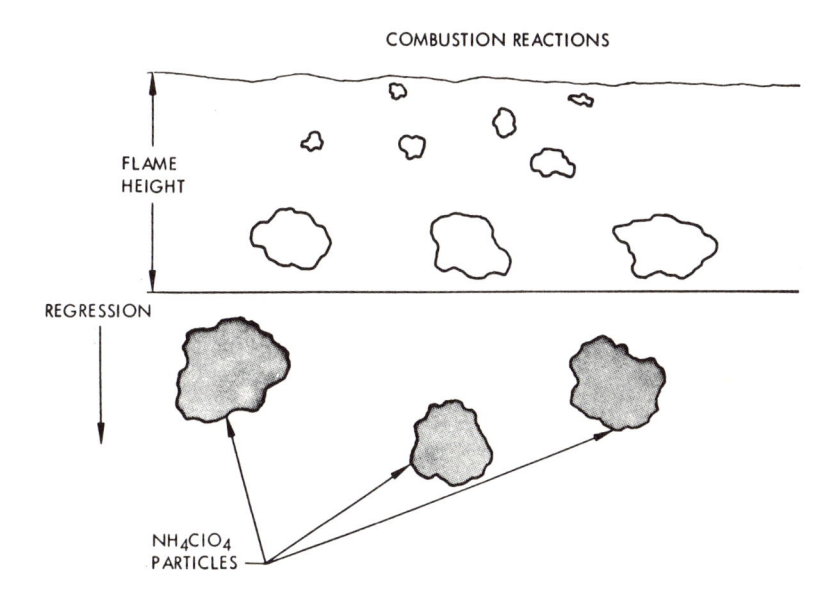

FIG. 18. Granular diffusion-flame model (S7).

Summerfield writes the energy balance at the propellant surface as

$$\dot{m}[c(T_s - T_o) - Q_v] = k_g(T_f - T_s)/L_f \quad (33)$$

where L_f is the flame thickness and Q_v the heat of vaporization. In this equation, an average surface temperature is assumed for the purpose of simplicity, even though it is recognized that the assumption may not be valid. However, the temperature difference $(T_f - T_s)$ in Eq. (33) is usually reasonably large, so that variations in surface temperature will have little effect on the final results.

Summerfield then argues that at low pressures, the combustion process should behave as a premixed flame. If the reaction is second-order, the flame thickness is given by the equation

$$L_{LP} = \frac{\dot{m}}{\rho_g^2 \exp(-E/RT_f)} \quad (34)$$

where L_{LP} is the low-pressure flame thickness.

At the other extreme, i.e., at high pressures, the mass of gas μ in a pocket determines the pocket size:

$$\mu = (P_c/RT_g)d_p^3 \quad (35)$$

where μ is the factor relating oxidizer particle size to gas pocket size and d_p is the oxidizer particle diameter. Because the diffusional relaxation time for the pocket and the flame thickness are given by the expressions

$$\tau_i \approx d_p/D \quad (36a)$$

and

$$L = m/\rho_g \tau_i \quad (36b)$$

where τ_i is the diffusion time, these equations can be combined with Eq. (33) to yield

$$L_{HP} = \frac{k_g^{1/2} \mu^{1/3}}{D_g^{1/2} \rho_g^{5/6}} \left[\frac{(T_f - T_s)}{c_g(T_s - T_o) - Q_v} \right]^{1/2} \quad (37)$$

where L_{HP} is the high-pressure flame thickness.

Summerfield then suggests that the flame thickness for the intermediate pressure case should be some weighted average between these two extremes, such as

$$L = d_1 L_{LP} + d_2 L_{HP} \quad (38)$$

where d_1 and d_2 are correlating factors. Combining Eqs. (33), (34), and (37) yields the final expression for the effect of pressure on the burning rate:

$$\frac{1}{\dot{r}} = \frac{b_1}{P_c} + \frac{b_2}{P_c^{1/3}} \quad (39)$$

where b_1 and b_2 are correlating factors.

In this equation, Summerfield has shown that the parameter b_1 should be very sensitive to the flame temperature of the propellant. At the same time, the factor b_2 should be strongly dependent on oxidizer particle size. To check these predictions, Summerfield prepared four propellants using 120 and 16 μ oxidizer particles at 75 and 80% loadings. Correlation of the burning-rate data with Eq. (39) yields the values for the parameters given in Table I. The experimentally observed trends are consistent with predicted effects.

TABLE I

Constants for Summerfield Combustion Equation

Propellant		Constants	
Loading (%)	Particle size (μ)	b_1	b_2
75	102	365	39.0
75	16	400	19.8
80	120	245	27
80	16	160	17.3

To test the theoretical development, Most et al. (M9) have noted that Eq. (39) has the burning rate increasingly dependent on pressure as the pressure decreases. At the same time, the effect of oxidizer particle size should decrease with decreasing pressure. Their experimental data show, however, that the particle-size effect persists at subatmospheric pressures. They also observed a lower-pressure deflagration limit which depends upon the particular propellant formulation. Finally, and perhaps most interestingly, they observed that certain propellant formulations would burn without the presence of a visible gas-phase combustion zone. Most et al. attribute the flameless combustion either to exothermic solid-phase reactions involving fuel and oxidizing species or else to continued exothermic deflagration of the ammonium perchlorate.

Barrère (B3) has recently presented an alternate approach to the analysis of Summerfield, but with essentially identical conclusions. Penner (P3) has offered a slightly different approach to the granular diffusion model. His expression becomes

$$\dot{r} = \frac{P_c^{1/2}}{[b^3 P_c^{\varepsilon''} + b_4/P_c]^{1/2}} \quad (40)$$

where ε'', b_3, and b_4 are correlating factors. Penner shows that ε'' can have a variety of values, depending on the nature of the mixing process. If fuel and oxidizer vaporize at the surface and interdiffuse, as in the sandwich model, $\varepsilon'' = 1$ is appropriate. If solid particles are torn off the surface to vaporize and react further in the gas phase, ε'' ranges from 0.25 to 0.6, as in droplet combustion. If the process is controlled by reactive gas pockets, as Summerfield suggests, then $\varepsilon'' = 0.33$ in Eq. (40), giving about the same accuracy as the equation derived by Summerfield [Eq. (39)]. Penner has further shown that with slight modification in the development, Eq. (40) can be rederived to yield Eq. (39).

Taking a completely different approach, Smith (S4) has proposed a model based on exothermic reactions between a gaseous molecule and a solid component occurring at the solid–gas interface. He assumes that the heat transfer from the hot gaseous combustion products is controlled by convective transport. He then considers the primary propellant reaction as occurring at the propellant surface, followed by the diffusion of reaction products away from the solid surface to undergo secondary combustion in the gas phase. The results yield a complex expression in terms of reaction rates and transport parameters. Using certain approximations, Smith shows a reasonable correlation of the experimental data reported by Summerfield, purely on the basis of surface reactions and convective energy transport.

A third alternative has been proposed by Anderson and Brown (A6, A9) as an outgrowth of their research on the ignition of composite propellants. Their ignition studies suggest significant contributions to the overall combustion process from the solid phase. Two exothermic reaction zones contributing to combustion are considered, as shown schematically in Fig. 19.

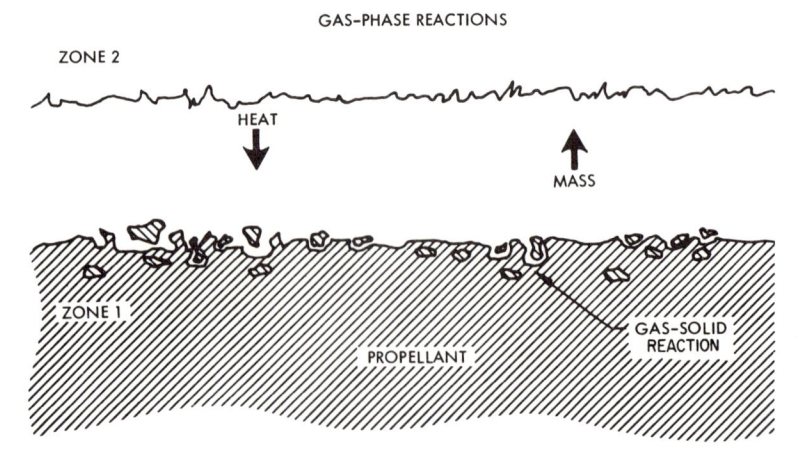

FIG. 19. Solid-propellant reaction zones (A6).

These zones are: (1) the ignition front zone at the propellant surface and in the subsurface layers, and (2) the luminous combustion zone in the gas phase.

In zone 1, the combustion processes considered are interfacial reactions between reactive decomposition species of the solid oxidizer and the solid fuel. The discussion of the decomposition characteristics of ammonium perchlorate indicate that intermediate species are formed which are highly reactive with normal propellant binders. The hypergolic ignition studies reported by these investigators have shown that both the highly reactive intermediate decomposition products, such as $HClO_4$, and the less reactive final products of decomposition, such as O_2, are both capable of chemically attacking typical binders. In addition, the analysis of the ignition response mechanisms indicate that such interfacial reactions could be important at typical propellant ignition temperatures. Because these interfacial or heterogeneous reactions may occur at temperatures of 200° to 300°C, much below the surface temperatures characteristic of steady-state combustion (400° to 700°C), it seems reasonable to expect that these reactions can be important in steady-state combustion as well. Based on this concept, these investigators picture that around each oxidizer crystal, near the propellant surface, exothermic chemical reactions liberate heat.

Figure 20, a magnitude schematic view of zone 1 in Fig. 19, depicts this effect. These exothermic oxidative reactions in zone 1 can release sufficient heat to expel partially combusted products, pyrolysis products, and fuel and oxidizer fragments into the gas phase, where they can intermix and burn completely. The maximum flame temperature will then be reached in the luminous zone, where the largest portion of the heat is released. However, a relatively

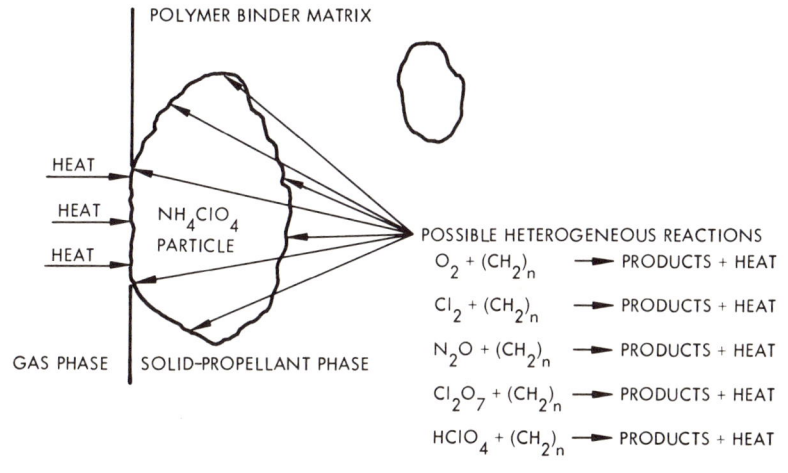

FIG. 20. Heterogeneous reactions occurring at oxidizer-particle-polymeric-binder interface (A6).

small fraction of the heat released in the luminous zone is transferred back to the propellant surface to supplement the heat generated by the heterogeneous reactions. Thus, the energy for the propagation of the ignition front in the propellant and for controlling the burning rate is supplied by the exothermic reactions (partially controlled by oxidizer decomposition) in zone 1, and by the gas-phase reactions in zone 2.

Based on this approach, an approximate energy balance around the solid-phase reaction zone is given as

$$\dot{m}[Q_v + C_p(T_s - T_o)] = \frac{k_g(T_f - T_s)}{L} + \dot{Q}_s + \dot{Q}_{ss} \qquad (41)$$

where $\dot{Q}_s$ is the heat generation at the propellant surface and $\dot{Q}_{ss}$ the heat generation below the propellant surface. Since

$$\dot{Q}_s = \left(\frac{P_c}{RT} X_o\right)^{n^+} Z \exp(-E/RT_s) \qquad (42)$$

where X_o is the mole fraction of oxidizing species, and since $\dot{Q}_{ss}$ can be approximated by the expression

$$\dot{Q}_{ss} \approx f\left(\frac{P_c}{RT} X_o\right) Z \exp(-E/RT) \qquad (43)$$

where Z depends on the binder structure, then

$$\dot{m}[Q_v + c(T_s - T_o)] = \frac{(T_f - T_s)}{L_f} + (K'P^{n^+}\exp(-E/RT_s) \\ + (K''P)^{n^+} \exp(-E/RT) \qquad (44)$$

Because the flame thickness in this equation is the same parameter discussed and analyzed by Summerfield (S7), Penner (P3), and Nachbar (N1), this equation suggests that the pressure effect on burning rate should approximate

$$\dot{m} = (a_2/L) + a_3 P_c + a_4 P_c^{a_5} \qquad (45)$$

where a_2, a_3, a_4, and a_5 are correlating factors.

The possibility of significant heat release below the surface suggests the importance of a number of variables which do not enter into analyses based on purely gas-phase processes. Among these are: (1) binder mechanical properties, and (2) coatings on the solid oxidizer, since they affect the burning rate and the pressure exponent.

Hightower and Price (H8) have conducted studies on burned and quenched ammonium perchlorate single crystals and two-dimensional propellant sandwiches prepared by laminating a thin binder layer between two ammonium perchlorate single crystals. The effect of various ballistic modifiers on the burning ammonium perchlorate was noted, as was evidence that some liquid may be present on the burning surface. Microscopic examination of the quenched sample disclosed a smooth burning surface across the oxidizer–binder interface, indicating that significant interface reactions are not occurring between the ammonium perchlorate crystal and the binder layer. These results suggest the relative unimportance of oxidative heterogeneous attack on the binder for this model configuration, but do not completely eliminate such reactions as important factors in propellant burning, where part of the surface regression involves pyrolysis of the binder layer separating successive oxidizer particles.

Powling (P7) recently reported on the results of an extensive study of the combustion characteristics of ammonium perchlorate-based composite propellants. The nature of the chemical processes taking place at the solid–gas interface and the possibility of heat release in the condensed phase were considered. Although the evidence is that some heat release is likely to occur within the solid surface, Powling found that the combustion in all pressure regions appears to be dominated by gas-phase reactions.

Most theoretical studies have concentrated on the analysis of the combustion zone of nonaluminized propellant systems. In actual practice, propellants containing aluminum are used in many applications. One study in which aluminum has been included has recently been published by Dunlop and Crowe (D2). In this study, the combustion zone is idealized to consist of four regions, as shown in Fig. 21. The results of these simplified one-dimensional analyses suggest that the combustion of aluminum particles in the gas

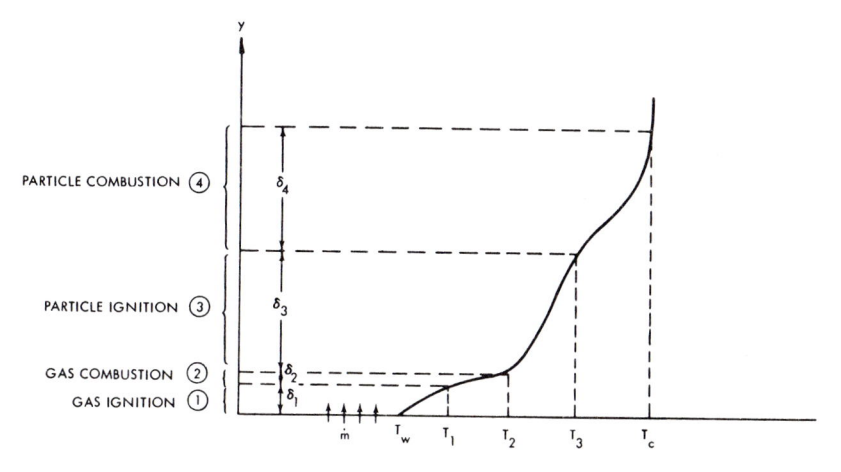

FIG. 21. Simplified model for temperature distribution in the combustion zone near a metallized solid propellant (D2).

phase should have little effect on the propellant burning rate. However, effects on the solid surface were not considered. Though many investigators have studied the combustion of metal particles in detail, little further effort has been applied to the incorporation of these processes into the analysis of propellant combustion phenomena.

C. Erosive Burning

The preceding discussions on the nature of the solid-propellant combustion zone universally agree that heat transfer from the gas phase to the solid surface contributes significantly to the overall combustion process. Because this contribution is primarily controlled by the heat-transfer characteristics of the gaseous layers over the propellant surface, the flow characteristics of the combustion gases within the chamber could have a significant effect on the combustion products through the motor. If the mass flow rate of combustion products through the motor port is sufficiently high, the resulting turbulence could alter the structure of the gas-phase flame zone and hence the burning rate. The result would be motor mass-flow rates and thrust levels which differ significantly from the ballistic design values based on nonflame burning-rate measurements. Thus, the designer is very concerned with these erosive-burning effects, since the maximum mass flow rate permitted determines the port diameter.

The effect on the propellant burning rate of gas flow parallel to the propellant surface has received considerable experimental and theoretical attention. These studies have generally shown that there is little effect of gas flow parallel to the propellant surface, provided the flow rate is below a certain critical level. However, once this critical value has been exceeded, the burning rate increases with increasing mass flow rate.

Lenoir and Robillard (L2) have developed a theoretical analysis based on this approach which yields the following expression:

$$\dot{m}[Q_v + c(T_s - T_o)] = \dot{Q}_{\text{static}} + \dot{Q}_{\text{erosive}} \tag{46}$$

where $\dot{Q}_{\text{static}}$ is the heat transfer under non-erosive conditions, and $\dot{Q}_{\text{erosive}}$ is the heat transfer contribution due to the gas velocity. Then, using the correlations for heat transfer to a transpiring surface developed by Mickley (M6), they derived an expression for the burning rate which includes the erosive effects:

$$\dot{r} = aP^n + \frac{a_5 G^{0.8}}{L^{0.2} \exp(a_6 \dot{r} P/G)} \tag{47}$$

where a_6 is a constant. Using this expression, they were able to correlate the increase in propellant burning rate along the surface of a cylindrical charge by curve-fitting to obtain the appropriate values for a_5 and a_6 suggested by the behavior of Eq. (47) shown in Fig. 22. Marklund and Lake (M2) also found that their blast-tube data were correlated by Eq. (47).

Green (G3) has proposed an alternate approach based on the concept of a critical mass-velocity required to produce a Mach number of 1 in a constant-area channel. Green showed this approach was able to correlate the erosive-burning data he obtained for both a double-base propellant and a composite propellant.

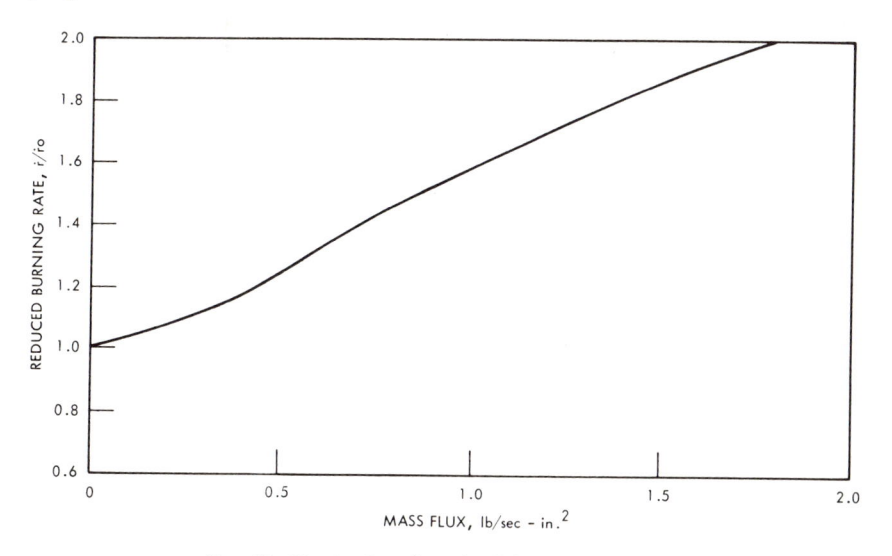

FIG. 22. Erosive burning of solid propellants (L2).

More recently, Zucrow *et al.* (Z1) have run experiments which show that in the region of low flow rates the burning rates of certain propellants actually decrease with increasing gas flow. As the gas flow rate increases, the burning rate is observed to go through a minimum and then increase with further increases in gas flow rate. The decrease in burning rate was attributed to undefined mass-transfer processes. Eventually, the convective heat-transfer processes overcome this effect to give results similar to those obtained by others.

Generally speaking, these studies of erosive burning have been able to correlate the observed effects. Until the structure of the combustion zone is defined and quantitatively characterized in detail, it would appear that the currently available bases for correlating erosive-burning data are adequate.

IV. Combustion Instability

The ballistic design derived for a solid-propellant rocket motor is based on the assumption that the burning rate in the actual motor will behave identically to the rate observed in laboratory and small-motor testing. Unfortunately, this is not always the case. During the testing of a significant number of developmental and operational design motors, the chamber pressure is observed to deviate from the values predicted from the ballistic equations and the normal burning rate, and these deviations can take several forms. In some cases, the chamber pressure is observed to oscillate at frequencies greater than 500 cps and to have amplitudes as high as 50% of the design pressure. On other occasions, the chamber pressure will grow to two and three times the design pressure at some point during the operating cycle; this behavior may be sustained during the remainder of the operation, or may be damped out. At still other times, motors have been observed to burn in a nonequilibrium manner for a short period and then apparently estinguish, only to "reignite" after a pause of several seconds; this cycle is repeated until all the propellant has been consumed.

Because any such behavior causes the motor to fail in its mission objective, these peculiar operational effects have received considerable research attention. The results of these research studies have shown that these various forms of instability result from a coupling between the transient combustion characteristics of the propellant and the transient ballistics of the combustion chamber. These instabilities are termed pressure-coupled, velocity-coupled, and bulk-coupled, and will be described below.

A. Pressure-Coupled Instability

In this mode, acoustic-pressure oscillations are similar to those established in a closed organ pipe. The resulting pressure oscillations then couple with the pressure-sensitive combustion processes to further excite the oscillating pressure and thus produce the high-pressure amplitudes.

Although this form of unstable combustion was observed as early as 1942 (C2), the problem has only recently received significant analytical and experimental attention. The combustion-chamber cavity is considered to be an acoustical oscillator, and the stability of the system is determined by the relative magnitudes of the acoustical energy sources and the acoustical energy losses. The energy sources for the oscillations are the reactions occurring in the combustion zone of the solid propellant. Acoustical losses in the system are caused by viscous and inertial drag effects in the gas phase, and by viscoelastic, mechanical, and case-bonding effects in the solid-propellant grain. Therefore, the response of each of these effects to acoustical oscillations must be known in order to accurately calculate the system's acoustic admittances and attenuations; these are needed so that a solution of the wave equations governing the acoustical behavior of the gas cavity can be obtained, and hence the stability of the combustor (chamber plus propellant) determined.

The most important parameter in the analysis of pressure-coupled combustion instability is the acoustic admittance Y, which is the ratio of the amplitude of the acoustic velocity $\tilde{V}$ to the amplitude of the acoustic pressure amplitude of the acoustic velocity $\tilde{V}$ to the amplitude of the acoustic pressure $\tilde{P}$:

$$Y = -\tilde{V}/\tilde{P} \tag{48}$$

Strittmater (S6) has presented a solution to the damped acoustic wave equation, and shown that the acoustic pressure has the form

$$P = A f(x) \exp(-ibt) \tag{49}$$

where A and b are constants. He goes on to show that

$$ib = -\rho C_s^2 Y/L_c g_c \tag{50}$$

where C_s is the sonic velocity and L_c the length of the acoustic cavity.

Thus, the exponential growth constant of the pressure oscillation is directly related to the acoustic admittance of the propellant. Hence, the acoustic admittance can be evaluated directly from the growth rate of the pressure amplitude. Ryan (R5) has also desired this espression on the basis of acoustic-energy considerations.

Hart and McClure (H2) have considered the combustion aspects of the problem and have shown the acoustic admittance may be written in the form

$$Y = -\frac{V}{P_c}\left[\frac{\mu'}{\varepsilon'} + \frac{1}{\gamma} - \sigma'\right] \tag{51}$$

where σ' is a factor to account for temperature oscillations, $\mu' = \tilde{m}/m$ (with $\tilde{m}$ the acoustic mass-burning rate), and $\varepsilon' = \tilde{P}/P_c$.

When the propellant burning rate is espressed by Eq. (5b), the parameter μ'/ε' can be considered as the transient sensitivity of the burning rate to pressure. This parameter depends on the transient combustion characteristics, and its evaluation depends on the particular model of the combustion process. Thus, the acoustic admittance provides the link between experimental observation and theoretical prediction.

The most notable theoretical analysis of the instability problem has been presented by McClure and Hart (M5). These investigators postulated a generalized combustion zone that includes a temperature-dependent and pressure-independent solid-phase reaction zone, and a temperature- and pressure-dependent gas-phase reaction zone. From this general model, Hart and McClure derived an espression for the acoustic admittance of the combustion zone, and showed the significance of this parameter in determining the stability of the combustor.

These studies have indicated that the independent parameters controlling the postulated solid-phase reactions significantly affect the resulting acoustic admittance of the combustion zone, even though these reactions were assumed to be independent of the pressure in the combustion zone. In this combustion model, the pressure oscillations cause the flame zone to move with respect to the solid surface. This effect, in turn, causes oscillations in the rate of heat transfer from the gaseous-combustion zone back to the solid surface, and hence produces oscillations in the temperature of the solid surface. The solid-phase reactions respond to these temperature oscillations, producing significant contributions to the acoustical response of the combustion zone.

Williams (W2) has recently modified the analysis of Hart and McClure by considering in more detail the effect of diffusional processes on the gas-phase reaction zone. The results of his study show that the diffusional processes tend to stabilize the gas-phase combustion process, indicating that the postulated solid-phase reactions are probably the underlying cause of the instability.

Friedly (F4) expanded the theoretical analysis of Hart and McClure and included second-order perturbation terms. His analysis shows that the linear response of the combustion zone (i.e., the acoustic admittance) is not signficantly altered by the incorporation of second-order perturbation terms. However, the second-order perturbation terms predict changes in the propellant burning rate (i.e., transition from the linear to nonlinear behavior) consistent with experimental observation. The analysis including second-order terms also shows that second-harmonic frequency oscillations of the combustion chamber can become important.

In his analysis, Friedly assumed a laminar premixed-gas-flame zone containing pressure- and temperature-dependent homogeneous chemical reactions coupled with pressure- and temperature-dependent surface reactions. His computations show that the gas-phase reactions are extremely fast and are always in phase with pressure oscillations typical of high-frequency combustion instability. However, the time lags in the solid are significant, and actions coupled with pressure- and temperature-dependent surface reactions. Friedly concluded that surface reactions control, to a large degree, the acoustic response of the combustion zone in the 1000- to 10,000-cps frequency range. In addition, he showed that if the surface reactions are sensitive to pressure oscillations, the effects of surface reactions are even more significant.

Horton and Price (H11) have obtained acoustic-admittance data for a series of double-base and composite propellants with different burning-rate characteristics. They examined the effects of pressure at various frequencies and made preliminary comparisons between their data and the theory of Hart and McClure. Exact comparisons are difficult because of the numerous independent parameters which are hard to measure. These comparisons do show qualitative agreement, but the measured admittances are significantly higher than expected from the theory.

Horton (H9, H10) has obtained additional acoustic-admittance data for a series of composite propellants. At a given frequency, decreasing the mean oxidizer particle size increases the acoustic admittance and thereby the tendency for instability. Horton also investigated the effects on the acoustic admittance of the incorporation of traces of copper chromite, a known catalyst, for the decomposition of ammonium perchlorate, lithium fluoride (a burning-rate depressant), and changes in binder; these data are difficult to analyze because of experimental errors.

A comparison of Horton's data for composite propellants with the theoretical results of Hart and Friedly is difficult. The theoretical studies are based on premixed flames, which are more appropriate for double-base propellants. The applicability of premixed flames to composite propellants is open to question, as indicated in Section II. Brown et al. (B13) have indicated that the data are consistent with the expected contributions of surface reactions in the transient combustion process. These comparisons are preliminary, however, and more research is required to study these observations in detail.

B. Velocity-Coupled Instability

Together with the acoustic pressure oscillations which can develop in a solid-propellant motor, energy-conservation considerations require that acoustic velocity oscillations be present. If the gas oscillates in a direction parallel to the burning propellant surface, these oscillations can cause significant changes in the local burning rate by changing the heat-transfer flux from the gas-phase combustion zone. This process can be viewed as transient erosive burning, and most studies have approached the analysis of this phenomenon in that light.

The velocity oscillations can occur along the motor axis (in the longitudinal mode), or rotational flows can occur (giving rise to tangential modes). Crump and Price (C4) have studied the effects of longitudinal-mode velocity-coupled instability and have shown that the maximum changes in burning rate correspond to the location of the velocity anti-mode. Povinelli (P6) has studied tangential modes experimentally; his preliminary results showed some effect of propellant burning rate on the tendency for the excitation of transverse or tangential modes. He has also conducted theoretical studies (P4) to show the effect of the erosion constant on the stability of a cylindrical burning grain. Watermeier (W1) has also reported experiments to determine the effect of the longitudinal-velocity amplitude on the mean burning; his data show the same trend observed by Crump. At the higher amplitudes, however, the burning rate goes through a maximum and then tends to decrease slowly with further increases in velocity amplitude.

C. Bulk-Coupled Instability

During the operation of some solid-propellant motors, several investigators have observed oscillations occurring at low frequences (0–500 cps), as shown in Fig. 23. These oscillations cannot be associated with any of the acoustic modes of the combustion chamber. Angelus (A11) was one of the first to investigate these low-frequency oscillations; later, Yount and Angelus (Y1) observed that the amplitude of the oscillations decreased and the frequency increased with increasing mean chamber pressure. They correlated

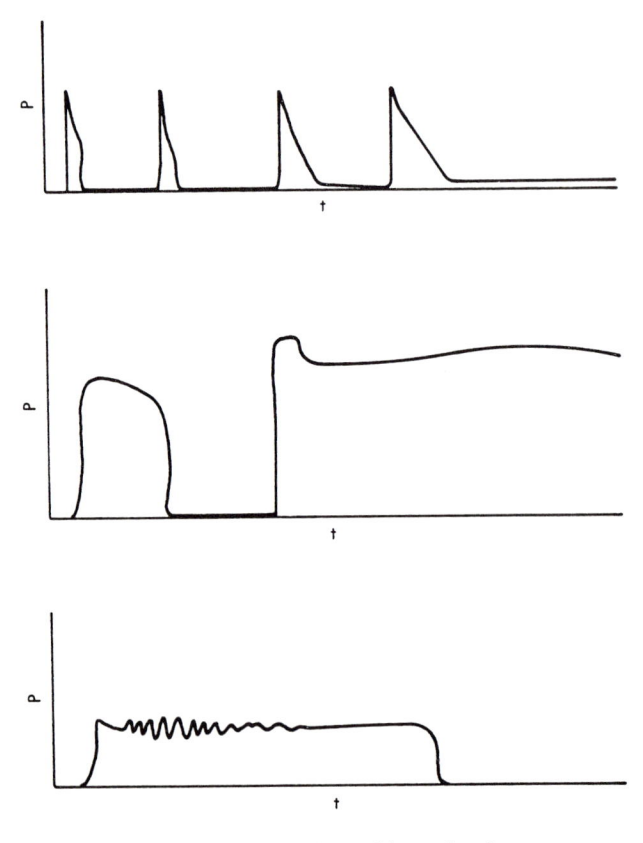

Fig. 23. Examples of unstable combustion.

their results with an analysis based on the thermal-explosion theory, thus suggesting that the problem is related to heat release within the sloid phase thus suggesting that the problem is related to heat release within the solid phase (at least for the double-base propellant they studied).

Similar phenomena have also been observed in the combustion of composite propellants. Eisel (E1) has observed that there is a unique frequency-pressure relation in a low-pressure region where nonacoustic instability results. He speculates that this preferred frequency is related to the periodic appearance and depletion of the aluminum particles on the propellant surface. High-speed pictures confirm the periodic sluffing of aluminum, but its relation to the preferred frequency is still not clear.

In an effort to determine the processes responsible for this type of behavior, Akiba and Tanno (A3), Sehgal and Strand (S2), and Beckstead (B6) have studied the coupling between the dynamics of the combustion process and the dynamic ballistics of the combustion chamber as described by Eq. (7). Each of these investigators has postulated admittedly simplified but slightly different combustion models to couple with the transient ballistic equations. Each has examined the combined equations for regions of instability. The results of these studies suggest a correlation between the L^* of the motor (the ratio of combustion-chamber volume to nozzle throat area) and the frequency of the oscillations.

Akiba and Sehgal have show that there is a critical relation between the L^* of the combustion chamber and the pressure which defines the region of stable operation. Sehgal showed that the critical relation is of the form

$$L^* \propto P_c^{-2n} \qquad (52)$$

This expression seemed to correlate the data of Anderson (A5) for non-aluminized propellants, but did not work for aluminized propellants. In later work, Sehgal (S2) has studied the aluminum effect in greater detail. He reports that the effect of aluminum appears to cause incomplete combustion. Price (P10) has reported essentially the same observation. Beckstead derived an expression between the frequency of the oscillations and the L^* of the combustion chamber. The resulting equations were then shown to correlate experimental data.

V. Combustion Termination

Often, an important consideration in the definition of a missile's mission is the ability to stop the rocket motor upon demand. Such capability provides the opportunity to exercise control over the flight path and to make corrections where necessary. Solid propellants do not readily lend themselves to start–stop applications because the oxidizer and fuel are already mixed, and no direct control exists over the mass flow into the combustion chamber. For this reason, liquid engines are often preferred over the simpler solid motors. Hence, there is considerable motivation to develop methods for terminating the combustion of solid propellants upon demand.

Research studies over the past several years have shown that least three possible methods exist for terminating propellant combustion—rapid depressurization of the combustion chamber, the L^* method, and rapid injection of a vaporizable fluid. Each of these methods initiates pressure and temperature disturbances within the combustion zone which disrupt the balance between the rate of heat generation by chemical reactions and the rate of heat loss. If the disturbances cause the heat loss to exceed the heat input, combustion will be extinguished. These three methods for achieving termination merely differ in the mechanism by which the pressure and temperature disturbances are created.

A. Rapid-Depressurization Termination

Ciepluch (C3) was the first to demonstrate that solid propellants could be extinguished by the rapid venting of gases from the combustion chamber. This was accomplished by suddenly opening a secondary nozzle to achieve the needed venting rate. If the depressurization rate was above a critical value, extinguishment could be achieved; if below it, the pressure would seek a new steady state determined by the new chamber ballistics.

Ciepluch (C3) and Povinelli (P5) investigated the effects of changes in the propellant formulation on the depressurization rate required to achieve extinguishment. The data, shown in Fig. 24, indicate that small changes in the oxidizer and aluminum content can significantly alter the critical depressurization rate. Brown has obtained data (previously unpublished), shown in Fig. 25, which indicate that the binder material has a significant effect on the critical venting rate even though the burning-rate behavior of the two propellants are nearly identical.

The combustion processes which control the critical depressurization rate are not understood. Landers (L1) and Von Elbe (V1) have tired to derive an expression for the critical depressurization rate, but the transient combustion model they used is far too simplified to predict the effects shown in Figs. 24 and 25. One possible explanation for these large variations would be that heat-release processes within the solid phase are important. From light-emission measurements during depressurization, Ciepluch observed that it was much easier to eliminate light emission than to terminate combustion (i.e., approximately 12,000 psi/sec produced light emission, compared with 100,000 psi/sec for termination).

IGNITION AND COMBUSTION OF SOLID ROCKET PROPELLANTS 59

Reactions within the solid phase could also account for the strong effect of binder properties, as shown in Fig. 25. This effect might be explained as follows: If exothermic chemical reactions occur on and within the surface layers of the propellant, as shown in Fig. 26, the subsurface reactions are confined by the surrounding binder. Heat and combustion products generated by these reactions produce significant internal pressure in the subsurface layers. As the external pressure is decreased, a pressure differential is generated between the luminous combustion zone and the ignition front. If the

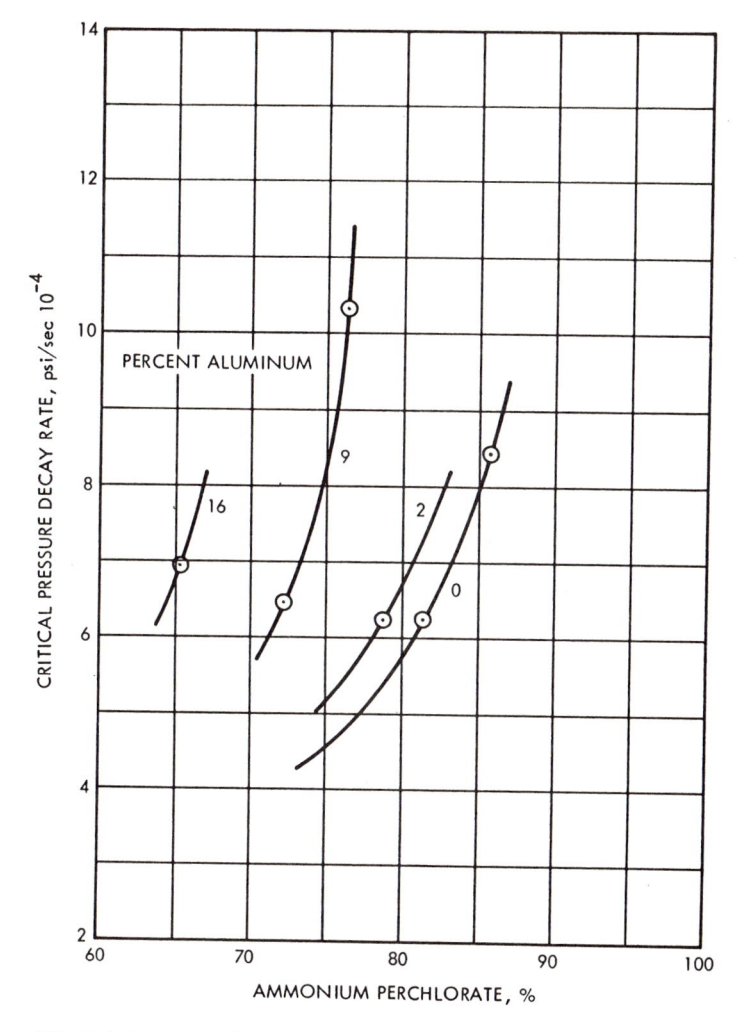

Fig. 24 Critical depressurization rate for various propellant compositions (C1, P5).

rate of pressure decrease is high enough, the combustion processes cannot follow the pressure decrease, and the pressure differential causes failure or rupture of the confining binder so that the reacting surface layers are torn away. Hence, there is no energy or reignition source remaining within the propellant to continue the ignition front, and combustion is terminated.

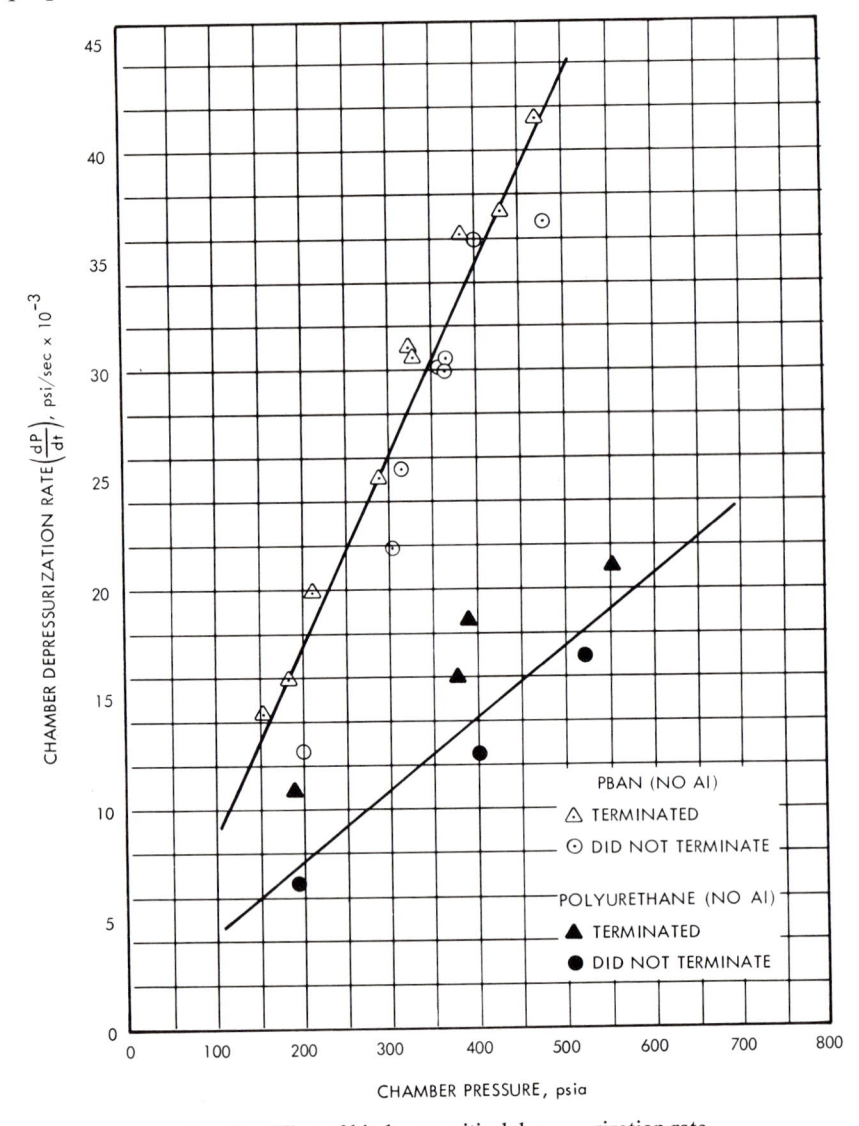

FIG. 25. Effect of binder on critical depressurization rate.

GAS PHASE
REACTIONS

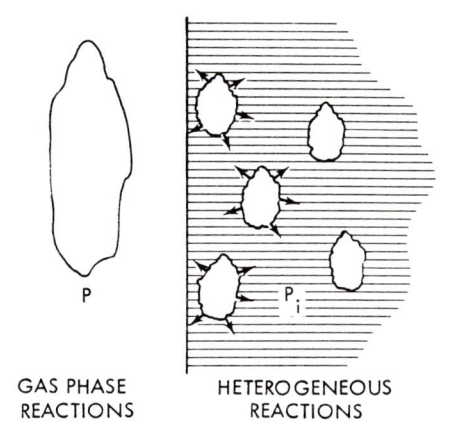

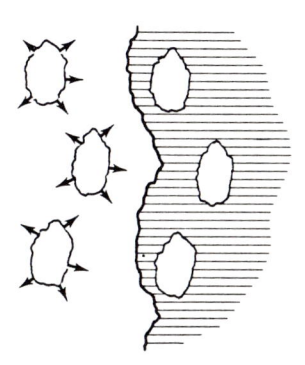

HETEROGENEOUS
REACTIONS

Fig. 26. Combustion zone during combustion termination.

When pressure-decay rates less than critical are employed, the gas-phase combustion zone is removed from the propellant surface and extinguished, but not the ignition from within the condensed phase. Therefore, the temperature of the surface material will be above the autoignition temperature, and steady-state combustion will eventually be initiated. This mechanism is consistent with the observation that the luminosity of the combustion zone can vanish without combustion having been completely terminated.

This approach offers interesting possibilities for explaining termination by rapid depressurization. However, it does depend on a special approach for modeling the combustion zone, an approach which is not understood in detail and which is not satisfactorily verified.

B. L^* Termination

In the investigation of low-frequency nonacoustic instability, Anderson (A5) found that the combustion of solid propellants in actual test motors could terminate as a result of the instability. Beckstead (B6) has also observed that chuffing (one form of low-frequency instability) sometimes precedes the extinguishment of combustion. Anderson has correlated his experimental data by showing a critical relation between L^* and chamber pressure, as derived by Sehgal, to separate the regions of stable operation from the regions of unstable operation and subsequent termination. As shown in Fig. 27, if the motor operates in the region above the critical line, normal operation is expected. On the other hand, if the operation should somehow be to the left of the critical line, termination will occur.

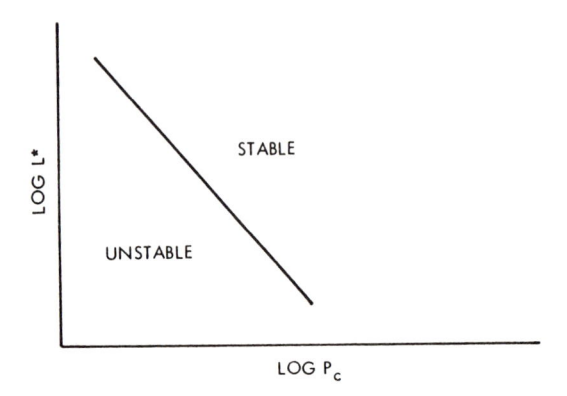

FIG. 27. Correlation of L^*-termination data (S2).

The connection between instability and termination is not clearly understood. Some investigators have suggested that the oscillating pressures resulting from the instability induce depressurization rates. Their argument says that by rearrangement of the transient ballistic mass balance [Eq. (7)], the depressurization rate is

$$\frac{dP_c}{dt} = \frac{\Gamma^2 c^* P_c}{L^*} [\rho_p K_n c^* P_c^{n-1} - 1] \qquad (53)$$

If the relation between the critical depressurization rate and pressure is then substituted into this expression,

$$L^*_{\text{crit}} = f[aP_c^{n-1} - b] \qquad (54)$$

where L^*_{crit} is the value of L^* required to sustain combustion. Because n is normally less than unity, the dynamics of the combustion chamber alone show that the inverse relationship between L^* and the pressure required to sustain combustion is consistent with the concept of a critical depressurization rate as observed in high-pressure termination studies.

Actually, the system capacitance and the pressure-decay rate measure the rate of capacitance discharge of the system. Because the combustion process is known to be controlled by pressure, the pressure-decay rate will disturb the combustion process. If the decay rate is greater than the intrinsic pressure-growth rate of the controlling reaction, the combustion process will not recover. This suggests that the pressure-decay rate dP/dt is the intrinsic term, rather than the capacitance term (L^*).

Other investigators feel that L^* termination is related to the low-pressure deflagration limit of the propellant. Their arguments are built on the fact that most L^*-extinguishment tests were conducted under conditions where the test apparatus exhausted into a vacuum. Because of the low combustion pressures involved, the experimental test motors all exhausted into a vacuum to maintain sonic flow through the nozzle. Under many extinguishment conditions, however, there is significant chuffing before extinguishment is achieved. In the rather long periods between chuffs, the chamber pressure drops to the nozzle exhaust pressure, which in many cases is of the order of the minimum deflagration pressure. To further substantiate their arguments, this school of thought cites evidence which suggests that the exhaust pressure affects the critical L^*–P_c correlation.

In actual fact, both approaches have considerable merit, and it would appear that the two schools are describing the actual physical mechanism from two different points of view. Certainly, a steady-state condition exists in which the rate of heat generation does not exceed the rate of heat loss from the combustion zone. There are also purely dynamic conditions related to the creation of the same imbalance between heat generation and heat loss. These purely static and purely dynamic conditions can be considered as the end points for a whole range of combined static (i.e., minimum-pressure) and dynamic (depressurization) conditions by which termination can be achieved. L^*-termination is probably one of these intermediate conditions.

C. Fluid-Injection Termination

If the injection of a readily vaporized liquid into a motor could be shown to terminate the combustion of the propellant, the development of such a system for operational motors would offer a number of significant advantages. In fact, several studies (A12, J1, M8, N2) have been conducted to demonstrate the applicability of combustion termination by fluid injection in a variety of motors using a wide range of propellants. One proposed mechanism for such an effect has the liquid fog which is formed acting as an evaporative cooler in the combustion chamber, rapidly reducing the temperature of the combustion gases, decreasing the heat transferred to the solid propellant surface, and thus depressing the steady-state combustion process. Concurrently, the rapid cooling of the gases produces a rapid pressure decay in the combustion chamber which further depresses the solid-propellant combustion process. If the heat generation in both gas- and solid-phase reaction zones is reduced sufficiently, combustion is terminated.

Another contributing mechanism is the direct cooling of hot propellant surface by contact with the injected fluid. The fluid should cause the decomposing surface to reduce its pyrolysis rate to a point where combustion cannot be sustained. In addition, the presence of water on the surface would obstruct heat transfer from the gas-phase reaction zones to the solid surface, thus augmenting the cooling of the surface. Proponents of these two approaches have correlated the injection data on the basis of mass of fluid required per unit area of surface, but theoretical justifications for the use of this particular correlating parameter have not been presented.

A third approach has been suggested by Jaroudi (J1), who points out that one necessary condition to prevent reignition of the propellant is to ensure that the gas temperature resulting from thermal equilibrium between the injected fluid and the combustion products is below the propellant autoignition temperature. This approach leads to the conclusion that the ratio of coolant mass flow to propellant mass flow is the critical correlating parameter.

In an effort to rationalize the basic mechanism, Brown and Jensen (B12) have solved the dynamic energy- and mass-flow equations, allowing for a finite rate of vaporization of the injected fluid. The results of these calculations have shown that both mechanisms can be important. For propellants which require relatively low depressurization rates (such as polyurethane types), the evaporative-cooling mechanism can develop sufficient depressurization rates. For PBAN propellants, direct surface-cooling is the only mechanism whereby estinguishment can be accomplished.

Based on these conclusions, Nielsen, Harris, and Kilgroe (N2) have been conducting a development program to design a water termination system for the 120-in. solid booster used on the Titan III-C vehicle. In the small-motor tests conducted as part of the program, they found that the total water required to achieve extinguishment could be correlated with the rate of fluid injection, and concluded from these preliminary results that the basic mechanism of fluid termination lies in "deigniting" (or cooling) the solid propellant phase.

Nomenclature

A	A constant	L_{LP}	Low-pressure flame thickness
a	Burning-rate factor	L_{HP}	High-pressure flame thickness
A_B	Area of burning propellant surface	$L^* = V_c/A_t$	
		L^*_{crit}	Value of L^* required to sustain combustion
A_p	Port area	$\dot{m}$	Mass flow rate
A_t	Area of nozzle throat	$\tilde{m}$	Acoustic mass burning rate
$a_1, \ldots, a_6$	Correlating factors	$\dot{m}_i$	Mass flow rate of igniter
B	Preexponential factor	$\dot{m}_x$	Mass flow rate in x direction
$b_1, \ldots, b_4$	Correlating factors	$\dot{m}_y$	Mass flow rate in y direction
C	Concentration	m_w	Molecular weight
C_f	Concentration of fuel species	N_{Nu}	Nusselt number
C_o	Concentration of oxidizing species	N_{Pr}	Prandtl number
		N_{Re}	Reynolds number
C_{oi}	Initial concentration of oxidizing species	n	Pressure exponent of propellant burning rate
C_o'	Concentration of oxidizing species at propellant surface	n'	Stoichiometric factor
		n^+	Order of surface reaction
c^*	Characteristic velocity	P	Pressure
c	Specific heat	$\tilde{P}$	Acoustic pressure
c_g	Specific heat of gas	P_a	Absolute pressure of atmosphere
C_s	Sonic velocity		
D	Diffusivity	P_c	Absolute pressure in combustion chamber
D_p	Diameter of motor port		
d, d_1, d_2	Correlation factors	P_e	Absolute pressure at nozzle exit plane
d_p	Oxidizer particle diameter		
E	Activation energy	P_i	Chamber pressure due to igniter
E_s	Activation energy of surface reaction	P_s	Pressure during ignition of motor
F	Thrust		
f_m	Ratio of mass of propellant to total mass of motor	Q	Heat of reaction
		$\dot{Q}$	Rate of heat transfer to propellant surface
G	Mass velocity		
G_n	Constants in equation for entry heat-transfer coefficients	Q_v	Heat of vaporization
		$\dot{Q}_s$	Heat generation at propellant surface
g	Exponent of mass flow rate effect on heat-transfer coefficient	$\dot{Q}_{ss}$	Heat generation below propellant surface
		$\dot{Q}_{static}$	Heat transfer under nonerosive conditions
g_c	Gravity constant		
h	Heat-transfer coefficient	$\dot{Q}_{erosive}$	Heat transfer contribution due to gas velocity
I_{sp}	Specific impulse of propellant		
$K_n = A_B/A_t$		R	Gas constant
k	Thermal conductivity of propellant	$\dot{r}$	Linear propellant burning rate
		S	Slope on plot of $\log \tau^{1/2}$ versus $\log \dot{Q}$
k_c	Mass-transfer coefficient		
k_g	Thermal conductivity of gas	T	Propellant temperature
L	Length of propellant grain	T_{ai}	Propellant-surface ignition temperature
L_c	Length of acoustic cavity		
L_f	Flame thickness		

T_e Absolute temperature at nozzle exit plane
T_f Absolute adiabatic flame temperature of propellant
T_g Gas temperature
T_i Initial gas temperature
T_o Initial propellant temperature
T_s Propellant surface temperature
t Time
V Linear gas velocity
$\tilde{V}$ Acoustic velocity
V_b Boost velocity
V_c Volume of combustion chamber

V_e Exhaust velocity
W Wetted perimeter of port
X_o Mole fraction of oxidizing species
x Distance from propellant surface
Y Acoustic admittance
$\tilde{Y}_k$ Mole fraction of Kth species
y Distance along propellant surface
y' Distance from impingement point
Z Arrhenius preexponential factor

GREEK LETTERS

α Propellant thermal diffusivity
γ Ratio of specific heats
$\Gamma = \gamma^{1/2}\left(\dfrac{2}{\gamma+1}\right)^{[(\gamma+1)/2(\gamma-1)]}$
ε Emissivity
ε' P/P_c
ε'' Correlating factor
λ A factor to account for the expansion angle of the nozzle
λ_n Eigenvalue

μ Factor relating oxidizer particle size to gas pocket size
μ' $\dot{m}/\dot{m}_o$
ρ Gas density
ρ_p Propellant density
σ Boltzmann constant
σ' A factor to account for temperature oscillations
τ ignition delay time
τ_i Diffusion time
$\dot{\omega}$ Heat generation rate in sandwich model

References

A1. Adams, G. K., et al., "Selected Combustion Problems: Fundamentals and Aeronautical Applications," p. 277. Butterworths, London, 1954.
A2. Aerojet General Corp., Final Rept. on Phase II, Contract AF 04(611)-8012 (1964).
A3. Akiba, R., and Tanno, M., *Proc. Intern. Symp. Rockets and Astronautics, 1st, Tokyo, 1959* p. 74.
A4. Altman, D., and Nichols, P., JPL Rept. 20-85 (1954) (ASTIA No. Ad 98466).
A5. Anderson, F. A., Strehlow, R. A., and Strand, L. D., TM 33-134, Jet Propulsion Lab., Pasadena, California (1963).
A6. Anderson, R., and Brown, R. S., paper presented at *ICRPG (Interagency Chem. Rocket Propulsion Group) Combust. Instability Conf., 1st, Orlando, Florida, 1964*.
A7. Anderson, R., Brown, R. S., and Shannon, L. J., Tech. Rept. TM 34-63-U2, United Technology Center (1963).
A8. Anderson, R., Brown, R. S., Shannon, L. J., *Chem. Eng. Prog. Symp. Ser.* **62**, p. 29 (1966).
A9. Anderson, R., Brown, R. S., and Shannon, L. J., Tech. Note AIAA J.**2**, 179 (1964).
A10. Anderson, R., Brown, R. S., Thompson, G. T., and Ebeling, R. W., paper presented at *AIAA Heterogeneous Combust. Conf., Palm Beach, Florida, 1963*.
A11. Angelus, T. A., *Symp. Combust. 8th, Pasedena, California, 1960* p. 921. Williams and Wilkins, Baltimore, Maryland, 1962.

A12. Auble, C. M., Brown, S. A., and Westphal, W. R., paper presented at *AIAA Solid Propellant Rocket Conf.*, *Washington, D.C., 1965*.
B1. Baer, A. D., and Ryan N. W., *AIAA J.* **3**, 884 (1965).
B2. Baer, A. D., Ryan, N. W., and Salt, D. L., " Progress in Astronautics and Rocketry," Vol. 1, "Solid Propellant Rocket Research," pp. 653–672. Academic Press, New York, 1960.
B3. Barrère, M., Williams, F. A., "Analytical and Experimental Studies of the Steady State Combustion Mechanism of Solid Propellants," 1965.
B4. Bastress, E. K., Allan, D. S., and Richardson, D. L., presented at *2nd ICRPG Combustion Conference, Cocoa Beach, Florida, 1965*.
B5. Beachell, H. C., and Nemphos, S. P., *J. Polymer Sci.* **21**, 133 (1956).
B6. Beckstead, M., Non-acoustic Instability of Solid Propellants, Ph.D. Thesis, Univ. of Utah, Salt Lake City, Utah, 1965.
B7. Beyer, R. B., and Fishman, N., " Progress in Astronautics and Rocketry," Vol. 1, "Solid Propellant Rocket Research," pp. 673–692. Academic Press, New York, 1960.
B8. Beyer, R. B., Anderson, R., MacLaren, R. O., and Corcoran, W. J., Final Rept. UTC-2079-FR, Contract No. AF 04(611)-9701. United Technology Center, Sunnyvale, California (1965).
B9. Bircumshaw, L. L., and Newman, B. H., *Proc. Roy. Soc. (London)* **A227**, 115 (1954); **A227**, 228 (1955).
B10. Bircumshaw, L. L., and Phillips, T. R., *J. Chem. Soc.* p. 4741 (1957).
B11. Brown, R. S., Final Rept., Contract NAS 7-156, United Technology Center (1965).
B12. Brown, R. S., and Jensen, G. E., *AIAA J.* **5**, 1917 (1967).
B13. Brown, R. S., Muzzy, R. J., and Steinlc, M. E., *AIAA J.* **5**, 1718 (1967).
B14. Brown, R. S., Wirrick, T. K., and Anderson, R., paper presented at *AIAA Solid Propellant Rocket Conf., Palo Alto, California, 1964*.
C1. Carlson, L. W., and Seader, J. D., Final Rept. AFRPL-TR-65-158, Rocketdyne Canoga Park, California (1965).
C2. Carpenter, C., and Longwell, P. A., Calif. Inst. of Technol. Rept. CIT JDC (1942).
C3. Ciepluch, C. C., *ARS Journal* 1584 (1961).
C4. Crump, J. E., and Price, E. W., paper presented at *AIAA Solid Propellant Rocket Conf., Palo Alto, California, 1964*.
D1. DeSoto, S., and Friedman, H. A., presented at *AIAA Solid Propellant Rocket Conf., Palo Alto, California, 1964*.
D2. Dunlop, R., and Crowe, C. T., Third Quart. Progr. Rept., Contract NOw 65-0222f, United Technology Center, Sunnyvale, California (September 1965).
E1. Eisel, J. L., Horton, M. D. and Price, E. W., paper presented at *AIAA Solid Propellant Rocket Conf., Palo Alto, California, 1964*.
F1. Fishman, N., Final Rept., Contract AF 04(611)-10534 (November 1965).
F2. Frank-Kamentskii, D. A., "Diffusion and Heat Exchange in Chemical Kinetics" (translation). Princeton Univ. Press, Princeton, New Jersey, 1955.
F3. Frazer, J. H., and Hicks, B. L., *J. Phys. Colloid Chem.* **54**, 872 (1950).
F4. Friedly, J. C., Unstable combustion of solid propellant rockets, *AIAA J.* **4**, 1604, 1932 (1966).
F5. Friedman, R., Nugent, R. G., Rumbel, K. E., and Scurlock, A. C., *Symp. Combust., 6th, Yale Univ., 1965* p. 612. Reinhold, New York, 1957.
F6. Fullman, C. H., and Nielsen, F. B., Final Rept., Contract AF 04(611)-7559 (May 1963).
G1. Geckler, R. D., "AGARD Selected Combustion Problems," p. 289. Butterworths, London, 1954.
G2. Grassie, N., and Weir, N. A., *J. Appl. Polymer Sci.* **9**, 963 (1965).

G3. Green, L., *Jet Propulsion* **24**, 386 (1954).
H1. Hansen, R. H., Bell Lab. Rept., New Jersey (1964).
H2. Hart, R. W., and McClure, F. T., *J. Chem. Phys.* **30**, 1501 (1959).
H3. Heath, G. A., and Majer, R. J., *Trans. Faraday Soc.* **60**, 1783 (1964).
H4. Hermance, C. E., Shinnar, R., and Summerfield, M., *AIAA J.* **3**, 1584 (1965).
H5. Hermance, C. E., Shinnar, R., and Summerfield, M., *Astronaut. Acta* **12**, 95 (1966).
H6. Hicks, B. L., *J. Chem. Phys.* **22**, 414 (1954).
H7. Hightower, J. D., and Price, E. W., *Symp. Combust., 11th, Berkeley, California, 1966*, p. 463. The Combustion Institute, Pittsburgh, Pennsylvania, 1967.
H8. Hightower, J. D., and Price, E. W., presented at *ICRPG Interagency Chem. Rocket Propulsion Group) Combust. Conf., 2nd, Los Angeles, California, 1965*.
H9. Horton, M. D., and Rice, D. W., *Comb. and Flame* **8**, 21 (1964).
H10. Horton, M. D., and McGie, M. R., *AIAA J.* **1**, 1319 (1963).
H11. Horton, M. D., and Price, E. W., *Symp. Combust., 9th, Cornell Univ., Ithaca, N.Y., 1962* p. 303 Academic Press, New York, 1963.
H12. Huggett, C., Bartley, C. E., and Mills, N. M., "Solid Propellant Rockets." Princeton Univ. Press, Princeton, New Jersey, 1960.
I1. Irwin, O. R., Salzman, P. K., and Anderson, W. H., *Symp. Combust., 9th, Cornell Univ., Ithaca, N.Y., 1962* Academic Press, New York, 1963.
J1. Jaroudi, R., and McDonald, A. J., Injection thrust modulation and modulation in solid rockets, *AIA J.* **2**, 2036–2037 (1964).
J2. Jensen, G. E., Final Rept., Contract NAS 7-329 (June 1966).
J3. Jensen, G. E., Brown, R. S., Cose, D. A., and Anderson, R., presented at *AIAA Propulsion Joint Specialist Conf., 2nd, Colorado Springs, Colorado, 1966*.
J4. Johnson, W. E., and Nachbar, W., *Symp. Combust., 8th, Pasedena, California, 1960* p. 678. Williams and Wilkins, Baltimore, Maryland, 1962.
K1. Kays, W. H., "Convective Heat and Mass Transfers," p. 125. McGraw-Hill, New York, 1966.
L1. Landers, L. C., Final Report, Aerojet-General, Contract AF 04(611)-9889 (1965).
L2. Lenoir, J. M., and Robillard, G., *Symp. Combust., 6th, Yale Univ. 1956* p. 663. Reinhold, New York, 1957.
M1. Madorsky, S. L., "Thermal Degradation of Organic Polymers." Wiley (Interscience), New York, 1964.
M2. Marklund, T., and Lake, A., *ARS J.* **30**, 173 (1960).
M3. McAlevy, R. F. and Summerfield, M., Tech. Rept. AFOSR TN1220, Princeton Univ. Princeton, New Jersey (June 1961), Contract No. USAF-AFSOR 49(638)-411.
M4. McAlevy, R. F., III, Cowan, P. L., and Summerfield, M., "Progress in Astronautics and Rocketry," Vol. 1, "Solid Propellant Rocket Research," pp. 623–652. Academic Press, New York, 1960.
M5. McClure, F. T., Hart, R. W., and Bird, J. F., *J. Appl. Phys.* **31**, 884 (1960).
M6. Mickley, H. S., Ross, R. C., Squyers, A. C., and Steward, W. E., NACA TN 3208 (1954).
M7. Miller, C. L., Final Rept., Contract NOw 64-0209-c, United Technology Center, Sunnyvale, California (1966).
M8. Morash, R. T., paper presented at *CPIA (Chemical Propulsion Information Agency), 22nd Meeting, Philadelphia, Pennsylvania, 1964*.
M9. Most, W. J., Wenograd, J., Summerfield, M., and 17th Progr. Rept., Contract Nor 1838(32), Princeton Univ., Princeton, New Jersey (1965).
M10. Mullis, B. G., Final Tech. Rept., Contract No. NAS 7-302, United Technology Center, Sunnyvale, California (September 1965).

N1. Nachbar, W., "Progress in Astronautics and Rocketry," Vol. 1, "Solid Propellant Rocket Research." Academic Press, New York, 1960.
N2. Nielsen, F. B., Rept. AFRPL-TR-66-9, prepared under Contract AF 04(695)-845 (January 1966).
N3. Notely, T. N., *Trans. Faraday Soc.* **58**, 66 (1962).
P1. Parker, D. B. V. Tech. Note Chem. 1284, Royal Establishment, Farnborough (1956).
P2. Parr, R. G., and Crawford, B. L., Jr., *J. Phys. Chem.* **54**, 929 (1950).
P3. Penner, S. S., "Chemical Rocket Propulsion and Combustion Research," Gordon and Breach, New York, 1962.
P4. Povinelli, L., paper presented at *ICRPG (Interagency Chem. Rocket Propulsion Group) Combust. Instability Conf., 2nd, Los Angeles, California, 1965*.
P5. Povinelli, L. A., and Ciepluch, C. C., Surface phenomena in solid propellant combustion, presented at *JANAF/ARPA/NASA Solid-Propellant Group Meeting, Pittsburgh, Pennsylvania, 1962*.
P6. Povinelli, L., and Heidmann, M. F., paper presented at *ICRPG (Interagency Chem. Rocket Propulsion Group) Combust. Instability Conf., 1st, Orlando, Florida, 1964*.
P7. Powling, J., presented at *Symp. Combust. 11th, Berkeley, California, 1966*.
P8. Price, E. W., Bradley, H. H., Jr., Dehority, G. L., and Ibiricu, M. M., *AIAA J.* **4**, 1153 (1966).
P9. Price, E. W., Bradley H. H., Jr., and Fleming, R., presented at *Western States Sect. Combust. Inst. Meeting, San Diego, California, 1963* (paper No. WSS/CI 63-6).
P10. Price, E. W., Rice, D. W., and Crump, J. E., Naval Ordinance Test Station, TP 3524 (July 1964).
R1. Rice, O. K., Office Sci. Res. Develop. Rept. 5224 (1945).
R2. Rice, O. K., and R. Ginnell, *J. Phys. Chem.* **54**, 885 (1950).
R3. Rosen, G., *J. Chem. Phys.* **32**, 89 (1960).
R4. Ryan, N. W., AFOSR Rept. 40–64 (1964).
R5. Ryan, N. W., Coates, R. L., and Baer, A. D., *Combust., 9th, Cornell Univ., Ithaca, N.Y., 1962* p. 328. Academic Press, New York, 1963.
S1. Schultz, R., Green, L., Jr., and Penner, S. S., paper presented at *AGARD Combust. Propulsion Colloq., 3rd, Palermo, Italy, 1958*.
S2. Sehgal, R., and Strand, L., *AIAA J.* **2**, 696 (1964).
S3. Shannon, L. J., and Anderson, R., Ann. Summ. Rept. 1, Contract AF 49(638)-1557 (May 1966).
S4. Smith, J. M., *A.I.Ch.E. (Am. Inst. Chem. Eng.) J.* **6**, 299 (1960).
S5. Spalding, D. B., *Combust. Flame* **4**, 59 (1960).
S6. Strittmater, R., Watermeier, L., and Pfaff, S., *Symp. 9th, Cornell Univ., Ithaca, N.Y., 1962* p. 311. Academic Press, New York, 1963.
S7. Summerfield, M., Sutherland, G. S., Webb, M. J., Taback, H. J., and Hall, K. P., "Progress in Astronautics and Rocketry," Vol. 1, "Solid-Propellant Rocket Research," pp. 141–182. Academic Press, New York, 1960.
V1. Von Elbe, G., paper presented at *CPIA Meeting, 21st, Seattle, Washington, 1963*.
W1. Watermeier, L. A., Aungst, W. P., and Strittmater, R. C., paper presented at *ICRPG (Interagency Chem. Rocket Propulsion Group) Combust. Instability Conf., 1st., Orlando, Florida, 1964*.
W2. Williams, F. A., *J. Appl. Phys.* **33**, 3153–3166 (1962).
Y1. Yount, R. A., and Angelus, T. A., paper presented at *AIAA Solid Propellant Rocket Conf., Palo Alto, California, 1964*.
Z1. Zucrow, M. J., Oxborn, J. R., Murphy, P., paper presented at *AIAA Solid Propellant Rocket Conf., Palo Alto, California, 1964*.

GAS–LIQUID–PARTICLE OPERATIONS IN CHEMICAL REACTION ENGINEERING*

Knud Østergaard

Department of Chemical Engineering
The Technical University of Denmark, Copenhagen, Denmark

I. Introduction.	71
II. Gas–Liquid–Particle Processes	73
A. Processes with Chemical Reaction between Gas, Liquid, and Solid	73
B. Processes with Chemical Reaction between Gas and Solid	76
C. Other Processes.	78
III. Gas–Liquid–Particle Operations	79
A. Fixed-Bed Operations	79
B. Suspended-Bed Operations	80
IV. Theoretical Models of Gas–Liquid–Particle Operations	81
A. Process Steps.	81
B. Models for Isothermal Operations	83
V. Transport Phenomena in Gas–Liquid–Particle Operations.	90
A. Fixed-Bed Trickle-Flow Operation	90
B. Fixed-Bed Bubble-Flow Operation	104
C. Bubble-Column Slurry Operation	108
D. Stirred-Slurry Operation	120
E. Gas–Liquid Fluidization	123
VI. Summary and Conclusions	130
Nomenclature	131
References	133

I. Introduction

A number of three-phase processes (processes in which contact is established between a gaseous phase, a liquid phase, and a solid-particle phase in order to promote chemical conversion and the transfer of momentum, heat, and mass) are becoming increasingly important in the process industries.

---

* Parts of this paper were prepared at the Department of Chemical Engineering, Massachusetts Institute of Technology, where the author was a guest in the academic year 1964–1965.

The catalytic hydrogenation of fatty oils, the desulfurization of liquid petroleum fractions by catalytic hydrogenation, Fischer–Tropsch-type synthesis in slurry reactors, and the manufacture of calcium bisulfite acid are familiar examples of this type of process, for which the term "gas–liquid–particle process" will be used in the following.

Several different types of industrial operation may be employed in order to obtain the desired contact between the three phases. They may be grouped into two main classes, depending upon the state of motion of the solid particles:

(1) In the first class, the particles form a fixed bed, and the fluid phases may be in either cocurrent or countercurrent flow. Two different flow patterns are of interest, trickle flow and bubble flow. In trickle-flow reactors, the liquid flows as a film over the particle surface, and the gas forms a continuous phase. In bubble-flow reactors, the liquid holdup is higher, and the gas forms a discontinuous, bubbling phase.

(2) In the second class, the particles are suspended in the liquid phase. Momentum may be transferred to the particles in different ways, and it is possible to distinguish between bubble-column slurry reactors (in which particles are suspended by bubble movement), stirred-slurry reactors (in which particles are suspended by bubble movement and mechanical stirring), and gas–liquid fluidized reactors (in which particles are suspended by bubble movement and cocurrent liquid flow).

All these gas–liquid–particle operations are of industrial interest. For example, desulfurization of liquid petroleum fractions by catalytic hydrogenation is carried out, on the industrial scale, in trickle-flow reactors, in bubble-column slurry reactors, and in gas–liquid fluidized reactors.

These operations are characterized by different reaction engineering properties. The transport of momentum, heat, and mass take place by different rates in the different operations, and the yield and selectivity obtained for a given chemical reaction will depend upon the type of operation employed. The operations also differ with respect to more loosely defined characteristics, such as ease of operation, and it can be noted in particular that some operations have been studied with considerably more thoroughness than others, and may consequently be designed with greater accuracy and reliability.

A considerable amount of information regarding the properties of various gas–liquid–particle operations is available in the chemical engineering literature. This information is of importance both with respect to the design of gas–liquid–particle operations and as a basis for the comparison of and choice between the various types of operation. It is of particular importance with regard to these uses that the available information be compiled and reviewed as comprehensively as possible, and it is intended in the present paper to attempt such a review.

The area of interest covered by this paper is limited to processes in which chemical conversion occurs, as in the processes noted above. Gas–liquid–particle processes in which a gaseous phase is created by the chemical reaction between a liquid and a solid (for example, the production of acetylene by the reaction between water and carbide) are excluded from the review. Also excluded are physical separation processes, such as flotation by gas–liquid–particle operation. Gas absorption in packed beds, another gas–liquid–particle operation, is not treated explicitly, although certain results for this operation must necessarily be referred to.

II. Gas–Liquid–Particle Processes

The gas–liquid–particle processes considered in this paper may be grouped into two major classes. In the first, components of all three phases participate in the chemical reaction. In the second, components of only the gaseous and the solid phases participate in the chemical reaction, the liquid phase functioning as a chemically inactive medium for the transfer of momentum, heat, and mass. Important examples of these two types of processes are described, respectively, in Sections II,A and II,B.

A. Processes with Chemical Reaction between Gas, Liquid, and Solid

1. *Catalytic Hydrogenation of Liquid Petroleum Fractions*

The catalytic hydrogenation of petroleum fractions may be carried out either in gas–particle operation, in which the petroleum fraction is completely vaporized, or in gas–liquid–particle operation, in which the petroleum fraction is only partly vaporized. Complete vaporization can be achieved by using a feed of sufficiently high hydrogen-to-oil ratio or by using a sufficiently high operating temperature. The first method may be too costly, requiring large reactor volumes and recirculation of large volumes of hydrogen, and the second may promote undesirable secondary reactions, such as thermal cracking and catalyst deactivation. Generally speaking, gas–particle operation may be advantageous for fractions of low molecular weight, such as refinery gases and naphthas, whereas gas–liquid–particle operation, although more complex and considerably less well understood than the former, may be advantageous for fractions of high molecular weight, such as gas oils and fuel oils. (Exceptions to this general rule are known, however, and one is given in the first example below.)

A number of industrially important processes are carried out by gas–liquid–particle operation, as illustrated by the following four examples.

a. Hydrogenation of C_4-*Hydrocarbons.* Krönig (K23) has described several processes for the selective catalytic hydrogenation of the C_4-acetylenes and the butadiene occurring in C_4-fractions obtained in the production of olefins by the pyrolysis of liquid or liquefiable hydrocarbons.

Because of the high pyrolysis temperature, the C_4-fraction contains quantities of vinyl acetylene and ethyl acetylene, the removal of which prior to the recovery of butadiene is necessary in certain cases, particularly if butadiene of low acetylene content is desired. Similar considerations apply to C_4-fractions obtained by the dehydrogenation of *n*-butane and *n*-butenes.

The removal of C_4-acetylenes may be effected by catalytic gas-phase hydrogenation in a gas–particle operation by a process similar to that widely used for removing acetylene from ethylene streams. However, in view of the strong polymerization tendency of the C_4-fractions, it is desirable in this case to work at the lowest possible temperature.

A gas–liquid–particle process termed "cold hydrogenation" has been developed for this purpose. The hydrogenation is carried out in fixed-bed operation, the liquefied hydrocarbon feed trickling downwards in a hydrogen atmosphere over the solid catalyst, which may be a noble metal catalyst on an inert carrier. Typical process conditions are a temperature of 10°–20°C and a pressure of 2.5–7 atm gauge. The hourly throughput is as high as 20-kg hydrocarbon feed per liter of catalyst volume.

The process is characterized by high yield (nearly complete hydrogenation of acetylenes) and high selectivity (only a small loss of butadiene by hydrogenation). The process does not lead to polymerization, which might otherwise cause catalyst deactivation, and only infrequent regeneration of catalyst is necessary.

Cold hydrogenation may also be used for the selective hydrogenation of butadiene and for the selective hydrogenation of methyl acetylene and propadiene in propylene feedstocks (K22).

b. Catalytic Desulfurization of Petroleum Fractions. The current emphasis on the desulfurization of petroleum fractions is due in part to the necessity of using crude oils of relatively high sulfur content, while at the same time low-sulfur content is preferred in intermediate refinery streams, where sulfur may act as a catalyst poison, and in final products, where sulfur may cause corrosion and pollution.

Processes other than catalytic hydrogenation may be used for desulfurization—extraction, for example. An important reason for the widespread use of catalytic hydrogenation for this purpose is the availability of hydrogen as a byproduct of catalytic reforming. The growing demand for gasoline of increasing octane number has caused a large, continuing expansion of reforming capacity. Since the hydrogen produced in reforming processes was in many cases (at least in the earlier stages of this development) available at fuel value, its use in desulfurization processes was advantageous (H11).

Catalytic desulfurization is at present carried out industrially by at least three of the major types of gas–liquid–particle operations referred to in Section I: trickle reactors, bubble-column slurry reactors, and gas–liquid fluidized reactors.

Trickle-bed operation is the oldest and the most commonly used; its development is described in a recent publication (V1). Cobalt–molybdenum catalysts may be used at a temperature of 360°C and a pressure of 57 atm for the hydrogenation of straight-run gas oils.

Bubble-column slurry reactors are used in Germany and in Eastern European countries for catalytic desulfurization as well as for the hydrocracking of crude oils and residues. The solid catalyst is charged to the gas–liquid mixture as a paste, is transferred in suspension in the reacting mixture through the reactors, and is recovered at a later process stage. Typical reaction conditions are 300 atm and 470°C in the Scholven process for the hydrogenation of vacuum residues (U1), and 60 atm and 440°C in the Varga process for the hydrogenation of topped crude oil (K1).

Gas–liquid fluidization is employed in the H-Oil process developed in the United States (H6). Cobalt–molybdenum catalyst particles of $\frac{1}{32}$-in. diameter may be used at a reaction pressure of 100 atm or more and a temperature of about 400°C (V4).

c. Liquid-Phase Hydrocracking. Hydrocracking is a process of rapidly increasing importance to the petroleum industry (H9). In this process, the molecular weight of a heavier petroleum fraction is reduced by catalytic hydrogenation. Hydrocracking may be carried out by gas–liquid–particle operation in, for example, a fixed bed, a bubble-column slurry reactor, or gas–liquid fluidized reactors. The Scholven and Varga processes referred to in the previous section thus combine desulfurization and hydrocracking. A modification of the H-Oil process, the Hy-C process, has been developed for hydrocracking (C8).

d. Hydrogenation of Lubricating Oils. The use of trickle-bed operation for the catalytic hydrogenation of lubricating oils has been reported; a main purpose of the process is improvement in product appearance.

2. *Catalytic Hydrogenation of Unsaturated Fats*

Hydrogenation of unsaturated fats and fatty oils is one of the oldest heterogeneous catalytic processes of industrial significance, and is carried out exclusively by gas–liquid–particle operation, the vaporization of the fats being impracticable. Stirred-slurry operation is the normal mode of operation, the suspended catalyst being finely divided by Raney nickel (B2).

3. Production of Calcium Acid Sulfite

The calcium bisulfite acid used in the manufacture of sulfite cellulose is the product of reaction between gaseous sulfur dioxide, liquid water, and limestone. The reaction is normally carried out in trickle-bed reactors by the so-called Jenssen tower operation (E3). The use of gas–liquid fluidized beds has been suggested for this purpose (V7). The process is an example of a noncatalytic process involving three phases.

4. Synthesis of Butynediol

2-Butyne-1,4-diol and propargylalcohol are produced by reaction between formaldehyde in aqueous solution and gaseous acetylene in the presence of a copper acetylide catalyst supported on nickel. The process is carried out by trickle-flow operation (B10, S4).

5. Production of Sorbitol

Sorbitol is produced by a gas–liquid–particle process in which a solution of glucose is hydrogenated in the presence of a solid catalyst consisting of nickel on diatomaceous earth carrier (B6).

6. Hydrogenation of Carboxylic Acids

The production of alcohols by the catalytic hydrogenation of carboxylic acids in gas–liquid–particle operation has been described. The process may be based on fixed-bed or on slurry-bed operation. It may be used, for example, for the production of hexane-1,6-diol by the reduction of an aqueous solution of adipic acid, and for the production of a mixture of hexane-1,6-diol, pentane-1,5-diol, and butane-1,4-diol by the reduction of a reaction mixture resulting from cyclohexane oxidation (C10).

B. Processes with Chemical Reaction between Gas and Solid

Gas-phase reactions catalyzed by solid catalysts are normally carried out in gas–particle operation in either fixed or fluidized beds. The possibility of using gas–liquid–particle operations for such reactions is, however, of interest in certain cases, particularly if the presence of a liquid medium for the transfer of heat or mass is desirable.

1. The Fischer–Tropsch Process

In the Fischer–Tropsch process, carbon monoxide reacts with hydrogen in the presence of a solid catalyst, with the formation of a mixture of hydrocarbons. The composition of the product varies considerably with the catalyst and the operating conditions. The mixture may include (in addition to hydrocarbons) alcohols, aldehydes, ketones, and acids.

A number of such processes were established before the second World War in Germany, Japan, and France for the production of hydrocarbon mixtures in the liquid fuel range (P2). This way of manufacturing automotive fuels is now uneconomical in most areas, but related processes may be utilized for the production of various chemicals, such as paraffinic waxes or oxygenated compounds. (The manufacture of methanol from carbon monoxide and hydrogen, usually by catalytic reaction in fixed-bed gas–particle operation, is an important process of this type.)

Commercial Fischer–Tropsch processes have been based exclusively on gas–particle operations, mainly in fixed beds (P2). The chemical reactions are highly exothermic, however, and accurate temperature control is therefore difficult to achieve in a fixed bed. Good temperature control is important because of the temperature sensitivity of the chemical reactions taking place, and several attempts have therefore been made to develop processes based on other types of operation.

In a number of these processes, a liquid heat- and mass-transfer medium is in direct contact with the catalyst and the reaction mixture. The main function of the liquid is to act as a heat sink and as a medium for convective heat transfer. However, since the liquid may be assumed to cover the solid particles and in this way act as a barrier between the gaseous and the solid phases, it must therefore also function as a mass-transfer medium.

Fixed-bed processes, in which a cooling oil is passed cocurrently with the synthesis gas through the catalyst bed and recirculated to the bed after cooling in an external heat exchanger, have been developed in Germany and the United States (B7, C12, K2, K9).

Slurry processes, in which a finely divided catalyst is suspended in a liquid medium which is cooled either by means of heat-transfer surfaces in the reactor or by recirculation of the liquid through external heat exchangers, have been under investigation for this purpose in Germany, the United States, and the United Kingdom (C12, F3, H1, K5, K7, K8, K12).

2. *Catalytic Oxidation of Olefins*

Epoxides such as ethylene oxide and higher olefin oxides may be produced by the catalytic oxidation of olefins in gas–liquid–particle operations of the slurry type (S7). The finely divided catalyst (for example, silver oxide on silica gel carrier) is suspended in a chemically inactive liquid, such as dibutylphthalate. The liquid functions as a heat sink and a heat-transfer medium, as in the three-phase Fischer–Tropsch processes. It is claimed that the process, because of the superior heat-transfer properties of the slurry reactor, may be operated at high olefin concentrations in the gaseous process stream without loss with respect to yield and selectivity, and that propylene oxide and higher epoxides which are difficult to produce by direct catalytic vapor-phase oxidation may be produced by this method.

It has been pointed out (S2) that this type of operation might be widely applicable for organic oxidation processes, provided suitable inert carrier liquids can be found. It may be noted in this connection that the liquid must be reasonably resistant against oxidation and that it must not cause catalyst deactivation—for example, by chemisorption.

3. *Catalytic Hydration of Olefins*

The heterogeneous catalytic hydration of light olefins, such as ethylene, propylene, and butenes, may be carried out under process conditions (higher pressure and high water-to-olefin ratio) such that both liquid and vapor phases are present in the reactor (Z1). Experiments have shown that much higher conversions may be obtained in such mixed-phase processes than in conventional direct vapor-phase hydration. (The relatively low conversions obtainable by direct vapor-phase hydration where the equilibrium relationship is unfavorable is a serious disadvantage of that type of process.) It seems likely that this difference is due to the solubilities of the product alcohols in the liquid phase, which consists largely of water. Presumably, the equilibrium conversion is higher in the mixed-phase system than in the vapor-phase system because the product alcohols are removed from the catalyst surface by solution in the aqueous liquid phase.

The liquid phase in this type of process seems to function as a sink and transfer medium for mass, in analogy to its function as a heat sink and heat-transfer medium in the processes referred to in Sections II,B,1 and II,B,2.

4. *Polymerization of Ethylene*

As a final example of the application of gas–liquid–particle operation to a process involving a gaseous reactant and a solid catalyst, the possibility of polymerizing ethylene in, for example, a slurry operation employing a metal or metal oxide catalyst can be cited. It has been suggested that the good control of reaction conditions obtained in a slurry-type operation may be of importance in the production of certain types of polyethylene (R1).

C. Other Processes

A number of important gas–liquid–particle processes fall outside the categories of Sections II,A and II,B. The review of gas–liquid–particle operations in the following sections is written with particular regard to applications in processes of the types already referred to, but may also be of some significance with regard to other types. A few examples of such processes will be briefly mentioned below.

Gas absorption in packed beds may be described as a gas–liquid–particle process involving reacting gas and liquid phases and an inert particle phase, the latter functioning mainly as a momentum-transfer medium.

An example of a gas–liquid–particle process involving reacting liquid and solid phases and an inert gaseous phase is the cleaning of sand filters in water-treating plants. The sand filter is fluidized by an upward flow of cleaning water, and gas is injected into the liquid fluidized bed in order to provide stirring. The gas functions as a momentum-transfer medium in the resulting gas–liquid fluidized bed (H2).

Processes in which two phases react and result in the formation of a third form an important group of gas–liquid–particle processes. In the production of acetylene, a gaseous phase is formed by reaction between a liquid and a particle phase: water and carbide. In the production of gas hydrates in desalination processes, a particle phase is formed by reaction between a liquid and a gaseous phase: sea water and, for example, propane. In the melting of gas-hydrate or ice crystals a liquid phase is formed when gaseous and particle phases are brought in contact.

It may finally be pointed out that certain separation processes in addition to packed-bed gas absorption are gas–liquid–particle operations. Examples are flotation and a special type of fluidized crystallization process (Z2).

III. Gas–Liquid–Particle Operations

The operations considered in this paper may be classified under the headings of fixed-bed operations or suspended-bed operations, depending on the state of motion of the solid particles.

A. Fixed-Bed Operations

Two types of fixed-bed operations, characterized by distinctly different flow patterns, are in current industrial use. These are usually described as trickle-flow operation and bubble-flow operation. In both cases, a lower limit exists for the particle size, usually about $\frac{1}{8}$ in.

1. *Trickle-Flow Operation*

In trickle-flow operation, the liquid phase flows downwards, and may or may not cover the solid particles as a film. The gaseous phase moves in either co- or countercurrent, continuous flow.

Trickle-flow operation is widely used for large-scale gas–liquid–particle processes, as noted in Section II.

2. *Bubble-Flow Operation*

In bubble-flow operation, the gaseous phase moves upwards as discrete bubbles. The liquid phase may be in either co- or countercurrent flow. The liquid holdup is relatively high.

B. SUSPENDED-BED OPERATIONS

In this group of operations, the solid particles are kept in a suspended state by momentum transfer from the liquid phase. Momentum may be transferred to the liquid phase by different means, and distinction will be made on this basis between three different types of operations. An upper limit exists for the particle size that can be used in suspended bed operations, and is of the order of $\frac{1}{4}$ in.

1. *Bubble-Column Slurry Reactors*

In bubble-column slurry reactors, momentum is transferred to the liquid phase by the movement of gas bubbles. The liquid medium is stationary in most cases. Finely divided solids with particle diameters of the order of 0.01 mm are used. The operation is usually carried out in columns with high height-to-diameter ratios. The operation may be employed for batchwise conversion of a liquid reactant, or for continuous reaction between gaseous reactants.

The use of this type of operation for Fischer–Tropsch synthesis or similar large-scale processes has been referred to in Section II.

2. *Stirred-Slurry Reactors*

In stirred-slurry reactors, momentum is transferred to the liquid phase by mechanical stirring as well as by the movement of gas bubbles. Small particles are used in most cases, and the operation is usually carried out in tank reactors with low height-to-diameter ratios. The operation is in widespread use for processes involving liquid reactants, either batchwise or continuous—for example, for the batchwise hydrogenation of fats as referred to in Section II.

3. *Gas–Liquid Fluidization*

In gas–liquid fluidization, the liquid flows upwards through a bed of solid particles which is fluidized by the flowing liquid, while the gaseous phase moves as discrete bubbles through the liquid-fluidized bed. Relatively large particles may be employed in this operation. The operation may be used for the continuous processing of liquid as well as gaseous reactants. It has been used for the industrial-scale hydrogenation of petroleum fractions.

If the liquid flow rate is increased above the value required for conveying the solid particles, they will be carried out of the reactor with the liquid. This effect may be of interest in cases where frequent catalyst regeneration is necessary.

IV. Theoretical Models of Gas–Liquid–Particle Operations

The purpose of this section is to present a general theoretical model of gas–liquid–particle operations, with a number of simplifying assumptions that make possible, at least in principle, the calculation of the conversion and yield from a specified amount of information regarding transport phenomena and reaction kinetics.

Gas–liquid–particle operations are of a comparatively complicated physical nature: Three phases are present, the flow patterns are extremely complex, and the number of elementary process steps may be quite large. Exact mathematical models of the fluid flow and the mass and heat transport in these operations probably cannot be developed at the present time. Descriptions of these systems will be based upon simplified concepts.

It seems probable that a fruitful approach to a simplified, general description of gas–liquid–particle operation can be based upon the film (or boundary-resistance) theory of transport processes in combination with theories of backmixing or axial diffusion. Most previously described models of gas–liquid–particle operation are of this type, and practically all experimental data reported in the literature are correlated in terms of such conventional chemical engineering concepts. In view of the so far rather limited success of more advanced concepts (such as those based on turbulence theory) for even the description of single-phase and two-phase chemical engineering systems, it appears unlikely that they should, in the near future, become of great practical importance in the description of the considerably more complex three-phase systems that are the subject of the present review.

In this section, a number of important elementary process steps into which a gas–liquid–particle process can be subdivided will be mentioned. Several theoretical models proposed in the literature will be discussed, and a slightly more comprehensive model will be described.

A. Process Steps

It is assumed in this section and the next that the solid particles are completely wetted by the liquid, and, as a consequence, that the gaseous phase is not in direct contact with the solid. Systems may conceivably exist in which the particles are not completely wetted by liquid, either because of poor liquid distribution or because of the surface properties of liquid and solid, but such systems are not adequately described by the models proposed in the following.

The most complex type of gas–liquid–particle process is one in which gaseous components participate in a heterogeneous catalytic reaction, with the formation of gaseous products. The following elementary steps must occur in a process of this type:

1. *Absorption of Gaseous Reactants*

The gaseous components must be transferred from the bulk gaseous phase to the bulk liquid phase. The components are transferred to the gas–liquid interface by convection and diffusion in the gas and from the interface by diffusion and convection in the liquid.

2. *Transfer of Reactants to the Exterior Particle Surface*

The absorbed components must be transferred from the bulk liquid phase to the exterior surface of the catalyst by convection and diffusion.

3. *Catalytic Reaction*

The catalytic reaction can be subdivided into pore diffusion and chemisorption of reactants, chemical surface reaction, and desorption and pore diffusion of products, the number of steps depending upon the nature of the catalyst and the catalytic reaction.

4. *Transfer of Products from the Exterior Particle Surface*

The products of the catalytic reaction are transferred by convection and diffusion to the bulk liquid phase.

5. *Desorption of Gaseous Products*

The absorbed products are transferred across the gas–liquid interface by convective and diffusive transport.

The process steps mentioned are not of importance in all gas–liquid–particle processes. In particular, the last step does not occur in processes in which a liquid product is formed by reaction between gaseous and liquid reactants, as may be the case, for example, in the catalytic hydrogenation of liquid petroleum fractions.

In addition to the process steps described above involving mass transfer to and across interfaces, mass transfer by fluid flow through the reactor must also be taken into account.

Finally, in the case of nonisothermal processes, the overall heat transfer in the process must be analyzed, preferably in terms of elementary process steps similar to those discussed for mass transfer.

B. Models for Isothermal Operations

A number of theoretical models of gas–liquid–particle operations have recently appeared in the literature. Those considered most significant will be discussed briefly in the following, as introduction to a somewhat more general model that will be proposed in this paper.

Calderbank *et al.* (C6) examined the Fischer–Tropsch reaction in a slurry operation with a view to establishing the rate-determining process steps. It was assumed that the diffusion of hydrogen in the liquid phase away from the gas–liquid interface was, together with the catalytic reaction rate-determining, while the gas-side resistance to hydrogen absorption was assumed negligible, as was the resistance to hydrogen transfer to the solid catalyst particle. The experimental data suggested a pseudo-first-order dependence on hydrogen concentration in the gas phase. Consequently the molal rate of disappearance of hydrogen per unit volume would be:

$$k_0 c_g \varepsilon_g = k_l A_g (c_l^* - c_l) = k_c c_p (1 - \varepsilon_g) Q_1 \approx k_c c_p Q_1 \tag{1}$$

where k_0 is the pseudo-first-order reaction rate constant, k_l the liquid-side film coefficient for absorption, k_c the rate constant of the surface reaction, c_g the hydrogen concentration in the gas phase, ε_g the fractional gas holdup, A_g the specific gas–liquid interfacial area, Q_1 the holdup of catalyst, and c_l^* the equilibrium concentration of hydrogen in the liquid phase. Under the assumption that Henry's law holds and that the surface concentration c_p is in equilibrium with the bulk liquid concentration c_l, to which it is related by a linear adsorption isotherm, the following relationships are obtained:

$$c_l^* = RT c_g / m \tag{2}$$

and

$$c_p = K_s c_l \tag{3}$$

m being the Henry law constant and K_s the adsorption equilibrium constant.

Substitution of Eqs. (2) and (3) in Eq. (1) yields

$$\frac{RT}{m} \frac{1}{k_0} = \frac{\varepsilon_g}{k_l A_g} + \frac{\varepsilon_g}{k_c Q_1 K_s} \tag{4}$$

This equation, however, does not adequately represent the experimental data when used to correlate measured values of k_0, A_g and Q_1. It was modified empirically, but the modified equation does not account for observed variations of conversion with bed height and pressure. The proposed model would therefore appear to rest on oversimplified assumptions.

Kölbel and Maennig (K11, K18, M1) studied the hydrogenation of ethylene catalyzed by Raney-nickel suspended in hydrogenated Kogasin II. A theoretical model of the process was based on two assumptions: (1) That the process can be described by an overall kinetic expression of the form

$$r_{\text{eff}} = [k_0 \exp(-E_{\text{eff}}/RT)][x_{H_2}^k x_{C_2H_4}^l] \tag{5}$$

where $k_0 \exp(-E_{\text{eff}}/RT)$ is the effective rate constant and x_{H_2} and $x_{C_2H_4}$ are mole fractions in the gas phase of hydrogen and ethylene, respectively, and (2) that the overall process rate can be derived from experimental measurements by use of the expression

$$r_{\text{eff}} = \frac{u_g p_{j0}}{v_j RT} \frac{dX_j}{dz} \tag{6}$$

where u_g is the superficial gas velocity, p_{j0} the partial pressure of component j at the reactor entrance v_j the stoichiometric coefficient of component j, X_j the conversion of component j, and z the linear distance from the reactor entrance. When used for interpreting experimental data, the rather empirical nature of the model is reflected in the fact that neither rate constant, activation energy, nor reaction order with respect to hydrogen are true constants. In particular, these parameters increase with increasing gas flow rate.

In a later publication, Kölbel et al. (K16) have proposed a less empirical model based on the assumption that the rate-determining steps for a slurry process are the catalytic reaction and the mass transfer across the gas–liquid interface. When used for the hydrogenation of carbon monoxide to methane, the process rate is expressed as moles carbon monoxide consumed per hour and per cubic meter of slurry:

$$r = q_{CO} \frac{dX_{CO}}{dz} \tag{7}$$

where q_{CO} is the flow rate of carbon monoxide, X_{CO} conversion with respect to carbon monoxide, and z distance from the reactor entrance. In the stationary state, the process rate is identical to the rate of the catalytic reaction:

$$r = (1 - \varepsilon_g) k c_c c_l^\alpha \tag{8}$$

and to the rate of the mass-transfer step:

$$r = k_l A_g (HP x_{H_2} - c_l) \frac{1}{v_{H_2}} \tag{9}$$

whereby it is assumed that the reaction is of first order with respect to hydrogen concentration at the catalyst surface and that mass transfer of hydrogen is rate-determining. Here, k is a reaction rate constant, c_c the concentration of solids in the slurry, c_l the hydrogen concentration in the liquid, H the proportionality constant from Henry's law (which is assumed applicable),

P the total pressure, x_{H_2} the mole fraction of hydrogen in the gas phase, and v_{H_2} the stoichiometric coefficient of hydrogen. It is assumed that the hydrogen concentration at the catalyst surface is in equilibrium with the hydrogen concentration in the liquid and is related to this through a Freundlich isotherm with the exponent α. The quantity x_{H_2} is related to X_{CO} by stoichiometry, and ε_g and A_g are related to X_{CO} because the reaction is accompanied by reduction of the gas-phase volume. The corresponding relationships are introduced into Eqs. (7)–(9), and these equations are solved by analog computation.

When experimentally determined values of ε_g, A_g, and k_l are used, the analog computation yields graphical relationships between X_{CO} and z. Effects of axial mixing are not accounted for in the analog computer program. It is reported that the results of the computation are corrected for these effects by the graphical method of Schoenemann (S4), and that the data of Siemes and Weiss (S14) for axial mixing in the liquid phase are used if the catalytic reaction is the slowest step, whereas the data of Kölbel et al. (K17) for axial mixing in the gas phase are used if the absorption step is slowest. The effects can, alternatively, be compensated for by introducing an effective reaction order, as done in the theoretical model proposed by Kölbel and Maennig and described above. Experimental and computed values of conversion versus gas velocity and temperature are reported and are in reasonable agreement.

Farkas and Sherwood (F1, S5) have interpreted several sets of experimental data using a theoretical model in which account is taken of mass transfer across the gas–liquid interface, of mass transfer from the liquid to the catalyst particles, and of the catalytic reaction. The rates of these elementary process steps must be identical in the stationary state, and may, for the catalytic hydrogenation of α-methylstyrene, be expressed by:

$$r = k_l A_g(c^* - c_l) = k_s A_p(c_l - c_p) = k_r A_p c_p \tag{10}$$

where c_p is the hydrogen concentration at the catalyst surface, A_p the external (geometric) surface area of the particles, k_s the coefficient of hydrogen transfer from liquid to particles, and k_r the rate constant of the chemical reaction (which is assumed to be of first order with respect to hydrogen and of zero order with respect to α-methylstyrene). The equilibrium concentration of hydrogen in the liquid phase, c^*, is related to the partial pressure of hydrogen by Henry's law:

$$c^* = \frac{P_{H_2}}{m} \tag{11}$$

Equation (10) can be rewritten:

$$r = \frac{P_{H_2}}{m} \left(\frac{1}{k_l A_g} + \frac{1}{k_s A_p} + \frac{1}{k_r A_p} \right) \tag{12}$$

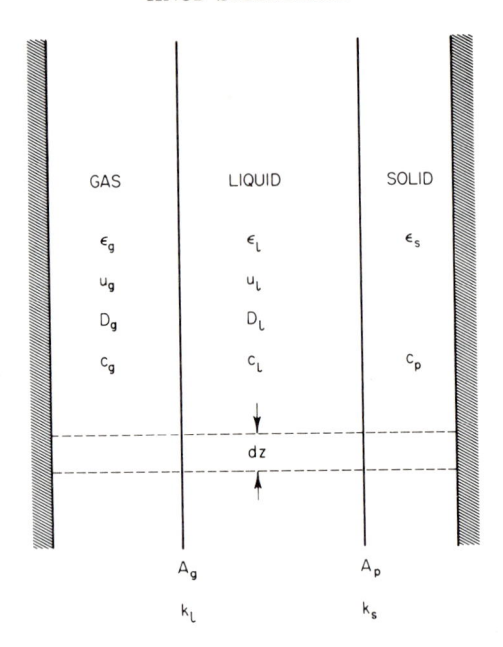

FIG. 1. Model of gas–liquid–particle operation.

where D_g and D_l are the superficial axial dispersion coefficients (based on the empty cross section of the column) of gas and liquid, respectively, u_g and u_l the nominal velocities of gas and liquid, respectively, m the ratio between equilibrium concentrations in gas and liquid, and $A_p f(c_p)$ the rate of the catalytic reaction; (it is assumed that the latter is proportional to the exterior particle surface area and that the reaction is of zero order with respect to components other than the one considered in the mass-transfer steps).

Equation (15) is derived under the assumption that the amount of adsorbed component transferred by flow or diffusion of the solid phase may be neglected. This assumption is clearly justified in cases of fixed-bed operation, and it is believed to be permissible in many cases of slurries or fluidized beds, since the absolute amount of adsorbed component will probably be quite low due to its low diffusivity in the interior of the catalyst pellet. The assumption can, however, be waived by including in Eq. (15) the appropriate diffusive and convective terms.

Equations similar to Eqs. (13)–(15) can be derived for any other components of which the concentrations appear in the rate expression for the catalytic reaction.

The systems represented by Eqs. (13)–(15) are of interest partly because analytical solutions may be derived for certain simple cases, for example, when the catalytic reaction is of zero or first order.

For a zero-order reaction, $f(c_p)$ is equal to k_r, and Eq. (15) may be solved for c_p:

$$c_p = c_l - \frac{k_r}{k_s} \quad (16)$$

This expression may be substituted in Eq. (14).

For a first-order reaction, $f(c_p)$ is equal to $k_r c_p$, and Eq. (15) may be solved for c_p with the following result:

$$c_p = \frac{k_s c_l}{k_s + k_r} \quad (17)$$

This expression may be substituted in Eq. (14).

For both of these cases, Eqs. (13)–(15) constitute a system of two linear ordinary differential equations of second order with constant coefficients. The boundary conditions are similar to those used by Miyauchi and Vermeulen, which are identical to those proposed by Danckwerts (D1). The equations may be transformed to a dimensionless form and solved analytically. The solutions may be recorded in dimensionless diagrams similar to those constructed by Miyauchi and Vermeulen. The analytical solutions in the present case are, however, considerably more involved algebraically.

The above approach to a theoretical description of gas–liquid–particle processes should be particularly useful in cases characterized by a moderate deviation from plug flow, such as a trickle-bed process. For a system with a high degree of mixing in both phases, such as a mechanically-stirred slurry reactor, a model based on the concept of perfectly mixed stages in cascade may have wider applicability. For some systems, it may be advantageous to employ a combination of the two models, a diffusion model being used for one of the fluid phases and a cascade model for the other. It is believed that models of these types in which the influence of mixing is included by the two generally accepted methods of mathematical description will be useful for the design of gas–liquid–particle processes and for the comparison of the several types of operations.

However, the models represent only crude approximate descriptions of the complex physical systems involved. Probably the most important phenomenon excluded is that of heat transfer. Suspended-bed operations are characterized by a high effective thermal conductivity, and thus represent a good approximation to isothermal behavior, and the above models should provide an adequate description of these systems. Fixed-bed operations will probably in many cases depart significantly from isothermal conditions, and in such cases models should be constructed that take heat transfer into account. Fairly simple models of heat transfer will probably often suffice—for example, those that include particle-to-liquid heat transfer and heat transfer between the liquid and the reactor wall.

Discussed in the following section will be such data and other information regarding the elementary process steps in gas–liquid–particle operations as have appeared in the chemical engineering literature.

V. Transport Phenomena in Gas–Liquid–Particle Operations

A comparison of the operations discussed in Section III with regard to applications in a particular chemical process should be based, at least in part, on the analysis of a theoretical model of the type discussed in Section IV. At the present stage of development, only an approximate estimate of reaction conversion and selectivity will be obtained in this way, and the analysis must in most cases be supplemented with qualitative considerations. The analysis is necessary, however, if optimum choice of operation and optimum design of the chosen operation are to be achieved.

The use of even the very simple models for isothermal operation described in Section IV,B requires a substantial amount of information regarding the elementary rate processes occurring in a gas–liquid–particle operation, as discussed in Section IV,A. While a considerable amount of information of this kind is available in the chemical engineering literature, it is widely scattered. It will be attempted in this section to present a comprehensive review of this information in order to facilitate its use. It is hoped that this review will be of value not only to those chemical engineers directly interested in the practical applications of gas–liquid–particle operations, but also, by pointing to the several areas characterized by very limited information, to those interested in research in this field.

The experimental and theoretical work reported in the literature will be reviewed for each of the five major types of gas–liquid–particle operation under the headings: Mass transfer across gas–liquid interface; mass transfer across liquid–solid interface; holdup and axial dispersion of gas phase; holdup and axial dispersion of liquid phase; heat transfer; reaction kinetics.

A. Fixed-Bed Trickle-Flow Operation

Trickle-flow operation is probably the most widely used operation for large-scale industrial gas–liquid–particle processes. It has been the subject of a large number of investigations, and is, as a result, relatively well described.

It may be noted that trickle-flow operations is not always clearly distinguished in the literature from fixed-bed bubble-flow operation. The two operations differ with respect to several important characteristics, and are treated separately in this review.

1. *Mass Transfer across Gas–Liquid Interface*

The absorption of reactants (or desorption of products) in trickle-bed operation is a process step identical to that occurring in a packed-bed absorption process unaccompanied by chemical reaction in the liquid phase. The information on mass-transfer rates in such systems that is available in standard texts (N2, S6) is applicable to calculations regarding trickle beds. This information will not be reviewed in this paper, but it should be noted that it has been obtained almost exclusively for the more efficient types of packing material usually employed in absorption columns, such as rings, saddles, and spirals, and that there is an apparent lack of similar information for the particles of the shapes normally used in gas–liquid–particle operations, such as spheres and cylinders.

By far the major portion of the available gas-absorption data have been obtained for countercurrent flow, which is the normal mode of operation for packed-bed absorbers. Special mention may be made of the results of Dodds *et al.* (D6), who examined mass transfer by the absorption of gas in liquid under cocurrent downward flow at flow rates higher than those corresponding to the flooding point for countercurrent operation.

Experiments were carried out with 1- and $1\frac{1}{2}$-in. Berl and Intalox saddles and 2-in. steel rings for liquid flow rates of 9330 and 32,600 lb/ft^2-hr and gas flow rates of 22.7 and 56.6 lb moles/ft^2-hr.

The results are of interest partly with respect to the design of certain types of trickle-flow operation and partly because they demonstrate that higher mass-transfer coefficients may be obtained for cocurrent than for countercurrent operation.

2. *Mass Transfer across Liquid–Solid Interface*

A considerable amount of information has been reported regarding mass transfer between a single fluid phase and solid particles (such as those of spherical and cylindrical shape) forming a fixed bed. A recent review has been presented by Norman (N2). The applicability of such data to calculations regarding trickle-flow processes is, however, questionable, due to the fundamental difference between the liquid flow pattern of a fixed bed with trickle flow and that of a fixed bed in which the entire void volume is occupied by one fluid.

Information regarding mass transfer between liquid and solid in fixed beds operated under trickle-flow conditions has apparently not appeared in the literature.

3. Holdup and Axial Dispersion of Gas Phase

Only a few investigations concerned with the measurement of gas holdup and residence-time distribution have been reported. The information regarding liquid holdup, which will be discussed in the following section, is considerably more abundant; in some cases, values of gas holdup can be deduced from the reported data on liquid holdup and total voidage.

De Maria and White (D4) measured gas holdup and residence-time distribution in a column of 4-in. diameter packed with unglazed porcelain Raschig rings of $\frac{1}{4}$-, $\frac{3}{8}$-, and $\frac{1}{2}$-in. nominal diameter. Column heights were 3 and 4 ft. The fluid media were atmospheric air flowing upward and purified water flowing downward; helium, or a mixture of helium and nitrogen, was used as tracer in either step or pulse inputs. The gas holdup was calculated from the mean residence-time determined from the response curves and the volumetric flow rate. It was found to depend mainly upon the voidage of a corresponding dry bed, packing diameter, and liquid flow rate, and could be correlated by the equation:

$$\varepsilon_g/\varepsilon_{g0} = 0.90 \times 10^{-3.43 \times 10^{-6}(d_p/d_t)^{2.31} N_{Re_l}} \tag{18}$$

where ε_g is the gas holdup of the trickle bed (volume fraction of empty column), ε_{g0} the voidage of the corresponding dry bed, d_p the nominal packing diameter, d_t the column diameter (this was not varied in the experiments), and N_{Re_l} the liquid Reynolds number (based on nominal packing diameter and average nominal liquid velocity).

The axial dispersion in the gas phase was expressed as the Peclet number, or the height of a perfectly mixed stage. The analysis was carried out by the method of moments, assuming perfect inputs and correcting for end effects. The extent of axial mixing was found to depend upon the following variables (listed in order of decreasing importance): liquid flow rate, packing diameter, and gas flow rate; the Peclet number could be correlated by the equation:

$$N_{Pe} = 2.4 N_{Re_g}^{-0.20} \times 10^{-(0.013 - 0.088 d_p/d_t) N_{Re_l}} \tag{19}$$

with N_{Re_g} the gas Reynolds number (based on nominal packing diameter and average nominal velocity). The column diameter d_t was not varied in the experiments.

By the injection of tracer step-inputs, de Waal and van Mameren (D5) measured gas residence-time distribution in a 10-ft high column of 1-ft diameter packed with 1-in. ceramic Raschig rings. The fluid media were air and water, and butane was used as a tracer. Little tailing was observed, and the height of a perfectly mixed stage was calculated from the slope of the residence-time distribution curve at the average residence time. The radial mixing was good. The height of a perfectly mixed stage was independent of the gas flow rate for the range 1500–4200 kg/m²-hr, but increased with liquid flow rate, from 1.8 packing diameters at zero liquid flow rate to 4 packing diameters at a flow rate of 80,000 kg/m²-hr. It was concluded that the effect of gas residence-time distribution is unimportant in tall packed columns, but that it may be of importance in measurements of gas-phase resistance to mass transfer.

Sater and Levenspiel (S1) measured the axial dispersion of gas in a column of 4-in. diameter and 12-ft height packed with ½-in. ceramic Berl saddles or ½-in. Raschig rings. The fluid media were atmospheric air and tap water, and the tracer used in the gas phase was argon-41. The imperfect-pulse technique and the analysis of moments were the experimental and mathematical methods employed. Use of a gamma-ray-emitting tracer avoided any disturbance of the flow pattern, such as would have resulted from the insertion of probes in the interior of the column. The results could be fitted to the following equation:

$$N_{Pe_g} = \frac{u_g d_p}{D_g} = 0.0585 (a_p d_p)^\beta N_{Re_g}^\gamma \times (10)^\delta \qquad (20)$$

where $\beta = 2.58 \pm 0.78$, $\gamma = -0.668 \pm 0.184$, $\delta = -(0.00259 \pm 0.00053) N_{Re_l}$, and D_g is the axial-dispersion coefficient for the gas phase. The factor $(a_p d_p)^{2.58 \pm 0.78}$ was included in order to account for the influence of packing geometry; its form is strictly arbitrary, however, since only two geometries were used in the experiments. The limits included in the correlation are 95% confidence limits.

Dunn et al. (D7) measured axial dispersion in the gas phase in the system referred to in Section V,A,4, using helium as tracer. The data were correlated reasonably well by the random-walk model, and reproducibility was good, characterized by a mean deviation of 10%. The degree of axial mixing increases with both gas flow rate (from 300 to 1100 lb/ft²-hr) and liquid flow rate (from 0 to 11,000 lb/ft²-hr), the following empirical correlations being proposed:

For 1-in. Berl saddles:

$$N_{Pe_g} = (0.822 - 4.73 \times 10^{-4} G_g) \times 10^{-3.85 \cdot 10^{-5} G_l} \qquad (21)$$

For 1-in. Raschig rings:

$$N_{Pe_g} = (0.665 - 3.83 \times 10^{-4} G_g) \times 10^{-3.85 \cdot 10^{-5} G_l} \qquad (22)$$

For 2-in. Raschig rings:

$$N_{Pe_g} = (0.756 - 1.875 \times 10^{-4} G_g) \times 10^{-1.61 \cdot 10^{-5} G_l} \qquad (23)$$

The Peclet number is based on the equivalent packing diameter (see Section V,A,4) and G_g and G_l are the gas and liquid flow rates, respectively.

Figure 2 shows a graphical comparison of the correlations derived by De Maria and White and by Sater and Levenspiel for ½-in. Raschig rings, and the results obtained by de Waal and van Mameren for 1-in. Raschig rings. It is apparent that the disagreement between these results is quite large. The results of Sater and Levenspiel show the greatest deviation from piston flow, and the use of this correlation in design calculations will therefore probably result in a conservative performance estimate.

Further work regarding the axial dispersion of gas in irrigated packed beds seems needed, and it may be noted, with particular regard to gas–liquid–particle processes, that no results have been reported for beds of cylindrically or spherically shaped packing materials.

4. *Holdup and Axial Dispersion of Liquid Phase*

These aspects of trickle-flow operation have been studied quite extensively. The available information will be reviewed in near-chronological order, the

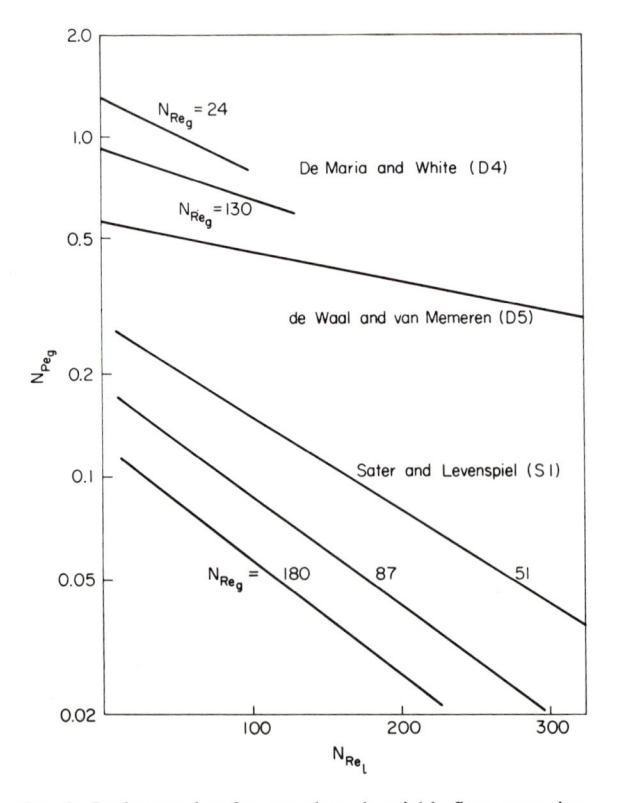

FIG. 2. Peclet number for gas phase in trickle-flow operation.

earlier publications being concerned with holdup alone, the later with residence-time distribution as well.

It may be noted here that the expression "trickle-bed operation" covers a number of quite distinct flow patterns; some studies of these that have been published in recent papers will be reviewed at the end of the section.

Liquid holdup has been studied by Elgin and Weiss (E2), Piret *et al.* (P3), Jesser and Elgin (J2), Shulman *et al.* (S9), and Otake and Okada (O9). The results of these studies are in substantial agreement.

Piret *et al.* measured liquid holdup in a column of 2½-ft diameter and 6-ft packed height, packed with graded round gravel of 1¾-in. size, the total voidage of the bed being 38.8%. The fluid media, air and water, were in countercurrent flow. The liquid holdup was found to increase markedly with liquid flow rate, but was independent of gas flow rate below the loading point. Above the loading point, an increase of liquid hold-up with gas flow rate was observed.

Shulman *et al.* measured liquid holdup for downward flow of water with countercurrent air flow in a column of 10-in. diameter and 3-ft height packed with ½-, 1- and 1½-in. unglazed porcelain Raschig rings, ½- and 1-in. unglazed porcelain Berl saddles, and 1-in. carbon Raschig rings. The total holdup was defined as the volume fraction of packed bed occupied by liquid under operating conditions; the static holdup, as the volume fraction occupied by liquid which does not drain from the column after the liquid supply has been interrupted; and the operating or dynamic holdup, as the difference between the two. It was concluded that total holdup is largely independent of gas flow rate below the loading point, increases with liquid flow rate, and is approximately proportional to the number of pieces of packing per unit volume. Static holdup is independent of gas and liquid flow rate, but is dependent upon the nature of the surface of the packings, whereas operating holdup is independent of the packing surface. The air flow rate in these experiments was varied from 100 to 1000 lb/ft²-hr, and the liquid flow rate from 1000 to 10,000 lb/ft²-hr.

Otake and Okada correlated measurements of total holdup by the equation

$$\varepsilon_l = (ad)1.295\, N_{Re_l}{}^{0.676} N_{Ga_l}{}^{-0.44} \tag{24}$$

where N_{Re_l} and N_{Ga_l} are the liquid-phase Reynolds and Galileo numbers, respectively, d the nominal particle diameter, and a the packing surface area per unit volume.

More than a dozen studies of liquid axial dispersion in trickle-flow operation have been published, but the results are not in complete agreement. More experimental work on the subject is certainly necessary, both to resolve these disagreements and to extend the range of operating variables for which information is available.

Kramers and Alberda (K20) have reported some data in graphical form for the residence-time distribution of water with countercurrent air flow in a column of 15-cm diameter and 66-cm height packed with 10-mm Raschig rings. It was concluded that axial mixing increased with increasing gas flow rate and decreasing liquid flow rate, and that the results were not adequately represented by the diffusion model.

Schoenemann (S4) examined the liquid holdup and the residence-time distribution in trickle-flow reactors used for butynediol synthesis. Responses to step changes of liquid-phase concentrations are reported for a technical reactor of 16-meter height and 0.8-meter diameter and for a pilot reactor of 3.5-meter height and 32-mm diameter. Cylindrical silica catalyst pellets of 4-mm diameter were used in both experiments. The total liquid holdup of the technical reactor was 50% at a nominal liquid velocity of 5.95 meters/hr, and that of the pilot reactor was 41% at a nominal liquid velocity of 0.99 meters/hr. The residence-time distribution curves correspond to approximately four ideal mixing stages. The residence-time distribution is reportedly unaffected by variations of gas velocity in co- and countercurrent operation and of liquid velocity in cocurrent operation (within the range 2 to 10 meters/hr). It is influenced by the characteristics of the packing materials (for example—porosity). It is reported that a narrower residence-time distribution is observed at high liquid velocities in countercurrent operation.

Lapidus (L1) described liquid residence-time distribution studies for air–water and air–hydrocarbon in cocurrent, downward flow through a column of 2-in. diameter and 3-ft height. Spherical glass beads of 3.5. mm diameter and cobalt molybdate catalyst cylinders of $\frac{1}{8}$-in. diameter were used as packing materials.

The results obtained for the nonporous glass beads indicate a close approximation to plug flow in the liquid phase. For the porous catalyst pellets, the results of pulse-tracer experiments likewise indicate a good approximation to plug flow, whereas much wider residence time-distribution curves were obtained by step-tracer experiments. This difference is explained as an effect of intraparticle mass diffusion in the porous material. An approximate method eliminates the influence of intraparticle mass diffusion, and the results from step-function experiments for porous packing, when corrected by this method, are in good agreement with the results for nonporous packing. This is also the case for similar data from other experiments in towers of diameters ranging from $\frac{3}{4}$ in. to 6 ft. It is suggested that residence-time distribution experiments may be used for the determination of effective diffusion coefficients for porous catalysts under operating conditions. The results seem to indicate further that liquid–catalyst contacting is poor. The total liquid holdup was approximately 12% for nonporous packing and approximately 30% for porous packing, and calculations show that only approximately 50% of the intraparticle voids are filled with liquid.

Schiesser and Lapidus (S3), in later studies, measured the liquid residence-time distribution for a column of 4-in. diameter and 4-ft height packed with spherical particles of varying porosity and nominal diameters of $\frac{1}{4}$ in. and $\frac{1}{2}$ in. The liquid medium was water, and as tracers sodium chloride or methyl orange were employed. The specific purposes of this study were to determine radial variations in liquid flow rate and to demonstrate how pore diffusivity and pore structure may be estimated and characterized on the basis of tracer experiments. Significant radial variations in flow rate were observed; methods are discussed for separating the hydrodynamic and diffusional contributions to the residence-time curves.

Otake and Kuniguta (O8) measured liquid-phase axial mixing for countercurrent flow in columns filled with 7.85- and 15.5-mm Raschig rings. The results, which were correlated in terms of liquid phase Reynolds and Galileo numbers, fit the correlation proposed by Sater and Levenspiel which will be referred to later in this section. Hofmann (H8) mentions that the correlation of Otake and Kuniguta has been found inapplicable in certain experiments and Weber (W1) refers to the same experiment as yielding much higher values for the dispersion coefficient than those calculated from the correlation. The correlation is of the form:

$$\frac{D_l}{v} = 0.527 N_{Re_{l_r}}^{0.5} N_{Ga_l}^{0.33} \tag{25}$$

Hofmann (H8) has proposed, on the basis of similarity theory, a correlation for the liquid Peclet number of the following form:

$$N_{Pe_l} = f(N_{Re_l}, N_{Ga_l}, d_p/d_t) \tag{26}$$

where N_{Ga_l} is the liquid-phase Galileo number, i.e., gd_p^3/v_l^2, where v_l is the kinematic viscosity. This is of the form employed by Otake and Kuniguta, although those authors did not include the diameter ratio. Other factors, such as the distribution of liquid and the method of packing, do, however, influence axial mixing.

Harrison et al. (H5) measured the axial dispersion of liquid flowing over a vertical string of 128 touching spheres of 3.8-cm diameter, this system functioning as a somewhat simplified model of a packed column. They concluded that the simple diffusion model provides only an approximate representation of the axial dispersion of the liquid. Tailing was observed, which is taken as supporting a theory that mass transfer from regions of low velocity or stagnation influences the axial dispersion.

Dunn et al. (D7) measured axial dispersion in a packed column of 2-ft diameter and 6-ft packed height. The fluid media were air flowing upward and water flowing downward, and as tracer sodium nitrate solution was used. A step change in tracer concentration was effected at the inlet to the column, the downstream tracer concentration being monitored continuously. One- and 2-in. Raschig rings and 1-in. Berl saddles were used as packing elements, the voidages of the dry column being 0.682, 0.690 and 0.740, respectively, and the equivalent diameter (diameter of a sphere with the same surface-to-volume ratio) being 0.376, 0.749, and 0.237 in., respectively.

The dispersion data were analyzed in terms of the diffusion model, which gave a better (although not an exact) fit to the results than the random-walk model, mixing-cell model, or segmented-laminar-flow model. The reproducibility of the data was poor, characterized by a mean deviation of 30%. The axial mixing was found to decrease with increasing liquid flow rate, but no significant variation with gas flow rate (for the range from 0 to 1100 lb/ft²-hr) was observed. The following empirical correlations were proposed:

For 1-in. Raschig rings:

$$N_{Pe_l} = 0.038 \times 10^{4.93 \times 10^{-5} G_l} \tag{27}$$

For 2-in. Raschig rings

$$N_{Pe_l} = 0.051 \times 10^{4.93 \times 10^{-5} G_l} \tag{28}$$

For 1-in. Berl saddles

$$N_{Pe_l} = 0.033 \times 10^{4.93 \times 10^{-5} G_l} \tag{29}$$

where the Peclet number is based upon the equivalent diameter and G_l is the liquid flow rate (which varied from 2000 to 11,000 lb/ft²-hr). It is suggested in that paper that a reasonable fit to the data may also be obtained for a square root dependence of Peclet number on G_l.

Glaser and Lichtenstein (G3) measured the liquid residence-time distribution for cocurrent downward flow of gas and liquid in columns of ¾-in., 2-in., and 1-ft diameter packed with porous or nonporous $\tfrac{1}{16}$-in. or ⅛-in. cylindrical packings. The fluid media were an aqueous calcium chloride solution and air in one series of experiments and kerosene and hydrogen in another. Pulses of radioactive tracer (carbon-12, phosphorous-32, or rubidium-86) were injected outside the column, and the effluent concentration measured by Geiger counter. Axial dispersion was characterized by "variability" (defined as the standard deviation of residence time divided by the average residence time), and corrections for end effects were included in the analysis. The experiments indicate no effect of bed diameter upon variability. For a packed bed of porous particles, variability was found to consist of three components: (1) Variability due to bulk flow through the bed channels, which had the approximately constant value of 0.20 to 0.25 over the range of liquid flow rates 100–10,000 lb/ft^2-hr; (2) variability due to diffusion to and from stagnant pools of liquid in the bed (such pools tend to disappear at liquid flow rates above 3000 lb/ft^2-hr due to turbulence); and (3) variability due to pore diffusion, which is large at low liquid flow rates but masked at higher flow rates.

Glaser and Litt (G4) have proposed, in an extension of the above study, a model for gas–liquid flow through a bed of porous particles. The bed is assumed to consist of two basic structures which influence the fluid flow patterns: (1) Void channels external to the packing, with which are associated dead-ended pockets that can hold stagnant pools of liquid; and (2) pore channels and pockets, i.e., continuous and dead-ended pockets in the interior of the particles. On this basis, a theoretical model of liquid-phase dispersion in mixed-phase flow is developed. The model uses three bed parameters for the description of axial dispersion: (1) Dispersion due to the mixing of streams from various channels of different residence times; (2) dispersion from axial diffusion in the void channels; and (3) dispersion from diffusion into the pores. The model is not applicable to turbulent flow nor to such low flow rates that molecular diffusion is comparable to Taylor diffusion. The latter region is unlikely to be of practical interest. The model predicts that the reciprocal Peclet number should be directly proportional to nominal liquid velocity, a prediction that has been confirmed by a few determinations of residence-time distribution for a wax desulfurization pilot reactor of 1-in. diameter packed with 10–14 mesh particles.

Ross (R2) measured liquid-phase holdup and residence-time distribution by a tracer-pulse technique. Experiments were carried out for cocurrent flow in model columns of 2- and 4-in. diameter with air and water as fluid media, as well as in pilot-scale and industrial-scale reactors of 2-in. and 6.5-ft diameters used for the catalytic hydrogenation of petroleum fractions. The columns were packed with commercial cylindrical catalyst pellets of $\frac{3}{16}$-in. diameter and length. The liquid holdup was from 40 to 50% of total bed volume for nominal liquid velocities from 8 to 200 ft/hr in the model reactors, from 26 to 32% of volume for nominal liquid velocities from 6 to 10.5 ft/hr in the pilot unit, and from 20 to 27% for nominal liquid velocities from 27.9 to 68.6 ft/hr in the industrial unit. In that work, a few sets of results of residence-time distribution experiments are reported in graphical form, as tracer-response curves.

Hoogendoorn and Lips (H10) carried out residence-time distribution experiments for countercurrent trickle flow in a column of 1.33-ft diameter and 5- and 10-ft height packed with $\frac{1}{2}$-in. porcelain Raschig rings. The fluid media were air and water, and ammonium chloride was used as tracer. The total liquid holdup was calculated from the mean residence time as found from the response curves and the liquid flow rate. Expressed as a percentage of void volume, it varied from 12 to 17.5% for nominal liquid velocities (based on the empty column) from 0.007 to 0.017 ft/sec. It was not affected by the nominal gas velocity in the range of 0.03 to 0.3 ft/sec. The static holdup was nearly independent of liquid velocity at a value of 6.5%, in agreement with the results of Shulman et al.

The pulse-response curves showed strong tailing, and analysis based on the diffusion model using van der Laan's equation (V2) was impossible. The effect of gas flow rate was negligible, whereas the liquid flow rate had a small effect upon the curves. The flow was approximately plug flow, especially for the front of the response curves. These observations are in general agreement with those of Lapidus and Schiesser and Otake and Kuniguta. In a discussion of possible causes for the observed tailing the following phenomena are ruled out: (1) Adsorption on packing; (2) end effects; and (3) uneven liquid distribution. It is concluded that the most likely cause is liquid retention in static holdup, a phenomenon which was also discussed by Glaser and Lichtenstein. A theoretical analysis of the proposed mechanism is also presented in that paper.

De Waal and van Mameren (D5) measured liquid holdup and residence-time distribution for the experimental system described in section V,A,3 using as tracer a common salt solution. Liquid holdup, measured by draining the column, is reported for gas flow rates from 0 to 5000 kg/m^2-hr and liquid flow rates from 5000 to 80,000 kg/m^2-hr. The maximum holdup was approximately 15%, corresponding to the flooding point, the holdup being markedly influenced by both gas and liquid flow rates. The average liquid residence time, as calculated from the data on holdup and flow rate, was found to pass through a minimum which was more pronounced at high gas flow rates. This phenomenon is referred to as a reason for limiting the gas flow rate: A fluctuation in liquid flow rate may significantly change the liquid residence time, which may be undesirable if, for example, a chemical reaction takes place. The static liquid holdup measured after 30-min draining was 1.17% and was independent of gas and liquid flow rate.

Tailing was absent in the pulse-response curves, and the height of a perfectly mixed stage was calculated from the slope of the residence-time distribution curve at the average residence time. The height of a perfectly mixed stage was independent of gas flow rate from 1340 to 3750 kg/m^2-hr and liquid flow rate from 27,000 to 69,000 kg/m^2-hr, the average value being 2.5 cm, i.e., one nominal packing diameter. The effect of liquid residence-time distribution can thus be neglected in tall columns.

De Waal and van Mameren also examined liquid distribution at the column outlet, and concluded that the liquid remains evenly distributed in the column if it is evenly distributed at the inlet.

Sater and Levenspiel (S1), using the experimental and theoretical methods referred to in Section V,A,3, measured liquid holdup and axial dispersion. Iodine-131 in the form of a sodium iodide solution was used as tracer. Liquid holdup increased with liquid flow rate in the range from 600 to 35.000 lb/ft²-hr from 6 to 20% of total bed volume, the results being in good agreement with those of Elgin and Weiss, Jesser and Elgin, and Shulman *et al.* No significant variation of holdup with gas flow rate was observed. The results were well correlated by the equation proposed by Otake and Okada. The dispersion coefficients could be correlated by the equation:

$$N_{Pe} = \frac{u_l d_p}{D_l} = 7.58 \times 10^{-3} N_{Re_l}^{0.703} \tag{30}$$

the 95% confidence limit for the exponent being ±0.238. These results agree reasonably well with those of Otake and Kuniguta. A correlation is also proposed which includes the liquid-phase Galileo number and the product of the nominal particle diameter and the surface area per unit volume of packed bed, but the experimental results did not permit the exponents of these dimensionless groups to be reliably determined.

In this section, four other papers may be finally referred to, which, in addition to liquid holdup, deal with various other aspects of liquid flow in trickle-bed operation.

Larkins *et al.* (L2) visually observed flow patterns and measured pressure drop and liquid holdup for cocurrent downflow of gas and liquid through beds of spheres, cylinders, and Raschig rings of diameters from 3 mm to ⅜ in. in experimental columns of 2- and 4-in. diameter, as well as in a commercial unit several feet in diameter. The fluid media were air, carbon dioxide, or natural gas; and water, water containing methylcellulose, water containing soap, ethylene glycol, kerosene, lubricating oil, or hexane.

The liquid holdup is recorded as the "saturation," i.e., the fraction of void bed volume occupied by the liquid phase. Two basic flow patterns, homogenous and heterogenous, were observed. In the homogenous mode, each local area in the bed is the same, and there is no large-scale transient behavior; in the heterogenous mode, slugs representing sharp local increases in liquid saturation travel down the column. The wide range of experimental results were satisfactorily correlated, and it was shown that design correlations can be based on the knowledge of friction losses for gas and liquid when they flow alone in the bed. The packing characteristics and fluid properties thus do not appear directly in the two-phase correlation. Packing porosity varied from 0.357 to 0.520, nominal gas velocity from 0 to 26.4 ft/sec, and nominal liquid velocity from 0 to 0.87 ft/sec.

Prost and Le Goff (P6) reported measurements of liquid holdup for co- and countercurrent flow of air and a 70% aqueous solution of saccharose in a column of 10-cm diameter and 3-meter height packed with 1-cm Raschig rings. The downward liquid flow rate was varied from 1.12 kg/m^2-sec to 10.3 kg/m^2-sec (corresponding to liquid holdup from 7 to 35% at zero gas flow rate) and the gas flow rate from 0 to 2.5 kg/m^2-sec. Liquid holdup decreased slightly with increasing gas flow rate at cocurrent flow, but increased sharply at countercurrent flow until the flooding point was reached.

These authors also measured the electrical conductivity of the irrigated bed in the horizontal and vertical directions. The ratio between the liquid holdup multiplied by the conductivity of the liquid and the effective conductivity of the bed was assumed to be a measure of the tortuosity of the liquid flow.

Weekman and Myers (W2) examined the fluid-flow characteristics of cocurrent downward flow of gas and liquid. The pulsing effect first noted by Larkins *et al.* was also observed in this work. Pressure-drop data could be correlated satisfactorily by a relation similar to those used for two-phase flow in pipes. Surface-active agents were observed to have a pronounced influence upon flow regime transition and pressure drop.

Kolár and Brož (K4) have described a theoretical analysis of countercurrent flow of liquid and gas through a packed bed. A relationship has been derived between holdup of liquid, flow rates of fluids, and physical properties of fluids. The relationship contains three parameters, the values of which must be determined by experiment. Experimental data are not presented.

In conclusion, one can state that the available information on holdup is in general agreement, although some difference exists with respect to the influence of gas flow rate and to the magnitude of static holdup.

More disagreement exists with respect to axial dispersion—for example, regarding the applicability of the diffusion model, and regarding the influence of gas and liquid flow rates. More work on these aspects and on the influence of fluid distribution and method of packing is required. Some of the available results are compared in Fig. 3.

Some of the later papers referred to have pointed to the existence of distinctly different flow patterns under conditions normally characterized as trickle-flow operation. The pulsing flow pattern observed may be of particular interest, and this mode of operation could be a fertile area for research.

5. *Heat Transfer*

A certain amount of information on particle-to-liquid and bed-to-wall heat transfer is available for single-phase fluid flow through a packed bed. It is not clear, however, to what extent this information can be applied to

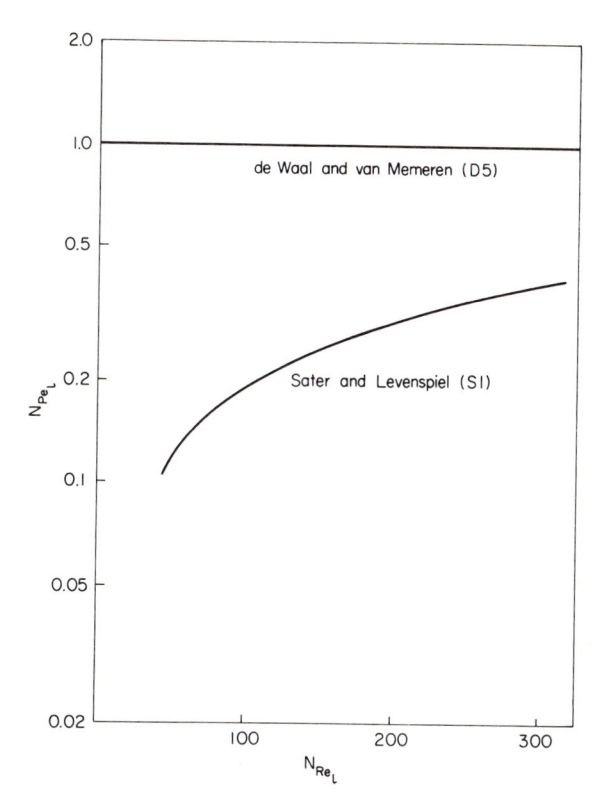

FIG. 3. Peclet number for liquid phase in trickle-flow operation.

calculations regarding trickle-bed operations, since very few systematic data on heat transfer in these systems appear to have been published.

It is well known that trickle-flow operation is characterized by comparatively poor heat-transfer properties, this being one of the disadvantages of this type of operation. Schoenemann (S4), for example, refers to the difficulties of controlling temperature in trickle-bed reactors.

Weekman and Myers (W3) measured wall-to-bed heat-transfer coefficients for downward cocurrent flow of air and water in the column used in the experiments referred to in Section V,A,4. The transition from homogeneous to pulsing flow corresponds to an increase of several hundred percent of the radial heat-transfer rate. The heat-transfer coefficients are much higher than those observed for single-phase liquid flow. Correlations were developed on the basis of a radial-transport model, and the penetration theory could be applied for the pulsing-flow pattern.

6. Reaction Kinetics

Comparatively few kinetic experiments in trickle-bed reactors have been described in the literature.

Babcock et al. (B1) examined the hydrogenation of α-methylstyrene catalyzed by palladium and platinum catalysts in a reactor of $1\frac{1}{2}$-in. diameter under countercurrent flow. Flow rates were above 1500 kg/m^2-hr for the liquid phase and above 15 kg/m^2-hr for the gas, and it was concluded from the experimental results that mass transfer was not of rate-determining influence under these conditions.

Zabor et al. (Z1) have described studies of the catalytic hydration of propylene under such conditions (temperature 279°C, pressure 3675 psig) that both liquid and vapor phases are present in the packed catalyst bed. Conversions are reported for cocurrent upflow and cocurrent downflow, it being assumed in that paper that the former mode corresponds to bubble flow and the latter to trickle-flow conditions. Trickle flow resulted in the higher conversions, and conversion was influenced by changes in bed height (for unchanged space velocity), in contrast to the case for bubble-flow operation. The differences are assumed to be effects of mass transfer or liquid distribution.

Ross (R2) reported measurements of desulfurization efficiency of fixed-bed pilot and commercial units operated under trickle-flow conditions. The percentage of retained sulfur is given as a function of reciprocal space velocity, and the curve for a 2-in. diameter pilot reactor was found to lie below the curves for commercial units; it is argued that this is proof of bad liquid distribution in the commercial units. The efficiency of the commercial units increased with increasing nominal liquid velocity. This may be an effect either of mass-transfer resistance or liquid distribution.

B. Fixed-Bed Bubble-Flow Operation

Fixed-bed bubble-flow operation has been the subject of considerably less experimental work than has trickle-flow operation. This reflects the fact that bubble-flow operation has been of much more limited industrial importance.

1. Mass Transfer across Gas–Liquid Interface

No information on gas absorption in fixed-bed bubble-operation has come to the author's attention.

2. Mass Transfer across Liquid–Solid Interface

No information on this subject has come to the author's attention.

3. Holdup and Axial Dispersion of Gas Phase

Weber (W1) has reported measurements of gas holdup for the experimental system described in Section V,B,4. The following empirical relationships can be derived from the graphical correlations:

For 6-mm Raschig rings, $\varepsilon_l + \varepsilon_g = 0.71$; liquid phase, water:

$$\varepsilon_g = 0.114 u_g^{0.30} \tag{31}$$

For 5-mm spheres, $\varepsilon_l + \varepsilon_g = 0.33$; for 4 × 10 mm cylinders, $\varepsilon_l + \varepsilon_g = 0.42$; liquid phase, water:

$$\varepsilon_g = 0.079 u_g^{0.30} \tag{32}$$

For 2-mm spheres, $\varepsilon_l + \varepsilon_g = 0.32$; liquid phase, water:

$$\varepsilon_g = 0.078 u_g^{0.21} \tag{33}$$

For 2-mm spheres; liquid phase, water + 0.2% ethanol:

$$\varepsilon_g = 0.113 u_g^{0.20} \tag{34}$$

For 5-mm spheres; liquid phase, water + 0.2% ethanol:

$$\varepsilon_g = 0.152 u_g^{0.20} \tag{35}$$

where ε_g is the fraction of total bed volume occupied by gas and u_g is the nominal gas velocity based on the empty-column cross section and is measured in centimeters per second. The correlations are valid for the approximate range 0.15 cm/sec $\leq u_g \leq$ 3 cm/sec. The significantly higher gas holdup measured in the experiments with water containing ethanol is explained as an effect of the reduction in surface tension from 72.9 dyn/cm for water at 18°C to 64.5 dyn/cm for water containing 0.2% ethanol. Gas holdup was not affected by changes in liquid flow rate.

Some data on gas holdup are also reported by Stemerding (S16).

Hoogendoorn and Lips (H10) have reported gas-holdup data for countercurrent bubble flow in the experimental system described in Section V,A,4. Gas holdup was not influenced by changes of liquid flow rate, but increased with nominal gas velocity in the range from 0.03 ft/sec to 0.3 ft/sec. The results are somewhat lower than those obtained by Weber, the difference being explained as due to the difference in gas distributor. Weber used a porous plate and Hoogendoorn and Lips a set of parallel nozzles.

No data on gas-phase residence-time distribution appear to have been published.

A study of interest has been reported by Gorring and Katz (G5) who measured drag coefficients for single air bubbles rising in packed beds of 1-in. glass spheres filled with stagnant glycerine solutions, water, or n-heptane. It was found that the drag coefficients in the hexagonal and random arrays were from two to three times those in clear liquids, whereas the drag in a cubic array was the same as that for clear liquids. The data were correlated by use of the minor bubble diameter in the drag coefficient, and the major diameter in the Reynolds number. This basis improved the correlation for bubble rise in clear liquids.

4. Holdup and Axial Dispersion in Liquid Phase

Schoenemann (S4) reported qualitatively that the liquid residence-time distribution for cocurrent upward bubble flow was narrower than that observed in trickle-flow operation.

Stemerding (S16) has reported dispersion measurements in a column filled with 13-mm Raschig rings with water and air in countercurrent flow. The dispersion coefficient was observed to be essentially independent of the water flow rate and dependent on the air flow rate only. For increasing air flow rates, the dispersion coefficient passed through a maximum.

On the basis of similarity theory Hofmann (H8) derived the following relationship describing axial dispersion in the liquid phase:

$$N_{Pe} = f(N_{Re_l}, N_{We}, d_t/d_p, u_l/u_g) \tag{35a}$$

where N_{Re_l} is the liquid-phase Reynolds number; N_{We} the Weber number $[\sigma/\rho d_b^2]$, where σ is the surface tension, ρ the density, and d_b the bubble diameter]; d_t the column diameter; d_p the packing diameter; and u_l and u_g the nominal velocities of liquid and gas, respectively, based on the voidage of the packed bed.

Weber (W1) has carried out an extensive study of axial dispersion in fixed-bed bubble-flow operations. The experimental column was of 50-mm diameter and 1-meter height, and used as packings were spheres of 2.0- and 5.0-mm diameter, cylinders of 3.8-mm diameter and 10-mm length (equivalent sphere diameter, 6.2-mm), and 6.2-mm Raschig rings (equivalent sphere diameter, 6-mm), all made of glass. The bed voidage was 0.32, 0.33, 0.42, and 0.71, respectively. Demineralized water and atmospheric air were the fluid media, and the dilute hydrochloric acid used as tracer was introduced as approximately perfect pulse or step inputs. All experiments were carried out at cocurrent flow. The dispersion measurements could be satisfactorily correlated by the following expressions:

For 5-mm spheres:

$$N_{Pe_l} = 0.012 \left(\frac{u_l}{u_g} N_{Re_l} N_{Sc_l}\right)^{0.48} \tag{36}$$

For 4 × 10-mm cylinders ($d_p = 6.2$ mm):

$$N_{Pe_l} = 0.024 \left(\frac{u_l}{u_g} N_{Re_l} N_{Sc_l}\right)^{0.46} \tag{37}$$

For 6.2-mm Raschig rings ($d_p = 6.0$ mm):

$$N_{Pe_l} = 0.017 \left(\frac{u_l}{u_g} N_{Re_l} N_{Sc_l}\right)^{0.43} \qquad (38)$$

where N_{Sc_l} is the Schmidt number.

These expressions have been found valid for the range $1 < [(u_l/u_g)N_{Re_l}N_{Sc_l}] < 1000$. For higher values, the axial mixing is identical to that observed for the single-phase liquid flow for which the following correlation was found for the experimental data:

$$N_{Pe_l}(\varepsilon_l + \varepsilon_g) = 0.4 \qquad (39)$$

N_{Pe_l} and N_{Re_l} are based on the equivalent sphere diameters and on the nominal velocities u_g and u_l, which in turn are based on the holdup of gas and liquid. The Schmidt number is included in the correlation partly because the range of variables covers part of the laminar-flow region ($N_{Re_l} < 1$) and the transition region ($1 < N_{Re_l} < 100$) where molecular diffusion may contribute to axial mixing, and partly because the kinematic viscosity (changes of which were found to have no effect on axial mixing) is thereby eliminated from the correlation.

Measurements for water containing 0.2% ethanol, the addition of which was found to influence markedly the gas holdup (see Section V,B,3), indicate that variation of surface tension has no significant effect upon axial mixing. The results for 2-mm spheres could not be correlated by a similar expression. It is proposed in that work that the flow mechanism in this case is significantly different because of the higher ratio between bubble size and particle size.

Hoogendoorn and Lips (H10), in the experimental system described in Section V,A,4, measured liquid axial dispersion under bubble-flow conditions with countercurrent flow. The axial-dispersion coefficient was calculated from van der Laan's equation (V2), the diffusion model being in excellent agreement with the experimental data. Axial dispersion was found to be largely independent of liquid flow rate (for nominal liquid velocities based on the empty column from 0.003 to 0.017 ft/sec) and to depend only slightly on gas flow rate. The dispersion coefficients measured are about 10 times the values obtained by Weber, the difference being explained partly as due to the difference in packing size (Weber used 6.2-mm Raschig rings and Hoogendoorn and Lips ½-in. Raschig rings) and partly as due to the difference in bubble velocity (cf. the gas holdup data referred to in Section V,B,3). It is suggested in (H10) that the experimental data be correlated in the form of a Peclet number based on the average bubble velocity, and this method produces a good agreement between the two sets of data.

Some entrance effects were observed at high air flow rates, the dispersion coefficient being higher for a 5- than for a 10-ft-high bed. This is presumably due to a greater degree of mixing in the lower part of the bed. It was further observed that nonuniform gas distribution could result in a large increase in dispersion coefficients.

5. *Heat Transfer*

No data on heat transfer in fixed-bed bubble-flow operation appear to have been published.

6. *Reaction Kinetics*

The only data on chemical conversion in fixed-bed bubble-flow operation that have come to the author's attention are the few results referred to in Section V,A,6 obtained by Zabor *et al.* (Z1) for catalytic mixed-phase hydration of propylene.

C. BUBBLE-COLUMN SLURRY OPERATION

Bubble-column slurry operation is the most widely used suspended-bed operation for large-scale continuous processes. It has been the subject of a considerable number of investigations, and it may be noted in particular that many studies of the overall reaction kinetics in such operations have been published.

Bubble-column slurries have much in common with two-phase bubble columns containing no solid particles, which have also been studied in great detail. Reference will be made in the following to a number of those studies considered to be of relevance with respect to the analysis and design of corresponding three-phase systems containing suspended solids.

Bubble-column slurry operations are usually characterized by zero net liquid flow, and the particles are held suspended by momentum transferred from the gas phase to the solid phase via the liquid medium. The relationships between solids holdup and gas flow rate is of importance for design of bubble-column slurries, and some studies of this aspect will be reviewed prior to the discussion of transport phenomena.

Kölbel *et al.* (K18) have studied catalyst distribution in a water–air system containing 5% by weight of silicon dioxide particles of 0.1- to 0.125-mm diameter. Large deviations from the average catalyst concentration were observed for a nominal gas velocity of 3.5 cm/sec, more uniform distributions being observed for higher gas velocities. The range of gas velocity over which a bubble-column slurry reactor may be operated decreases with increasing catalyst concentration. The lower limiting gas velocity increases because of an increasing tendency towards sedimentation, and the upper limiting gas velocity decreases because of an increasing tendency towards bubble coalescence and the corresponding increase in bubble size.

Kato (K3) measured so-called critical gas velocities corresponding to the complete suspension of solids, and presents a graphical correlation of the results for glass spheres (diameters from 0.074 to 0.295 mm), magnetite particles (particle size from 0.038 to 0.175 mm), and sand particles (particle size 0.147 to 0.295 mm).

Roy et al. (R3) define the critical solids holdup as the maximum quantity of solids that can be held in suspension in an agitated liquid. They present measurements of this factor for various values of gas velocity, gas distribution, solid-particle size, liquid surface tension, liquid viscosity, and a solid–liquid wettability parameter, and they propose the following two correlations in terms of dimensionless groups containing these parameters:

$$H_s = 6.84 \times 10^{-4} N_{Re_g} N_b^{-0.23} (u_t/u_b)^{-0.18} \gamma^{-3.0} C_\mu \tag{40}$$

for $N_{Re_g} < 600$, and

$$H_s = 0.1072 N_{Re_g}^{0.2} N_b^{-0.23} (u_t/u_b)^{-0.18} \gamma^{-3.0} C_\mu \tag{41}$$

for $N_{Re_g} > 600$. Here, H_s is the critical solids holdup (weight solids/weight suspension); N_{Re_g} is a superficial gas Reynolds number based on the empty-column cross section; u_t is Stokes' free settling velocity; u_b is the bubble velocity, N_b is a bubble flow number, the ratio between the bubble Reynolds and Weber numbers, $\sigma_l/(u_b \mu_l)$ [where σ_l and μ_l are the liquid surface tension and viscosity, respectively]; γ is a relative wettability factor; and C_μ is a viscosity correction factor related to the viscosity (measured in cP) by the following expression:

$$C_\mu = 1 - 0.5892 \log \mu_l + 0.1026 \log^2 \mu_l \tag{42}$$

Cova (C11) has examined the vertical distribution of catalyst concentration as a function of gas and liquid flow rates for systems with finite net liquid flow. A theoretical model is presented which predicts the catalyst profile as a function of physical properties and operating conditions, and which adequately represents observations for both laboratory and pilot-scale operations.

1. Mass Transfer across Gas–Liquid Interface

Only one publication on gas–liquid mass transfer in bubble-column slurry reactors has come to the author's attention. However, a relatively large volume of information regarding mass transfer between single bubbles or bubble swarms and pure liquid containing no suspended solids is available, and this information is probably of some relevance to the analysis of systems containing suspended particles. It is conceivable that the information obtained for solids-free systems may be used for systems containing suspended solids if suitable correction factors or functions can be developed, and some significant contributions from the solids-free field will therefore be reviewed in the present section.

Kato (K3) determined the overall volumetric absorption coefficient $k_g A_b$ for the absorption of oxygen from air in aqueous sodium sulfite solutions containing cupric ions as catalyst, and with the types of particles referred to in the preceding section suspended in the liquid phase. The coefficient $k_g A_b$ increased with gas velocity towards an upper limiting value of about 0.8 kg-mole/m²-hr-atm. At low gas velocities, the absorption coefficient was observed to decrease with increasing particle size, increasing amount of solids in the column, and increasing density difference between solid and liquid, whereas at higher velocities it was nearly independent of these variables. The gas velocity above which $k_g A_b$ is nearly constant is related by analytical and graphical correlations to the amount of solids in the column and to the critical gas velocity corresponding to complete suspension of the solids. The interfacial area A_b is reported to be nearly constant at a value of 1000 m²/m³ for gas velocities corresponding to a constant volumetric absorption coefficient.

The remaining studies reviewed in this section are concerned with gas–liquid mass transfer for single bubbles or bubble swarms in clear liquids.

Shulman and Molstad (S8) examined the mass-transfer characteristics of bubble-columns in which carbon dioxide absorption or desorption, or hydrogen desorption, took place, with downflowing water as the liquid phase. Two distinctly different regions were observed: At low gas velocities (labeled the streamline region), the rate of mass transfer increases with increasing gas velocity, whereas at high gas velocities (labeled the turbulent region) it is independent of gas velocity. The transition between the two regions is characterized by a decrease in mass-transfer rate. Mass-transfer rates also varied with liquid flow rate, liquid temperature, column height, and the liquid diffusion coefficient of the absorbed component, but were independent of column diameter (columns of 1-, 2-, and 4-in. diameter were used), gas distributor plate porosity, and direction of mass transfer.

The results, in terms of the height of a liquid-film transfer unit $(HTU)_L$ measured in feet, could be correlated by the following two expressions:

For the streamline region:

$$(HTU)_L = 5.1 \times 10^{-6}(G_l + 10{,}730)\left(\frac{M}{G_l}\right)^{0.74} Z^{0.26} N_{\text{Sc}}^{0.25} \qquad (43)$$

For the turbulent region:

$$(HTU)_L = 4.0 \times 10^{-8}(G_l + 5530)t^{0.48} Z^{0.26} N_{\text{Sc}}^{0.86} \qquad (44)$$

where G_l is the liquid flow rate (lb/ft²-hr), M the molecular weight of the gas entering the column, Z the column operating height, N_{Sc} the Schmidt number, and t the water temperature (°C). The Schmidt number should be evaluated at the liquid temperature when Eq. (43) is used and at 15°C when Eq. (44) is used. A graphical correlation between gas and liquid flow rates for the transition region is included.

It is concluded that gas bubble-columns exhibit mass-transfer rates of the same order of magnitude as packed columns at low liquid flow rates, and much higher mass-transfer rates at high liquid flow rates. The pressure drop across a bubble-column is much greater than that across packed columns of the same height.

Coppock and Meiklejohn (C9) determined liquid mass-transfer coefficients for the absorption of oxygen in water. The value of k_l was observed to vary markedly with variations of bubble velocity, from 0.028 to 0.055 cm/sec for a velocity range from 22 to 28 cm/sec. These results appear to be in general agreement with the results obtained by Datta et al. (D2) and by Guyer and Pfister (G9) for the absorption of carbon dioxide by water.

Pasveer (P1) studied oxygen absorption from air bubbles by water and found that the experimental results could be adequately correlated in terms of a penetration or surface-renewal theory.

Houghton et al. (H12) examined the effects of various operating variables on the behavior of bubble-column water scrubbers for carbon dioxide. It was found that the bubble-column is superior to the packed tower for the absorption of carbon dioxide in water, where the liquid film resistance is believed to control the rate of absorption. Increase of gas velocity causes an increase of absorption rate in the region where gas holdup increases with increasing gas velocity. It was observed, in a separate study (H13), that the average gas bubble volume in water increases only slightly as the gas velocity is increased, but that the bubble number and thus the interfacial area increase. Gas absorption rate becomes independent of gas velocity at a point corresponding to maximum gas holdup, further increase in gas velocity causing formation of slugs. The volumetric absorption coefficient increases with increasing downward liquid flow rate, but the absorber efficiency decreases. The volumetric absorption coefficient was observed to decrease with increasing bed height, and this is explained as being caused by the decrease in interfacial area corresponding to the depletion of the gas phase.

Grassman (G7) has proposed a simplified theoretical treatment of heat and mass transfer between two fluid phases, as, for example between a dispersed gas phase and a continuous liquid phase; von Bogdandy et al. (V8) measured the rate of absorption of carbon dioxide by water and by decalin, and found that the absorption rate approximated that predicted by Grassmann in the laminar region but was above the theoretical values in the turbulent region. A theoretical and experimental study of mass transfer from gas bubbles to liquids has been published by Griffith (G8), who also discusses the effect of surface-active agents on the mass-transfer rate.

Most of the studies have been concerned with bubbles of from 0.1 to 0.8-cm equivalent diameter (the diameter of a sphere of identical volume), which is the size range usually met with in industrial operations. Baird and Davidson (B3) have measured the rate of absorption of carbon dioxide by water from single bubbles of equivalent diameter from 0.8 to 4.2 cm and compared the experimental results with a theoretical model for absorption from the upper surface of a spherical cap bubble. The absorption rate in tap water is about 50% greater than is predicted by the theory; it is assumed that this is due to absorption by the rippling rear surface of the bubble. The ripples can be suppressed by the addition of n-hexanol. Bubbles whose equivalent diameters exceed about 2.5 cm absorb at an unsteady rate which decreases with time; it is assumed that this is due to the gradual saturation of the liquid carried as the bubble wake. In a theoretical and experimental study of the mass transfer and the rise velocity of small and large single bubbles, Leonard and Houghton (L4) compare their results with, among others, those of Baird and Davidson and of Hammerton and Garner (H3).

Recent publications on mass transfer from single bubbles include a theoretical study by Ruckenstein (R4) and theoretical and experimental studies by Lochiel and Calderbank (C4, L7). Bowman and Johnson (B8, J3) and Li *et al.* (L5) have studied mass transfer from streams of bubbles rising in water.

Among the more recent publications on mass transfer in bubble-columns, the following may be mentioned:

Kölbel and Langemann (K10) measured rates of oxygen absorption and desorption in gas–liquid bubble-columns at zero net liquid flow. The mass-transfer rate increases with liquid velocity in the so-called "laminar" region, while at higher gas velocities it is approximately independent thereof. A maximum value of about 0.03 cm/sec is reached for the absorption coefficient at a gas velocity corresponding to the transition from "laminar" to "turbulent" flow. The mass-transfer rates were observed to decrease with column height, and it is suggested in that paper that this may be due to oscillations of gas bubbles in the entrance region and to bubble coalescence. The findings are in general agreement with those of Shulman and Molstad and Houghton *et al.*

Tadaki and Maeda (T1) examined the desorption of carbon dioxide from water in a bubble-column and analyzed the experimental results under the assumption that while the gas phase moves in piston flow, the liquid undergoes axial mixing that can be characterized by the diffusion model. (Shulman and Molstad, in contrast, assumed piston flow for both phases.) Only poor agreement was obtained between the theoretical model and the experimental results; the neglect of end effects and of the possible variation of the volumetric absorption coefficient with column height are assumed likely reasons for this difference.

Yoshida and Akita (Y1) determined volumetric mass-transfer coefficients for the absorption of oxygen by aqueous sodium sulfite solutions in countercurrent-flow bubble-columns. Columns of various diameters (from 7.7 to 60.0 cm) and liquid heights (from 90 to 350 cm) were used in order to examine the effects of equipment size. The volumetric absorption coefficient reportedly increases with increasing gas velocity over the entire range investigated (up to approximately 30 cm/sec nominal velocity), and with increasing column diameter, but is independent of liquid height. These observations are somewhat at variance with those of other workers.

Braulick et al. (B9) studied mass transfer in bubble-columns at high gas velocities (from 0.09 to 1.10 ft/sec). Mass-transfer rates increased with increasing gas velocity and with increasing liquid height-to-column diameter ratio. In electrolyte solutions, a large number of very small ionic bubbles were observed which contribute significantly to mass transfer because of their high interfacial area and long residence time.

The data on mass transfer in bubble-columns are in general agreement as to the magnitude of the volumetric mass-transfer coefficient in water, as well as to the fact that the absorption rate increases with increasing gas flow rate over some range of the latter. Considerably more work appears necessary, however, in order to establish the effect of several important process parameters, notably, the presence of electrolytes and surface-active agents. The qualitative agreement between the results of Kato for bubble-columns containing suspended solids and those obtained in several studies of corresponding solids-free systems indicate that the former may be analyzed by regarding the liquid–solid phase as a continuous liquid phase. More work on gas–liquid mass-transfer rates in bubble-column slurry operations is obviously required.*

2. *Mass Transfer across Liquid–Solid Interface*

Mass transfer between a liquid and suspended solids in mechanically agitated systems has been widely studied, and a number of important investigations will be referred to in Section V,D,2.

Yoshitome et al. (Y2) examined mass transfer from single samples of benzoic acid suspended in an air–water bubble-column. Spherical, cylindrical, and disk-shaped samples of diameters from 25 to 75 mm were used,

---

* *Note added in proof:* A comprehensive review on gas absorption from bubbles has recently been published by P. H. Calderbank, *Trans. Inst. Chem. Engrs. (London)*, Chem. Eng. 209 (1967).

and the results have probably only limited relevance for calculations regarding bubble-column slurry systems where the particles are of much smaller size. Other studies have apparently not been published.

3. *Holdup and Axial Dispersion of Gas Phase*

Information on gas holdup and axial dispersion in bubble-columns containing suspended solid particles is scarce; reference will therefore also be made to significant studies of bubble-columns with no particles present, results obtained for these systems being probably of some relevance to the understanding of bubble-column slurry operations.

Kato (K3) reported gas holdup as a function of gas velocity, particle size, amount of solids and liquid in the bed, as well as of density of solids, for the system described in Section V,C. The holdup, defined as the ratio between the gas volume and the sum of gas and liquid volumes, increased with increasing nominal gas velocity to a maximum value ranging from 0.40 to 0.75 reached for gas velocities of from 10 to 20 cm/sec. The gas holdup decreased with increasing particle size and with increasing amounts of solids in the bed.

Various methods may be used for the determination of gas holdup—for example, displacement measurements and tracer experiments. Farley and Ray (F2) have described the use of gamma-radiation absorption measurement for the determination of gas holdup in a slurry reactor for the Fischer–Tropsch synthesis.

Numerous studies of the formation and flow of gas bubbles in liquids have appeared in the literature, and a complete review will not be attempted. Attention is drawn to two recently published reviews, that of Jackson (J1) on the formation and coalescence of drops and bubbles in liquids, and that of Govier (G6) on developments in the understanding of the vertical flow of two phases.

Verschoor (V5) studied the motion of swarms of gas bubbles formed at a porous glass gas distributor. Gas holdup was observed to increase approximately linearly with nominal gas velocity up to a critical point (corresponding to a nominal gas velocity of about 4 cm/sec), whereupon it decreased to a minimum and then increased again on further increase of the gas velocity. Higher holdup was observed for a water–glycerine mixture than for water.

A series of studies of the properties of gas bubble swarms have been published by Siemes, Kölbel, and co-workers. Siemes (S10) studied holdup and flow patterns in air–water bubble-columns as a function of gas flow rate and gas distributor design. The existence of a " laminar " flow region in which gas holdup increases with increasing gas velocity, similar to that observed by Verschoor and by Shulman and Molstad, was demonstrated. At higher gas velocities, holdup was approximately independent of gas velocity. Gas distributor design had some influence on gas holdup. Siemes and Gunther (S13) studied the dispersion of gas from a jet injected into a liquid. Bubble size-distribution curves were obtained for varying values of orifice diameter, gas flow rate, liquid viscosity, and surface tension. Viscosity was observed to have a pronounced effect on bubble size distribution: For viscosities above 20 cp, large spherical cap bubbles formed, and it was concluded that mechanical stirring is required in such cases in order to obtain satisfactory interfacial areas. Siemes and Borchers (S11) examined the effect of porous sintered metal distributors on bubble size for varying gas velocity. They concluded that the pore size has little effect on bubble size, whereas the pressure drop increases markedly with decreasing pore size, and that distributors with large pores should therefore be employed. Bubble coalescence can be prevented by the addition of small amounts of surface-active agents or electrolytes. Similar conclusions were reached in a later publication (S12) in which, as in a paper by Kölbel et al. (K13), it is demonstrated that the mean bubble diameter increases with increasing gas velocity up to a limiting value which is reached at a nominal gas velocity of approximately 1 cm/sec. In the latter publication, the effect of pressure on bubble size distribution and gas holdup is also examined; it is demonstrated that this effect can be eliminated from the correlations when the nominal gas velocity is calculated at the column pressure.

Houghton et al. (H13) have reported data on the size, number, and size-distribution of bubbles. Distinction is made between " bubble beds," in which bubble diameter and gas holdup tend to become constant as the gas velocity is increased (these observations being in agreement with those of other workers previously referred to), and " foam beds," in which bubble diameter increases and bubble number per unit volume decreases for increasing gas velocity. Pore characteristics of the gas distributor affect the properties of foam beds, but not of bubble beds. Whether a bubble bed or a foam bed is formed depends on the properties of the liquid, in particular on the stability of bubbles at the liquid surface, foam beds being more likely to form in solutions than in pure liquids.

Hughmark and Pressburg (H14) studied holdup and pressure drop for cocurrent gas–liquid flow, and correlated holdup with a function of gas and liquid flow rates, surface tension, densities of gas and liquid, viscosities of gas and liquid, and total mass velocity.

A theory of two-phase bubble flow has been developed by Nicklin (N1), who shows that the motion of bubbles arises partly from buoyancy and partly from the nominal velocity caused by the entry of the two phases into the tube. Theoretical and semiempirical studies of bubble flow have also been presented by Azbel (A2) and by Azizyan and Smirnov (A3), and further experimental data on holdup have been recently reported by Yoshida and Akita (Y1), by Braulick et al. (B9) and by Towell et al. (T3).

Axial mixing of the gas phase in two-phase bubble-columns has apparently not been studied extensively, and only two publications on this subject have come to the author's attention: Kölbel et al. (K17) and Kölbel and Langemann (K10) carried out residence-time distribution experiments using step inputs of tracer and effluent analysis. The results indicate a characteristic variation of axial mixing with gas velocity: At low gas velocities (1 cm/sec nominal gas velocity in a nitrogen–water bubble-column of 92-mm diameter and 2000-mm height, with carbon monoxide as the displacing gas), considerable axial mixing is observed, the residence-time distribution curve being similar to that corresponding to laminar flow; however, for gas velocities increasing up to about 4 cm/sec (corresponding to the end of the "laminar" flow region), axial mixing is reduced until piston flow is finally approached. Further increase of gas velocity causes an increase of axial mixing. This variation, which is illustrated in Fig. 4, is explained as an effect of variation of bubble size with gas flow rate. Some experiments (such as those of Siemes et al. previously referred to in this section) have shown that the average bubble

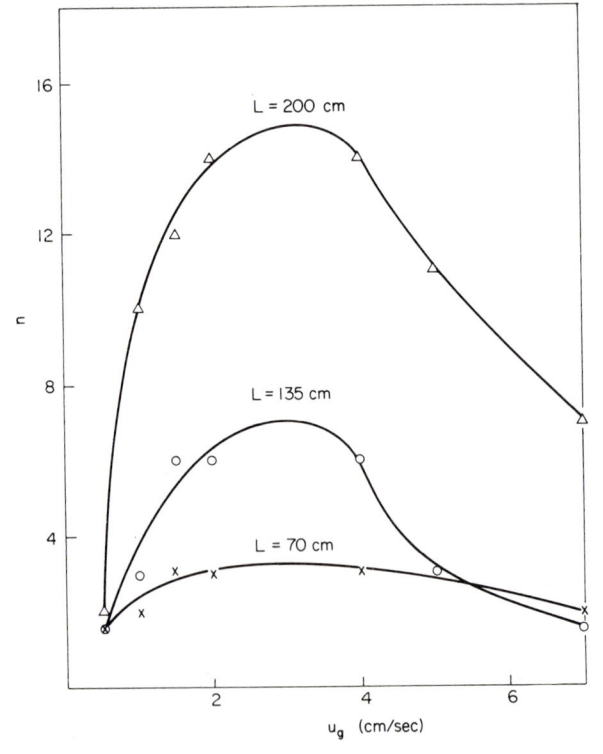

FIG. 4. Number of perfectly mixed stages–for gas flow in a bubble-column as a function of bed height L and superficial gas velocity u_g [Kölbel et al. (K17)].

diameter changes rapidly (from 1 to 3 mm) with gas velocity in the low-velocity range (up to 0.5 cm/sec nominal velocity). In this range of bubble size, the bubble rise-velocity changes rapidly with bubble size; a wide bubble-velocity distribution is therefore probable, and the residence-time distribution may be expected to be similar to that observed for a well-mixed vessel. For the larger bubbles (3 to 4 mm average equivalent diameter) obtained at higher gas velocities (1 to 5 cm/sec) the rise-velocity changes considerably more slowly with bubble size, and a residence-time distribution approaching that of piston flow may be expected. At still higher gas velocities, turbulence is assumed to cause a renewed increase of axial mixing in the gas phase.

Bubble coalescence may considerably influence holdup, residence-time distribution, and other properties of bubble-columns. Reference is made to the review by Jackson and to a recent study by Calderbank et al. (C7).

4. *Holdup and Axial Dispersion of Liquid Phase*

Siemes and Weiss (S14) investigated axial mixing of the liquid phase in a two-phase bubble-column with no net liquid flow. Column diameter was 42 mm and the height of the liquid layer 1400 mm at zero gas flow. Water and air were the fluid media. The experiments were carried out by the injection of a pulse of electrolyte solution at one position in the bed and measurement of the concentration as a function of time at another position. The mixing phenomenon was treated mathematically as a diffusion process. Diffusion coefficients increased markedly with increasing gas velocity, from about 2 cm^2/sec at a superficial gas velocity of 1 cm/sec to from 30 to 70 cm^2/sec at a velocity of 7 cm/sec. The diffusion coefficient also varied with bubble size, and thus, because of coalescence, with distance from the gas distributor.

Tadaki and Maeda (T2) examined the axial mixing characteristics of a two-phase bubble-column with downflowing liquid. Oxygen and water were the fluid media, and as tracer pulses of a sodium chloride solution were injected. The experimental measurements produced a good fit to the diffusion model. The diffusion coefficient was independent of column height and of position in the column (in contrast to the findings of Siemes and Weiss), as well as of liquid flow rate. It varied markedly with gas flow rate in a manner similar to that observed by Siemes and Weiss, and was of the same order of magnitude. Some variation of diffusion coefficient with gas orifice diameter and column diameter was also observed.

Measurements of axial mixing in a hydrogen–α-methylstyrene bubble-column have been reported by Farkas (F1). Measurements of axial heat diffusion were carried out, the diffusion coefficient being of the order of 7.5 cm^2/sec at a nominal gas velocity of about 4 cm/sec, a value that is in good agreement with some of the results of Siemes and Weiss.

5. Heat Transfer

Kölbel *et al.* have published a series of studies of heat transfer from a cylindrical heating element suspended in bubble-columns with and without suspended solids (K14, K15, K19, M10).

Experiments with bubble-columns containing no suspended solids have demonstrated that the heat-transfer coefficient increases with increasing gas velocity to a maximum value of about 4500 kcal/m²-hr-°C, which is reached for a superficial gas velocity of 8 to 10 cm/sec. Bubble size and height above the gas distributor had no significant influence on heat transfer, whereas the reduction of surface tension had a diminishing influence at low gas velocities but no effect at high velocities, the former effect being explained as corresponding to a reduction in bubble rise-velocity. Viscosity variations from 1 to 900 centistokes correspond to a pronounced reduction of the heat-transfer coefficients. Variations of column diameter had no effect if the ratio between column diamater and bubble diameter was above 20. The results could be correlated by the following equations:

For $N_{Re} < 150$:

$$N_{Nu} = 22.4 N_{Re}^{0.355} \tag{45}$$

For $N_{Re} > 150$:

$$N_{Nu} = 43.7 N_{Re}^{0.22} \tag{46}$$

Nusselt and Reynolds numbers are based on the diameter of the heating element, the conductivity and viscosity of the liquid, and the nominal gas velocities. The heat-transfer coefficient is constant for nominal liquid velocities above 10 cm/sec. The results were obtained for Prandtl numbers from 5 to 1200, but no effect of this variation was observed.

In experiments with bubble-columns containing suspended sand particles with average diameter 0.12 mm, an increase in heat-transfer coefficient was observed with increasing sand concentration, maximum values of 6000 kcal/m²-hr-°C being measured for suspensions containing 50% sand (based on the liquid volume).

For suspensions of kieselguhr in water and oil, the following correlations were developed, in which the Nusselt and Reynolds numbers are based on the length of the heating element:

For the laminar region:

$$N_{Nu} = 227.5 \left(\frac{v_T}{v_W}\right)^{0.1} N_{Re}^{0.161} N_{Pr}^{-0.038} \tag{47}$$

For the turbulent region:

$$N_{Nu} = 454.0 \left(\frac{v_T}{v_W}\right)^{0.1} N_{Re}^{0.113} N_{Pr}^{-0.135} \tag{48}$$

where v_T and v_W are the kinematic viscosities of the liquid medium and of water, respectively.

Experiments with bubble-columns containing suspended solid particles of varying diameter have demonstrated an increase of heat-transfer coefficient with increasing particle size.

6. Reaction Kinetics

In this section, only those studies, all of relatively recent date, that particularly emphasize the determination of rate-determining process steps and the application of the relatively advanced theoretical models discussed in Section IV will be reviewed. For earlier studies of overall reaction kinetics, the reader is referred to the publications of Hall *et al.* (H1) and Kölbel (K6).

Maennig and Kölbel (K11, K18, M1) studied the hydrogenation of ethylene catalyzed by Raney-nickel suspended in hydrogenated Kogasin II. Experiments were carried out at atmospheric pressure and at temperatures ranging from 30 to 100°C. For the conversion of ethylene in an ethylene—hydrogen mixture of stoichiometric composition (50 mole % ethylene and 50 mole % hydrogen) the following observations were made for various values of gas flow rate and temperature:

In the low-temperature range, the conversion increased linearly with increasing temperature, and it is assumed that the chemical reaction is the rate-determining process step. (An activation energy of 13 kcal/mole was determined.) At higher temperatures, conversion is influenced only a little by temperature, the activation energy being from 4 to 8 kcal/mole, and it is assumed that mass transfer is the limiting step. The higher the gas flow rate, the higher the temperature at which mass transfer becomes rate-determining. Reaction orders were calculated on the basis of gas-phase concentrations, that for ethylene being near zero for a wide range of flow rates, and the reaction order for hydrogen varying from 1.5 at a superficial gas velocity of 1 cm/sec, to 3 at higher velocities. Among the effects included in this variation of apparent reaction order is the effect of change in axial mixing. It was found that the activation energy is lower for a catalyst concentration of 5 g/liter than for one of 2.5 g/liter, and it is concluded that gas–liquid mass transfer is rate-determining in the former case.

Calderbank *et al.* (C6) studied the Fischer–Tropsch reaction in slurry reactors of 2- and 10-in. diameters, at pressures of 11 and 22 atm, and at a temperature of 265°C. It was assumed that the liquid-film diffusion of hydrogen from the gas–liquid interface is a rate-determining step, whereas the mass transfer of hydrogen from the bulk liquid to the catalyst was believed to be rapid because of the high ratio between catalyst exterior surface area and bubble surface area. The experimental data were not in complete agreement with a theoretical model based on these assumptions.

A complicating factor in this process is the formation of finely divided carbon, which causes an increase of liquid viscosity and promotes bubble coalescence whereby the gas–liquid interfacial area is reduced. Also observed was a effect of reactor height, which may be attributed to bubble coalescence.

Farley and Ray (F3) have reported holdup and conversion data for the Fischer–Tropsch process carried out in a pilot-scale reactor.

Kölbel *et al.* (K16) examined the conversion of carbon monoxide and hydrogen to methane catalyzed by a nickel–magnesium oxide catalyst suspended in a paraffinic hydrocarbon, as well as the oxidation of carbon monoxide catalyzed by a manganese–cupric oxide catalyst suspended in a silicone oil. The results are interpreted in terms of the theoretical model referred to in Section IV,B, in which gas–liquid mass transfer and chemical reaction are assumed to be rate-determining process steps. Conversion data for technical and pilot-scale reactors are also presented.

Farkas and Sherwood (F1, S5) have experimentally and theoretically studied the catalytic hydrogenation of α-methylstyrene and of cyclohexene, and have reinterpreted the experimental data of Kölbel and Maennig for the catalytic hydrogenation of ethylene. They concluded that mass transfer to the catalyst particles was rate-controlling for the hydrogenation of α-methylstyrene; the activation energy of the reaction was 4 kcal/mole, of the same order of magnitude as that predicted by use of empirical correlations for diffusion in liquids, and the rate was of the same order of magnitude as the calculated rate of mass transfer to suspended solid particles. The analysis of the data for the hydrogenation of ethylene indicates that the chemical reaction is rate-determining, whereas for the hydrogenation of cyclohexene it seems probable that mass transfer to the particles is the controlling process step.

Slesser and Highet (S15) have reported on a theoretical and experimental study of hydrogenation of ethylene catalyzed by Raney-nickel.

D. Stirred-Slurry Operation

Stirred-slurry reactors are of considerable industrial importance in batchwise processing. The catalytic hydrogenation of fats and fatty acids is an example of a process that is carried out almost exclusively in mechanically stirred slurry reactors. The operation is of less significance with respect to continuous processing.

It is the oldest type of gas–liquid–particle operation, and has thus been the subject of a large number of investigations.

1. *Mass Transfer across Gas–Liquid Interface*

Morris (M9) has recently reviewed a number of studies of mass transfer across the gas–liquid interface in mechanically agitated systems containing suspended solid particles. These studies [Hixon and Gaden (H7), Eckenfelder

(E1), Oldshue (O1), and Johnson *et al.* (J4)] have been concerned with the determination of the volume transfer coefficient $K_1 A_b$ (liter/hr), where K_1 is the mass-transfer coefficient and A_b is the total gas–liquid interfacial area. The results obtained using a turbine impeller and an open pipe sparger can be correlated in terms of the nominal gas velocity u_g(ft/hr) and the horsepower input to the impeller HP by an expression of the following form:

$$K_1 A_b = u_g{}^x (HP)^y \qquad (49)$$

The various experimental data correspond to the values of x and y are listed in Table I.

TABLE I

EXPONENTS x AND y FOR EQ. (49)

System	x	y	References
Baker's yeast—air	0.68	—	H7
Sewage—air	0.80	1.54	E1
Cannery waste—air	0.70	—	E1
Dairy waste—air	0.39	0.54	O1
α-methylstyrene—hydrogen, palladium catalyst	0.75	0.56	J4
Flue gas—calcium hydroxide slurry	1.01	0	M9

Later publications have been concerned with mass transfer in systems containing no suspended solids. Calderbank measured and correlated gas–liquid interfacial areas (C1), and evaluated the gas and liquid mass-transfer coefficients for gas–liquid contacting equipment with and without mechanical agitation (C2). It was found that gas film resistance was negligible compared to liquid film resistance, and that the latter was largely independent of bubble size and bubble velocity. He concluded that the effect of mechanical agitation on absorber performance is due to an increase of interfacial gas–liquid area corresponding to a decrease of bubble size.

Calderbank and Moo-Young (C5) have studied gas–liquid mass transfer in systems characterized by high viscosities and high diffusion coefficients, and have on the basis of data obtained in this and other studies developed correlations for the mass-transfer coefficients.

Westerterp *et al.* (W5) measured interfacial areas in mechanically agitated gas–liquid contactors. The existence of two regions was demonstrated: At agitation rates below a certain minimum value, interfacial areas are unaffected by agitation and depend only on nominal gas velocity and the type of gas distributor, whereas at higher agitation rates, the interfacial areas are linearly dependent on the agitation rate and are independent of gas velocity and distributor type. In a following publication, Westerterp (W4) utilizes this and other information as a basis for design recommendations for agitated gas–liquid contactors.

Gal-Or and Resnick (G1) have developed a simplified theoretical model for the calculation of mass-transfer rates for a sparingly soluble gas in an agtitated gas–liquid contactor. The model is based on the average gas residence-time, and its use requires, among other things, knowledge of bubble diameter. In a related study (G2) a photographic technique for the determination of bubble flow patterns and of the relative velocity between bubbles and liquid is described.

2. *Mass Transfer across Liquid–Solid Interface*

Mass transfer across the liquid–solid interface in mechanically agitated liquids containing suspended solid particles has been the subject of much research, and the data obtained for these systems are probably to some extent applicable to systems containing, in addition, a dispersed gas phase. Liquid–solid mass transfer in such systems has apparently not been studied separately. Recently published studies include papers by Calderbank and Jones (C3), Barker and Treybal (B5), Harriott (H4), and Marangozis and Johnson (M3, M4). Satterfield and Sherwood (S2) have reviewed this subject with specific reference to applications in slurry-reactor analysis and design.

3. *Holdup and Axial Dispersion in Gas Phase*

Foust *et al.* (F4) measured gas holdup in mechanically stirred gas–liquid contactors of various diameters (from 1 to 8 ft) and various liquid contents (from 5 to 2250 gal). The nominal gas velocity varied from 1 to 5 ft/min and the power input from 0.01 to 6.5 hp. The contact time (sec/ft) could be correlated by the following expression:

$$\tau = k(HP)^n/(Vu_g) \tag{50}$$

where k and n are constants, V is the liquid content (ft^3), HP the horsepower input, and u_g the nominal gas velocity (ft/sec). Values of k and n are presented for various geometries.

Kramers *et al.* (K21) measured gas residence-time distribution in a mechanically agitated gas–liquid contactor of 0.6-m diameter for various gas velocities and agitator speeds. In the region where agitation has an effect on the gas–liquid interfacial area (cf. the study by Westerterp *et al.* (W5), Section V,D,1), the residence-time distribution was found to resemble closely that of a perfect mixer.

4. Holdup and Axial Dispersion in Liquid Phase

Liquid residence-time distributions in mechanically stirred gas–liquid–solid operations have apparently not been studied as such. It seems a safe assumption that these systems under normal operating conditions may be considered as perfectly mixed vessels. Van de Vusse (V3) have discussed some aspects of liquid flow in stirred slurry reactors.

5. Heat Transfer

Studies dealing specifically with heat transfer in mechanically stirred gas–liquid–particle operations have apparently not been published.

6. Reaction Kinetics

Among the earlier studies of reaction kinetics in mechanically stirred slurry reactors may be noted the papers of Davis *et al.* (D3), Price and Schiewitz (P5), and Littman and Bliss (L6). The latter investigated the hydrogenation of toluene catalyzed by Raney-nickel with a view to establishing the mechanism of the reaction and reaction orders, the study being a typical example of the application of mechanically stirred reactors for investigations of chemical kinetics in the absence of mass-transfer effects.

Johnson *et al.* (J4) investigated the hydrogenation of α-methylstyrene catalyzed by a palladium–alumina catalyst suspended in a stirred reactor. The experimental data have recently been reinterpreted in a paper by Polejes and Hougen (P4), in which the original treatment is extended to take account of variations in catalyst loading, variations in impeller type, and variations of gas-phase composition. Empirical correlations for liquid-side resistance to gas–liquid and liquid–solid mass transfer are presented.

E. Gas–Liquid Fluidization

This operation has been examined to a considerably lesser extent than most of the other gas–liquid–particle operations, presumably because it is of relatively recent origin and because it has, thus far, been applied on a commercial scale in only a few cases.

The expression "gas–liquid fluidization," as defined in Section III,B,3, is used for operations in which momentum is transferred to suspended solid particles by cocurrent gas and liquid flow. It may be noted that the expression "gas–liquid–solid fluidization" has been used for bubble-column slurry reactors (K3) with zero net liquid flow (of the type described in Sections III,B,1 and III,V,C). The expression "gas–liquid fluidization" has also been used for dispersed gas–liquid systems with no solid particles present.

1. *Mass Transfer across Gas–Liquid Interface*

Massimilla *et al.* (M5) measured the rate of absorption of carbon dioxide in water from a mixture of carbon dioxide and nitrogen. Used as solid phase were silica sand particles of average equivalent diameter 0.22 mm, or glass ballotini of average equivalent diameter 0.50 and 0.80 mm. Columns of 30- and 90-mm i.d. were used, and the column height was varied from 100 to 1200 mm.

The absorption rate increased with increasing nominal liquid velocity for all particle sizes and decreased with increasing particle size for all liquid velocities. The absorption rates were lower than those measured in an equivalent gas–liquid system with no solid particles present. The difference is explained as being due to a higher rate of bubble coalescence and, consequently, a lower gas–liquid interfacial area in the gas–liquid fluidized bed.

In a later publication, Massimilla *et al.* (M6) investigated the formation and coalescence of gas bubbles formed at a single orifice in a liquid fluidized bed. The experiments were carried out with atmospheric air and tap water and with silica sand particles, glass ballotini, and iron sand particles of average equivalent diameters from 0.22 to 1.09 mm. Orifices of 0.4 and 1.0 mm i.d. were used, and the gas flow rate was varied from 0.5 to 6.5 cm^3/sec. The average bubble size was determined from cinephotographs of the bed surface. A measure of the bubble size at the orifice was obtained from cinephotographs of the surface of a very shallow bed (the distance between orifice and bed surface being 2 in., and bubble coalescence being assumed negligible). In addition, the rise-velocities of isolated gas bubbles in a liquid fluidized bed were determined by the injection of a well-defined gas volume and measurement of the time interval between the instant of injection and the instant at which the bubble emerged through the bed surface.

The bubble size at formation varied with particle characteristics. It was further observed that the bubble size decreased with increasing fluidization intensity (i.e., with increasing liquid velocity). The rate of coalescence likewise decreased with increasing fluidization intensity; the net rate of coalescence had a positive value at distances from 1 to 2 ft above the orifice, whereas at larger distances from the orifice the rate approached zero. The bubble rise-velocity increased steadily with bubble size in a manner similar to that observed for viscous fluids, but different to that observed for water. An attempt was made to explain the dependence of the rate of coalescence on fluidization intensity in terms of a relatively high viscosity of the liquid fluidized bed.

Østergaard (O4), by somewhat similar experiments, measured the rate of growth of gas bubbles formed in a liquid fluidized bed at a single orifice of 3.0-mm i.d. for gas flow rates varying from 9 to 63 cm^3/sec. The experiments were carried out with tap water, atmospheric air, and sand particles of average equivalent diameter 0.64 mm. The bubble frequency at the orifice was measured by an electrical resistance probe connected to an oscilloscope, which produced a straight line at zero gas flow rates and a series of peaks at finite gas flow rates, each peak corresponding to the increase in electrical resistance resulting from the formation of a bubble. The bubble frequency at the bed surface was calculated from cinephotographs.

The measured bubble frequencies at the orifice did not generally deviate significantly from those measured in water with no solid particles present. Near incipient fluidization, however, the frequencies appeared to be lower than in water. The measured rate of coalescence was markedly dependent on bed porosity, having a relatively high value near the point of incipient fluidization and decreasing with increasing liquid velocity and bed porosity. This is in general agreement with the results of Massimilla *et al.*, as is the observation that the main change in bubble frequency occurs within a relatively short distance from the orifice. The rate of coalescence did not vary significantly with gas flow rate.

Observations of bubbles emerging through the bed surface show that bubble shape is markedly dependent on liquid velocity. This indicates the existence of a relationship between bed viscosity and liquid velocity. A bed near incipient fluidization is characterized by a high viscosity, and an emerging bubble is of nearly spherical shape, whereas a fluidized bed of high porosity is characterized by a viscosity not very much higher than that of water, so that an emerging bubble is of spherical cap shape.

Adlington and Thompson (A1) measured the gas–liquid interfacial area in beds of particles of from 0.3- to 3-mm diameter by oxygen absorption in a sodium sulfite solution. They found that the interfacial area decreased with decreasing bed porosity, and was less sensitive to changes in particle size.

The results of Massimilla *et al.*, Østergaard, and Adlington and Thompson are in substantial agreement on the fact that gas–liquid fluidized beds are characterized by higher rates of bubble coalescence and, as a consequence, lower gas–liquid interfacial areas than those observed in equivalent gas–liquid systems with no solid particles present. This supports the observations of gas absorption rate by Massimilla *et al.* It may be assumed that the absorption rate depends upon the interfacial area, the gas residence-time, and a mass-transfer coefficient. The last of these factors is probably higher in a gas–liquid fluidized bed because the bubble Reynolds number is higher, but the interfacial area is lower and the gas residence-time is also lower, as will be further discussed in Section V,E,3.

These results are, however, only valid for the particle sizes referred to. Lee (L3) has reported measurements of average bubble diameter and gas–liquid interfacial area for gas–liquid fluidized beds of glass beads of 6-mm diameter. The gas injection system was arranged to give fairly large bubbles at the base, and it was found that the bubble size decreased and the gas–liquid interfacial area increased with increasing height above the gas distributor. The breakup of bubbles occurred at a higher rate in beds of low expansion. Lee suggests that beds of larger particles may be of practical value because of the improved gas absorption which, presumably, may be obtained.

2. *Mass Transfer across Liquid–Solid Interface*

No work on mass transfer across the liquid–solid interface in gas–liquid fluidized beds has come to the author's attention.

3. *Holdup and Residence-Time Distribution of Gas Phase*

Adlington and Thompson (A1) observed that the presence of solids had little influence on gas holdup below nominal gas velocities of about 1.5 cm/sec. At higher gas velocities, the gas holdup was depressed by the solids, particularly in beds of low porosity. The gas holdup was only little influenced by changes in particle size.

Viswanathan *et al.* (V6) measured gas holdup in fluidized beds of quartz particles of 0.649- and 0.928-mm mean diameter and glass beads of 4-mm diameter. The fluid media were air and water. Holdup measurements were also carried out for air–water systems free of solids in order to evaluate the influence of the solid particles. It was found that the gas holdup of a bed of 4-mm particles was higher than that of a solids-free system, whereas the gas holdup in a bed of 0.649- or 0.928-mm particles was lower than that of a solids-free system. An attempt was made to correlate the gas holdup data for gas–liquid fluidized beds using a mathematical model for two-phase gas–liquid systems proposed by Bankoff (B4).

Østergaard and Gilliland (O6) measured the gas holdup in beds of sand particles of 40–60 and 60–80 mesh. The fluid media were nitrogen and water. The gas holdup was largely independent of particle size and liquid velocity, but increased linearly with nominal gas velocity. Comparison with an equivalent gas–liquid system free of solids showed that the gas holdup of such a system was higher than that of a gas–liquid fluidized bed.

The results reported for beds of small particles (1 mm diameter and less) are in substantial agreement on the fact that the presence of solid particles tends to decrease the gas holdup and, as a consequence, the gas residence-time. This fact may also support the observations of gas absorption rate by Massimilla *et al.* (Section V,E,1) if it is assumed that a decrease of absorption rate caused by a decrease of residence time outweighs the increase of absorption rate caused by increase of mass-transfer coefficient arising from the increase in bubble Reynolds number. These results on gas holdup are in agreement with the results on bubble coalescence in beds of small particles (Section V,E,1) if it is assumed that an increase in bubble volume causes an increase in bubble rise-velocity.

Based on the same assumption, the results by Viswanathan *et al.* on gas holdup in beds of larger particles are in agreement with the results of Lee (L3) on bubble breakup in beds of larger particles (O5). No work on the residence-time distribution of the gas phase in gas–liquid fluidized beds has come to the author's attention.

4. *Holdup and Residence-Time Distribution of Liquid Phase*

Measurement of the expansion of a gas–liquid fluidized bed provides a measure of the holdup of solids or of the corresponding combined holdup of gas and liquid. From such measurements, the holdup of liquid may be calculated if the gas holdup has been determined independently.

Measurements of bed expansion have been reported by Turner (T4), Østergaard (O2), Stewart and Davidson (S17), Adlington and Thompson (A1), and Østergaard and Theisen (O7). These workers are in agreement on the following points: The bed expansion (or the total holdup of gas and liquid) increases with increasing liquid velocity, as is also the case for liquid fluidized beds containing no gas bubbles. The bed expansion may, however, decrease with increasing gas velocity, and this is a unique feature of gas–liquid fluidized beds. This contraction of a liquid fluidized bed upon the injection of gas bubbles may be quite considerable (a reduction of bed height by 48% has been observed for a highly expanded bed of small particles), and may thus correspond to a very significant reduction of liquid holdup and residence-time. The reduction of bed expansion caused by gas injection increases with increasing bed expansion. The reduction decreases with increasing particle size.

Stewart and Davidon (S17) and Østergaard (O3) have proposed qualitative and semiquantitative explanations for this characteristic phenomenon. These explanations are based on the assumption that the gas bubbles in the bed are followed by bubble wakes in which liquid volume elements travel at velocities that are of the same magnitude as the bubble velocities and thus considerably higher than the average superficial liquid velocity in the bed. It then follows from the equation of mass conservation that the liquid velocity in the bed outside the bubble wakes must be lower than the average liquid velocity, and that the expansion of this part of the bed must be correspondingly reduced. In Fig. 5 are shown some of the data reported by Turner in comparison with theoretical curves computed from the semiempirical theory of Østergaard. This theory cannot at the present time be applied to systems of geometries and particle sizes different from those used in these experiments.

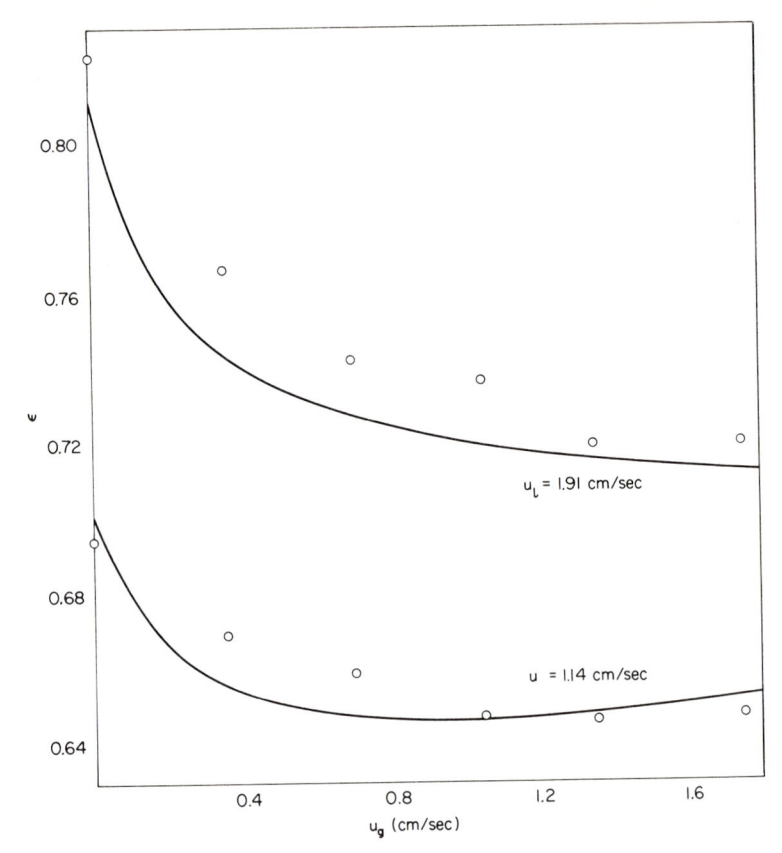

FIG. 5. Sum of gas and liquid holdup in gas–liquid fluidized bed. Experimental data of Turner (T4) and theoretical curves of Østergaard (O3).

Østergaard and Theisen (O7) have proposed a qualitative explanation for the observation that the reduction of bed expansion decreases with increasing particle size.

From the assumption that liquid volume elements travel as bubble wakes at velocities higher than the average liquid velocity, it follows that the bubble movement must influence the residence-time distribution of the liquid phase. However, no work on this subject has come to the author's attention.

5. Heat Transfer

Østergaard (O2) measured the wall-to-bed heat-transfer coefficient in a bed of 3-in. diameter. The media were air, water, and glass ballotini of 0.5-mm diameter. It was observed that the heat-transfer coefficient for a liquid fluidized bed near the point of incipient fluidization could be approximately doubled by the injection of gas, whereas for higher liquid velocities, only small increases in the heat-transfer coefficient would result.

Wall-to-bed heat-transfer coefficients were also measured by Viswanathan et al. (V6). The bed diameter was 2 in. and the media used were air, water, and quartz particles of 0.649- and 0.928-mm mean diameter. All experiments were carried out with constant bed height, whereas the amount of solid particles as well as the gas and liquid flow rates were varied. The results are presented in that paper as plots of heat-transfer coefficient versus the ratio between mass flow rate of gas and mass flow rate of liquid. The heat-transfer coefficient increased sharply to a maximum value, which was reached for relatively low gas–liquid ratios, and further increase of the ratio led to a reduction of the heat-transfer coefficient. It was also observed that the maximum value of the heat-transfer coefficient depends on the amount of solid particles in the column. Thus, for 0.928-mm particles, the maximum value of the heat-transfer coefficient obtained in experiments with 750-gm solids was approximately 40% higher than those obtained in experiments with 250- and 1250-gm solids.

Van Driesen and Stewart (V4) have reported temperature measurements for various locations in commercial gas–liquid fluidized reactors for the large-scale catalytic desulfurization and hydrocracking of heavy petroleum fractions (2500 barrels per day capacity). The hydrogenation was carried out in two stages; the maximum and minimum temperatures measured were 774° and 778°F for the first stage and 768° and 770°F for the second. These results indicate that gas–liquid fluidized reactors are characterized by a high effective thermal conductivity.

Measurements of heat-transfer coefficients and effective thermal conductivity for gas–liquid fluidized beds have also been carried out by Manchanda (M2).

6. Reaction Kinetics

Adlington and Thompson (A1) investigated catalytic desulfurization by hydrogenation in a gas–liquid fluidized bed of 1-in. diameter under typical process conditions. From these experiments, calculations were made of the mass-transfer driving force across the gas–liquid interface, i.e., of the difference in hydrogen concentration. Two extreme situations were examined in experiments with (1) low liquid-feed space velocity, high gas flow rate, and high liquid recycle rate, and (2) high liquid-feed space velocity, low gas flow rate, and no liquid recycle. Calculations based on the experimental data showed that the ratio between the mass-transfer driving force and the saturation concentration of hydrogen in the feed was less than 0.1, and it was concluded that desulfurization can be carried out in a fluidized bed under conditions such that diffusion restrictions in the bulk phase are not important.

D_g, D_l Superficial axial dispersion coefficients for gas and liquid phases, respectively
d_b Diameter of bubble
d_p Nominal diameter of packing
d_t Diameter of column
G_g Rate of flow of gas
G_l Rate of flow of liquid
g Acceleration due to gravity
H Henry's Law constant
H_s Critical holdup of solids, the limiting weight fraction of solids in suspension
(H.t.u.)$_L$ Height of liquid-film transfer unit
K_s Equilibrium constant for adsorption
K_1 Mass transfer coefficient
k Reaction-rate constant
k_c Rate constant of surface reaction
k_g Gas-to-liquid mass transfer coefficient
k_1 Liquid-side film coefficient for absorption
k_0 Pseudo first order reaction rate constant
$k_0 \exp(-E_{eff}/RT)$ Effective reaction rate constant
k_r A reaction rate constant
k_s Mass transfer coefficient from liquid to particle
M Molecular weight
m Ratio of equilibrium concentration in gas phase to that in liquid phase
P Total pressure
P_{H_2} Partial pressure of hydrogen
P_{j0} Partial pressure of component j at inlet
Q_1 Holdup of catalyst
q_{CO} Rate of flow of crabon monoxide
R Gas constant
r Reaction rate
r_{eff} Effective reaction rate
T Absolute temperature
t Temperature
u_b Velocity of bubble
u_g Superficial gas velocity
u_l Superficial liquid velocity
u_t Stokes' Law free-settling velocity
V Total volume of liquid
X_j Conversion of component j
x_{H_2} Mole-fraction of hydrogen
$x_{C_2H_4}$ Mole-fraction of ethylene
z Distance from inlet
α Exponent in Freundlich isotherm
ε_g Fractional gas holdup
ε_{g0} Volumetric void fraction of dry packed-bed
μ_l Viscosity of liquid
ν_j Stoichiometric coefficient of component j
ν_l Kinematic viscosity of liquid
ν_T Kinematic viscosity of liquid
ν_W Kinematic viscosity of water
σ Surface tension
τ Contact time

DIMENSIONLESS GROUPS

N_b "Bubble-flow number," $\sigma_l/u_b\mu_l$
N_{Ga_1} Gallileo number for liquid phase, $g\,d_p^3/\nu_l^2$
N_{Nu} Nusselt number
N_{Pe} Peclet number
N_{Pe_g} Peclet number for gas phase
N_{Pr} Prandtl number
N_{Re_g} Reynolds number for gas phase
N_{Re_l} Reynolds number for liquid phase
N_{Sc} Schmidt number
N_{We} Weber number

References

A1. Adlington, D., and Thompson, E., *Proc. European Symp. Chem. React. Eng.*, 3rd p. 203. Pergamon Press, Oxford, 1965.
A2. Azbel, D. S., *Intern. Chem. Eng.* **3**, 319 (1963).
A3. Azizyan, A. G., and Smirnov, N. I., *Intern. Chem. Eng.* **5**, 533 (1965).
B1. Babcock, B. D., Mejdell, G. T., and Hougen, O. A., *A.I.Ch.E. (Am. Inst. Chem. Eng.) J.* **3**, 366 (1957).
B2. Bailey, A. E., *in* "Encyclopedia of Chemical Technology" (R. E. Kirk and D. F. Othmer, eds.), 1st ed., Vol. 6, p. 140. Wiley (Interscience), New York, 1951.
B3. Baird, M. H. I., and Davidson, J. F., *Chem. Eng. Sci.* **17**, 87 (1962).
B4. Bankoff, S. G., *J. Heat Transfer* **82**, 265 (1960).
B5. Barker, J. J., and Treybal, R. E., *A.I.Ch.E. (Am. Inst. Chem. Eng.) J.* **6**, 289 (1960).
B6. Benson, F. R., *in* "Encyclopedia of Chemical Technology" (R. E. Kirk and D. F. Othmer, eds.), 2nd ed., Vol. 1, p. 569. Wiley (Interscience), New York, 1963.
B7. Benson, H. E., Field, J. H., Bienstock, D., and Storch, H. H., *Ind. Eng. Chem.* **46**, 2278 (1954).
B8. Bowman, C. W., and Johnson, A. I. *Can. J. Chem. Eng.* **40**, 139 (1962).
B9. Braulick, W. J., Fair, J. R., and Lerner, B. J., *A.I.Ch.E. (Am. Inst. Chem. Eng.) J.* **11**, 73 (1965).
B10. Brusie, J. P., and Hort, E. V., *in* "Encyclopedia of Chemical Technology" (R. E. Kirk and D. F. Othmer, eds.), 2nd ed., Vol. 1, p. 609. Wiley (Interscience), New York, 1963.
C1. Calderbank, P. H., *Trans. Inst. Chem. Engrs. (London)* **36**, 443 (1958).
C2. Calderbank, P. H., *Trans. Inst. Chem. Engrs. (London)* **37**, 173 (1959).
C3. Calderbank, P. H., and Jones, S. J. R., *Trans. Inst. Chem. Engrs. (London)* **39**, 363 (1961).
C4. Calderbank, P. H., and Lochiel, A. C., *Chem. Eng. Sci.* **19**, 485 (1964).
C5. Calderbank, P. H., and Moo-Young, M. B., *Chem. Eng. Sci.* **16**, 39 (1961).
C6. Calderbank, P. H., Evans, F., Farley, R., Jepson, G., and Poll, A., *Proc. Symp. Catalysis Pract.* p. 66. Inst. Chem Engrs. London, 1963.
C7. Calderbank, P. H., Moo-Young, M. B., and Bibby, R., *Proc. European Symp. Chem. React, Eng.*, 3rd p. 91. Pergamon Press, Oxford, 1965.
C8. Chervenak, M. C., Feigelman, S., Wolk, R., Byrd, C. R., Hellwig, L. R., and van Driesen, R. P., *Oil Gas J.* **61** (41), 227 (1963).
C9. Coppock, P. D., and Meiklejohn, G. T., *Trans. Inst. Chem. Engrs. (London)* **29**, 75 (1951).
C10. Corr, H., Haarer, E., and Hornberger, P., British Patent 921,447 (1963).
C11. Cova, D. R., *Ind. Eng. Chem. (Process)* **5**, 20 (1966).
C12. Crowell, J. H., Benson, H. E., Field, J. H., and Storch, H. H., *Ind. Eng. Chem.* **42**, 2376 (1950).
D1. Danckwerts, P. V., *Chem. Eng. Sci.* **2**, 1 (1953).
D2. Datta, R. L., Napier, D. H., and Newitt, D. M., *Trans. Inst. Chem. Engrs. (London)* **28**, 14 (1950).
D3. Davis, H. S., Thomson, G., and Crandall, G. S., *J. Am. Chem. Soc.* **54**, 2340 (1932).
D4. De Maria, F., and White, R. R., *A.I.Ch.E. (Am. Inst. Chem. Eng.) J.* **6**, 473 (1960).
D5. De Waal, K. J. A., and van Mameren, A. C., *Proc. Symp. Transport Phenomena* p. 60. Inst. Chem. Engrs., London, 1965.
D6. Dodds, W. S., Stutzman, L. F., Sollami, B. J., and McCarter, R. J., *A.I.Ch.E. (Am. Inst. Chem. Eng.) J.* **6**, 197 (1960).

D7. Dunn, W. E., Vermeulen, T., Wilke, C. R., and Word, T. T., Univ. Calif. Radiation Lab. Rept. 10394 (1963).
E1. Eckenfelder, W. W., Jr., *Chem. Eng. Progr.* **52**, 286 (1956).
E2. Elgin, J. C., and Weiss, E. B., *Ind. Eng. Chem.* **31**, 435 (1939).
E3. Ericsson, E. O., McCarthy, J. L., and Pearson, D. A., *in* "Pulp and Paper Science and Technology" (C. E. Libby, ed.), Vol. 1, p. 240. McGraw-Hill, New York, 1962.
F1. Farkas, E. J., A Study of the Bubble Column Slurry Reactor. Diss., Mass. Inst. Technol. Cambridge, Massachusetts, 1964.
F2. Farley, R., and Ray, D. J., *Brit. Chem. Eng.* **9**, 830 (1964).
F3. Farley, R., and Ray, D. J., *J. Inst. Petrol.* **50**, 27 (1964).
F4. Foust, H. C., Mack, D. E., and Rushton, J. H., *Ind. Eng. Chem.* **36**, 517 (1944).
G1. Gal-Or, B., and Resnick, W., *Chem. Eng. Sci.* **19**, 653 (1964).
G2. Gal-Or, B., and Resnick, W., *A.I.Ch.E. (Am. Inst. Chem. Eng.) J.* **11**, 740 (1965).
G3. Glaser, M. B., and Lichtenstein, I., *A.I.Ch.E. (Am. Inst. Chem. Eng.) J.* **9**, 30 (1963).
G4. Glaser, M. B., and Litt, M., *A.I.Ch.E. (Am. Inst. Chem. Eng.) J.* **9**, 103 (1963).
G5. Gorring, R. L., and Katz, D. L., *A.I.Ch.E. (Am. Inst. Chem. Eng.) J.* **8**, 123 (1962).
G6. Govier, G. W., *Can. J. Chem. Eng.* **43**, 3 (1965).
G7. Grassmann, P., *Chem. Ingr.-Tech.* **31**, 148 (1959).
G8. Griffith, R. M., *Chem. Eng. Sci.* **12**, 198 (1960).
G9. Guyer, A., and Pfister, X., *Helv. Chim. Acta* **29**, 1173 (1946).
H1. Hall, C. C., Gall, D., and Smith, S. L., *J. Inst. Petrol* **38**, 845 (1952).
H2. Halvorson, H. O., *in* "Encyclopedia of Chemical Technology" (R. E. Kirk and D. F. Othmer, eds.), 1st ed., Vol. 14, p. 946. Wiley (Interscience), New York, 1955.
H3. Hammerton, D., and Garner, F. H., *Trans. Inst. Chem. Engrs. (London) Symp. Gas Absorption* **32**, S18 (1954).
H4. Harriott, P., *A.I.Ch.E. (Am. Inst. Chem. Eng.) J.* **8**, 93 (1962).
H5. Harrison, D., Lane, M., and Walne, D. J., *Trans. Inst. Chem. Engrs. (London)* **40**, 214 (1962).
H6. Hellwig, L. R., van Driesen, R. P., Schuman, S. C., and Slyngstad, C. E., *Oil Gas J.* **60** (21), 119 (1962).
H7. Hixon, A. W., and Gaden, E. L., Jr., *Ind. Eng. Chem.* **42**, 1792 (1950).
H8. Hofmann, H., *Chem. Eng. Sci.* **14**, 193 (1961).
H9. Hoog, H., *Proc. Intern. Congr. Catalysis, 3rd.* p. 7. North-Holland Publ., Amsterdam, 1965.
H10. Hoogendoorn, C. J., and Lips, J., *Can. J. Chem. Eng.* **43**, 125 (1965).
H11. Horne, W. A., and McAfee, J., *Advan. Petrol. Chem. Refining* **3**, 193 (1960).
H12. Houghton, G., McLean, A. M., and Ritchie, P. D., *Chem. Eng. Sci.* **7**, 26 (1957).
H13. Houghton, G., McLean, A. M., and Ritchie, P. D., *Chem. Eng. Sci.* **7**, 40 (1957).
H14. Hughmark, G. A., and Pressburg, B. S., *A.I.Ch.E. (Am. Inst. Chem. Eng.) J.* **7**, 677 (1961).
J1. Jackson, R., *Trans. Inst. Chem. Engrs. (London) Chem. Eng.* p. 107 (1964).
J2. Jesser, B. W., and Elgin, J. C., *Trans. Am. Inst. Chem. Engrs.* **39**, 277 (1943).
J3. Johnson, A. I., and Bowman, C. W., *Can. J. Chem. Eng.* **36**, 253 (1958).
J4. Johnson, D. L., Saito, H., Polejes, J. D., and Hougen, O. A., *A.I.Ch.E. (Am. Inst. Chem. Engrs.) J.* **3**, 411 (1957).
K1. Károlyi, J., Zalai, A., Birthler, R., and Spitzner, H., *Intern Chem. Eng.* **3**, 597 (1963).
K2. Kastens, M. L., Hirst, L. L., and Dressler, R. G., *Ind. Eng. Chem.* **44**, 450 (1952).
K3. Kato, Y., *Kagaku Kogaku* (Abr. Ed. English) **1**, 3 (1963).
K4. Kolár, V., and Brož, Z., *Collection Czech. Chem. Commun.* **30**, 2527 (1965).
K5. Kölbel, H., *in* "Chemische Technologie" (K. Winnacher and L. Küchler, eds.), Vol. 3, p. 439. Hauser, Munich, 1959.

K6. Kölbel, H., *Chem. Eng. Sci.* **14**, 151 (1961).
K7. Kölbel, H., and Ackermann, P., *World Petrol. Congr.*, *Proc.* 4*th*, *Rome*, 1955 Sect. IV C, p. 228 (1955).
K8. Kölbel, H., and Ackermann, P., *Chem. Ingr.-Tech.* **28**, 381 (1956).
K9. Kölbel, H., and Ackermann, P., *in* "Ullmanns Encyklopädie der technischen Chemie" (W. Foerst, ed.), 3rd ed., Vol. 9, p. 716. Urban und Schwarzenberg, Munich, 1957
K10. Kölbel, H., and Langemann, H., *Dechema Monograph.* **49**, 253 (1964).
K11. Kölbel, H., and Maennig, H.-G., *Z. Elektrochem.* **66**, 744 (1962).
K12. Kölbel, H., Ackermann, P., and Engelhardt, F., *Erdoel Kohle* 153, 225, 303 (1956).
K13. Kölbel, H., Borchers, E., and Langemann, H., *Chem. Ingr.-Tech.* **33**, 668 (1961).
K14. Kölbel, H., Borchers, E., and Martins, J., *Chem. Ingr.-Tech.* **32**, 84 (1960).
K15. Kölbel, H., Borchers, E., and Müller, K., *Chem. Ingr.-Tech.* **30**, 729 (1958).
K16. Kölbel, H., Hammer, H., and Meisl, U., *Proc. European Symp. Chem. React. Eng.*, *3rd* p. 115. Pergamon Press, Oxford, 1965.
K17. Kölbel, H., Langemann, H., and Platz, J., *Dechema Monograph.* **41**, 225 (1962).
K18. Kölbel, H., Hammer, H., Henne, H. J., and Maenning, H.-G., *Dechema Monograph.* **49**, 277 (1964).
K19. Kölbel, H., Siemes, W., Maas, R., and Müller, K., *Chem. Ingr.-Tech.* **30**, 400 (1958).
K20. Kramers, H., and Alberda, G., *Chem. Eng. Sci.* **2**, 173 (1953).
K21. Kramers, H., Hanhart, J., and Westerterp, K. R., *Chem. Eng. Sci.* **18**, 503 (1963).
K22. Krönig, W., *Erdoel Kohle* **15**, 176 (1962).
K23. Krönig, W., *World Petrol. Congr.*, *Proc. 6th* Sect. IV, paper 7 (1963).
L1. Lapidus, L., *Ind. Eng. Chem.* **49**, 1000 (1957).
L2. Larkins, R. P., White, R. R., and Jeffrey, D. W., *A.I.Ch.E.* (*Am. Inst. Chem. Eng.*) *J.* **7**, 231 (1961).
L3. Lee, *Proc. European Symp. Chem. React. Eng.*, *3rd* p. 211. Pergamon Press, Oxford, 1965.
L4. Leonard, J. H., and Houghton, G., *Chem. Eng. Sci.* **18**, 133 (1964).
L5. Li, P.-S., West, F. B., Vance, W. H., and Moulton, R. W., *A.I.Ch.E.* (*Am. Inst. Chem. Eng.*) *J.* **11**, 581 (1965).
L6. Littman, H., and Bliss, H., *Ind. Eng. Chem.* **51**, 659 (1959).
L7. Lochiel, A. C., and Calderbank, P. H., *Chem. Eng. Sci.* **19**, 471 (1964).
M1. Maennig, H.-G., Heterogen-katalytische Gasreaktionen im Blasensäulereaktor. Diss., Tech. Univ. Berlin-Charlottenburg, 1963.
M2. Manchanda, K. D., Diss., Indian Inst. of Technol., Bombay, India, 1963.
M3. Marangozis, J., and Johnson, A. I., *Can. J. Chem. Eng.* **40**, 231 (1962).
M4. Marangozis, J., and Johnson, A. I., *Can. J. Chem. Eng.* **41**, 133 (1962).
M5. Massimilla, L., Majuri, N., and Signorini, P., *Ric. Sci.* **29**, 1934 (1959).
M6. Massimilla, L., Solimando, A., and Squillace, E., *Brit. Chem. Eng.* **6**, 232 (1961).
M7. Miyauchi, T., and Vermeulen, T., *Ind. Eng. Chem. Fundamentals* **2**, 113 (1963).
M8. Miyauchi, T., and Vermeulen, T., *Ind. Eng. Chem. Fundamentals* **2**, 304, (1963).
M9. Morris, R. M., *Chem. Ind.* (*London*) p. 1836 (1964).
M10. Müller, K., Der Wärmeübergang und das hydrodynamische Verhalten in Blasensäulen. Diss., Tech. Univ. Berlin-Charlottenburg, 1962.
N1. Nicklin, D. J., *Chem. Eng. Sci.* **17**, 693 (1962).
N2. Norman, W. S., "Absorption, Distillation and Cooling Towers." Longmans, London, 1961.
O1. Oldshue, J. Y., *Ind. Eng. Chem.* **48**, 2194 (1956).
O2. Østergaard, K., *in* "Fluidization," p. 58. Soc. Chem. Ind., London, 1964.
O3. Østergaard, K., *Chem. Eng. Sci.* **20**, 165 (1965).
O4. Østergaard, K., *Chem. Eng. Sci.* **21**, 470 (1966).

O5. Østergaard, K., *Chem. Eng. Sci.* **21**, 837 (1966).
O6. Østergaard, K., and Gilliland, E. R., to be published.
O7. Østergaard, K., and Theisen, P. I., *Chem. Eng. Sci.* **21**, 413 (1966).
O8. Otake, T., and Kuniguta, E., *Kagaku Kogaku* **22**, 144 (1958).
O9. Otake, T., and Okada, K., *Kagaku Kogaku* **17**, 176 (1953).
P1. Pasveer, A., *Sewage Ind. Wastes* **27**, 1130 (1955).
P2. Pichler, H., and Hector, A., *in* "Encyclopedia of Chemical Technology" (R. E. Kirk. and D. F. Othmer, eds.), 2nd ed., Vol. 4, p. 446. Wiley (Interscience), New York, 1964.
P3. Piret, E. L., Mann, C. A., and Wall, T., Jr., *Ind. Eng. Chem.* **32**, 861 (1940).
P4. Polejes, J. D., and Hougen, O. A., to be published (O. A. Hougen, private communication).
P5. Price, R. H., and Schiewetz, D. B., *Ind. Eng. Chem.* **49**, 807 (1957).
P6. Prost, C., and Le Goff, P., *Genie Chim.* **91**, 6 (1964).
R1. Raff, R. A. V., and Allison, J. B., "Polyethylene." Wiley (Interscience), New York, 1956.
R2. Ross, L. D., *Chem. Eng. Progr.* **61** (10), 77 (1965).
R3. Roy, N. K., Guha, D. K., and Ras, M. N., *Chem. Eng. Sci.* **19**, 215 (1964).
R4. Ruckenstein, E., *Chem. Eng. Sci.* **19**, 131 (1964).
S1. Sater, V. E., and Levenspiel, O., *Ind. Eng. Chem. Fundamentals* **5**, 86 (1966).
S2. Satterfield, C. N., and Sherwood, T. K., "The Role of Diffusion in Catalysis." Addison-Wesley, Reading, Massachusetts, 1963.
S3. Schiesser, W. E., and Lapidus, L., *A.I.Ch.E. (Am. Inst. Chem. Eng.) J.* **7**, 163 (1961).
S4. Schoenemann, K., *Dechema Monograph.* **21**, 203 (1952).
S5. Sherwood, T. K., and Farkas, E. J., *Chem. Eng. Sci.* **21**, 573 (1966).
S6. Sherwood, T. K., and Pigford, R. L., "Absorption and Extraction." McGraw-Hill, New York, 1952.
S7. Shingu, H., U.S. Patent 2,985,668 (1961).
S8. Shulman, H. L., and Molstad, M. C., *Ind. Eng. Chem.* **42**, 1058 (1950).
S9. Shulman, H. L., Ullrich, C. F., and Wells, N., *A.I.Ch.E. (Am. Inst. Chem. Eng.) J.* **1**, 247 (1905).
S10. Siemes, W., *Chem. Ingr.-Tech.* **26**, 614 (1954).
S11. Siemes, W., and Borchers, E., *Chem. Ingr.-Tech.* **28**, 783 (1956).
S12. Siemes, W., and Borchers, E., *Chem. Eng. Sci.* **12**, 77 (1960).
S13. Siemes, W., and Günther, K., *Chem. Ingr.-Tech.* **28**, 389 (1957).
S14. Siemes, W., and Weiss, W., *Chem. Ingr.-Tech.* **29**, 727 (1957).
S15. Slesser, C. G. M., and Highet, J., *Brit. Chem. Eng.* **11**, 247 (1966).
S16. Stemerding, *Chem. Eng. Sci.* **14**, 209 (1961).
S17. Stewart, P. S. B., and Davidson, J. F., *Chem. Eng. Sci.* **19**, 319 (1964).
T1. Tadaki, T., and Maeda, S., *Kagaku Kogaku* (Abr. ed. English) **2**, 69, (1964).
T2. Tadaki, T., and Maeda, S., *Kagaku Kogaku* (Abr. ed. English) **2**, 195 (1965).
T3. Towell, G. D., Strand, C. P., and Ackerman, G. H., *Proc. Symp. Mixing* p. 97. Inst. Chem. Engrs., London, 1965.
T4. Turner, R., *in* "Fluidisation," p. 47, Soc. Chem. Ind., London, 1964.
U1. Urban, W., *Erdoel Kohle* **8**, 780 (1955).
V1. Van Deemter, J. J., *Proc. European Symp. Chem. React. Eng.*, 3rd p. 215. Pergamon Press, Oxford, 1965.
V2. Van der Laan, E. T., *Chem. Eng. Sci.* **7**, 187 (1958).
V3. Van de Vusse, J. G., *Proc. Symp. Scaling-Up* p. 41. Inst. Chem. Engrs., London, 1957.
V4. Van Driesen, R. P., and Stewart, N. C., *Oil Gas J.* **62** (20), 100 (1964).
V5. Verschoor, H., *Trans. Inst. Chem. Engrs. (London)* **28**, 52 (1950).

V6. Viswanathan, S., Kakar, A. S., and Murli, P. S., *Chem. Eng. Sci.* **20**, 903 (1965).
V7. Volpicelli, G., and Massimilla, L., *Pulp Paper Mag. Can.* **66**, T512 (1965).
V8. Von Bogdandy, L., Rutsch, W., and Stranski, J. N., *Chem. Ingr.-Tech.* **31**, 580 (1959).
W1. Weber, H. H., Untersuchungen über die Verweilzeitverteilung in Aufstromkolonnen. Diss., Tech. Hochschule, Darmstadt, Germany, 1961.
W2. Weekman, V. W., Jr., and Myers, J. E., *A.I.Ch.E. (Am. Inst. Chem. Eng.) J.* **10**, 951 (1964).
W3. Weekman, V. W., Jr., and Myers, J. E., *A.I.Ch.E. (Am. Inst. Chem. Eng.) J.* **11**, 13 (1965).
W4. Westerterp, K. R., *Chem. Eng. Sci.* **18**, 495 (1963).
W5. Westerterp, K. R., van Dierendonck, L. L., and de Kraa, J. A., *Chem. Eng. Sci.* **18**, 157 (1963).
Y1. Yoshida, F., and Akita, K., *A.I.Ch.E. (Am. Inst. Chem. Eng.) J.* **11**, 9 (1965).
Y2. Yoshitome, H., Kakihara, M., and Tsuchiya, Y., *Kagaku Kogaku* (Abr. ed. English) **2**, 186 (1965).
Z1. Zabor, R. C., Odiozo, R. C., Schmidt, B. K., and Kaiser, J. R., *Actes Congr. Intern. Catalyse*, 2^e, *Paris*, 1960 p. 2601. Editions Technip, Paris, 1961.
Z2. Zenz, F. A., and Othmer, D. F., "Fluidization and Fluid-Particle Systems." Reinhold, New York, 1960.

THERMODYNAMICS OF FLUID-PHASE EQUILIBRIA AT HIGH PRESSURES

J. M. Prausnitz

*Department of Chemical Engineering,
University of California, Berkeley, California*

I. Introduction	140
II. Fugacities in Gas Mixtures: Fugacity Coefficients	144
A. Lewis Fugacity Rule	144
B. Virial Equation of State for Moderate Densities	145
C. Empirical Equation of State for High Densities	149
D. Pseudocritical Hypothesis	152
E. Conclusion	154
III. Fugacities in Liquid Mixtures: Activity Coefficients	154
A. Standard States and Normalization of Activity Coefficients	155
B. Constant-Pressure Activity Coefficients and the Gibbs–Duhem Equation	158
IV. Effect of Pressure on Activity Coefficients: Partial Molar Volumes	160
A. Dilute Solutions	161
B. Partial Molar Volumes in Saturated Liquids from an Equation of State	162
V. Dilute Solutions of Gases in Liquids at High Pressures	166
A. Krichevsky–Kasarnovsky Equation	166
B. Krichevsky–Ilinskaya Equation	169
VI. Concentrated Solutions: High-Pressure Vapor–Liquid Equilibria	170
A. Equation of State for Both Phases	171
B. Corresponding States for Mixtures	172
C. Semiempirical Models for Activity Coefficients	173
D. Thermodynamic Consistency	179
VII. Liquid–Liquid Equilibria in Binary Systems	184
VIII. Gas–Gas Equilibria in Binary Systems	190
IX. Liquid–Liquid Equilibria in Ternary Systems Containing One Supercritical Component	194
A. Effect of Third Component on Critical Solution Temperature	195
B. Ternary Phase Diagram from Binary Data	196
C. Thermodynamic Criteria for Separation-Process Application	203
References	203

I. Introduction

It is now nearly 100 years since Gibbs first discussed the use of thermodynamics to obtain a quantitative description of phase behavior. Although the importance of Gibbs' work was recognized by leading European scientists toward the end of the 19th century, it did not receive much attention in America until after his death in 1903. Since that time, hundreds of books and thousands of articles in the chemical literature have made use of Gibbs' ideas to elucidate our understanding of the equilibrium behavior of multiphase systems. However, with few exceptions, these publications on thermodynamics have restricted their attention to systems which are at low or moderate pressures (in the vicinity of 1 atm). Little attention has been given to the thermodynamics of fluid mixtures at high pressures, despite the abundance of experimental studies of such mixtures, especially of those containing hydrocarbons. In this paper, we discuss some current ideas on the thermodynamics of high-pressure fluid-phase mixtures. No attempt is made to present an exhaustive review; instead, it is our purpose here to summarize a few of those contributions which appear to be most promising for chemical-engineering applications.

Phase equilibria at high pressures are of interest to a variety of chemical engineers, especially to those in the petroleum, cryogenic, and related industries. Although there are some important exceptions, calculations of high-pressure phase behavior before about 1960 were primarily based on empirical correlations such as the well-known K-charts. Only in recent years has there been a serious tendency to extend to high-pressure equilibria those thermodynamic concepts which for many years have been accepted for the description of equilibria at ordinary pressures. This tendency is to be welcomed, for such an extension is an essential prerequisite for further progress in our understanding of high-pressure phenomena. Two important factors may be cited in support of this conclusion. First, deeper insight into the properties of fluids can be obtained only through progress in molecular science and statistical mechanics. Much effort is currently being expended in these areas by leading scientists; since the results of such efforts are necessarily expressed in thermodynamic terms, we can make use of these results only if we are able to describe high-pressure phase behavior with thermodynamic functions (fugacity, activity coefficient, partial molar volume, etc.) rather than with empirical *ad hoc* expressions such as K-factors. Thermodynamic analysis is the essential step which links molecular physics with practical phase-equilibrium problems.

Second, the number of binary fluid systems which are (or may soon become) of interest in technical processes is already large, and if one thinks of the possible number of ternary, quarternary, etc., fluid systems which may find their place in the chemical industry, the number becomes extremely large. To investigate experimentally even a small fraction of these systems under varying conditions of temperature and pressure represents an effort which is prohibitive in both cost and time. Therefore, the only reasonable long-range approach to increasing our quantitative knowledge of phase behavior for engineering purposes is to restrict experimental study to a few representative systems; on the basis of such studies, we must then use thermodynamic analysis as a basic tool for correlating and generalizing the experimental data. From these generalizations, reasonable predictions can be made for the phase behavior of previously investigated systems under new conditions, and perhaps for the behavior of new systems which have not been investigated at all. Such correlations are of special utility for predicting the properties of multicomponent systems from data obtained for binary mixtures; these predictions are frequently required for rational engineering design of separation operations.

The fundamental equilibrium relation for multicomponent, multiphase systems is most conveniently expressed in terms of fugacities; for any component i, the fugacity must be the same in all equilibrated phases. However, this equilibrium relation is of no use until we can relate the fugacity of a component to directly measurable properties. The task of phase-equilibrium thermodyanamics, therefore, is to describe the effects of temperature, pressure, and composition on the fugacity of each component in each phase of the system of interest. For any component i in a system containing N components, the total differential of the logarithm of the fugacity f_i is given by

$$d \ln f_i = \left(\frac{\partial \ln f_i}{\partial T}\right)_{P,x} dT + \left(\frac{\partial \ln f_i}{\partial P}\right)_{T,x} dP$$

$$+ \sum_{j=1}^{N-1} \left(\frac{\partial \ln f_i}{\partial x_j}\right)_{P,T,x_k} dx_j \quad k = 1, \ldots j-1, j+1 \ldots, N-1 \quad (1)$$

Thermodynamics gives limited information on each of the three coefficients which appear on the right-hand side of Eq. (1). The first term can be related to the partial molar enthalpy and the second to the partial molar volume; the third term cannot be expressed in terms of any fundamental thermodynamic property, but it can be conveniently related to the excess Gibbs energy which, in turn, can be described by a solution model. For a complete description of phase behavior we must say something about each of these three coefficients for each component, in every phase. In high-pressure work, it is important to give particular attention to the second coefficient, which tells us how phase behavior is affected by pressure.

When analyzing typical experimental data, it is often difficult to isolate the effect of pressure because, more often than not, a change in the pressure is accompanied by a simultaneous change in either the temperature or the composition, or both. A striking example of such simultaneous changes is given by the experimental results in Fig. 1, which show the effect of pressure on the melting temperature of solid argon (M6). Curve (a) gives the melting curve for pure argon as reported by Clusius and Weigand (C5); the melting temperature rises with pressure, as predicted by the Clapeyron equation, since argon expands upon melting. Curve (b) gives results of Mullins and Ziegler (M6) for the melting temperature of argon in the presence of high-pressure helium gas; these results are very similar to those for pure argon, and again the melting temperature rises with increasing pressure. Finally, curve (c) gives the melting temperature for argon in the presence of high-pressure hydrogen gas; these data were also reported by Mullins and Ziegler. One may at first be struck and perhaps puzzled by the fact that curve (c) shows completely different behavior: the melting temperature now falls with rising pressure.

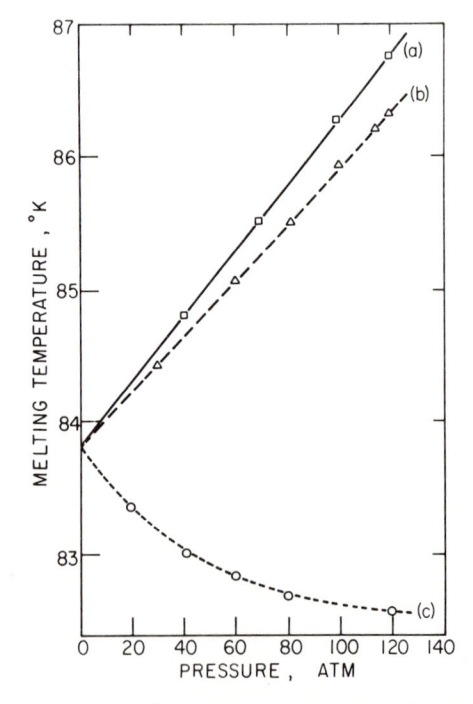

FIG. 1. Effect of pressure on the melting temperature of argon. (a) Pure argon, (b) argon-helium, (c) argon-hydrogen. Qualitative difference between curves (b) and (c) is due to the effect of composition on the liquid-phase fugacity of argon (M6).

The qualitative difference between curves (b) and (c) can be clarified by thermodynamic analysis if we note that the three-phase equilibrium temperature is determined by two separate effects: first, there is the effect of pressure on the fugacities of solid and liquid argon which is essentially the same for the three cases (a), (b), and (c); and second, there is the effect of liquid composition on the fugacity of liquid argon. It is the effect of composition which for case (b) is very much different from that for case (c). At a given pressure, the solubility of hydrogen in liquid argon is much larger than that of helium; since pressure and composition simultaneously influence the fugacity of liquid argon, we find that the presence of hydrogen alters the equilibrium temperature in a manner qualitatively different from that found in the presence of helium. The fugacity of liquid argon is raised by pressure but lowered by the solubility of another component. In the case of helium (low solubility), the pressure effect dominates, but for hydrogen (high solubility), the effect of composition is more important. We need not here go into a more detailed analysis of this particular system, but merely want to emphasize that the fugacity of a component is determined by the three variables temperature, pressure, and composition; that in a typical experimental situation these influences operate in concert; and that whatever success we may expect in explaining the behavior of a multicomponent, multiphase system is determined directly by the extent of our knowledge of the three coefficients given in Eq. (1).

At present we know least about the first coefficient, which is rigorously related to the enthalpy by

$$\left(\frac{\partial \ln f_i}{\partial T}\right)_{P,x} = \frac{-[\bar{h}_i - h_i^{\circ}]}{RT^2} \tag{2}$$

where $\bar{h}_i$ is the partial molar enthalpy of i. While we would like to use information on $\bar{h}_i$ to help us toward calculating f_i, we almost never can do so in high-pressure work because so little is known about enthalpies of fluid mixtures at high pressures. (In fact, in typical chemical-engineering work, it is more common to reverse the procedure, viz., to differentiate fugacity data with respect to temperature, in order to estimate enthalpies.) As a result, the best we can do at present is to analyze experimental phase-equilibrium data as a function of pressure and composition along an isotherm, and to allow any empirical parameters obtained from such analysis to vary with temperature as dictated by the experimental results.

We have a considerable body of knowledge to help us to say something about the third coefficient, the variation of fugacity with composition. Many empirical and semiempirical expressions (e.g., Margules, Van Laar, Scatchard–Hildebrand) have been investigated toward that end. Most of our experience in this regard is limited to liquid mixture at low pressures, where we can safely neglect the second coefficient which relates the fugacity of the pressure according to

$$\left(\frac{\partial \ln f_i}{\partial P}\right)_{T,x} = \frac{\bar{v}_i}{RT} \tag{3}$$

where $\bar{v}_i$ is the partial molar volume. It is this coefficient which is of primary concern in high-pressure phase equilibria, and no significant progress in high-pressure thermodynamics can be expected until we achieve improved quantitative understanding of partial molar volumes in liquid mixtures. We return to this key problem in Section IV after our discussion of fugacities in gas mixtures.

II. Fugacities in Gas Mixtures: Fugacity Coefficients

The fugacity of a component i in a gas mixture is related to the total pressure P and to its mole fraction y_i through the fugacity coefficient φ_i defined by

$$\varphi_i \equiv f_i/y_i P \tag{4}$$

The fugacity coefficient is a function of pressure, temperature, and gas composition. It has the useful property that for a mixture of ideal gases $\varphi_i = 1$ for all i. The fugacity coefficient is related to the volumetric properties of the gas mixture by either of the exact relations (B3, P5, R6):

$$RT \ln \varphi_i = \int_0^P \left[\left(\frac{\partial V}{\partial n_i}\right)_{T,V,n_j} - \frac{RT}{P} \right] dP \tag{5}$$

or

$$RT \ln \varphi_i = \int_V^\infty \left[\left(\frac{\partial P}{\partial n_i}\right)_{T,V,n_j} - \frac{RT}{V} \right] dV - RT \ln z \tag{6}$$

where V is the total volume of the gas mixture containing n_1 moles of component 1, n_2 moles of component 2, etc., and z is the compressibility factor of the gas mixture at total pressure P and temperature T. While Eq. (5) is the one most commonly quoted in textbooks, Eq. (6) is preferable for most practical calculations. Whenever volumetric properties are expressed in the form of an equation of state, Eq. (6) is likely to be more useful, since almost all equations of state are explicit in pressure rather than in volume.

A. Lewis Fugacity Rule

In order to simplify Eq. (5), it was suggested many years ago by Lewis (L3) that Amagat's law be used, viz., to assume that the partial molar volume of component i at any temperature and pressure is equal to the molar volume of pure (gaseous) i at the same temperature and pressure. When this assumption is incorporated into Eq. (5) we obtain a very simple relation, known as the Lewis fugacity rule:

$$\varphi_i = \varphi_{\text{pure } i} \qquad \text{(at same } T \text{ and } P) \tag{7}$$

The limits of the Lewis fugacity rule are not determined by pressure but by composition; the Lewis rule becomes exact at any pressure in the limit as $y_i \to 1$, and therefore it always provides a good approximation for any component i which is present in excess. However, for a component with small mole fraction in the vapor phase, the Lewis rule can sometimes lead to very large errors (P5, R3, R10).

B. Virial Equation of State for Moderate Densities

The chemical literature is rich with empirical equations of state and every year new ones are added to the already large list. Every equation of state contains a certain number of constants which depend on the nature of the gas and which must be evaluated by reduction of experimental data. Since volumetric data for pure components are much more plentiful than for mixtures, it is necessary to estimate mixture properties by relating the constants of a mixture to those for the pure components in that mixture. In most cases, these relations, commonly known as mixing rules, are arbitrary because the empirical constants lack precise physical significance. Unfortunately, the fugacity coefficients are often very sensitive to the mixing rules used.

The only equation of state with a firm theoretical foundation is the virial equation which gives the compressibility factor as a power-series expansion in the density. It is discussed in more detail elsewhere (H6, P5, P9, P10, R8, R10). For chemical-engineering calculations on mixtures, the virial equation has as a great advantage the fact that the various constants in the equation (the virial coefficients) can be related exactly to the composition of the mixture by simple polynomial expansions of the mole fractions. In these expansions, new constants (cross coefficients) appear; these constants have a precise physical significance and, in principle, can be evaluated from fundamental knowledge (potential functions) of the intermolecular forces between dissimilar molecules. The main disadvantage of the virial equation comes from the fact that while we have much information on second virial coefficients and some on third virial coefficients, very little is known about fourth and higher virial coefficients. As a result, application of the virial equation is limited to moderate pressures, since higher coefficients become increasingly important as the pressure becomes large. The virial equation is

$$z = \frac{Pv}{RT} = 1 + \frac{B}{v} + \frac{C}{v^2} + \cdots \tag{8}$$

The composition dependences of the virial coefficients B, C, ..., are given by

$$B = \sum_i \sum_j y_i y_j B_{ij} \qquad (9)$$

$$C = \sum_i \sum_j \sum_k y_i y_j y_k C_{ijk} \qquad (10)$$

etc.

When the virial expansion (truncated after the third term) is substituted into Eq. (6), we obtain

$$\ln \varphi_i = \frac{2}{v}\sum_j y_j B_{ij} + \frac{3}{2v^2}\sum_j \sum_k y_j y_k C_{ijk} - \ln z \qquad (11)$$

where B_{ij} is the second virial coefficient characteristic of the interaction between molecules i and j, and C_{ijk} is the third virial coefficient characteristic of the interactions between the three molecules i, j, and k. The molar volume v and the compressibility factor z of the gas mixture are determined from the virial equation with the pressure, temperature, and composition as independent variables.

An application of Eq. (11) is shown in Fig. 2, which gives the solubility of solid carbon dioxide in compressed air at a low temperature. The solubility is calculated from the equation of equilibrium

$$y_i = \frac{f_i^S}{\varphi_i P} \qquad (12)$$

where f_i^S is the fugacity of pure solid carbon dioxide at the temperature and pressure of the system. The fugacity coefficient φ_i was found from Eq. (11) with the various virial coefficients calculated from statistical-mechanical expressions using the Kihara potential (C8, K2, K3, P11, S3). The predicted solubilities are in good agreement with experimental results, and the virial equation correctly predicts the minimum in the solubility curve. Also shown are calculated results using only second virial coefficients (third virial coefficients were neglected), and while this calculation correctly predicts the minimum, it fails at pressures just beyond that corresponding to minimum solubility. For contrast, Fig. 2 also shows calculations based on the ideal gas assumption, i.e., $\varphi_i = 1$.

While virial coefficients can be calculated from statistical-mechanical formulas, for practical work it is usually more convenient to employ semiempirical correlations. Most of these correlations are based on the principle of corresponding states and as a result their applicability is limited to normal fluids, i.e., to fluids which are nonpolar (or slightly polar) and which do not exhibit any specific chemical interactions such as hydrogen bonding. For engineering purposes, useful correlations have been given by Pitzer and Curl (P1), McGlashan, Potter, and Wormald (M1, M2), Black (B5), Prausnitz (P5, P8), O'Connell (O2), and Chueh (C2, C3). These correlations are based on pure-component data, and some assumptions must be made to apply them to mixtures (P5, P8, P9, P11, R8, R10).

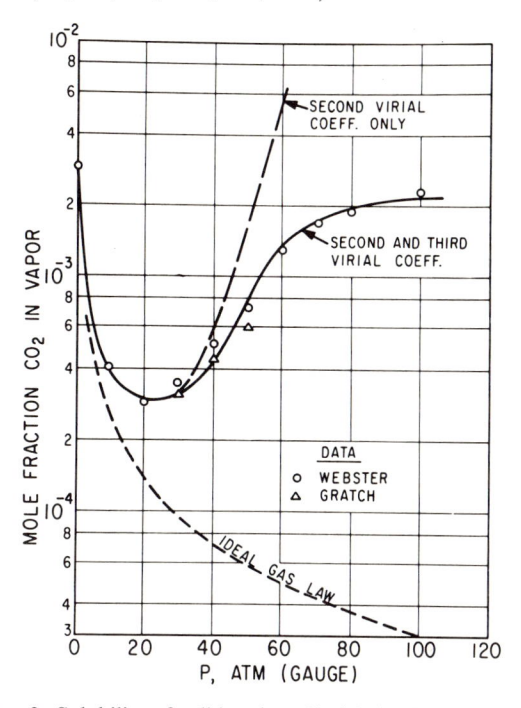

FIG. 2. Solubility of solid carbon dioxide in air at 143°K.

In Fig. 2, we showed the solubility of a solid component in a compressed gas as calculated from Eq. (11). A similar calculation can be made for the solubility of a liquid component in a compressed gas which is only slightly soluble in the liquid. For the liquid component, when $x_i \approx 1$, the equation of equilibrium is

$$y_i = \frac{x f^L_{\text{pure } i}}{\varphi_i P} \tag{13}$$

where $f^L_{\text{pure } i}$ is the fugacity of pure liquid i at the temperature and pressure of the system.

Figure 3 shows the solubility of n-decane in nitrogen at 50°C. Calculations were made with the virial equation [Eq. (11)] neglecting third and higher virial coefficients. Second virial coefficients for decane and nitrogen were taken from the equations of Pitzer and Curl (P2), and the cross-coefficient which characterizes the interaction between decane and nitrogen was taken from the correlation of Benson (P8). Figure 3 also shows the experimental results (P8) as well as calculations based on the assumption of ideal-gas behavior and on the Lewis rule (Eq. 7). Both of the simplified calculations lead to appreciable error; whereas solubilities calculated from the Lewis rule are too large, those calculated from the ideal-gas law are too low. The directions of these deviations from the data follow from the assumptions made. The Lewis rule, in effect, assumes that the vaporized decane molecules are surrounded only by other decane molecules; the Lewis rule is " blind " to the presence of other molecules (in this case, nitrogen), and since the attractive forces between decane–decane are considerably larger than those between decane–nitrogen, the calculated mole fraction of decane in the vapor is too high. On the other hand, the ideal-gas law assumes that the vaporized decane molecules are, essentially, in a vacuum, experiencing no attractive forces

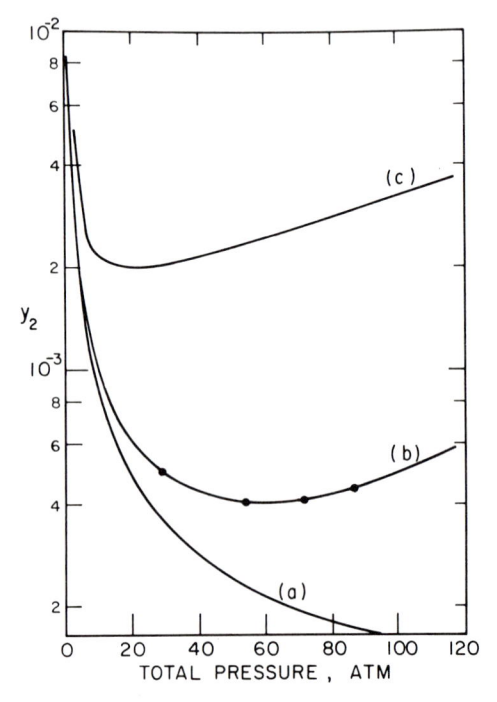

FIG. 3. Vapor-phase solubility of n-decane in nitrogen at 50°C. (a) Ideal gas, (b) virial equation, (c) Lewis rule, (●) experimental.

whatsoever; the ideal gas law is "blind" to the presence of any molecules at all, and as a result of this assumption the calculated vapor-phase solubility is too low. If the gaseous solvent had been helium (or hydrogen) instead of nitrogen, calculations based on the ideal-gas law would provide a good approximation; at normal temperatures, the intermolecular forces between decane and helium (or hydrogen) are so small that the vaporized decane molecules, in effect, "feel" as if they are alone. Solubility of a heavy component in a dense gas is strongly dependent on the intermolecular forces between solute and solvent, and the virial coefficients take these forces into account. For example, the virial equation correctly predicts that at 75°C and 60 atm the solubility of decane in carbon dioxide is about three times that in nitrogen and about seven times that in hydrogen (P8).

C. Empirical Equation of State for High Densities

When the density of the gas mixture approaches or exceeds a critical value, the virial equation is not useful and an empirical equation of state must be used to compute the fugacity coefficients. For this purpose it is most convenient to utilize an analytical equation of state. However, if the volumetric properties are given in numerical (or graphical) form, the calculations can also be performed as discussed elsewhere (M7); in particular, it is possible to calculate fugacity coefficients of components in high-pressure mixtures using generalized (corresponding-states) correlations (L2, P2). Such calculations are always based on assumptions regarding the variation of effective critical constants with composition. These assumptions comprise what is frequently called the pseudocritical hypothesis (see Section II,D).

To illustrate how an empirical equation may be used to calculate fugacity coefficients, we use the equation of Redlich and Kwong (R2), which generally gives satisfactory results for nonpolar fluids. This equation is

$$P = \frac{RT}{(v-b)} - \frac{a}{T^{1/2}v(v+b)} \tag{14}$$

where a and b are constants for each gas. These constants are related to the critical temperature T_c and critical pressure P_c by

$$a = \frac{0.4278 R^2 T_c^{2.5}}{P_c} \tag{15}$$

$$b = \frac{0.0867 RT_c}{P_c} \tag{16}$$

For a mixture, we assume the essentially arbitrary mixing rules

$$a = \sum_i \sum_j y_i y_j a_{ij} \tag{17}$$

$$b = \sum_i y_i b_i \tag{18}$$

Substitution into Eq. (11) gives

$$\ln \varphi_i = \ln \frac{v}{v-b} + \frac{b_i}{v-b} - \frac{2\sum_j y_j a_{ij}}{RT^{3/2}b} \ln \frac{v+b}{v}$$
$$+ \frac{ab_i}{RT^{3/2}b^2}\left[\ln\frac{v+b}{v} - \frac{b}{v+b}\right] - \ln\frac{Pv}{RT} \tag{19}$$

The molar volume v is found from the equation of state [Eq. (14)] when the pressure, temperature, and composition are the independent variables.

In their original paper, Redlich and Kwong also used the mixing rules given by Eqs. (17) and (18); in addition, however, they made the important simplification $a_{ij} = (a_i a_j)^{1/2}$. This simplification may introduce appreciable error (C3b, J2), and we do not use it here. Instead, we first rewrite Eq. (15) in the form

$$a = \frac{0.4278 RT_c^{1.5} v_c}{z_c} \tag{20}$$

where v_c and z_c are, respectively, the critical volume and the compressibility factor at the critical point, and then generalize Eq. (20) to read

$$a_{ij} = \frac{0.4278 RT_{c_{ij}}^{1.5} v_{c_{ij}}}{z_{c_{ij}}} \tag{21}$$

where $T_{c_{ij}}$, $v_{c_{ij}}$, and $z_{c_{ij}}$ are characteristic parameters for the interaction between molecules i and j. For two nonpolar molecules i and j, not too dissimilar in size, we suggest the rules

$$v_{c_{ij}}^{1/3} = \tfrac{1}{2}(v_{c_i}^{1/3} + v_{c_j}^{1/3}) \tag{22}$$

$$z_{c_{ij}} = \tfrac{1}{2}(z_{c_i} + z_{c_j}) \tag{23}$$

$$T_{c_{ij}} = (T_{c_i} T_{c_j})^{1/2}[1 - k_{ij}] \tag{24}$$

where k_{ij} is a constant characteristic of the interactions between two unlike molecules. As a first approximation, we may set $k_{ij} = 0$, and, provided molecules i and j are similar in size, shape, and chemical nature, good results are usually obtained with this simplification. In general, however, k_{ij} is a small number (usually positive and of the order 10^{-2} to 10^{-1}) which may not be neglected if accurate results are desired. Subject to several simplifying assumptions, it is possible to predict k_{ij} from the theory of intermolecular forces (London or Kirkwood–Muller), but such predictions are often not reliable. For small molecules, k_{ij} is, to a good approximation, independent of density, composition, and temperature; therefore, even fragmentary volumetric data may be used to provide a good estimate. A particularly good source of k_{ij} is provided by an analysis of second-virial-coefficient data as shown by Gunn (P9) and also by Eckert (E2). Characteristic constants k_{ij} for 10 binary systems are given in Table 1. A more complete list is available elsewhere (C3b, P8a).

TABLE I

CHARACTERISTIC BINARY CONSTANTS k_{ij} OBTAINED FROM VOLUMETRIC DATA

System	k_{ij}
Propane–methane	0.02
n-Butane–methane	0.04
n-Pentane–methane	0.06
Isopentane–ethane	0.02
Benzene–propane	0.02
Hydrogen sulfide–methane	0.05
Carbon tetrafluoride–methane	0.07
Argon–methane	0.02
Argon–krypton	0.03
n-Butane–Carbon dioxide	0.18

An application of Eq. (19) is shown in Fig. 4, which gives the solubility of solid naphthalene in compressed ethylene at three temperatures slightly above the critical temperature of ethylene. The curves were calculated from the equilibrium relation given in Eq. (12). Also shown are the experimental solubility data of Diepen and Scheffer (D4, D5) and calculated results based on the ideal-gas assumption ($\varphi_i = 1$). The ordinate scale is logarithmic and it is evident that very large errors are incurred when corrections for gas-phase nonideality are neglected.

The solubility of solids in compressed gases has attracted the attention of geologists (M4) and of mechanical engineers concerned with the design of high-pressure steam plants. Nonelectrolytes as well as salts have large solubilities in supercritical steam, and when these solutes precipitate upon expansion of the vapor, they can do serious mechanical damage to the turbine. The high solubilities of condensed components follow from their very small fugacity coefficients in the dense gas phase (F1, F2, R10).

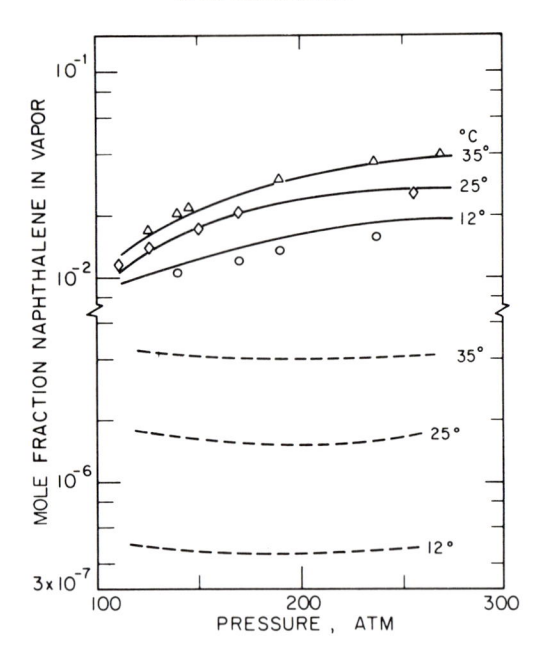

FIG. 4. Vapor-phase solubility of naphthalene in ethylene. Data points from G. A. M. Diepen and F. E. C. Scheffer, *J. Am. Chem. Soc.* **70**, 4085 (1948); vapor-phase fugacities from (——) Redlich-Kwong equation; (– – –) Ideal gas law.

D. Pseudocritical Hypothesis

A somewhat different method of calculating vapor-phase fugacity coefficients is based on an extension to mixtures of the theorem of corresponding states (B6, J1, L2, P2). Assuming the validity of a three-parameter corresponding-states theory, it can readily be shown that the fugacity coefficient for a pure component is a function of reduced pressure, reduced temperature, and one other characterizing parameter such as the acentric factor (P3). If we consider a mixture to be equivalent to a pseudopure substance whose characteristic parameters are averages of the pure-component parameters of the mixture's constituents, we then have for φ_M, the fugacity coefficient of the mixture,

$$\ln \varphi_M = f(T_R, P_R, \omega_M) \tag{25}$$

where $\varphi_M = f_M/P$; $T_R = T/T_{c_M}$ the reduced temperature; $P_R = P/P_{c_M}$ the reduced pressure; and ω_M is the acentric factor. The subscript M refers to the mixture and the generalized function f is the same as that obtained from pure-component properties.[1]

---

[1] The function f is given by Pitzer and co-workers (see C9) and can also be found in Appendix 1 of Lewis and Randall (L4).

To obtain the fugacity coefficient of a component i in the mixture, we must differentiate $\ln \varphi_M$ with respect to composition. For example, for a binary solution having composition y_1 we have the rigorous relation:

$$\ln \varphi_1 = \ln \varphi_M + (1 - y_1)\left(\frac{\partial \ln \varphi_M}{\partial y_1}\right)_{T,P} \tag{26}$$

Since φ_M depends on y_1 through T_{c_M}, P_{c_M}, and ω_M, we have

$$\left(\frac{\partial \ln \varphi_M}{\partial y_1}\right)_{T,P} = \left(\frac{\partial \ln \varphi_M}{\partial T_R}\right)_{P_R,\omega} \left(\frac{\partial T_R}{\partial y_1}\right)_T + \left(\frac{\partial \ln \varphi_M}{\partial P_R}\right)_{T_R,\omega} \left(\frac{\partial P_R}{\partial y_1}\right)_P$$
$$+ \left(\frac{\partial \ln \varphi_M}{\partial \omega_M}\right)_{P_R,T_R} \left(\frac{d\omega_M}{dy_1}\right) \tag{27}$$

The variation of $\ln \varphi_M$ with T_R, P_R, and ω_M can be determined from the generalized function f. The variation of T_{c_M}, P_{c_M}, and ω_M with composition is arbitrary and must be fixed by some mixing rule. For example, Pitzer and Hultgren (P2) proposed that for a binary mixture

$$T_{c_M} = y_1 T_{c_1} + y_2 T_{c_2} + 2y_1 y_2 [2T_{c_{1/2}} - T_{c_1} - T_{c_2}] \tag{28}$$

$$P_{c_M} = y_1 P_{c_1} + y_2 P_{c_2} + 2y_1 y_2 [2P_{c_{1/2}} - P_{c_1} - P_{c_2}] \tag{29}$$

$$\omega_M = y_1 \omega_1 + y_2 \omega_2 \tag{30}$$

where $T_{c_{1/2}}$ and $P_{c_{1/2}}$ are the pseudocritical temperature and pressure of the equimolar mixture; these parameters must be found from equilibrium data for the binary mixture. It is tempting to assume that

$$T_{c_{1/2}} = \tfrac{1}{2}(T_{c_1} + T_{c_2}) \tag{31}$$

$$P_{c_{1/2}} = \tfrac{1}{2}(P_{c_1} + P_{c_2}) \tag{32}$$

in which case Eqs. (28) and (29) reduce to the familiar linear rules first proposed by Kay (K1). Kay's rules often provide good approximations for mixtures whose components are similar in size and chemical nature, but they can lead to large errors for other mixtures (R5). The parameters $T_{c_{1/2}}$ and $P_{c_{1/2}}$ have been reported by Pitzer and Hultgren from volumetric data for twelve binary systems. Although Eqs. (26)–(30) are written for binary mixtures, they are readily generalized to mixtures containing more than two components.

Corresponding-states correlations for pure components rest upon several fundamental assumptions which are discussed elsewhere (G3, H4, H6). To extend such correlations to mixtures, it is necessary to make an additional fundamental assumption, viz., that the characterizing parameters chosen (T_{c_M}, P_{c_M}, and ω_M) are independent of temperature and density and are functions of composition only. Also, as a practical matter, one must assume particular relations such as Eqs. (28)–(30) to express the variation of the characterizing parameters with composition. These relations are arbitrary, and for accurate results it is necessary that at least some binary equilibrium data be available to establish them. While no systematic studies have been made, it is likely that the accuracy of fugacity coefficients in mixtures as obtained from corresponding-states correlations is similar to that obtained from the Redlich–Kwong equation as discussed above. In both cases, the accuracy of prediction is very much improved if at least a few experimental binary data are used to establish the composition dependence of the characterizing parameters or equation-of-state constants.

Vapor-phase fugacity coefficients are needed not only in high-pressure phase equilibria, but are also of interest in high-pressure chemical equilibria (D6, K7, S4). The equilibrium yield of a chemical reaction can sometimes be strongly influenced by vapor-phase nonideality, especially if reactants and products have small concentrations due to the presence in excess of a suitably chosen nonreactive gaseous solvent (S4).

E. Conclusion

With currently available techniques, we are able to make good estimates of gas-phase fugacity coefficients for components in nonpolar mixtures; however, some judgment and experience are required, since no single prescription is optimum for all cases. Even fragmentary binary data can be helpful in increasing accuracy, and are essential for mixtures which contain one or more polar components. At high temperatures, when the effect of polarity is small, good predictions can sometimes be made from pure-component data alone, as shown, for example, by Franck's study of the steam–carbon dioxide system (F3). However, at normal temperatures, especially for mixtures containing strongly polar or associating components, neither theoretical nor empirical tools are sufficiently developed for consistently good results. New research is badly needed for understanding dense gas mixtures containing one or more polar components.

III. Fugacities in Liquid Mixtures: Activity Coefficients

For any component i in a liquid phase, the fugacity of i is most conveniently related to the mole fraction x_i by use of the activity coefficient, γ_i, according to

$$f_i = \gamma_i x_i f_i^0 \qquad (33)$$

where f_i^0 is the fugacity of i in its standard state. The choice of a standard state is arbitrary, and is dictated by convenience; as shown below, it is frequently convenient to choose a standard state for supercritical components which is different from that used for subcritical components. The independent variables which determine the standard state are the temperature of the system, a fixed composition, and a specified pressure. The choice of standard state seriously influences the dependence of the activity coefficient on composition, and, in particular, determines the normalization of the activity coefficient, i.e., the condition at which γ_i attains a fixed value, usually taken as unity.

For liquid mixtures at low pressures, it is not important to specify with care the pressure of the standard state because at low pressures the thermodynamic properties of liquids, pure or mixed, are not sensitive to the pressure. However, at high pressures, liquid-phase properties are strong functions of pressure, and we cannot be careless about the pressure dependence of either the activity coefficient or the standard-state fugacity.

A. Standard States and Normalization of Activity Coefficients

The most frequently used standard state is the pure liquid ($x_i = 1$) at the system temperature and pressure. When this standard state is used for all components in the mixture, the activity coefficients are said to be symmetrically normalized, because in this case, for every component i,

$$\gamma_i \to 1 \quad \text{as} \quad x_i \to 1 \tag{34}$$

It is sometimes preferable to define the standard state as the pure liquid at the system temperature and at its own saturation pressure. For any component i for which this convention is used, the normalization is also given by Eq. (34).

Whenever we are dealing with a supercritical component, there is a serious disadvantage in using a pure-liquid standard state. If the system temperature T is larger than T_{c_i} (the critical temperature of component i), pure liquid i cannot exist at T, and we are faced with the problem of finding the fugacity of a physically unattainable state. In the past, this problem has been a major obstacle retarding the thermodynamics of high-pressure vapor-liquid equilibria. The fugacity of a hypothetical liquid state can only be found by one of several arbitrary methods of extrapolation; while some extrapolations are more popular than others, workers in this field have not been able to agree on a single unambiguous way to compute the fugacity of any supercritical "liquid." Nor is it likely that such agreement will be reached; the main difficulty here derives from the fact that if we use hypothetical pure liquid i as our standard state, then we must have the limiting relation $\gamma_i \to 1$ as $x_i \to 1$. However, experience with experimental data for mixtures has repeatedly shown that once we have calculated f_i^0 by a particular extrapolation, this limiting relation for γ_i may hold for a few solutions of i in various solvents, only to fail badly in others. Such failure is serious because unless we know how γ_i is normalized, we cannot write thermodynamically significant relations for the variation of γ_i with composition, nor can we meaningfully use binary data to estimate the properties of multicomponent mixtures. The failure of the limiting relation that $\gamma_i \to 1$ as $x_i \to 1$, regardless of the nature of the other component, is especially apparent whenever T is very much larger than T_{c_i} (e.g., solutions of hydrogen in common solvents near room temperature), since in that case the value of f_i^0 is extremely sensitive to the details of extrapolation. On the other hand, when T is only slightly larger than T_{c_i}, the use of hypothetical liquid i for the standard state often provides a useful thermodynamic procedure, since then f_i^0 is only slightly sensitive to the method of extrapolation.

1. *Unsymmetric Normalization in Binary Solutions*

The difficulties engendered by a hypothetical liquid standard state can be eliminated by the use of unsymmetrically normalized activity coefficients. These have been used for many years in other areas of solution thermodynamics (e.g., for solutions of electrolytes or polymers in liquid solvents) but they have only recently been employed in high-pressure vapor–liquid equilibria (P7).

In a binary liquid solution, we differentiate between the subcritical component 1, called the solvent, (for which $T < T_{c_1}$), and the supercritical component 2, called the solute ($T > T_{c_1}$). For component 1, we normalize γ_1 in the usual way:

$$\gamma_1 \to 1 \quad \text{as} \quad x_1 \to 1 \tag{35}$$

For component 2, however, we normalize γ_2 according to

$$\gamma_2^* \to 1 \quad \text{as} \quad x_2 \to 0 \tag{36}$$

Equations (35) and (36) constitute what is called the unsymmetric convention of normalization, because γ_1 and γ_2^* go to unity in different ways. The asterisk serves as a reminder that the activity coefficient so designated is normalized in a manner different from the customary one. Separate notation for activity coefficients normalized according to Eq. (36) is psychologically useful because the effect of composition on γ^* is radically different from that on γ.

For the solvent (component 1), f_1^0 is the fugacity of pure saturated liquid 1 at the system temperature. However, the standard-state fugacity for the solute (component 2) is given by

$$f_2^0 = \lim_{x_2 \to 0} (f_2/x_2) = H_{2,1}^{(P_1^s)} \tag{37}$$

where $H_{2,1}^{(P_1^s)}$ stands for Henry's constant of solute 2 in solvent 1 at the system temperature and at P_1^s, the saturation pressure of pure liquid 1.

The standard state given by the unsymmetric convention for normalization has one very important advantage: it avoids all arbitrariness about f_2^0, which is an experimentally accessible quantity; the definition of f_2^0 given by Eq. (37) assures that the activity coefficient of component 2 is unambiguously defined as well as unambiguously normalized. There is no fundamental arbitrariness about f_2^0 because $H_{2,1}^{(P_1^s)}$ can be determined from experimental measurements.

There appears to be a certain operational disadvantage in the use of Henry's constant as the standard-state fugacity for a supercritical component which in the past has prevented its widespread acceptance. For a given solute (component 2), Henry's constant depends not only on temperature but also on the solvent (component 1). It appears, then, that whenever we change the solvent, we must also change the standard-state fugacity of the solute. However, this disadvantage is more apparent than real, and is only a consequence of our narrowminded habits of normalization. For example, if we are concerned with a solution of solute 2 in some solvent 3, we may still use $H_{2,1}^{(P_1^s)}$ as our standard-state fugacity if we wish to avoid changing standard states as we go from one solvent to another. In fact, this is an entirely reasonable procedure for the multicomponent case where solute 2 is dissolved in a mixed solvent containing components 1 and 3 (O1) (see next section). If we use $H_{2,1}^{(P_1^s)}$ for f_2^0 in a binary solution of 2 in 3, the normalization for γ_2^* becomes

$$\gamma_2^* \to \frac{H_{2,3}^{(P_3^s)}}{H_{2,1}^{(P_1^s)}} \quad \text{as} \quad x_2 \to 0 \tag{38}$$

2. Unsymmetric Normalization in Multicomponent Solutions

The unsymmetric convention of normalization is readily applicable to multicomponent solutions, but care must be taken to specify exactly the conditions that give $\gamma_2^* \to 1$. Whereas Eq. (35) is immediately applicable to solutions containing any number of components, Eq. (36) is not complete for a solution containing components in addition to 1 and 2. For a solute 2 dissolved in a mixture of solvents 1 and 3, the normalization conditions are completely specified if we write, for a fixed ratio $x_1/(x_1 + x_3)$,

$$\gamma_1 \to 1 \quad \text{as} \quad x_1 \to 1 \tag{39}$$

$$\gamma_3 \to 1 \quad \text{as} \quad x_3 \to 1 \tag{40}$$

$$\gamma_2^* \to 1 \quad \text{as} \quad x_2 \to 0 \tag{41}$$

As in Eq. (38), it is frequently convenient to use consistently one of the solvents (say, component 1) as the reference solvent; in that event, the fixed ratio is set equal to unity and $\gamma_2^* \to 1$ as $x_2 \to 0$ and $x_1 \to 1$. However, the fixed ratio $x_1/(x_1 + x_3)$ may be chosen to have any convenient value between zero and unity.

It is important to remind ourselves that thermodynamics does not provide any rigid conventions for the choice of standard states and for the normalization of activity coefficients. These conventions are determined strictly by convenience; while one particular convention may be most useful for a particular class of solutions, it may be awkward for another class. Indeed, one of the main advantages of the activity-coefficient concept is its flexibility. The purpose of the activity coefficient is no more and no less than to relate the fugacity of a component at some condition of composition and pressure to what it is at some other (reference) condition of composition and pressure where we accurately know its numerical value. Although it is advantageous to choose a standard state which gives activity coefficients of the order of unity in the concentration ranges of interest, the primary criterion must be that the fugacity of the standard state is known precisely and without ambiguity.

B. Constant-Pressure Activity Coefficients and the Gibbs–Duhem Equation

At constant temperature, the activity coefficient depends on both pressure and composition. One of the important goals of thermodynamic analysis is to consider separately the effect of each independent variable on the liquid-phase fugacity; it is therefore desirable to define and use constant-pressure activity coefficients which at constant temperature are independent of pressure and depend only on composition. The definition of such activity coefficients follows directly from either of the exact thermodynamic relations

$$\left(\frac{\partial \ln \gamma_i}{\partial P}\right)_{T,x} = \frac{\bar{v}_i}{RT} \qquad (42)$$

$$\left(\frac{\partial \ln \gamma_i}{\partial P}\right)_{T,x} = \frac{\bar{v}_i - v_i^0}{RT} \qquad (43)$$

where v_i^0 is the molar volume of i in the standard state. Equation (42) holds for the case where the standard-state fugacity is defined at a fixed pressure P^r while Eq. (43) holds for the case where the standard-state fugacity is defined at system pressure P. Equations (42) and (43) are independent of the type of normalization used for γ_i; identical equations hold for γ_i^*. Using Eq. (42), we write for the fugacity of any component i in the liquid phase

$$f_i = \gamma_i^{(P^r)} x_i f_i^{0(P^r)} \exp \int_{P^r}^{P} \frac{\bar{v}_i \, dP}{RT} \qquad (44)$$

where $f_i^{0(Pr)}$ is the standard-state fugacity of i at the system temperature T and the arbitrary reference pressure P^r. Equation (44) also holds for the unsymmetrically normalized activity coefficient $\gamma_i^{*(Pr)}$.[2] At a fixed temperature, Eq. (44) has the advantage over Eq. (33) that, whereas γ_i in Eq. (33) depends on both pressure and composition, $\gamma_i^{(Pr)}$ in Eq. (44) depends only on composition. Since it is our aim to relate activity coefficients to composition through some physical or mathematical model, it is essential that we deal with $\gamma_i^{(Pr)}$ rather than with γ_i when we attempt to test, interpret, and correlate high-pressure vapor–liquid equilibrium data.

The advantages of constant-pressure activity coefficients also become clear when we try to relate to one another the activity coefficients of all the components in a mixture through the Gibbs–Duhem equation (P6, P7). For variable-pressure activity coefficients at constant temperature we obtain

$$\sum_i x_i \, d \ln \gamma_i = \frac{v \, dP}{RT} \qquad (45)$$

or

$$\sum_i x_i \, d \ln \gamma_i = \frac{\Delta v \, dP}{RT} \qquad (46)$$

where Eq. (45) corresponds to all standard states at fixed pressure P^r and Eq. (46) to all standard states at system total pressure P, with v the molar volume of the liquid mixture, and

$$\Delta v \equiv v - \sum_i x_i v_i^0 \qquad (47)$$

For constant-pressure activity coefficients, however, we obtain the much simpler relation

$$\sum_i x_i \, d \ln \gamma_i^{(Pr)} = 0 \qquad (48)$$

Equations (45), (46), and (48) are all independent of the normalization used for the activity coefficients.

It is desirable to use activity coefficients which satisfy Eq. (48) rather than Eqs. (45) or (46) because all well-known mixture models (e.g., Van Laar,

---

[2] For a supercritical component, we have $f_i^0 = H_{i,r}^{(Pr)}$, Henry's constant of i in reference solvent r (which may be a solvent mixture) at the system temperature and at reference pressure P^r. Experimentally, Henry's constant is obtained at pressure P_r^s, the saturation pressure of the reference solvent. To obtain the desired $H_{i,r}^{(Pr)}$, we use the exact relation

$$H_{i,r}^{(Pr)} = H_{i,r}^{(Prs)} \exp \int_{Pr^s}^{Pr} \frac{\bar{v}_{i,r}^\infty \, dP}{RT}$$

where $\bar{v}_{i,r}^\infty$ is the partial molar volume of i in solvent r at infinite dilution.

Margules, and Scatchard–Hildebrand) are particular mathematical solutions to Eq. (48); these models do not satisfy Eqs. (45) and (46), except in the limiting case where the right-hand sides of these equations vanish. This limiting case provides a good approximation for mixtures at low pressures but introduces serious error for mixtures at high pressures, especially near critical conditions.

IV. Effect of Pressure on Activity Coefficients: Partial Molar Volumes

In Section I, we indicated that significant progress in understanding high-pressure thermodynamics of mixtures requires a quantitative description of the variation of fugacity with pressure as given by Eq. (3). To obtain the effect of pressure on activity coefficient we substitute as follows:

$$\left(\frac{\partial \ln f_i}{\partial P}\right)_{T,x} = \left(\frac{\partial \ln \gamma_i x_i f_i^0}{\partial P}\right)_{T,x} = \frac{\bar{v}_i}{RT} \qquad (49)$$

which simplifies to

$$\left(\frac{\partial \ln \gamma_i}{\partial P}\right)_{T,x} + \left(\frac{\partial \ln f_i^0}{\partial P}\right)_{T,x} = \frac{\bar{v}_i}{RT} \qquad (50)$$

If we define the standard-state fugacity f_i^0 at a fixed pressure, then the second term on the left side of Eq. (50) vanishes and we obtain Eq. (42). However, if we define f_i^0 at the total pressure of the system, we obtain Eq. (43).

Equations (42), (43), and (50) are independent of the convention used for the normalization of activity coefficients; they apply equally to γ_i and to γ_i^* (see Section III).

For simplicity, it is tempting to assume $\bar{v}_i = v_i^0$, since we can then readily predict the effect of pressure on the activity coefficient. For mixtures of liquids at pressures near 1 atm, this is a valid assumption; for dilute solutions of gases in liquids at high pressures, it is also valid provided we use the unsymmetric convention for normalization as shown in Section III. However, for concentrated solutions of gases in liquids, the assumption is poor and at conditions approaching critical, it is highly erroneous. Partial molar volumes may be positive or negative, and near critical conditions they are usually strong functions of the composition. For example, in the saturated liquid phase of the system carbon dioxide–n-butane at 160°F, $\bar{v}$ for butane is $+1.8$ ft^3/lb-mole when the mole fraction of CO_2 is very small, and is -2.4 ft^3/lb-mole at the critical composition $x_{CO_2} = 0.71$.

It is difficult to measure partial molar volumes, and, unfortunately, many experimental studies of high-pressure vapor–liquid equilibria report no volumetric data at all; more often than not, experimental measurements are confined to total pressure, temperature, and phase compositions. Even in those cases where liquid densities are measured along the saturation curve, there is a fundamental difficulty in calculating partial molar volumes as indicated by the exact relations between partial molar volumes $\bar{v}_1$ and $\bar{v}_2$ and saturated molar volume v^s in a binary system:

$$\bar{v}_1 = v^s - x_2 \left[\left(\frac{\partial v^s}{\partial x_2} \right)_T + v^s \beta \left(\frac{\partial P}{\partial x_2} \right)_T \right] \tag{51}$$

$$\bar{v}_2 = v^s + x_1 \left[\left(\frac{\partial v^s}{\partial x_2} \right)_T + v^s \beta \left(\frac{\partial P}{\partial x_2} \right)_T \right] \tag{52}$$

where the compressibility β is defined by

$$\beta = -\frac{1}{v^s} \left(\frac{\partial v^s}{\partial P} \right)_{T,x} \tag{53}$$

Experimental data for v^s are sometimes available, but experimental compressibilities for mixtures are rare.

A. Dilute Solutions

For dilute solutions, the technical literature contains some direct (dilatometric) measurements of $\bar{v}_2$, the partial molar volume of the more volatile component, but the accuracy of these measurements is usually not high. A survey was made by Lyckman and Eckert (L6) and they established the rough correlation shown in Fig. 5. On the ordinate, the partial molar volume is

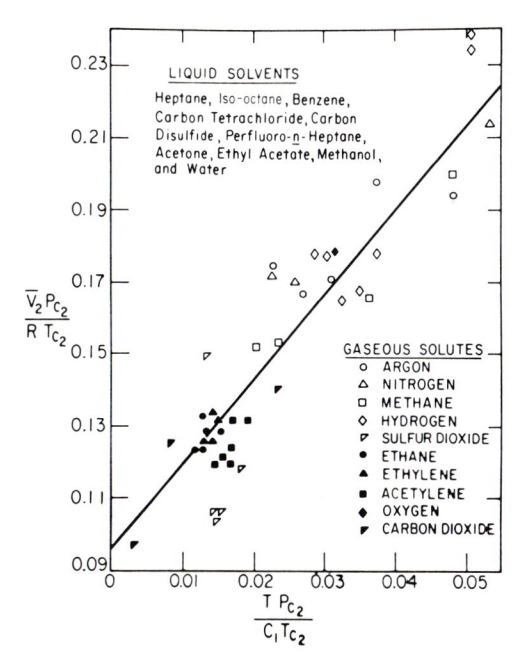

FIG. 5. Partial molar volumes of gases in dilute liquid solutions.

nondimensionalized with the critical temperature and pressure of component 2; the abscissa is also dimensionless and includes C_1, the cohesive energy density of the solvent, component 1.[3] Figure 5 is useful for rough approximations in systems remote from critical conditions. For expanded solvents, i.e., for solvents at temperatures approaching T_{c_1}, the partial molar volume of the solute tends to be much larger than that suggested by the correlation, as indicated in Fig. 6.

B. Partial Molar Volumes in Saturated Liquids from an Equation of State

If we can write an equation of state for liquid mixtures, we can then calculate partial molar volumes directly by differentiation. For a pressure-explicit equation, the most convenient procedure is to use the exact relation

$$\bar{v}_i = -\frac{(\partial P/\partial n_i)_{T,V,n_j}}{(\partial P/\partial V)_{T,\text{all } n}} \tag{54}$$

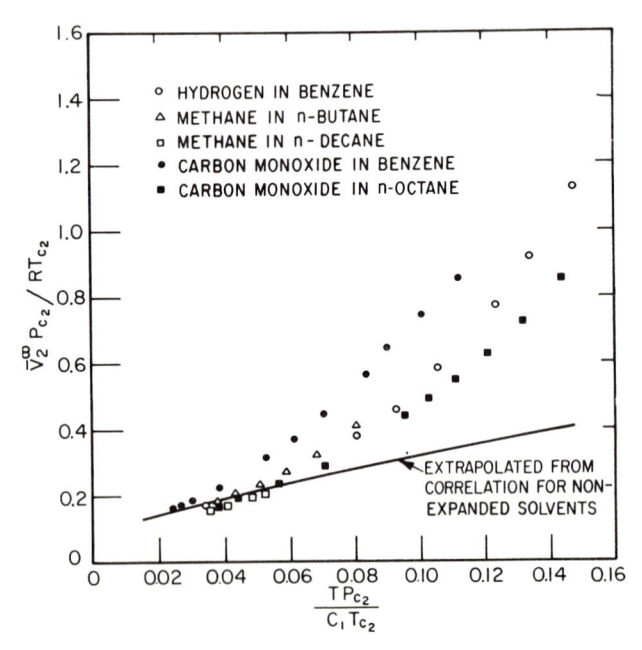

Fig. 6. Partial molar volumes of gaseous solutes at infinite dilution in expanded solvents.

[3] Cohesive energy density is the energy of isothermal vaporization per unit volume to the ideal-gas state. It is the square of the Hildebrand solubility parameter.

where V is the total volume of the mixture containing n_1 moles of component 1, etc.

When a pressure-explicit equation of state for a liquid mixture is substituted into Eq. (54), we obtain an expression of the form

$$\bar{v}_i = f(T, x_1, \ldots, v) \tag{55}$$

where v is the molar volume and $x_1, \ldots,$ the composition of the liquid mixture. Given the pressure, temperature, and composition, we can, in principle, use the equation of state to calculate the molar volume of the mixture; but this is not a good procedure, since we rarely have an equation of state sufficiently accurate for that calculation. Instead, it is better to use in Eq. (55) an independent correlation for mixture volumes, as shown by Chueh (C3). Confining his attention to saturated liquid mixtures, Chueh extends to mixtures the corresponding-states correlation for molar volumes of pure saturated liquids given by Lyckman and Eckert (L7). The correlation is of the form

$$\frac{v(T)}{v_c} = v_R^{(0)}\left(\frac{T}{T_c}\right) + \omega v_R^{(1)}\left(\frac{T}{T_c}\right) + \omega^2 v_R^{(2)}\left(\frac{T}{T_c}\right) \tag{56}$$

where v_c is the critical volume, T_c is the critical temperature, and ω is the acentric factor. The three generalized functions $v_R^{(0)}$, $v_R^{(1)}$, and $v_R^{(2)}$ depend only on the reduced temperature T/T_c and are represented by empirical algebraic equations. For mixtures, Chueh uses empirical mixing rules to calculate the composition-dependent parameters v_c, T_c, and ω.

To obtain an analytic function f in Eq. (55), Chueh uses the Redlich–Kwong equation; however, since the application is intended for liquids, the two constants in that equation were not evaluated (as is usually done) from critical data alone, but rather from a fit of the pure-component saturated-liquid volumes. The constants a and b in the equation of Redlich and Kwong are calculated from the relations

$$a = \Omega_a R^2 T_c^{5/2}/P_c \tag{57}$$

$$b = \Omega_b R T_c/P_c \tag{58}$$

where Ω_a and Ω_b are characteristic dimensionless constants determined from volumetric data for each saturated liquid. Table II gives Ω_a and Ω_b for 19 liquids. When a and b are evaluated from critical data alone, $\Omega_a = 0.4278$ and $\Omega_b = 0.0867$.

By adopting mixing rules similar to those given in Section II, Chueh showed that Eq. (55) can be used for calculating partial molar volumes in saturated liquid mixtures containing any number of components. Some results for binary systems are given in Figs. 7 and 8, which compare calculated partial molar volumes with those obtained from experimental data.

TABLE II

DIMENSIONLESS CONSTANTS FOR SATURATED LIQUIDS IN THE REDLICH–KWONG EQUATION OF STATE

Liquid	Ω_a	Ω_b
Methane	0.4546	0.0872
Ethane	0.4347	0.0827
Ethylene	0.4289	0.0185
Acetylene	0.4230	0.0802
Propane	0.4138	0.0802
Propylene	0.4130	0.0803
n-Butane	0.4184	0.0794
Isobutane	0.4100	0.0820
1-Butene	0.4000	0.0780
n-Pentane	0.3928	0.0767
Isopentane	0.3870	0.0748
n-Hexane	0.3910	0.0752
n-Heptane	0.3900	0.0740
n-Nonane	0.3910	0.0738
Cyclohexane	0.4060	0.0787
Benzene	0.4100	0.0787
Nitrogen	0.4593	0.0882
Hydrogen sulfide	0.4220	0.0823
Carbon dioxide	0.4184	0.0794

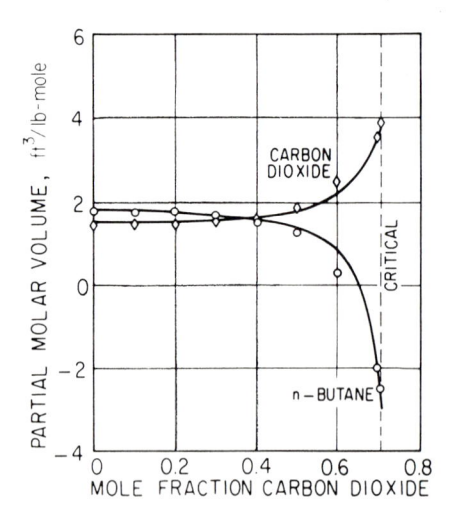

FIG. 7. Partial molar volumes in the saturated liquid phase of the n-butane–carbon dioxide system at 160°F. (———) Calculated. (○) (◊) From volumetric data of B. H. Sage and W. N. Lacey, "Some Properties of the Lighter Hydrocarbons, Hydrogen Sulfide, and Carbon Dioxide." American Petroleum Institute, New York 1955.

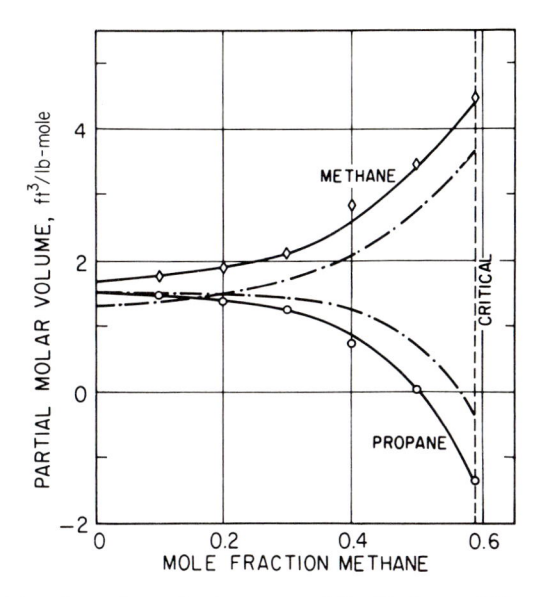

FIG. 8. Partial molar volumes in the saturated liquid phase of the propane–methane system at 100°F. (○) (◇) Data of B. H. Sage and W. N. Lacey, "Some Properties of the Lighter Hydrocarbons, Hydrogen Sulfide, and Carbon Dioxide." American Petroleum Institute, New York 1955. (———) Calculated with Ω_a and Ω_b from pure saturated liquid. (—·—) Calculated with universal constants in Redlich–Kwong equation.

Chueh's method gives consistently good results for mixtures except in the immediate vicinity of the critical region ($T/T_{c_{mix}} > 0.93$). For the critical region, his procedure was modified by using true critical constants, rather than pseudocritical constants in Eq. (56). For this purpose, he has established a separate correlation of true critical volumes and temperatures (C3).

Chueh's method for calculating partial molar volumes is readily generalized to liquid mixtures containing more than two components. Required parameters are Ω_a and Ω_b (see Table II), the acentric factor, the critical temperature and critical pressure for each component, and a characteristic binary constant k_{ij} (see Table I) for each possible unlike pair in the mixture. At present, this method is restricted to saturated liquid solutions; for very precise work in high-pressure thermodynamics, it is also necessary to know how partial molar volumes vary with pressure at constant temperature and composition. An extension of Chueh's treatment may eventually provide estimates of partial compressibilities, but in view of the many uncertainties in our present knowledge of high-pressure phase equilibria, such an extension is not likely to be of major importance for some time.

In the past, it has been customary to assume that partial molar volumes depend only on temperature and are independent of composition and pressure (C1, P13). This assumption is very poor in the critical region. Primarily because of this assumption, previous attempts to correlate high-pressure vapor–liquid equilibria have failed as the critical region is approached. Chueh's work provides much improvement, since it permits us to calculate the joint effects of composition and pressure on partial molar volumes along an isothermal saturation curve. With these partial molar volumes, it is then possible to make good estimates of the effect of pressure on activity coefficients.

V. Dilute Solutions of Gases in Liquids at High Pressures

In the previous sections, we emphasized that at constant temperature, the liquid-phase activity coefficient is a function of both pressure and composition. Therefore, any thermodynamic treatment of gas solubility in liquids must consider the question of how the activity coefficient of the gaseous solute in the liquid phase varies with pressure and with composition under isothermal conditions.

The simplest assumption one can make for a dilute solution at low pressure is to assert that the activity coefficient of the solute is a constant, independent of both pressure and composition. This assumption leads directly to Henry's law; for the solute (component 2), we obtain

$$f_2 = Hx_2 \tag{59}$$

where, for a given solute–solvent system, Henry's constant, H, is a function of temperature only. The molecular theory of fluids suggests that Henry's law is always valid for sufficiently dilute solutions of nondissociating solutes; in the highly dilute range, each solute molecule is completely surrounded by solvent molecules, and therefore the activity of the solute is proportional to its mole fraction, or equivalently, the activity coefficient of the solute is independent of the mole fraction. Also, at moderate pressures and at conditions far from critical, the properties of the liquid phase are not significantly influenced by the pressure; under these conditions, therefore, the activity coefficient of the solute is essentially independent of pressure as well as composition.

A. Krichevsky–Kasarnovsky Equation

For a dilute solution at high pressure, the variation of activity coefficient with pressure cannot be neglected. But when x_2 is small, it is often a good approximation to assume, as above, that the activity coefficient is not significantly affected by composition. If we also assume that $\bar{v}_2$, the partial molar volume of the solute, is independent of both pressure and composition we obtain the equation of Krichevsky and Kasarnovsky (see D7, K6)[4]

$$\ln \frac{f_2}{x_2} = \ln H + \frac{\bar{v}_2^\infty (P - P_1^s\cdot)}{RT} \tag{60}$$

where Henry's constant, H, is defined by

$$H \equiv \lim_{x_2 \to 0} (f_2/x_2)_T \tag{61}$$

and where P_1^s is the saturation pressure of the solvent (component 1). Infinite dilution is designated by the superscript ∞.

The Krichevsky–Kasarnovsky equation is a two-parameter isothermal equation which successfully represents high-pressure solubility data for a sparingly soluble gas. For example, Figs. 9 and 10 give solubility data for hydrogen and nitrogen in water at several temperatures at pressures up to 1000 atm; it is evident that Eq. (60) gives an excellent representation. For contrast, Fig. 11 gives solubility data for nitrogen in liquefied ammonia, and while the equation of Krichevsky and Kasarnovsky gives a very good fit at 0°C for pressures up to 1000 atm, it fails at 70°C for pressures larger than about 500 atm. The reason for this failure becomes clear when we note that the liquid-phase equilibrium mole fraction of nitrogen at 1000 atm and 0°C is 0.0221, while that at 1000 atm and 70°C is 0.129. The solubility data at 0°C represent a truly dilute solution throughout, but the solubility data at 70°C do not. As a result, the assumptions leading to Eq. (60) are valid for a larger range of pressure at 0°C than at 70°C.

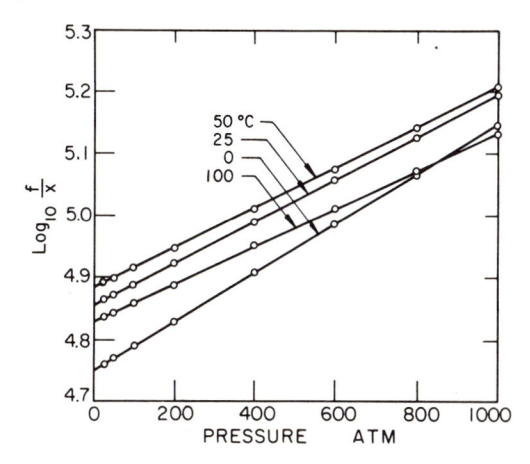

FIG. 9. Solubility of hydrogen in water at high pressures. Fugacity is in atmospheres.

[4] Equation (60) follows directly from the stated assumptions, from the definition of γ^* given by Eqs. (33) and (37), and from Eq. (42), which rigorously gives the effect of pressure on γ^*.

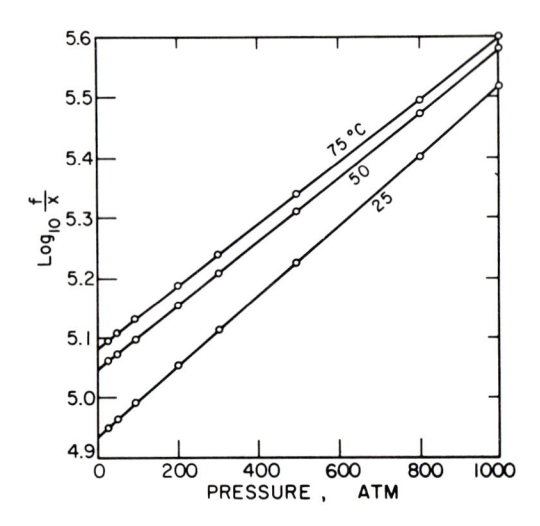

FIG. 10. Solubility of nitrogen in water at high pressures. Fugacity is in atmospheres.

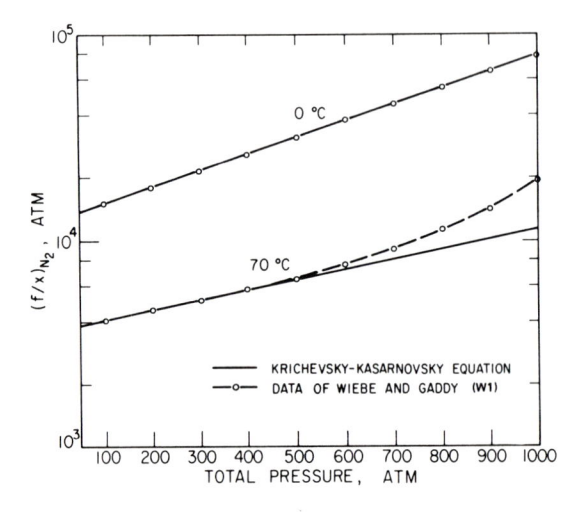

FIG. 11. Success and failure of the Krichevsky–Kasarnovsky equation. Solubility of nitrogen in liquid ammonia.

One interesting feature of the Krichevsky–Kasarnovsky equation is that it predicts a maximum in the solubility as the pressure becomes very large. If we consider the liquid solvent (component 1) to be involatile, differentiation of Eq. (60) with respect to pressure gives

$$\left(\frac{\partial \ln x_2}{\partial P}\right)_T = \frac{v_2^V - \bar{v}_2^\infty}{RT} \tag{62}$$

where v_2^V is the molar volume of component 2 in the vapor phase. According to Eq. (62), the solubility increases with pressure as long as $v_2^V > \bar{v}_2^\infty$, but it decreases when $v_2^V < \bar{v}_2^\infty$. Since a gas is more compressible than a liquid, the molar volume of gaseous component 2 at very high pressure may become smaller than the partial molar volume of infinitely dilute component 2 in the liquid solution. This prediction of the Krichevsky–Kasarnovsky equation has been verified experimentally by Basset and Dodé for the solubility of nitrogen in water at 18°C (B2); the solubility reaches a maximum near 3000 atm. in good agreement with the pressure predicted by Eq. (62) using independently determined data for v_2^V and $\bar{v}_2^\infty$ (K4).

B. Krichevsky–Ilinskaya Equation

For those dilute mixtures where the solute and the solvent are chemically very different, the activity coefficient of the solute soon becomes a function of solute mole fraction even when that mole fraction is small. That is, if solute and solvent are strongly dissimilar, the relations valid for an infinitely dilute solution rapidly become poor approximations as the concentration of solute rises. In such cases, it is necessary to relax the assumption (made by Krichevsky and Kasarnovsky) that at constant temperature the activity coefficient of the solute is a function of pressure but not of solute mole fraction. For those moderately dilute mixtures where the solute–solute interactions are very much different from the solute–solvent interactions, we can write the constant-pressure activity coefficients as Margules expansions in the mole fractions; for the solvent (component 1), we write at constant temperature and at reference pressure P^r:

$$RT \ln \gamma_1^{(P^r)} = A x_2^2 + B x_2^3 + \cdots \qquad (63)$$

where $A, B, \ldots$, are empirical coefficients. It follows from the Gibbs–Duhem equation [Eq. (48)] that for the solute (component 2) we have

$$RT \ln \gamma_2^{*(P^r)} = A(x_1^2 - 1) - \tfrac{1}{2}B(2x_1^3 - 3x_1^2 + 1) + \cdots \qquad (64)$$

The parameters $A, B, \ldots$, depend on temperature but not on pressure, and must be determined from experimental data for the binary mixture.

If we truncate Eqs. (63) and (64) after the first terms, and assume as before that the partial volume of the solute is a function of temperature only, we obtain the equation of Krichevsky and Ilinskaya (K5):[5]

$$\ln \frac{f_2}{x_2} = \ln H + \frac{\bar{v}_2^\infty (P - P_1^s)}{RT} + \frac{A}{RT}(x_1^2 - 1) \qquad (65)$$

---

[5] The reference pressure P^r for the activity coefficients is here taken as P_1^s, the saturation pressure of pure solvent 1.

As expected, this equation reduces to Eq. (60) for the case where $A = 0$.

For gas–liquid solutions which are only moderately dilute, the equation of Krichevsky and Ilinskaya provides a significant improvement over the equation of Krichevsky and Kasarnovsky. It has been used for the reduction of high-pressure equilibrium data by various investigators, notably by Orentlicher (O3), and in slightly modified form by Conolly (C6). For any binary system, its three parameters depend only on temperature. The parameter H (Henry's constant) is by far the most important, and in data reduction, care must be taken to obtain H as accurately as possible, even at the expense of lower accuracy for the remaining parameters. While H must be positive, A and $\bar{v}_2^\infty$ may be positive or negative; A is called the self-interaction parameter because it takes into account the deviations from infinite-dilution behavior that are caused by the interaction between solute molecules in the solvent matrix.

When solute and solvent are very dissimilar chemically, A is large. Therefore, deviations from infinitely-dilute-solution behavior are frequently observed for such mixtures at very small values of x_2. For example, solutions of helium in nonpolar solvents show deviations from dilute solution behavior at values of x_2 as low as 0.01. On the other hand, since both A and $\bar{v}_2^\infty$ are usually positive, it sometimes happens that the last two terms in Eq. (65) tend to cancel each other, with the fortuitous result that Henry's law provides a good approximation to unexpectedly high pressures and concentrations (M3).

The equation of Krichevsky and Ilinskaya can readily be extended to high-pressure solutions of a gas in a mixed solvent, as shown by O'Connell (O1) and discussed briefly by Orentlicher (O3). This extension makes it possible to predict the behavior of simple multicomponent systems using binary data only.

VI. Concentrated Solutions: High-Pressure Vapor–Liquid Equilibria

When we consider equilibrium between two phases at high pressure, neither phase being dilute with respect to one of the components, we can no longer make the simplifying assumptions made in some of the earlier sections. We must now realistically describe deviations from ideal behavior in both phases; for each phase, the effect of both pressure and composition must be seriously taken into account if we wish to describe vapor–liquid equilibria at high pressures for a wide range of conditions, including the critical.

In Section II, we discussed the fugacity coefficient, which relates the vapor-phase fugacity to the total pressure and to the composition. The fugacity coefficient can be calculated exactly from an equation of state and, therefore, the problem of calculating vapor-phase fugacities reduces to the problem of providing a good description of the volumetric (P–v–T–y) properties of the vapor mixture.

The technique for calculating vapor-phase fugacity coefficients is completely standardized by Eqs. (5) and (6); the only choice we have is in the particular equation of state we wish to use. Many equations of state have been proposed, and doubtless more will appear in the literature from time to time. As we have indicated in Section II, good estimates can usually be made of the fugacity coefficients of components in nonpolar mixtures at conditions away from the critical. The accuracy usually declines somewhat at near-critical conditions; however, since the fugacity coefficient as given in Eq. (6) is evaluated by an integration over the whole range from zero density to the density of interest, it frequently happens that the fugacity coefficients are satisfactory even when the equation of state provides poor estimates of P–v–T–y behavior in the critical region.

A. Equation of State for Both Phases

Since Eqs. (5) and (6) are not restricted to the vapor phase, they can, in principle, be used to calculate fugacities of components in the liquid phase as well. Such calculations can be performed provided we assume the validity of an equation of state for a density range starting at zero density and terminating at the liquid density of interest. That is, if we have a pressure-explicit equation of state which holds for mixtures in both vapor and liquid phases, then we can use Eq. (6) to solve completely the equations of equilibrium without explicitly resorting to the auxiliary-functions activity, standard-state fugacity, and partial molar volume. Such a procedure was discussed many years ago by van der Waals and, more recently, it has been reduced to practice by Benedict and co-workers (B4).

The fundamental idea of this procedure is as follows: For a system of two fluid phases containing N components, we are concerned with $N-1$ independent mole fractions in each phase, as well as with two other intensive variables, temperature T and total pressure P. Let us suppose that the two phases (vapor and liquid) are at equilibrium, and that we are given the total pressure P and the mole fractions of the liquid phase $x_1, x_2, \ldots, x_{N-1}$. We wish to find the equilibrium temperature T and the mole fractions of the vapor phase $y_1, y_2, \ldots, y_{N-1}$. The total number of unknowns is $N+2$: there are $N-1$ unknown mole fractions, one unknown temperature, and two unknown densities corresponding to the two limits of integration in Eq. (6), one for the liquid phase and the other for the vapor phase. To solve for these $N+2$ unknowns, we require $N+2$ equations of equilibrium. For each component i we have an equation of the form

$$f_i^V = f_i^L \quad (N \text{ such equations}) \tag{66}$$

In addition, we have the equation of state applied once to the vapor phase:

$$P = P(T, v^V, y_1, y_2, \ldots, y_{N-1}) \tag{67}$$

and once to the liquid phase:

$$P = P(T, v^L, x_1, x_2, \ldots x_{N-1}) \tag{68}$$

With a suitable equation of state, all the fugacities in each phase can be found from Eq. (6), and the equation of state itself is substituted into the equilibrium relations Eq. (67) and (68). For an N-component system, it is then necessary to solve simultaneously $N + 2$ equations of equilibrium. While this is a formidable calculation even for small values of N, modern computers have made such calculations a realistic possibility. The major difficulty of this procedure lies not in computational problems, but in our inability to write for mixtures a single equation of state which remains accurate over a density range that includes the liquid phase. As a result, phase-equilibrium calculations based exclusively on equations of state do not appear promising for high-pressure phase equilibria, except perhaps for certain restricted mixtures consisting of chemically similar components.

B. Corresponding States for Mixtures

In Section II we indicated that, subject to well-defined assumptions, corresponding-states correlations of thermodynamic properties of pure components may be extended to yield fugacities in vapor mixtures. Exactly the same procedure may be used to compute fugacities in liquid mixtures. We must first consider a liquid mixture to be equivalent to a pseudo-pure liquid whose characteristic parameters (e.g., T_{c_M}, P_{c_M}, and ω_M: see Section II) are averages of the characteristic parameters of the components in that mixture. We can then find f_M, the fugacity of the mixture, from the generalized fugacity correlation for pure liquids. Next, as discussed in Section II, we differentiate f_M with respect to composition to find f_i, the fugacity of any component i in the mixture. Variations on this procedure (two-liquid and three-liquid theory) have been discussed by Scott (P15, S2).

Several authors, notably Leland and co-workers (L2), have discussed vapor–liquid equilibrium calculations based on corresponding-states correlations. As mentioned in Section II, such calculations rest not only on the general assumptions of corresponding-states theory, but also on the additional assumption that the characterizing parameters for a mixture do not depend on temperature or density but are functions of composition only. Further, it is necessary clearly to specify these functions (commonly known as mixing rules), and experience has shown that if good results are to be obtained, these mixing rules must be flexible and must contain at least one or two adjustable constants which are determined from binary data. Application of corresponding-states correlations to liquid mixtures is rarely successful if based on characterizing parameters obtained exclusively from pure-component data.

C. Semiempirical Models for Activity Coefficients

In Section III, we discussed the relation between fugacities and activity coefficients in liquid mixtures, and we emphasized that we have a fundamental choice regarding the way we wish to relate the fugacity of a component to the pressure and composition. This choice follows from the freedom we have in choosing a convention for the normalization of activity coefficients.

The activity coefficient is related to the mole fraction and the fugacity by

$$f_i = \gamma_i x_i f_i^0 \tag{69}$$

If we use the symmetric convention for normalization, f_i^0 is the fugacity of pure liquid i at the temperature of the mixture and at some specified pressure, usually taken to be the total pressure of the system. Equation (69) presents no problem for subcritical components, where the pure liquid can exist at the system temperature. However, for supercritical components in the symmetric convention, f_i^0 is a fictitious quantity which must be evaluated by some arbitrary extrapolation.

In addition to deciding on the method of normalization of activity coefficients, it is necessary to undertake two additional tasks: first, a method is required for estimating partial molar volumes in the liquid phase, and second, a model must be chosen for the liquid mixture in order to relate γ to x. Partial molar volumes were discussed in Section IV. This section gives brief attention to two models which give the effect of composition on liquid-phase thermodynamic properties.

1. *Scatchard–Hildebrand Equation*

The Scatchard–Hildebrand theory of regular solutions is most attractive because of its simplicity, and it is of special interest here because it has been applied to hydrocarbon mixtures at high pressures (P13), leading to the correlation of Chao and Seader (C1).

In their correlation, Chao and Seader use the original Redlich–Kwong equation of state for vapor-phase fugacities. For the liquid phase, they use the symmetric convention of normalization for γ and partial molar volumes which are independent of composition, depending only on temperature. For the variation of γ with temperature and composition, Chao and Seader use the equation of Scatchard and Hildebrand for a multicomponent solution:

$$RT \ln \gamma_i = v_i [\delta_i - \bar{\delta}]^2 \tag{70}$$

where v_i is the molar volume of pure liquid i, δ_i is the solubility parameter of i, and $\bar{\delta}$ is the average solubility parameter for the *entire* mixture defined by

$$\bar{\delta} \equiv \sum_{\text{all } j} \Phi_j \delta_j \tag{71}$$

where Φ_j, the volume fraction of component j, is defined by

$$\Phi_j \equiv \frac{x_j v_j}{\sum_{\text{all } j} x_j v_j} \tag{72}$$

Although δ varies with temperature, the quantity $[\delta_i - \bar{\delta}]$ is insensitive to temperature; the solubility parameters used in Eq. (70) were therefore treated as constants. Table III gives some of the solubility parameters used by Chao and Seader. For supercritical components, the solubility parameters were back-calculated from binary-mixture data, as was also done by Shair (P2).

TABLE III

Solubility Parameters and Liquid Molar Volumes in the Chao–Seader Correlation

Liquid	δ (cal/cm³)$^{1/2}$	v (cm³/gm-mole)
Hydrogen	3.25	31
Methane	5.68	52
Ethane	6.05	68
Propane	6.40	84
n-Butane	6.73	101
i-Butane	6.73	106
n-Pentane	7.02	116
n-Hexane	7.27	132
n-Decane	7.72	196
Ethylene	6.08	61
Propylene	6.43	79
1-Butene	6.76	95
1, 3-Butadiene	6.94	88
1-Pentene	7.05	110
Cyclohexane	8.20	109
Benzene	9.16	89
Toluene	8.92	107
m-Xylene	8.82	124
Ethylbenzene	8.79	123

For components near or above their critical temperatures, the liquid volume v_i was evaluated by extrapolation with respect to temperature. For supercritical components, the fugacity f_i^0 was also evaluated by extrapolation; the effect of pressure was found from the Poynting relation using the previously extrapolated liquid molar volumes.

The correlation of Chao and Seader has been computerized and has been used extensively in the petroleum industry. It provides a useful method for estimating high-pressure vapor–liquid equilibria in hydrocarbon systems over a wide range of temperature, pressure, and composition, and presents a significant improvement over the previously used K-charts first introduced by W. K. Lewis, B. F. Dodge, G. G. Brown, M. Souders, and others (see D6) almost forty years ago. However, the Chao–Seader correlation is unreliable at conditions approaching the critical. Various extensions have been proposed (G2), especially for application at extreme temperatures.

The method of Chao and Seader is subject to certain limitations, and it is instructive briefly to consider the most serious ones:

(1) The Scatchard–Hildebrand equation is based on several important simplifying assumptions which are known to be approximations, as pointed out by Hildebrand and Scott (H5). The most serious of these is the geometric-mean assumption which gives the cohesive energy density of a mixture in terms of the cohesive energy densities of the pure components, without introducing any new parameters characteristic of the molecular interaction between dissimilar components. This assumption can sometimes lead to large errors because, unfortunately, the activity coefficients are very sensitive to even small deviations from the geometric-mean rule (E1, H5).

(2) Chao and Seader assume that the partial molar volumes are independent of composition; this assumption is equivalent to saying that at constant temperature and pressure there is no volume change upon mixing the pure liquid components, be they real or hypothetical. The term on the right-hand side of Eq. (46) is assumed to be zero for all temperatures, pressures, and compositions. This assumption is very poor near critical conditions, and is undoubtedly the main reason for the poor performance of the Chao–Seader correlation in the critical region.

(3) The temperature dependence of the activity coefficients is assumed to have a particularly simple form, and this can sometimes lead to serious error at temperatures far away from those used to evaluate the solubility parameters.

(4) While the arbitrary extrapolation methods used to evaluate v_i and f_i^0 for a supercritical component are partly compensated by evaluating δ_i from data for binary mixtures, such compensation cannot apply generally to mixtures containing supercritical components; i.e., for a supercritical component, δ_i found from data for solutions of i in one solvent may be quite different from that found from data for the same component i in another solvent.

2. The Dilated van Laar Equation

The difficulties encountered in the Chao–Seader correlation can, at least in part, be overcome by the somewhat different formulation recently developed by Chueh (C2, C3). In Chueh's equations, the partial molar volumes in the liquid phase are functions of composition and temperature, as indicated in Section IV; further, the unsymmetric convention is used for the normalization of activity coefficients, thereby avoiding all arbitrary extrapolations to find the properties of hypothetical states; finally, a flexible two-parameter model is used for describing the effect of composition and temperature on liquid-phase activity coefficients. The flexibility of the model necessarily requires some binary data over a range of composition and temperature; to obtain the desired accuracy, especially in the critical region, more binary data are required for Chueh's method than for that of Chao and Seader (C1). Fortunately, reliable data for high-pressure equilibria are now available for a variety of binary mixtures of nonpolar fluids, mostly hydrocarbons. Chueh's method, therefore, is primarily applicable to equilibrium problems encountered in the petroleum, natural-gas, and related industries.

Following the idea of van Laar, Chueh expresses the excess Gibbs energy per unit effective volume as a quadratic function of the effective volume fractions. For a binary mixture, using the unsymmetric convention of normalization, the excess Gibbs energy g^{E*} is found from[6]

$$\frac{g^{E*}}{RT(x_1 q_1 + x_2 q_2)} = -\alpha_{22} \Phi_2^2 \tag{73}$$

where $\Phi_2 = x_2 q_2/(x_1 q_1 + x_2 q_2)$, α_{22} is the self-interaction coefficient of the (light) component 2, and q_i is the effective molar volume of component i.

In the original equation of van Laar, the effective molar volume was assumed to be independent of composition; this assumption implies zero volume-change of mixing at constant temperature and pressure. While this assumption is a good one for solutions of ordinary liquids at low pressures, it is poor for high-pressure solutions of gases in liquids which expand (dilate) sharply as the critical composition is approached. The dilated van Laar model therefore assumes that

$$q_1 = v_{c_1}[1 + \eta_{12} \Phi_2^2] \tag{74}$$

$$q_2 = v_{c_2}[1 + \eta_{12} \Phi_2^2] \tag{75}$$

where v_{c_1} and v_{c_2} are the critical volumes of the pure components, and η_{12} is a positive dilation coefficient that reflects the tendency of the mixture to increase its molar volume as the concentration of solute rises to the critical composition. For a given binary system, the dilation coefficient is a function

---

[6] In the unsymmetric convention, $g^{E*}/RT = x_1 \ln \gamma_1 + x_2 \ln \gamma_2^*$.

of temperature only. An important and useful property of Eqs. (74) and (75) is that, while q_1 and q_2 are both dependent on composition, the ratio of q_1 to q_2 is independent of composition.

For dilute solutions, the ratio of q_1 to q_2 is given by the ratio of the pure-component critical volumes. This limiting relationship is somewhat arbitrary and is chosen primarily for convenience; any other convenient measure of molecular size could be used—for example, van der Waals' b or Lennard-Jones' σ^3.

From the dilated van Laar model, Chueh obtains two-parameter expressions for the activity coefficients. They are:

$$\ln \gamma_1^{(Pr)} = A\Phi_2^2 + B\Phi_2^4 \tag{76}$$

$$\ln \gamma_2^{*(Pr)} = A(v_{c_2}/v_{c_1})[\Phi_2^2 - 2\Phi_2] + B(v_{c_2}/v_{c_1})[\Phi_2^4 - \tfrac{4}{3}\Phi_2^3] \tag{77}$$

where $A \equiv \alpha_{22} v_{c_1}$ and $B \equiv 3\eta_{12}\alpha_{22}v_{c_1}$.

Equations (76) and (77) contain two constants, A and B, which, for any binary pair, are functions of temperature only. These equations appear to be satisfactory for accurately representing activity coefficients of nonpolar binary mixtures from the dilute region up to the critical composition. As examples, Figs. 12 and 13 present typical results of data reduction for two systems; in these calculations, the reference pressure P^r was set equal to zero.

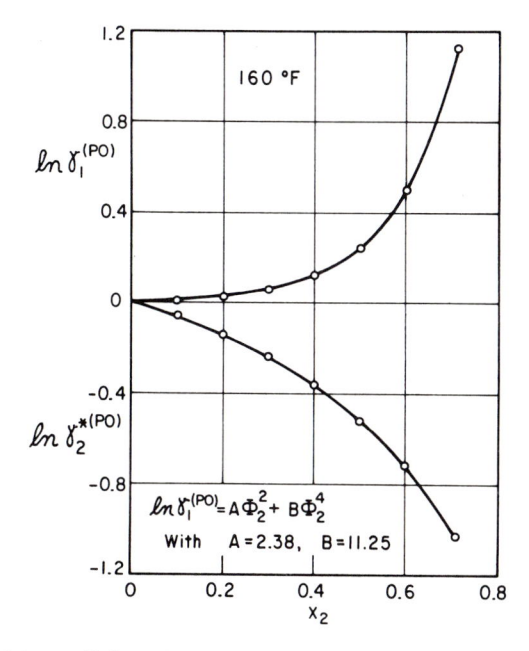

FIG. 12. Activity coefficients for the n-butane (1)–carbon dioxide (2) system.

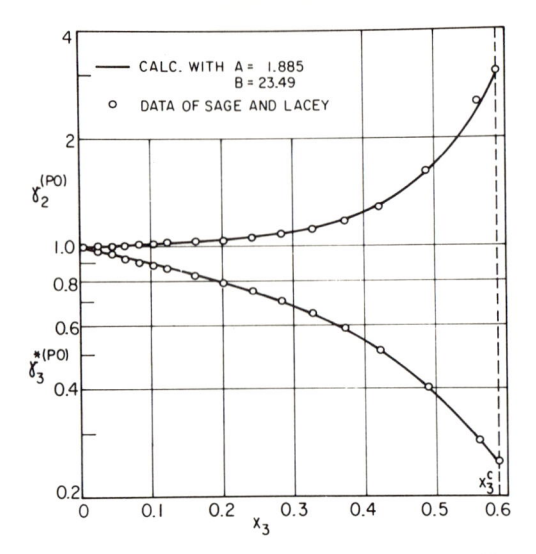

FIG. 13. Activity coefficients for the n-propane (2)–methane (3) system at 100°F. (○) Data of B. H. Sage and W. N. Lacey, "Some Properties of the Lighter Hydrocarbons, Hydrogen Sulfide, and Carbon Dioxide." American Petroleum Institute, New York, 1955.

It is easily possible to introduce refinements into the dilated van Laar model which would further increase its accuracy for correlating activity coefficient data. However, such refinements unavoidably introduce additional adjustable parameters. Since typical experimental results of high-pressure vapor–liquid equilibria at any one temperature seldom justify more than two adjustable parameters (in addition to Henry's constant), it is probably not useful for engineering purposes to refine Chueh's model further, at least not for nonpolar or slightly polar systems.

The dilated van Laar model is readily generalized to the multicomponent case, as discussed in detail elsewhere (C3, C4). The important technical advantage of the generalization is that it permits good estimates to be made of multicomponent phase behavior using only experimental data obtained for binary systems. For example, Fig. 14 presents a comparison of calculated and observed K-factors for the methane–propane–n-pentane system at conditions close to the critical.[7]

While the dilated van Laar model gives a reliable representation of constant-pressure activity coefficients for nonpolar systems, the good agreement between calculated and experimental high-pressure phase behavior shown in Fig. 14 is primarily a result of good representation of the partial molar volumes, as discussed in Section IV. The essential part of any thermodynamic description of high-pressure vapor–liquid equilibria must depend,

---

[7] By definition, $K_i = y_i/x_i$.

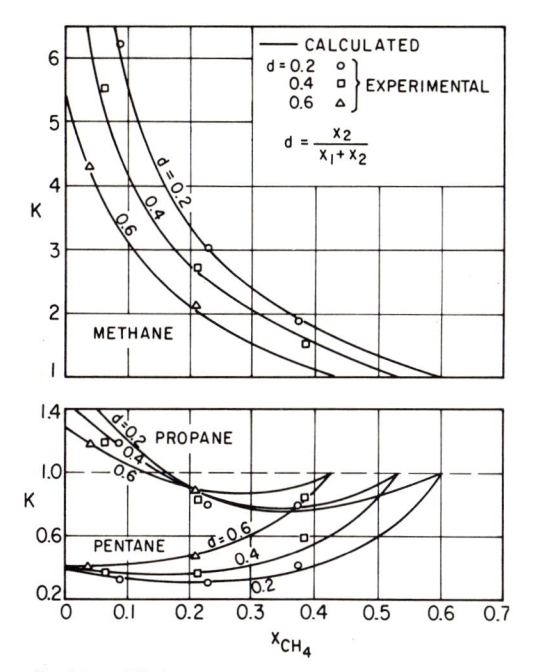

FIG. 14. Vapor–liquid equilibrium constants for the *n*-pentane (1)–propane (2)–methane (3) system at 220°F.

in one way or another, upon a realistic description of the variation of liquid fugacities with pressure as well as with composition. Therefore, future efforts to improve our understanding of high-pressure phase behavior must focus not only on better models for liquid mixtures, but even more on improved methods for interpreting, correlating, and predicting liquid-phase volumetric behavior in the critical region.

D. Thermodynamic Consistency

Vapor–liquid equilibrium data are said to be thermodynamically consistent when they satisfy the Gibbs–Duhem equation. When the data satisfy this equation, it is likely, but by no means guaranteed, that they are correct; however, if they do not satisfy this equation, it is certain that they are incorrect.

Thermodynamic consistency tests for binary vapor–liquid equilibria at low pressures have been described by many authors; a good discussion is given in the monograph by Van Ness (V1). Extension of these methods to isothermal high-pressure equilibria presents two difficulties: first, it is necessary to have experimental data for the density of the liquid mixture along the saturation line, and second, since the ideal gas law is not valid, it is necessary to calculate vapor-phase fugacity coefficients either from volumetric data for the vapor mixture or else from an equation of state. In an attempt to simplify the second requirement, Adler and co-workers (A1) used the Lewis fugacity rule to calculate vapor-phase fugacities; however, since the errors in this approximation are often very large, it is not possible to come to meaningful conclusions concerning the thermodynamic consistency of experimental equilibrium data when this approximation is used as an essential part of the consistency test.

A consistency test described by Chueh and Muirbrook (C4) extends to isothermal high-pressure data the integral (area) test given by Redlich and Kister (R1) and Herington (H2) for isothermal low-pressure data. [A similar extension has been given by Thompson and Edmister (T2)]. For a binary system at constant temperature, the Gibbs–Duhem equation is written

$$x_1 \, d \ln f_1 + x_2 \, d \ln f_2 = v^L \, dP/RT \tag{78}$$

When one uses the identity $x_1 \, d \ln x_1 + x_2 \, d \ln x_2 = 0$, Eq. (78) can be rearranged to read

$$\ln \frac{f_2/x_2}{f_1/x_1} dx_2 + \frac{v^L \, dP}{RT} = d\left(\ln \frac{f_1}{x_1} + x_2 \ln \frac{f_2/x_2}{f_1/x_1} \right) \tag{79}$$

Introducing fugacity coefficients φ and K-factors ($K_i = y_i/x_i$) into Eq. (79), one obtains

$$\left(\ln \frac{\varphi_2}{\varphi_1} + \ln \frac{K_2}{K_1} \right) dx_2 + \frac{v^L \, dP}{RT} = d\left[\ln K_1 + \ln \varphi_1 P + x_2 \left(\ln \frac{\varphi_2}{\varphi_1} + \ln \frac{K_2}{K_1} \right) \right] \tag{80}$$

In Eqs. (78)–(80), the subscript 2 refers to the light component. Equation (80) is integrated from $x_2 = 0$ to some arbitrary upper limit x_2. The following boundary condition applies:

when $x_2 = 0$: $\quad \varphi_1 = \varphi_1{}^s; \quad P = P_1{}^s; \quad K_1 = 1$

The integrated form of Eq. (80) can be most conveniently written as

Area I + Area II + Area III

$$= \left[\ln K_1 + \ln \frac{\varphi_1 P}{\varphi_1{}^s P_1{}^s} + x_2 \left(\ln \frac{\varphi_2}{\varphi_1} + \ln \frac{K_2}{K_1} \right) \right]_{\text{at } x_2} \tag{81}$$

where

$$\text{Area I} = \int_{x_2=0}^{x_2} \ln \frac{K_2}{K_1} \, dx_2 \tag{82}$$

$$\text{Area II} = \int_{x_2=0}^{x_2} \ln \frac{\varphi_2}{\varphi_1} \, dx_2 \tag{83}$$

$$\text{Area III} = \frac{1}{RT} \int_{x_2=0}^{x_2} v^L \, dP \tag{84}$$

The three areas are found by graphical integration. The thermodynamic consistency test consists of comparing the sum of the three areas [left-hand side of Eq. (81)] with the right-hand side of Eq. (81). The three areas depend upon equilibrium data for the composition range $x_2 = 0$ to $x_2 = x_2$. However, the right-hand side of Eq. (81) depends only on equilibrium data at the upper limit $x_2 = x_2$. The comparison indicated by Eq. (81) should be made for several values of x_2 up to and including the critical composition.

To illustrate this thermodynamic consistency test, Figs. 15, 16, and 17 show plots of the appropriate functions needed to calculate Areas I, II, and III, respectively, for the nitrogen–carbon dioxide system at 0°C; the data are taken from Muirbrook (M5). Fugacity coffiecients were calculated with the modified Redlich–Kwong equation (R4).

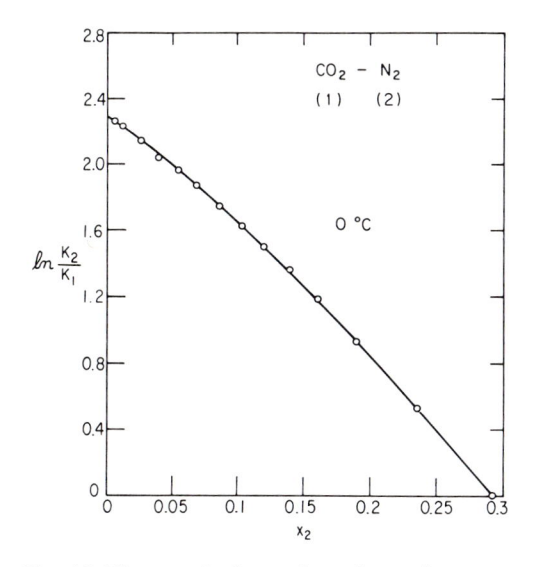

FIG. 15. First area in thermodynamic consistency test.

A comparison of the left-hand side (LHS) and the right-hand side (RHS) of Eq. (81) is given in Table IV. The comparison is made at three different values of x_2, including the critical point. In order to assess their relative importance, values of all the individual terms in Eq. (81) are reported in the table. It is apparent that all the terms contribute significantly and that none may be neglected (except that $\ln K_1$ must necessarily vanish at the critical mole fraction).

The final column in Table IV reports the absolute value of the difference between LHS and RHS divided by their arithmetic mean. For the system considered here, the percent inconsistency is always less than about 5%. In

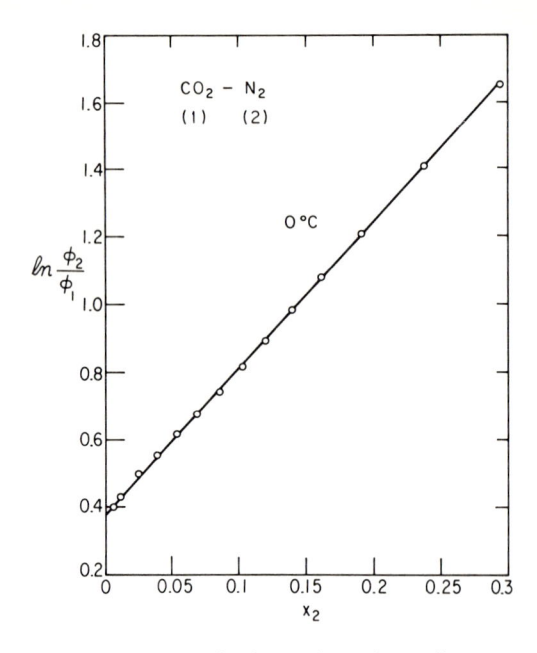

FIG. 16. Second area in thermodynamic consistency test.

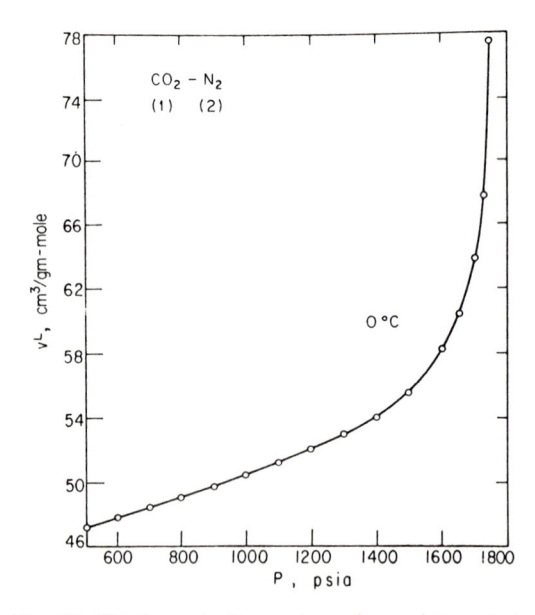

FIG. 17. Third area in thermodynamic consistency test.

TABLE IV
Thermodynamic Consistency Test for Carbon Dioxide (1)–Nitrogen (2) at 0°C

| x_2 | Left-hand side (LHS) of Eq. (81) ||||| Right-hand side (RHS) of Eq. (81) |||| Inconsistency between RHS and LHS (%) |
	Area I	Area II	Area III	Total, LHS	$\ln K_1$	$\ln \dfrac{\varphi_1 P}{\varphi_1^s P_1^s}$	$x_2(\ln K_2/K_1 + \ln \varphi_2/\varphi_1)$	Total, RHS	
0.1030	0.2024	0.0618	0.1042	0.369	−0.3496	0.4855	0.2505	0.386	4.5
0.1902	0.3142	0.1500	0.1703	0.635	−0.2562	0.5044	0.404	0.652	2.6
0.2926[a]	0.3618	0.2965	0.1968	0.855	0	0.3974	0.482	0.879	2.8

[a] Critical value.

view of the uncertainties in the fugacity coefficients (resulting from a good but still approximate equation of state) it is probably fair to judge the thermodynamic consistency of these data as very good.

The thermodynamic consistency test for binary systems described above can be extended to ternary (and higher) systems with techniques similar to those described by Herington (H3). The necessary calculations become quite tedious, and unless extensive multicomponent data are available, they are usually not worthwhile.

VII. Liquid–Liquid Equilibria in Binary Systems

In the previous sections we have been concerned with high-pressure equilibria in systems containing one liquid phase and one vapor phase. We now briefly consider the effect of pressure on equilibria between two liquid phases. In particular, we are concerned with the question of how pressure may be used to induce miscibility or immiscibility in a binary liquid system.

Two liquids are miscible in all proportions if Δg, the molar Gibbs energy of mixing at constant temperature and pressure, satisfies the relations

$$\Delta g < 0 \tag{85}$$

and

$$\left(\frac{\partial^2 \Delta g}{\partial x^2}\right)_{P,T} > 0 \tag{86}$$

Since Δg is a function of pressure, it follows that, under certain conditions, a change in pressure may produce immiscibility in a completely miscible system, or, conversely, such a change may produce complete miscibility in a partially immiscible system. The effect of pressure on miscibility in binary liquid mixtures is closely connected with the volume change on mixing, as indicated by the exact relation

$$\left(\frac{\partial \Delta g}{\partial P}\right)_{T,x} = \Delta v \tag{87}$$

To explain these concepts, let us consider a binary liquid mixture which at normal pressure is completely miscible and whose isothermal Gibbs energy of mixing is given by curve (a) in Fig. 18. Let us suppose that for this system Δv is positive; an increase in pressure raises Δg, and at some higher pressure, the variation of Δg with x_1 may be given by curve (b). As indicated in Fig. 18, Δg at the high pressure no longer satisfies Eq. (86), and the liquid mixture now has a miscibility gap in the composition interval $x'_1 < x_1 < x''_1$.

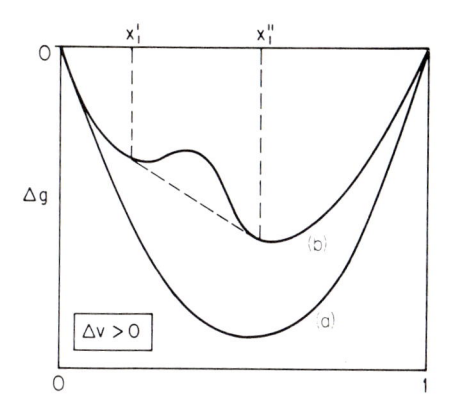

FIG. 18. Effect of pressure on miscibility. (a) Low pressure: no immiscibility. (b) High pressure: immiscible for $x'_1 < x_1 < x''_1$.

For contrast, let us also consider a binary liquid mixture which at normal pressures is incompletely miscible, as shown by curve (a) in Fig. 19. If Δv for this system is negative, then an increase in pressure lowers Δg, and at some high pressure the variation of Δg with x_1 may be given by curve (b), indicating complete miscibility. It follows from these simple considerations that the qualitative effect of pressure on the phase stability of binary liquid mixtures depends on the magnitude and sign of the volume change of mixing. In order to carry out quantitative calculations at some fixed temperature, it is necessary to have information on the variation of Δv with x and P, as well as information on the variation of Δg with x at one pressure.

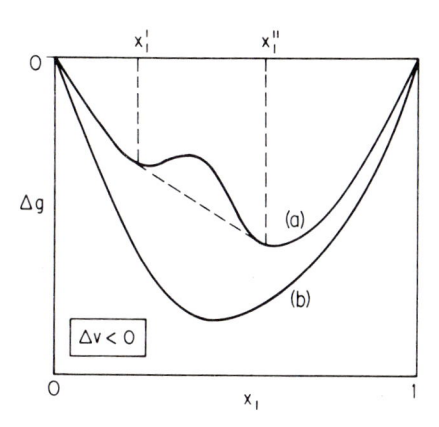

FIG. 19. Effect of pressure on miscibility. (a) Low pressure: immiscible for $x_1' < x_1 < x''_1$. (b) High pressure: no immiscibility.

To illustrate, let us consider a simple, symmetric binary mixture at some fixed temperature and 1-atm pressure. For this liquid mixture, let us assume that

$$\Delta g = RT(x_1 \ln x_1 + x_2 \ln x_2) + A x_1 x_2 \tag{88}$$

$$\Delta v = B x_1 x_2 \tag{89}$$

where A and B are experimentally determined constants. Further, let us assume that the liquid mixture is incompressible for all values of x; i.e., we assume that B is independent of pressure. At any pressure P then, we have for Δg

$$\Delta g = RT(x_1 \ln x_1 + x_2 \ln x_2) + [A + B(P-1)] x_1 x_2 \tag{90}$$

where P is in atmospheres. Substituting Eq. (90) into Eq. (86), we find that the mixture is partially immiscible when

$$\frac{A + B(P-1)}{RT} > 2 \tag{91}$$

Equation (91) tells us that if $A/RT < 2$ (complete miscibility at 1 atm) and if $B > 0$, then there is a certain pressure P (larger than 1 atm) at which immiscibility is induced. On the other hand, if $A/RT > 2$ (incomplete immiscibility at 1 atm) and if $B < 0$, then there is a certain pressure P (larger than 1 atm) at which complete miscibility is attained.

When two liquid phases exist, the compositions of the two phases α and β are governed by the equality of fugacities for each component:

$$f_1^\alpha = f_1^\beta \tag{92}$$

$$f_2^\alpha = f_2^\beta \tag{93}$$

or, equivalently, if the same standard-state fugacity is used for any component in both phases,

$$(\gamma_1 x_1)^\alpha = (\gamma_1 x_1)^\beta \tag{94}$$

$$(\gamma_2 x_2)^\alpha = (\gamma_2 x_2)^\beta \tag{95}$$

For the simple mixture described by Eqs. (88) and (89), we can substitute into Eqs. (94) and (95) and thus obtain

$$x_1^\alpha \exp \frac{[A + B(P-1)][1 - x_1^\alpha]^2}{RT} = x_1^\beta \exp \frac{[A + B(P-1)][1 - x_1^\beta]^2}{RT} \tag{96}$$

$$x_2^\alpha \exp \frac{[A + B(P-1)][1 - x_2^\alpha]^2}{RT} = x_2^\beta \exp \frac{[A + B(P-1)][1 - x_2^\beta]^2}{RT} \tag{97}$$

Simultaneous solution of these equilibrium relations (coupled with the conservation equations $x_1^\alpha + x_2^\alpha = 1$ and $x_1^\beta + x_2^\beta = 1$) gives the coexistence curve for the two-phase system as a function of pressure.

Experimental studies of liquid–liquid equilibria at high pressures were reported many years ago by Roozeboom (R7) and by Timmermans (T3) and Poppe (P4). More recently, experimental work has been reported by Schneider (S1) and by Winnick and Powers (see W2). These latter authors made a detailed study of the acetone–carbon disulfide system at 0°C; at normal pressure, this system is completely miscible, and, for an equimolar mixture, the volume increase upon mixing is of the order of 1 cm^3/gm-mole; which represents a fractional change of about 1.5%. Winnick and Powers measured the volume change on mixing as a function of both pressure and composition; these measurements, coupled with experimentally determined activity coefficients at low pressure, were then used to calculate the phase diagram at high pressures, using a thermodynamic procedure very similar to the one outlined above. The calculations showed that incomplete miscibility should be observed at pressures larger than about 75,000 psia. Winnick and Powers also made experimental measurements of the phase behavior of this system at high pressures, and they found (as shown in Fig. 20) that their observed results are in good agreement with those calculated from volumetric data. This is not surprising, since the calculations follow from rigorous thermodynamic relations and are not based on any physical or mathematical approximations. Perfect agreement between predicted and observed results was not obtained because the calculations are very sensitive to small inaccuracies in the data used.

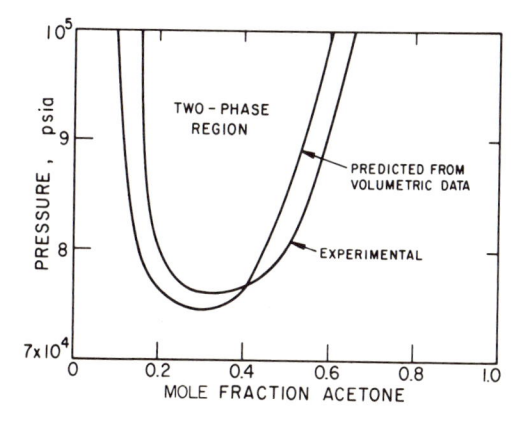

Fig. 20. Liquid–liquid equilibria for a system completely miscible at normal pressure. Calculated and observed behavior of the acetone–carbon disulfide system at 0°C [Winnick and Powers (W2)].

Schneider (S1) has presented a thorough review of the effect of pressure on liquid–liquid equilibria. Certain phenomena discussed in his review are of particular interest to chemical engineers, and are indicated here.

First, Schneider's article clearly shows that the behavior indicated in Fig. 20 illustrates just one of many types which have been observed. The behavior of each type depends on the sign and magnitude of the excess volume, as indicated by Eq. (87); in general, the excess volume depends not only on composition, but also on pressure and temperature, and as a result various types of phase behavior can be obtained. For example, at ordinary pressures, the system methyl ethyl ketone–water has an upper critical solution temperature and a lower critical solution temperature; between these two temperatures, the two liquids are incompletely miscible. As observed by Timmermans (T3), increasing pressure lowers the upper critical solution temperature and raises the lower critical solution temperature, as indicated in Fig. 21. As the pressure rises, the region of partial miscibility decreases, and at pressures beyond 1100 bar,[8] this region has disappeared. By contrast, the system triphenylmethane–sulfur exhibits the opposite behavior (S1); in this case, increasing pressure reduces the region of complete miscibility, as indicated by Fig. 22. Finally, Figs. 23 and 24 illustrate two additional types of phase behavior. Figure 23 shows that for the system 2-methylpyridine–deuterium oxide, increasing pressure first induces miscibility and then, at still higher pressures, again produces a region of incomplete miscibility. On the other hand, Fig. 24 shows that for the system 4-methylpiperidine–water, increasing pressure does not eliminate the region of partial miscibility; very high pressures, after first causing this region to shrink, bring about an increase in the two-phase

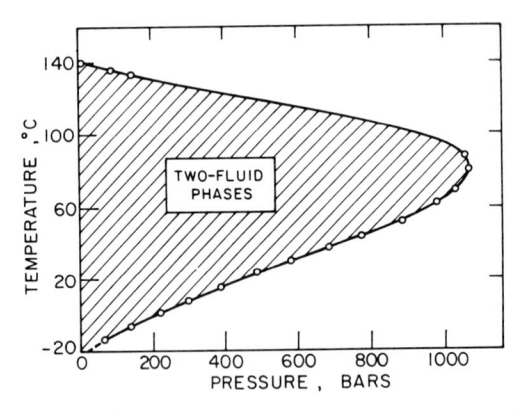

FIG. 21. Liquid–liquid phase behavior in the system methylethylketone–water [from Timmermans (T3)].

[8] 1 bar = 0.98692 atm.

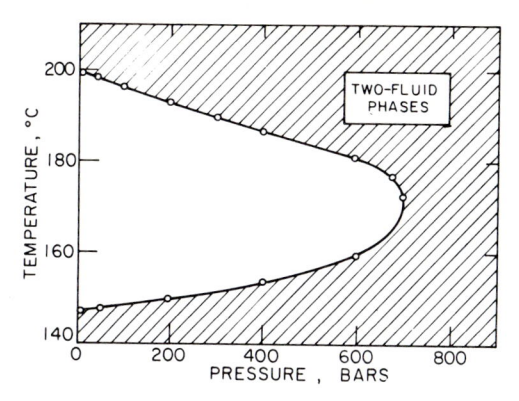

FIG. 22. Liquid–liquid phase behavior in the system triphenylmethane–sulfur [from Schneider (S1)].

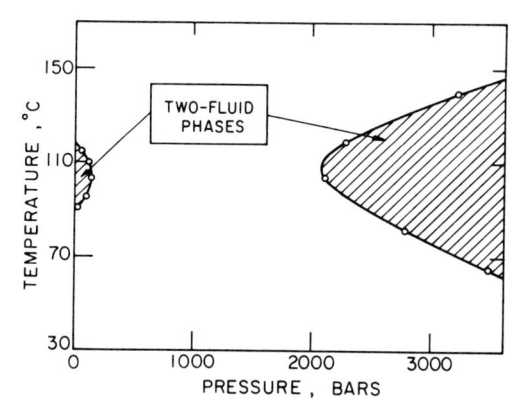

FIG. 23. Liquid–liquid phase behavior in the system 2-methylpyridine–deuterium oxide [from Schneider (S1)].

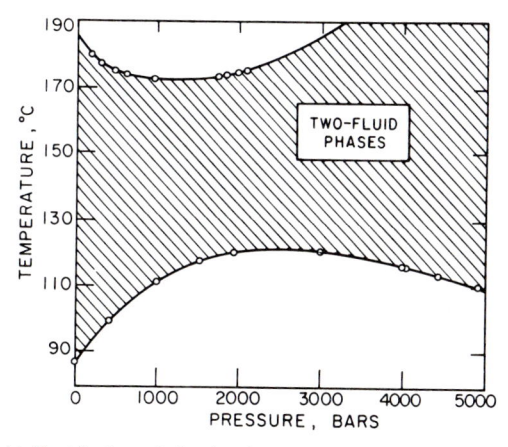

FIG. 24. Liquid–liquid phase behavior in the system 4-methylpiperidine–water [from Schneider (S1)].

region. Behavior similar to that shown in Fig. 24 has been observed by Connolly (C8) for some water–hydrocarbon systems.

Second, Schneider's article reviews recent work (notably by Rowlinson, Kohn and co-workers) on phase relations in binary liquid systems where one of the components is much more volatile than the other (D1, D2, E3, M8, R9). Such systems may have lower critical solution temperatures; for these systems, an increase in temperature (and, indirectly, pressure) causes precipitation of the heavy component, thereby providing a possible separation technique, e.g., for the fractionation of polymers.

Finally, Schneider's review clearly implies that the use of pressure as an independent process-variable bears much promise for bringing about desirable changes of state for new separation techniques. Such techniques may be applicable not only to fluids but also to solids and possibly even to materials having biological importance (L1).

VIII. Gas–Gas Equilibria in Binary Systems

It has been known since the beginning of recorded history that not all liquids are completely miscible with one another. But only in recent times have we learned that gases may also, under suitable conditions, exhibit limited miscibility. The possible existence of two gaseous phases at equilibrium was predicted on theoretical grounds by van der Waals as early as 1894, and again by Onnes and Keesom in 1907 (see R8). Experimental verification, however, was not obtained until about forty years later, primarily by Krichevsky, Tsiklis, and their co-workers in Russia (see G1, S1), by Lindroos and Dodge at Yale (L5), and, more recently, by de Swaan Arons and Diepen at Delft (D3).

Immiscibility of gases is observed only at high pressures when gases are at liquidlike densities. The term "gas–gas equilibria" is therefore somewhat misleading, because it refers to fluids whose properties are similar to those of liquids and are very different from those of gases under normal conditions. For our purposes, we define two-component gas–gas equilibria as the existence of two equilibrated, stable fluid phases at a temperature in excess of the critical of either pure component, both phases being at the same pressure but having different compositions. We cannot here treat this subject in detail, but shall restrict our discussion to a brief consideration of the pertinent thermodynamics. More detailed expositions are given elsewhere (G1, S1).

The conditions which lead a homogeneous fluid mixture to split into two separate fluid phases can be described by classical thermodynamic stability analysis as discussed in numerous textbooks.[9] Such analysis has often been

---

[9] E.g., I. Prigogine and R. Defay, "Chemical Thermodynamics" Chapters 15 and 16. Longmans, London, 1954.

applied to liquid mixtures, and exactly the same analysis is also applicable to gaseous mixtures. For gaseous mixtures, the analysis is a little more difficult in practice, because the effect of pressure on the thermodynamic properties of gases is large, whereas it is often negligible for liquids.

Thermodynamics analysis shows that a binary mixture is stable (with respect to the formation of a new phase) provided that

$$\left(\frac{\partial^2 g}{\partial y^2}\right)_{T,P} > 0 \qquad (98)$$

where g is the molar Gibbs energy of the mixture and y is the mole fraction. An alternate necessary (but not sufficient) criterion for phase stability is given by

$$\left(\frac{\partial^2 \mathbf{a}}{\partial y^2}\right)_{T,v} > 0 \qquad (99)$$

where $\mathbf{a}$ is the molar Helmholtz energy and v is the molar volume of the mixture.

If we have an equation of state for a gaseous mixture, then at any composition we can obtain either g (at any pressure) or $\mathbf{a}$ (at any volume) relative to g or $\mathbf{a}$ at some fixed state at the same temperature, by integrating either of the exact relations

$$dg = v\, dp \qquad (100)$$

or

$$d\mathbf{a} = -p\, dv \qquad (101)$$

Upon substitution into either one of the equations of stability [Eq. (98) or (99)], we can then determine whether the gas mixture exists in one or two stable phases. If two phases exist at some temperature and pressure, we can calculate the two phase compositions by utilizing the two equilibrium relations

$$f_1{}^\alpha = f_1{}^\beta \qquad (102)$$

$$f_2{}^\alpha = f_2{}^\beta \qquad (103)$$

where $f_1{}^\alpha$ stands for the fugacity of component 1 in phase α, etc. The fugacities are obtained from the equation of state in the manner discussed in Section II.

Since most equations of state are pressure-explicit, Eqs. (6) and (99) are often more convenient than Eqs. (5) and (98). With these equations, basing his calculations on van der Waals' equation of state, Temkin (T1) showed that gas–gas immiscibility may occur if the van der Waals constants a and b

of the pure components obey the inequalities

$$\frac{b_2}{b_1} > 0.42 \tag{104}$$

and

$$\frac{a_2}{a_1} < 0.053 \tag{105}$$

where component 1 is the heavier component, i.e., the critical temperature of component 1 exceeds that of component 2. These inequalities provide only a rough guide for nonpolar mixtures, since they are based on the simple mixing rules

$$b_{\text{mixture}} = y_1 b_1 + y_2 b_2 \tag{106}$$

$$a_{\text{mixture}}^{1/2} = y_1 a_1^{1/2} + y_2 a_2^{1/2} \tag{107}$$

Equation (105) is much more restrictive than Eq. (104); Eq. (105) is obeyed only by those systems having highly dissimilar components, e.g., systems containing helium.

Some experimental results for the helium–xenon system are shown in Fig. 25. (The critical temperature of xenon is 16.6°C.) At temperatures several degrees above the critical of xenon, the two phase-compositions are significantly different even at pressures as low as 200 atm. However, to obtain the same degree of separation at higher temperatures, much higher pressures are required.

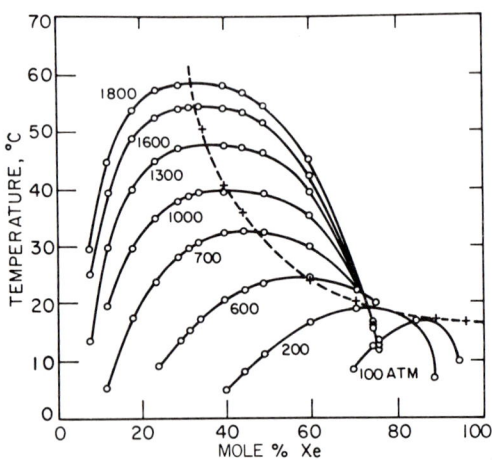

FIG. 25. Gas–gas equilibria in the helium–xenon system [from de Swaan Arons and Diepen (D3)].

A theoretical analysis of the helium–xenon system was reported by Zandbergen and Beenakker (Z1), who based their calculations on the Prigogine–Scott theory of corresponding states for mixtures (P15, S2). We cannot here go into the details of their analysis, but will merely indicate the essential elements. Zandbergen and Beenakker use the "three-liquid" theory to obtain an expression for the volumes of helium–xenon mixtures as a function of temperature, pressure, and composition. This expression is

$$v_{\text{mixture}} = y_1^2 v_1^0 + 2 y_1 y_2 v_{12}^0 + y_2^2 v_2^0 \tag{108}$$

where v_1^0 and v_2^0 are the pure-component volumes at the temperature and pressure of interest, and where v_{12}^0 is a function of pressure P'_1 and temperature T'_1 given by

$$v_{12}^0 = \left(\frac{\sigma_{12}}{\sigma_{11}}\right)^3 v'_1(P'_1, T'_1) \tag{109}$$

where v'_1 is the molar volume of pure component 1 at pressure P'_1 and temperature T'_1 defined by

$$P'_1 = P \frac{\varepsilon_{11}}{\varepsilon_{12}} \left(\frac{\sigma_{12}}{\sigma_{11}}\right)^3 \tag{110}$$

$$T'_1 = T \frac{\varepsilon_{11}}{\varepsilon_{12}} \tag{111}$$

In these equations, σ and ε are parameters in the Lennard–Jones potential function; for interactions between unlike molecules, the customary mixing rules were used:

$$\sigma_{12} = \tfrac{1}{2}(\sigma_{11} + \sigma_{22}) \tag{112}$$

$$\varepsilon_{12} = (\varepsilon_{11} \varepsilon_{22})^{1/2} \tag{113}$$

For helium, $\sigma = 2.56$ Å and $\varepsilon/k = 10.22°$K, where k is Boltzmann's constant. For xenon, $\varepsilon/k = 221°$K and $\sigma = 4.10$. Because of symmetry, the subscript 1 may refer to either helium or xenon; from the assumption of corresponding-states behavior, v_{12}^0 should be independent of the component chosen for subscript 1. Equation (108) is an equation of state for the binary mixture, and from it the phase behavior can be calculated without further assumptions.

Some results reported by Zandbergen and Beenakker are shown in Fig. 26. Considering the severe simplifying assumptions made, the calculated phase boundary is in gratifying agreement with that found experimentally. Because of symmetry with respect to mole fraction in the "three-liquid" model, the calculated T–y curve shown is necessarily a parabola whose maximum is at the composition midpoint. However, the vertical position and the width of the calculated parabola are subject to adjustment upon making small changes in the semiempirical mixing rules for Lennard–Jones parameters, Eqs. (112) and (113).

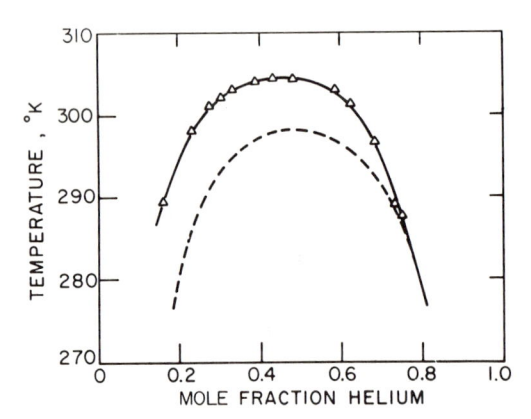

FIG. 26. Gas–gas equilibria for the system helium–xenon at 686 atm. (Δ) Data of de Swaan Arons (D3). (———) Calculated from "three-liquid" corresponding-states theory [from Zandbergen and Beenakker (Z1)].

IX. Liquid–Liquid Equilibria in Ternary Systems Containing One Supercritical Component

In the previous sections, we indicated how, under certain conditions, pressure may be used to induce immiscibility in liquid and gaseous binary mixtures which at normal pressures are completely miscible. We now want to consider how the introduction of a third component can bring about immiscibility in a binary liquid that is completely miscible in the absence of the third component. Specifically, we are concerned with the case where the added component is a gas; in this case, elevated pressures are required in order to dissolve an appreciable amount of the added component in the binary liquid solvent. For the situation to be discussed, it should be clear that phase instability is not a consequence of the effect of pressure on the chemical potentials, as was the case in the previous sections, but results instead from the presence of an additional component which affects the chemical potentials of the components to be separated. High pressure enters into our discussion only indirectly, because we want to use a highly volatile substance for the additional component.

The situation under discussion is very similar to the familiar salting-out effect in liquids, where a salt added to an aqueous solution serves to precipitate one or more organic solutes. Here we are considering the case where a gas is added instead of a salt; in addition, in order to dissolve an appreciable quantity of gas, the system must be at an elevated pressure. This separation method is called "high-pressure vapor extraction" (B1). In this method, two completely miscible liquids may be separated by solvent extraction using a high-pressure gas as the solvent; simplicity of solvent recovery is one of its main advantages. In order to understand the thermodynamic principles of high-pressure vapor extraction, we must first briefly consider how the presence of a third component induces immiscibility in a binary liquid mixture.

A. Effect of Third Component on Critical Solution Temperature

We consider a binary liquid mixture of components 1 and 3; to be consistent with our previous notation, we reserve the subscript 2 for the gaseous component. Components 1 and 3 are completely miscible at room temperature; the (upper) critical solution temperature T^c is far below room temperature, as indicated by the lower curve in Fig. 27. Suppose now that we dissolve a small amount of component 2 in the binary mixture; what happens to the critical solution temperature? This question was considered by Prigogine (P14), who assumed that for any binary pair which can be formed from the three components 1, 2 and 3, the excess Gibbs energy (symmetric convention) is given by

$$g^E_{ij} = \alpha_{ij} x_i x_j \qquad (i, j = 1, 2; 1, 3; \text{ or } 2, 3) \tag{114}$$

where α_{ij} is an empirical (Margules) coefficient determined by the properties of the ij binary. Prigogine has shown that the change in critical solution

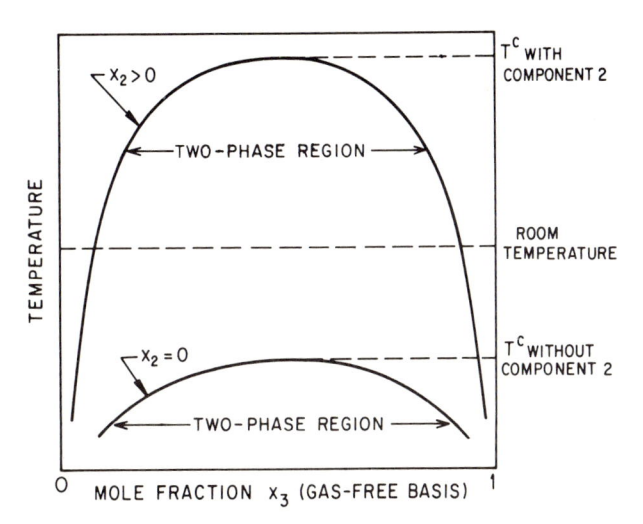

FIG. 27. Effect of gaseous component (2) on mutual solubility of liquids (1) and (3).

temperature which results upon adding a small amount of component 2 is given by

$$\frac{\partial T^c}{\partial x_2} = -\frac{1}{2R}\frac{(\alpha_{13} - \alpha_{12} + \alpha_{23})(\alpha_{13} - \alpha_{23} + \alpha_{12})}{\alpha_{13}} \tag{115}$$

In order to induce the desired immiscibility in the 1–3 binary at room temperature, we want $\partial T^c/\partial x_2$ to be positive and large.

Let us now focus attention on the common case where all three binaries exhibit positive deviations from Raoult's law, i.e., $\alpha_{ij} > 0$ for all ij pairs. If T^c for the 1–3 binary is far below room temperature, then that binary is only moderately nonideal and α_{13} is small. We must now choose a gas which forms a highly nonideal solution with one of the liquid components (say, component 3) while it forms with the other component (component 1) a solution which is only modestly nonideal. In that event,

$$\alpha_{23} \gg \alpha_{12} \tag{116}$$

and also

$$\alpha_{23} \gg \alpha_{13} \tag{117}$$

Equations (115)–(117), indicate that under the conditions just described, $\partial T^c/\partial x_2$ is both large and positive, as desired; i.e., dissolution of a small amount of component 2 in the 1–3 mixture raises the critical solution temperature, as shown in the upper curve of Fig. 27. From Prigogine's analysis, we conclude that if component 2 is properly chosen, it can induce binary miscible mixtures of components 1 and 3 to split at room temperature into two liquid phases having different compositions.

Basic thermodynamic considerations may be used to establish the phase diagram of a ternary system consisting of liquid components 1 and 3 and supercritical gas 2. In the subsequent discussion we follow the simplified treatment given by Balder (B1).

B. Ternary Phase Diagram from Binary Data

1. Coexistence Curve

In a ternary system with two liquid phases α and β, the coexistence curve is the locus of points satisfying the relationships

$$f_i^\alpha = f_i^\beta \quad \text{for each } i = 1, 2, \text{ or } 3 \tag{118}$$

If, for any component, the standard-state fugacity in phase α is the same as that in phase β, Eq. (118) can be rewritten in the more useful form

$$[\gamma_i^{(P)} x_i]^\alpha = [\gamma_i^{(P)} x_i]^\beta \quad \text{for each } i = 1 \text{ or } 3 \tag{119}$$

$$[\gamma_i^{*\,(P)} x_i]^\alpha = [\gamma_i^{*\,(P)} x_i]^\beta \quad \text{for } i = 2 \tag{120}$$

where $\gamma_i^{(P)}$ (or $\gamma_i^{*(P)}$) is the activity coefficient of component i at the system temperature and at the total pressure P. As discussed in Section III, the activity coefficient at pressure P is related at constant temperature and composition to that at some arbitrary reference pressure P^r through the partial molar volume $\bar{v}_i$ according to[10]

$$\frac{\gamma_i^{(P)}}{\gamma_i^{(P^r)}} = \exp \int_{P^r}^{P} \frac{\bar{v}_i \, dP}{RT} \tag{121}$$

The activity coefficient $\gamma_i^{(P^r)}$ is determined by differentiation of g^E, the molar excess Gibbs energy at reference pressure P^r,

$$RT \ln \gamma_i^{(P^r)} = \left(\frac{\partial n_T g^E}{\partial n_i} \right)_{T, P^r, n_j} \tag{122}$$

Therefore, if we have information on the partial molar volumes and the excess Gibbs energy of the ternary system, we can use Eqs. (119)–(122) to find the ends of the tie lines which comprise the coexistence curve.

To illustrate such a calculation, Balder (B1) considered a simple case wherein he assumed that the (symmetric) excess Gibbs energy of the ternary system is given by a two-suffix Margules expansion:

$$g_{123}^E = \alpha_{12} x_1 x_2 + \alpha_{13} x_1 x_3 + \alpha_{23} x_2 x_3 \tag{123}$$

where α_{ij} is an empirical coefficient determined by the properties of the ij binary. Further, Balder assumed that for each component i, $\bar{v}_i$ is independent of composition; as a result of this simplifying assumption, pressure has no effect on the coordinates of the coexistence curve. However, this does not imply that the phase diagram is independent of pressure; as discussed in the next section, pressure has a strong effect on the solubility of (gaseous) component 2, and, therefore, the region of composition in which two liquid phases can exist is very much affected by the total pressure of the system.

It has been shown previously (O1) that

$$\gamma_2^* = \gamma_2 \exp - \frac{\alpha_{12}}{RT} \tag{124}$$

From Eqs. (122)–(124), we obtain for the activity coefficients[11]

$$RT \ln \gamma_1 = \alpha_{12} x_2^2 + \alpha_{13} x_3^2 + (\alpha_{12} + \alpha_{13} - \alpha_{23}) x_2 x_3 \tag{125}$$

$$RT \ln \gamma_2^* = \alpha_{12}(x_1^2 - 1) + \alpha_{23} x_3^2 + (\alpha_{23} + \alpha_{12} - \alpha_{13}) x_1 x_3 \tag{126}$$

$$RT \ln \gamma_3 = \alpha_{13} x_1^2 + \alpha_{23} x_2^2 + (\alpha_{13} + \alpha_{23} - \alpha_{12}) x_1 x_2 \tag{127}$$

---
[10] Equation (121) also holds for the unsymmetrically normalized activity coefficient γ_i^*.
[11] The standard-state fugacity for γ_2^* is $H_{2,1}^{(P^r)}$, Henry's constant for solute 2 in solvent 1 at pressure P^r.

When the three coefficients α_{12}, α_{13}, and α_{23} are known, the coexistence curve can be found by simultaneous solution of Eqs. (119) and (120). A numerical iterative technique given by Hennico and Vermeulen (H1) was used by Balder for performing these calculations with a digital electronic computer.

2. Solubility Curve

In our discussion of the coexistence curve, we tacitly assumed that at least one liquid phase can exist for any composition. However, since component 2 is a supercritical gas, the physically allowable liquid compositions are limited by the solubility of the gas in the binary solvent mixture. As the pressure rises, the solubility of the gas increases, thereby enlarging the composition range which permits the existence of a liquid phase.

To simplify matters, let us assume that at the system temperature, solvents 1 and 3 have negligible volatility; the gas phase, then, contains only component 2. The solubility of component 2 in the liquid is governed by the equation of equilibrium

$$f^V_{\text{pure }2} = \gamma_2^* x_2 H^{(Pr)}_{2,1} \exp \int_{Pr}^{P} \frac{\bar{v}_2 \, dP}{RT} \tag{128}$$

In Eq. (128), the superscript V stands for the vapor phase; $\bar{v}_2$ is the partial molar volume of component 2 in the liquid phase; γ_2^* is the (unsymmetric) activity coefficient; and $H^{(Pr)}_{2,1}$ is Henry's constant for solute 2 in solvent 1 at the (arbitrary) reference pressure P^r, all at the system temperature T. Simultaneous solution of Eqs. (126) and (128) gives the solubility (x_2) of the gaseous component as a function of pressure P and solvent composition (x_1/x_3).

3. Construction of Ternary Phase Diagram

Once the solubility and coexistence curves have been determined, the complete ternary phase diagram may be constructed. This procedure is illustrated by the example shown in Fig. 28. The following parameters have been used:

$\alpha_{12}/RT = 0.5 \qquad \bar{v}_2 = 60 \text{ cm}^3/\text{gm-mole}$
$\alpha_{13}/RT = 1.0 \qquad H^{(Pr)}_{2,1} = 1500 \text{ psia}$
$\qquad\qquad\qquad\quad T = 15°C$
$\alpha_{23}/RT = 4.0 \qquad P^r = 0$
$f_2^V = P^{0.95} \qquad (P \text{ and } f_2^V \text{ have units of psia})$

Depending on the value of the total pressure, the following cases can arise.

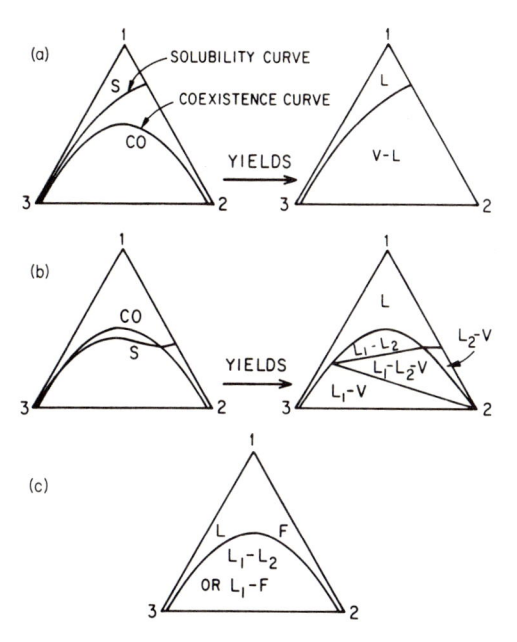

FIG. 28. Phase diagram from computer results. (a) No two-liquid region. (b) Two-liquid region. (c) No vapor region. F stands for fluid.

Case I. At sufficiently low pressures, the solubility curve does not intersect the coexistence curve. In this case, the gas solubility is too low for liquid–liquid immiscibility, since the coexistence curve describes only liquid-phase behavior. Stated in another way, the points on the coexistence curve are not allowed because the fugacity f_2^L on this curve exceeds the prescribed vapor-phase value f_2^V. The ternary phase diagram therefore consists of only the solubility curve, as shown in Fig. 28a where V stands for vapor phase.

Case II. Raising the pressure increases the solubility of the gas such that the solubility curve intersects a part of the coexistence curve. The stability criteria do not allow the existence of a single phase inside the coexistence curve; therefore, a liquid–liquid region and a vapor–liquid–liquid region are formed, as illustrated in Fig. 28b.

Case III. As the pressure increases still further, the solubility curve intersects larger liquid–liquid regions until the critical solution pressure of the system has been reached. Above this critical pressure, no vapor phase exists, and the phase diagram consists of only the coexistence curve, as shown in Fig. 28c. In Fig. 28, L_1 and L_2 stand for the two liquid phases and F stands for a fluid phase.

Figure 29 shows a comparison of results based on the calculation methods developed by Balder with those found experimentally by Elgin and Weinstock (E4). Quantitatively, the agreement is not good, as was to be expected, since the simple expressions for the activity coefficients based only on a total of three binary Margules constants cannot accurately describe highly polar systems. More significant, however, is the excellent qualitative agreement with respect to the presence of multiple-phase regions and the proper slope of the tie lines. It is precisely this qualitative picture which is necessary to investigate possible industrial applicability of a particular ternary system prior to detailed experimental study.

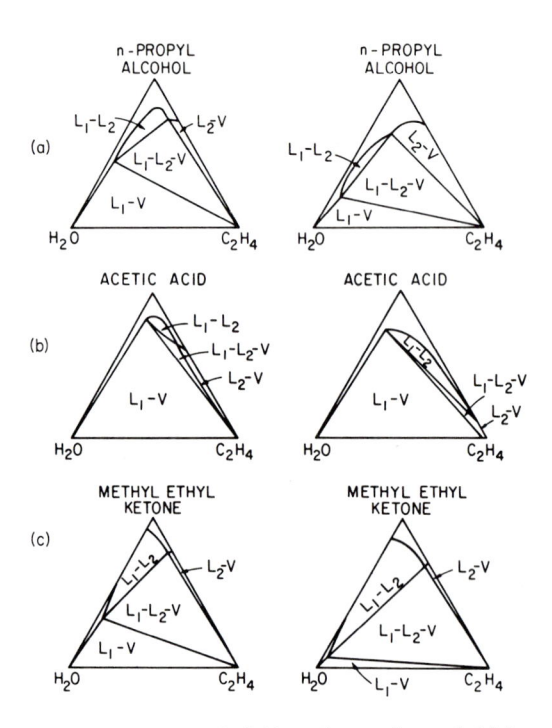

FIG. 29. Comparison of computed (left) and experimental (right) phase diagrams [experimental data by Elgin and Weinstock (E4)].

4. *Effect of Binary Margules Constants on Phase Behavior*

Since industrial separation processes operate in the L_1–L_2 region, it is important to determine how the Margules parameters affect the shape of the coexistence curve and the slope of the tie lines. For any liquid–liquid region to exist, at least one of the binary Margules constants must be greater than $2RT$ (only positive values are considered here); this is a consequence of the stability criteria (H4). Assume for illustrative purposes that the 2–3 binary satisfies this condition, i.e.,

$$\alpha'_{23} = \alpha_{23}/RT > 2 \qquad (129)$$

The effect of increasing α'_{23} is to increase the area enclosed by the liquid–liquid region, thereby providing greater separation at a given overall composition. This is demonstrated by Fig. 30, in which, for illustrative purposes, the Margules constants for the 1–2 and 1–3 binaries have been set equal to zero.

If the reduced 1–2 and 1–3 Margules parameters are less than 2, their primary effect is to determine the slope of the tie lines. Three cases can occur.

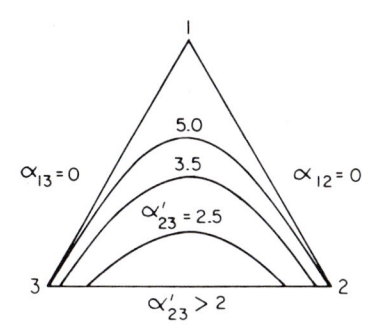

FIG. 30. Effect of 2–3 Margules constant on coexistence curve. Systems 1–2 and 1–3 are ideal. System 2–3 is highly nonideal.

Case I. $\alpha_{13} = \alpha_{12}$. The tie lines are parallel to the 2–3 binary base line.

Case II. $\alpha_{13} > \alpha_{12}$. In this case, the tie lines slope toward the 1–3 binary line. This could have been intuitively predicted by considering the limiting case of an immiscibility band across the phase diagram, as shown in Fig. 31C. Of necessity, the tie lines become parallel to either the 1–3 or the 2–3 binary lines in the limit of pure 1–3 binary or pure 2–3 binary, respectively.

Case III. $\alpha_{13} < \alpha_{12}$. (See Fig. 31D.) Following consistently the explanation of Case II, the tie lines slope toward the 1–2 binary line. In general, then, the effect of the relative values of the 1–2 and 1–3 Margules parameters is to cause the tie lines to slope toward the least ideal of these two binaries.

The value of α_{12} affects the total pressure required to achieve any specified degree of separation. This is illustrated in Fig. 32. A binary solvent separates into phases *a* and *b* at lower pressures as α_{12} increases. This follows from the fact that a large α_{12} indicates that the gas is much more soluble than predicted by Henry's law.

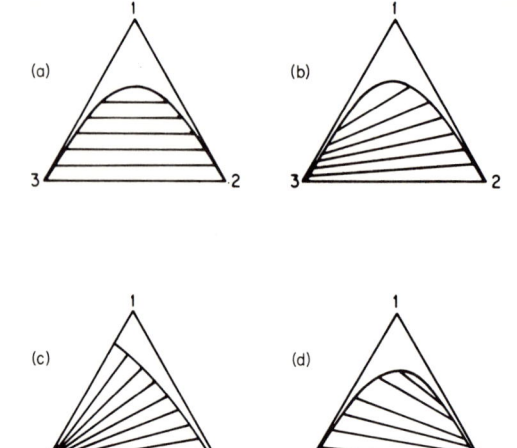

FIG. 31. Effect of 1–2 and 1–3 Margules constants on phase behavior.

(a): $\alpha'_{12} = 1.0$, $\alpha'_{13} = 1.0$, $\alpha'_{23} = 5.0$.
(b): $\alpha'_{12} = 0.5$, $\alpha'_{13} = 1.5$, $\alpha'_{23} = 5.0$.
(c): $\alpha'_{12} < 2$, $\alpha'_{13} > 2$, $\alpha'_{23} > 2$.
(d): $\alpha'_{12} = 1.5$, $\alpha'_{13} = 0.5$, $\alpha'_{23} = 5.0$.

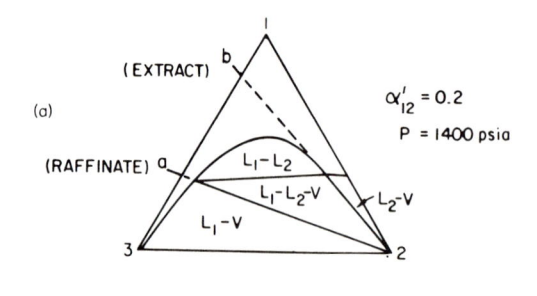

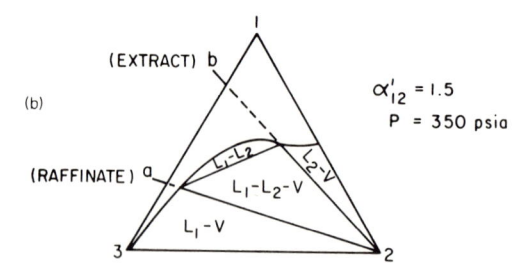

FIG. 32. Effect of 1–2 nonideality on pressure required for specified separation of 1 and 3. Lower pressure is sufficient for system with larger nonideality.

C. Thermodynamic Criteria for Separation-Process Application

The criteria which would be most desirable for industrial application of a separation process involving a supercritical gas may be established by comparing Figs. 31B, 31D, and 32. The largest cost in such a process is likely to be that of gas compression. Therefore, the maximum separation possible of the two solvents should occur for the addition of a given amount of gas, and the total pressure required to dissolve this gas should be small. This is the case if the tie lines slope toward the 1–3 binary line and if the gas is readily soluble. In terms of the Margules parameters and Henry's constant, these favorable criteria are:

(1) To form an immiscible region α_{23}/RT should be greater than 2.

(2) To obtain the proper slope of the tie lines, the 1–3 binary should be more nonideal than the 1–2 binary: $\alpha_{13} > \alpha_{12}$.

(3) To minimize the pressure requirement, $H_{2,1}$ should be small [gas (2) readily soluble in liquid (1)], and α_{12} should be large and positive (the 1–2 binary is highly nonideal with positive deviations from Raoult's law).

As an example, the *n*-propyl alcohol–water–ethylene system shown in Fig. 29 satisfies these criteria.

Balder's work indicates that a good semiquantitative picture may be obtained for phase behavior of ternary systems containing a supercritical gas by using thermodynamic considerations and a limited amount of binary data. A study of the effects of the empirical Margules constants on the shape of the coexistence curve has led to a set of criteria desirable in employing a supercritical gas in a separation process. The accuracy of Balder's treatment could be significantly improved by accounting for nonidealities introduced by the small quantities of heavy components in the vapor phase and by including higher-order terms in the empirical expressions for activity coefficients.

Acknowledgment

The author is grateful to T. Vermeulen for helpful comments and to the Miller Institute for Basic Science, University of California, Berkeley and to the National Science Foundation for financial support.

References

A1. Adler, S. B., Friend, L., Pigford, R. L., and Rosselli, G. M., *A.I.Ch.E. (Am. Inst. Chem. Eng.) J.* **6**, 104 (1960).
B1. Balder, J. R., and Prausnitz, J. M., *Ind. Eng. Chem. Fundamentals* **5**, 449 (1966).
B2. Basset, J., and Dodé, C. R., *Compt. Rend.* **203**, 775 (1936).
B3. Beattie, J. A., *Chem. Rev.* **44**, 141 (1949).
B4. Benedict, M., Webb, G. B., and Rubin, L. C., *Chem. Eng. Progr.* **47**, 419 (1951).
B5. Black, C., *Ind. Eng. Chem.* **50**, 391 (1958).
B6. Brown, W. B., *Phil. Trans. Roy. Soc. (London)* **A250**, 175 (1957).

C1. Chao, K. C., and Seader, G. D., *A.I.Ch.E. (Am. Inst. Chem. Eng.) J.* **7**, 598 (1961).
C2. Chueh, P. L., Dissertation, Univ. of Calif., Berkeley, 1967.
C3. Chueh, P. L., and Prausnitz, J. M., *A.I.Ch.E. (Am. Inst. Chem. Eng.) J.* **13**, 896 (1967).
C3a. Chueh, P. L., and Prausnitz, J. M., *A.I.Ch.E. J.*, **13**, 1099, 1107 (1967).
C3b. Chueh, P. L., and Prausnitz, J. M., *Ind. Eng. Chem. Fundamentals* **6**, 492 (1967).
C4. Chueh, P. L., Muirbrook, N. K., and Prausnitz, J. M., *A.I.Ch.E. (Am. Inst. Chem. Eng.) J.* **6**, 1097 (1965).
C5. Clusius, K., and Weigand, K., *Z. Phys. Chem.* **B46**, 1 (1940).
C6. Connolly, J. F., *J. Chem. Phys.* **35**, 2892 (1962).
C7. Connolly, J. F., *J. Chem. Eng. Data* **11**, 13 (1966).
C8. Connolly, J. F., and Kandalic, G. A., *Phys. Fluids* **3**, 463 (1960).
C9. Curl, R. F., and Pitzer, K. S., *Ind. Eng. Chem.* **50**, 265 (1958).
D1. Davenport, A. J., Freeman, P. I., and Rowlinson, J. S., *A.I.Ch.E. (Am. Inst. Chem. Eng.) J.* **8**, 428 (1962).
D2. Davenport, A. J., and Rowlinson, J. S., *Trans. Faraday Soc.* **59**, 78 (1963).
D3. DeSwaan Arons, J., and Diepen, G. A. M., *J. Chem. Phys.* **44**, 2323 (1966).
D4. Diepen, G. A. M., and Scheffer, F. E. C., *J. Am. Chem. Soc.* **70**, 4085 (1948).
D5. Diepen, G. A. M., and Scheffer, F. E. C., *J. Phys. Chem.* **57**, 575 (1953).
D6. Dodge, B. F., "Chemical Engineering Thermodynamics." McGraw-Hill, New York, 1944.
D7. Dodge, B. F., and Newton, R. H., *Ind. Eng. Chem.* **29**, 718 (1947).
E1. Eckert, C. A., and Prausnitz, J. M., *A.I.Ch.E. (Am. Inst. Chem. Eng.) J.* **11**, 886 (1965).
E2. Eckert, C. A., Renon, H., and Prausnitz, J. M., *Ind. Eng. Chem. Fundamentals* **6**, 52, 58, correction on p. 619, (1967).
E3. Ehrlich, P., and Graham, E. B., *J. Polymer Sci.* **45**, 246 (1960).
E4. Elgin, J. C., and Weinstock, J. J., *J. Chem. Eng. Data* **4**, 3 (1959).
F1. Franck, E. U., *Z. Phys. Chem. (Frankfurt)* **6**, 345 (1956).
F2. Franck, E. U., *Angew. Chem.* **73**, 309 (1961).
F3. Franck, E. U., and Tödheide, K., *Z. Phys. Chem.* **22**, 232 (1959).
G1. Gonikberg, M. G., "Chemical Equilibria and Reaction Rates at High Pressures" (translated from Russian). Office of Tech. Services, U.S. Dept. of Commerce, Washington, D.C. 1963.
G2. Grayson, H. G., personal communication.
G3. Guggenheim, E. A., *J. Chem. Phys.* **13**, 253 (1945).
H1. Hennico, A., and Vermeulen, T., Rept UCRL-9415, Univ. of Calif. Radiation Lab. (1961).
H2. Herington, E. F. G., *Nature* **160**, 610 (1947).
H3. Herington, E. F. G., *J. Appl. Chem. (London)* **2**, 11 (1952).
H4. Hildebrand, J. H., and Scott, R. L., "Solubility of Nonelectrolytes." Reinhold, New York, 1950.
H5. Hildebrand, J. H., and Scott, R. L., "Regular Solutions." Prentice-Hall, Englewood Cliffs, New Jersey, 1962.
H6. Hirschfelder, J. O., Curtiss, C. F., and Bird, R. B., "Molecular Theory of Gases and Liquids." Wiley, New York, 1954.
J1. Joffe, J., *Ind. Eng. Chem.* **40**, 1738 (1948).
J2. Joffe, J., and Zudkevitch, D., *Ind. Eng. Chem. Fundamentals* **5**, 245 (1967).
K1. Kay, W. B., *Ind. Eng. Chem.* **28**, 1014 (1936).
K2. Kihara, T., *Rev. Mod. Phys.* **25**, 831 (1953).
K3. Kihara, T., *in* "Advances in Chemical Physics" (I. Prigogine, ed.), Vol. 1, 1958; Vol. 5, 1963. Wiley (Interscience), New York.

K4. Krichevsky, I. R., *J. Am. Chem. Soc.* **59**, 596 (1937).
K5. Krichevsky, I. R., and Ilinskaya, *Zh. Fiz. Khim.* **19**, 621 (1945).
K6. Krichevsky, I. R., and Kasarnovsky, J. S., *J. Am. Chem. Soc.* **57**, 2168 (1935).
K7. Kuss, E., *Chimia (Aarau)* **18**, 75 (1964).
L1. Landau, J. V., *J. Cell Biol.* **24**, 332 (1965).
L2. Leland, T. W., Chappelear, P. S., and Gamson, B. W., *A.I.Ch.E. (Am. Inst. Chem. Eng.) J.* **8**, 482 (1962).
L3. Lewis, G. N., and Randall, M., "Thermodynamics." McGraw-Hill, New York, 1923.
L4. Lewis, G. N., and Randall, M., "Thermodynamics," 2nd. ed. (by Pitzer, K. S., and Brewer, L.). McGraw-Hill, New York, 1961.
L5. Lindroos, A. E., and Dodge, B. F., *Chem. Eng. Progr. Symp. Ser.* **48**, 10 (1952).
L6. Lyckman, E. W., Eckert, C. A., and Prausnitz, J. M., *Chem. Eng. Sci.* **20**, 685 (1965).
L7. Lyckman, E. W., Eckert, C. A., and Prausnitz, J. M., *Chem. Eng. Sci.* **20**, 703 (1965).
M1. McGlashan, M. L., and Potter, D. J. B., *Proc. Roy. Soc. (London)* **A267**, 478 (1962).
M2. McGlashan, M. L., and Wormald, C. J., *Trans. Faraday Soc.* **60**, 646 (1964).
M3. Miller, P., and Dodge, B. F., *Ind. Eng. Chem.* **32**, 434 (1940).
M4. Morey, G. N., *Econ. Geol.* **52**, 225 (1957).
M5. Muirbrook, N. K., and Prausnitz, J. M. *A.I.Ch.E (Am. Inst. Chem. Eng.) J.* **6**, 1092 (1965).
M6. Mullins, J. C., and Ziegler, W. T., *Intern. Advan. Cryogenic Eng.* **10**, 171 (1964).
M7. Myers, C. O., *Chem. Eng. Progr. Symp. Ser.* **46**, No. 7 (1953).
M8. Myrat, C. D., and Rowlinson, J. S., *Polymer* **6**, 645 (1965).
O1. O'Connell, J. P., and Prausnitz, J. M., *Ind. Eng. Chem. Fundamentals* **3**, 347 (1964).
O2. O'Connell, J. P., and Prausnitz, J. M., *Ind. Eng. Chem. Process Design Develop* **6**, 245 (1967).
O3. Orentlicher, M., and Prausnitz, J. M., *Chem. Eng. Sci.* **19**, 777 (1964).
P1. Pitzer, K. S., and Curl, R. F., Jr., *J. Am. Chem. Soc.* **79**, 1269 (1957).
P2. Pitzer, K. S., and Hultgren, G. O., *J. Am. Chem. Soc.* **80**, 4793 (1958).
P3. Pitzer, K. S., Lippman, D., Curl, R. F., Huggins, C. M., and Petersen, D. E., *J. Am. Chem. Soc.* **77**, 3427 (1955).
P4. Poppe, G., *Bull. Soc. Chim. Belges* **44**, 640 (1935).
P5. Prausnitz, J. M., *A.I.Ch.E. (Am. Inst. Chem. Eng.) J.* **5**, 3 (1959).
P6. Prausnitz, J. M., *A.I.Ch.E. (Am. Inst. Chem. Eng.) J.* **8**, 563 (1962).
P7. Prausnitz, J. M., *Chem. Eng. Sci.* **18**, 613 (1963).
P8. Prausnitz, J. M., and Benson, P. R., *A.I.Ch.E. (Am. Inst. Chem. Eng.) J.* **5**, 161 301 (1959).
P8a. Prausnitz, J. M., and Chueh, P. L., "Computer Calculations for High-Pressure Vapor-Liquid Equilibria." Prentice-Hall, Englewood Cliffs, New Jersey, to be published (1968).
P9. Prausnitz, J. M., and Gunn, R. D., *A.I.Ch.E. (Am. Inst. Chem. Eng.) J.* **4**, 430 (1958).
P10. Prausnitz, J. M., and Keeler, R. N., *A.I.Ch.E. (Am. Inst. Chem. Eng) J.* **7**, 399 (1961).
P11. Prausnitz, J. M., and Myers, A. L., *A.I.Ch.E. (Am. Inst. Chem. Eng.) J.* **9**, 5 (1963).
P12. Prausnitz, J. M., and Shair, F. H., *A.I.Ch.E. (Am. Inst. Chem. Eng.) J.* **7**, 682 (1961).
P13. Prausnitz, J. M., Chao, K. C., and Edmister, W. C., *A.I.Ch.E. (Am. Inst. Chem. Eng.) J.* **6**, 214 (1960).
P14. Prigogine, I., *Bull. Soc. Chim. Belges* **52**, 115 (1943).
P15. Prigogine, I., "The Molecular Theory of Solutions." North-Holland Publ. Co., Amsterdam, 1957.
R1. Redlich, O., and Kister, A. T., *Ind. Eng. Chem.* **40**, 345 (1948).
R2. Redlich, O., and Kwong, J. N. S., *Chem. Rev.* **44**, 233 (1949).

R3. Redlich, O., Kister, A. T., and Turnquist, C. E., *Chem. Eng. Progr. Symp. Ser.* **48**, 49 (1952).
R4. Redlich, O., Ackerman, F. J., Gunn, R. D., Jacobson, M., and Lau, S., *Ind. Eng. Chem. Fundamentals* **4**, 369 (1965).
R5. Reid, R. C., and Leland, T. W., *A.I.Ch.E.* (*Am. Inst. Chem. Eng.*) *J.* **11**, 228 (1965).
R6. Reuss, J., and Beenakker, J. J. M., *Physica* **22**, 869 (1956).
R7. Roozeboom, H. W., " Die heterogenen Gleichgewichte." Braunschweig, 1918.
R8. Rowlinson, J. S., "Liquids and Liquid Mixtures." Butterworths, London, 1959.
R9. Rowlinson, J. S., and Freeman, P. I., *Pure Appl. Chem.* **2**, (1961).
R10. Rowlinson, J. S., and Richardson, M. J., *in* "Advances in Chemical Physics" (I. Prigogine, ed.), Vol. 2. Wiley (Interscience), New York, 1959.
S1. Schneider, G., *Ber. der Bunsengesellschaft* **70**, 497 (1966).
S2. Scott, R. L., *J. Chem. Phys.* **25**, 193 (1956).
S3. Sherwood, A. E., and Prausnitz, J. M., *J. Chem. Phys.* **41**, 413, 429 (1964).
S4. Sorgato, I., and Angelin, L., *Chem. Eng. Sci.* **20**, 431 (1965).
S5. Sortland, L. D., and Prausnitz, J. M., *Chem. Eng. Sci.* **20**, 847 (1965).
T1. Temkin, M., *Zh. Fiz. Khim.* **9**, 2040 (1959).
T2. Thompson, R. E., and Edmister, W. C., *A.I.Ch.E.* (*Am. Inst. Chem. Eng.*) *J.* **11**, 457 (1965).
T3. Timmermans, J., *J. Chim. Phys.* **20**, 491 (1923).
V1. Van Ness, H. C., "Classical Thermodynamics of Nonelectrolyte Solutions." Pergamon, Oxford, 1964.
W1. Wiebe, R., and Gaddy, V. L., *J. Am. Chem. Soc.* **59**, 19, 84 (1937).
W2. Winnick, J., Dissertation, Univ. of Oklahoma, Norman, Oklahoma, 1963.
Z1. Zandbergen, P., Dissertation, Univ. of Leiden, 1966.

THE BURN-OUT PHENOMENON IN FORCED-CONVECTION BOILING

Robert V. Macbeth

*United Kingdom Atomic Energy Authority
Atomic Energy Establishment, Winfrith, England*

I. Introduction	208
II. The Meaning of Burn-Out	210
A. Temperature-Controlled Systems	210
B. Heat-Flux-Controlled Systems	212
C. Detection of the Burn-Out Point	214
D. Burn-Out in the Annular Flow Regime	218
E. The Connection between Burn-Out and Various Other Flow Patterns	222
F. The Post-Burn-Out Condition Giving Tolerable Wall Temperatures	223
III. The Important System-Describing Parameters	225
A. Distinction between a Hard and a Soft Inlet	226
B. The Influence of Gravity on Burn-Out	231
C. The Influence of Surface Conditions and Liquid Purity	232
D. The Influence of Test-Section Wall Thickness	233
IV. The Linear Relation between Burn-Out Flux and Inlet Subcooling	235
V. Possible Contractions among the System-Describing Parameters	238
VI. The Local-Conditions Concept of Burn-Out	241
A. The Barnett Local-Conditions Hypothesis	242
B. The Kirby Local-Conditions Hypothesis	245
VII. The Low-Velocity Burn-Out Regime	246
VIII. The Correlation of Water Data for Uniformly Heated Channels	249
A. Round Tubes	251
B. Rectangular Channels	258
C. Rod Bundles	260
D. Annular Test Sections	267
IX. The Equivalent-Diameter Hypothesis	273
X. The Effect on Burn-Out of Nonuniform Heating	274
XI. The Technique of Modeling Forced-Convection Burn-Out	280
A. The Scaling-Law Approach	280
B. The Scaling-Factor Approach	284
Nomenclature	286
References	287

I. Introduction

The burn-out phenomenon is a condition usually associated with a heat-flux-controlled boiling system and not with a temperature-controlled one. The condition arises when part of a boiling surface suddenly becomes blanketed with vapor, causing a local deterioration in heat transfer and a consequent rise in the surface temperature. The cause of vapor blanketing is primarily a physical crisis in the hydrodynamics of a system. The temperature rise may be more than a thousand degrees and lead to melting and the rupture of metal surfaces; this is physical burn-out. On the other hand, depending on the nature of a boiling system, the temperature rise may hardly be noticeable.

In most engineering and chemical processes involving boiling, the burn-out phenomenon is unlikely to be destructive, since heat-flux levels are much too low and many boiling systems are effectively temperature-controlled. The water-cooled nuclear reactor is an important exception; nearly all experimental work and theoretical and analytical effort relating to burn-out, starting about 1950, has been aimed at producing design information for these reactors. Thus, the majority of experimental data available refer to water. There is an increasing amount of important data appearing for the two fluids CCl_2F_2 and $CHFCl_2$ (generally known as "Freons" or "Arctons," to quote but two of their trade names), as a consequence of a basic research program being carried out by the United Kingdom Atomic Energy Authority (UKAEA) at Winfrith in England to determine, if possible, the scaling laws of forced-convection burn-out. These data have been compiled by Stevens *et al.* (S13), and they will be referred to later. Published data for other fluids with forced convection are too sparse and cover too limited a range of parameters to be of much general value. Some examples are: hydrazine (H7, N1), anhydrous ammonia (B7, D3, N2), polyphenyls (C7, R3, V1), liquid hydrogen and nitrogen (P1), liquid potassium (H8), nitrogen tetroxide (B21), ethylene glycol (G1), isopropanol (S9), and ethanol (S10).

The water data have come from nearly every country with a nuclear power program, and the combined data, totaling about 15,000 experimental results, represent an effort costing tens of millions of dollars and involving hundreds of research workers. The early data were for simple geometries such as uniformly heated round tubes and rectangular channels of short length. Experimentation can be difficult, and it has taken time for techniques to evolve; only since about 1960 have data become available for complex rodbundle systems and nonuniformly heated channels with wide ranges of system parameters.

It is found today that effective data handling is no longer possible except by computer, and several organizations now maintain systematic and up-to-date records of burn-out data on computer cards. Useful compilations of data for water have appeared in the literature, and the reader is referred to De Bortoli *et al.* (D2), Firstenberg *et al.* (F3), Silvestri (S4), Thompson and Macbeth (T1), Tong *et al.* (T5), Becker *et al.* (B13), Janssen and Kervinen (J3), Macbeth (M4), and Barnett (B6).

A proven theoretical basis for dealing with burn-out in forced-convection boiling has not yet been found, although numerous theories have been proposed, such as those of Randles (R1), Goldmann *et al.* (G2), Levy (L7), Tippets (T2), Carr (C1), Isbin *et al.* (I1), Grace (G3), Becker and Persson (B12), and Grace and Isbin (G4). These theories relate to a variety of physical models, and they have recently been systematically examined by Kirby (K3), and either rejected because they failed to conform reasonably well with experimental facts, or found unproven because they result in an equation with too many arbitrary constants. Almost any form of equation can be fitted to a set of experimental data if the equation has enough degrees of freedom in the form of arbitrary constants. When the numerical values of the arbitrary constants have to be found by referring to the experimental data, any success the equation may have cannot be ascribed with certainty to the complete or even partial correctness of the theory.

The main weakness with burn-out theory lies in the fact that the physical models used have been largely speculative, i.e., they have not started from what is actually seen or measured in a boiling channel at burn-out. This is not without good reason, however, as such measurements are difficult and entail trying to look quantitatively at the hydrodynamic conditions in the immediate vicinity of the burn-out area. One important step forward in this direction has recently been made by Hewitt *et al.* (H5) and Staniforth *et al.* (S8), who have shown very convincingly that at conditions of from about 10 to 100% quality, burn-out coincides with the vanishing to zero thickness of the thin film of liquid that exists on a heated surface at these qualities. Early indications of such a burn-out mechanism were obtained from experimental work on climbing film evaporators by Stroebe *et al.* (S16), Coulson and McNelly (C9), and Cathro and Tait (C3), as well as from experiments on annular flow by Sachs and Long (S1). The physical picture of the vanishing of a liquid film to zero thickness is the one on which Isbin *et al.* (I1) based their theory, and that theory may well prove to be correct. For its confirmation, however, still more quantitative experimental information is required on the deposition and entrainment mass-transfer rates to and from the liquid film. Without this detailed information, the theory of Isbin *et al.* leads to an equation with too many arbitrary constants for there to be complete satisfaction with its validity. For subcooled boiling conditions, Kirby (K4) has found evidence of a continuous liquid film existing on the heated surface underneath a bubble layer, although at very high subcooling this appears not to be the case. Thus, several different burn-out mechanisms may exist, and more than one theory is probably necessary to describe burn-out for all conditions.

While dimensional analysis is often a useful tool for dealing with a problem, it has not yet been successful for studying this phenomenon, mainly because the fluid properties of importance in forced-convection boiling have not been identified. Burn-out correlations based on dimensional analysis have appeared, e.g., Griffith (G5), Reynolds (R2), Zenkevitch (Z1), Ivashkevich (I2), Tong *et al.* (T6), but the fluid properties used in these cases have been chosen on the basis of various assumptions without any demonstration that the properties used were the correct ones. They have, in fact, been shown in recent work by Barnett (B5), (to be considered later) to be either incorrect or incomplete.

At the present time, our knowledge and understanding of the burn-out phenomenon in a boiling system must depend largely upon interpreting correctly the vast amount of experimental data; in the following pages this is attempted.

II. The Meaning of Burn-Out

Jens and Leppert (J4), and, more recently, Barnett (B2), have emphasized the need to distinguish between a temperature-controlled surface and a heat-flux-controlled surface when referring to boiling phenomena. Failure to observe the distinction has caused some confusion in the literature, particularly as further complications arise from differences that exist between pool boiling and forced-convection boiling.

A. Temperature-Controlled Systems

A perfect temperature-controlled heat-transfer surface is difficult to achieve, but it is closely simulated in practice by using a control fluid on one side of, for example, a metal tube. The tube wall should be thin and, ideally, the heat-transfer resistance comparatively large for the other fluid on the "working" side of the tube; the latter surface is then effectively temperature-controlled and responds only to changes in the control fluid.

With the above type of system, the boiling curve, giving the relationship between surface temperature at a local area and the heat flux, is of the form shown in Fig. 1. The precise values depend on the flow velocity over the surface, on surface geometry, etc. Experimental confirmation of Fig. 1 has been given, for example, for pool boiling by Berenson (B16) and Stock (S15) and for forced-convection boiling by Ellion (E2). Stock used $\frac{5}{8}$-in. diameter horizontal tubes of different material immersed in a test fluid, Freon 11, with high-pressure water flowing through the tubes for temperature control. Ellion used a $\frac{1}{4}$-in. diameter electrically heated tube inside a $2\frac{1}{2}$-in. diameter

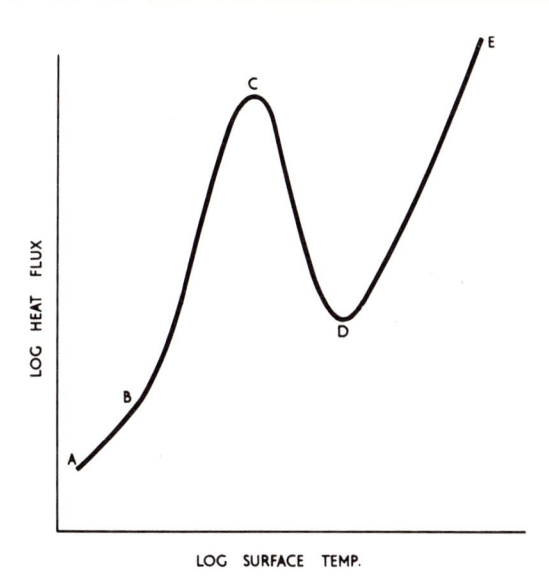

FIG. 1. Boiling curve of a temperature-controlled surface.

Pyrex jacket. The test fluid, distilled water, flowed vertically upwards through the annulus, while inside the heated tube a control fluid flowed which was either water or nitrogen gas, depending on the tube temperature required.

The well-known regions of the boiling curve in Fig. 1 are: AB—single-phase heat transfer, BC—nucleate boiling, CD—a transition region called "unstable film boiling," and DE—a region of stable film boiling. The heat flux at point C is called the critical heat flux, and point D corresponds to the Leidenfrost point.

Burn-out can have only one meaning with a temperature-controlled system, and that is physical burn-out, which occurs when the surface temperature is made high enough to result in a rupture. Physical burn-out is a function of the mechanical properties of the surface material and of any load stresses it may carry.

It must be pointed out that while the region BC in Fig. 1 is usually called the nucleate boiling region, this may not be an accurate description in the case of forced convection. Thus, it is possible with annular flow (which corresponds to a thin film of liquid moving cocurrently along a heated channel wall with vapor and entrained liquid in the main stream) for there to be no nucleate boiling. Heat transfer is by ordinary forced convection through the liquid film (which will be superheated) and by evaporation (not boiling) from the film interface to the main stream. This is the phase-change mechanism occurring in climbing film evaporators and has been investigated by Coulson and McNelly (C9) and more recently by Hewitt et al. (H4). Whether it occurs in a boiling channel depends on the system parameters, but it does mean that the region BC in Fig. 1 may comprise a nucleate boiling region and an evaporative region.

B. Heat-Flux-Controlled Systems

1. Pool Boiling

In a heat-flux-controlled system such as an electrically heated tube or wire, or a nuclear-reactor fuel element, the heat flux is an independent variable and the surface temperature a dependent quantity; with pool boiling, the boiling curve is typically as shown in Fig. 2. Nukiyama (N3) was the first to demonstrate the characteristics of Fig. 2. It has the regions AB—single-phase heat transfer, BC—nucleate boiling, and DE—stable film boiling similar to a temperature-controlled system (Fig. 1). However, on reaching point C, a further small increase in the heat flux causes the surface temperature to undergo a jump discontinuity along the dotted line CF and stabilize at the point F, provided physical burn-out does not occur on the way. It is the uncontrolled temperature rise of the transition CF that is customarily known as burn-out, and the point C is the burn-out point. Some proliferation of terminology exists in the literature, and the reader is reminded that terms such as departure from nucleate boiling (DNB), the first boiling crisis, dry out, and critical heat flux are, for all practical purposes, synonymous with burn-out or the burn-out point as defined above. It would appear that the

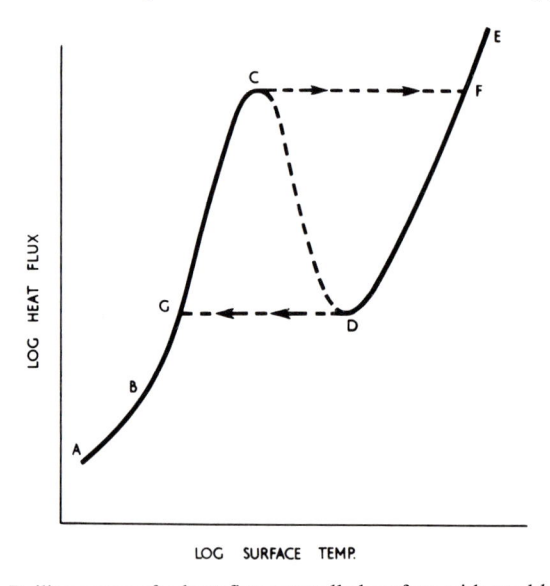

Fig. 2. Boiling curve of a heat-flux-controlled surface with pool boiling.

term DNB may be a misnomer in the case of forced convection, since burn-out may occur with annular flow, but without nucleate boiling, as described earlier. It is also considered a useful distinction to reserve the term "critical heat flux" for a temperature-controlled surface, since the latter does not show the uncontrolled temperature rise that is the essential characteristic of the burn-out point in a heat-flux-controlled system.

If point F in Fig. 2 is reached without physical burn-out occurring, then, as shown by Nukiyama, a further increase in heat flux will raise the surface temperature in the direction of E until physical burn-out does occur. If, however, the heat flux at point F is decreased, the surface temperature does not revert to the value at C, but moves along the curve towards D. On reaching D, it was observed by Nukiyama that the surface temperature undergoes another jump discontinuity along the dotted line DG, and stabilizes at G in the nucleate-boiling region. Both the transition lines CF and DG can be passed only in the direction shown by the arrows in Fig. 2.

Nukiyama was unable to establish a condition on the line CD, and in his original experimental curves he shows this line dotted. However, Farber and Scorah (F2) have reported that with special surface treatment and very careful experimentation, quasistable conditions can be set up on a curve such as CD in pool boiling. Their procedure was to first reach point D by way of ACFD and then to very carefully increase the heat flux. However, any disturbance or sudden change of heat flux when in the region CD caused the system to revert to the stable regions GC or DF.

2. *Forced-Convection Boiling*

A heat-flux-controlled surface with forced-convection boiling is the situation to which the majority of burn-out data correspond. An example of a set of experimental boiling curves obtained by Stevens *et al.* (S13) using Freon 12 in a uniformly heated tube with vertical upflow is shown in Fig. 3. The Freons are good fluids to use for getting a complete boiling curve, since the maximum tube-wall temperatures reached are not so high as to cause physical burn-out. In Fig. 3, the wall temperature is that recorded by thermocouples attached to the tube at the outlet end of the heated length, which is where burn-out normally first occurs with uniform heating. A jump discontinuity in wall temperature can be seen, and this jump indicates the burn-out flux in the same way as in a pool-boiling system. However, on decreasing the heat flux from a point in the stable film-boiling region of Fig. 3, Stevens (S11) has observed that the system follows the same jump discontinuity and returns to nucleate boiling by way of the burn-out point. McEwan *et al.* (M5) also found no signs of any hysteresis effect in tests with a heated annulus section using water at 1500 psia. On the other hand, Sterman *et al.* (S9, S10) found that in tests with a 0.63-in. diameter heated tube 3.54-in. long using

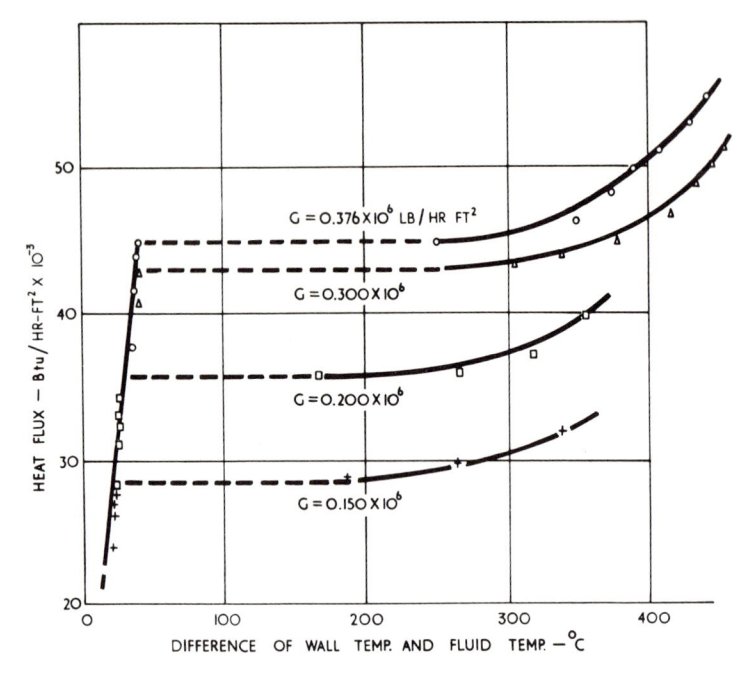

FIG. 3. Boiling curves for Freon 12 in a heat-flux-controlled uniformly heated tube [from Stevens et al. (S13)]. $L = 51$ in., $d = 0.633$ in., $P = 155$ psia, $\Delta h = 13$ Btu/lb.

isopropyl alcohol, ethyl alcohol, and distilled water at low pressures (2–7 atm), operation in the stable film-boiling region at heat fluxes well below their burn-out flux was possible over a wide range of mass velocities. In the case of the distilled water, however, it is stated in Sterman et al. (S10) that prolonged film boiling below the burn-out flux could not be maintained. Thus, it seems that in a forced-convection system, a region corresponding to FD in Fig. 2 for pool boiling may or may not be attainable, depending on the system parameters. Experimental data are too few to allow any further comment on this issue.

From here on, unless otherwise stated, we shall be considering only forced-convection boiling with a heat-flux-controlled surface.

C. Detection of the Burn-Out Point

A brief consideration of the methods used to determine the burn-out point and the various effects seen when using these methods will be helpful in indicating further exactly what is meant in practice by burn-out, and also in showing some difficulties of interpretation that exist.

The most direct way of detecting burn-out is to monitor by thermocouple the channel-wall temperature in the area where burn-out is likely to occur. With uniformly heated channels, burn-out usually occurs at the channel outlet, although the precise location to within about $\frac{1}{2}$ in. of the channel outlet or with respect to the heated perimeter is unpredictable. Consequently, since burn-out can sometimes be very localized, a number of thermocouples are needed. Stevens *et al.* (S13), for example, using uniformly heated round tubes, employed two sets of three-wire thermocouples spot-welded to the outer tube wall 180° apart, with the three wires in each set (chromel–alumel–chromel) spaced axially at $\frac{1}{4}$-in. intervals starting at a point $\frac{1}{4}$ in. from the outlet end. The boiling curves in Fig. 3 were obtained by Stevens *et al.* using this arrangement, and it has been found very reliable with Freons under all test conditions. The same is not generally true of water, however, particularly with subcooled boiling, for if localized burn-out were to occur midway between the two sets of thermocouples, the latter may not sense even a large temperature rise and a physical burn-out may occur. The essential difference here is that the burn-out heat flux with water tends to be an order of magnitude greater than with Freons, and the local temperature rise in the metal wall does not have time to spread before damage takes place. Further difficulties arise in the case of nonuniformly heated channels, for, depending on the heat-flux profile and other system parameters, burn-out may occur an appreciable distance before the channel outlet, and a multiplicity of thermocouples is necessary to give adequate coverage. Uncertainty with regard to burn-out location is also present with heated rod-bundle arrangements, since the rod to show first signs of burn-out is generally unpredictable and thermocouples must be placed on all the rods.

Disadvantages of the thermocouple method are overcome by the use of bridge-type burn-out detectors. In this device, a Wheatstone-bridge circuit is formed by wiring two portions of the test channel, usually of equal electrical resistance, with two variable resistors and putting an out-of-balance detector across the bridge diagonals. If burn-out then occurs on one of the portions of the test channel, the change in electrical resistance due to the local temperature rise can readily be detected as a small, but rapidly increasing out-of-balance signal. The detector is usually set to trip off the power on the test section when the signal attains a predetermined level, thus protecting the test section against possible damage. The trip action must be fast because the temperature rise at burn-out can reach a dangerous level in as little as 10 msec under certain conditions, such as subcooled boiling with water. Salt and Wintle (S2) describe a transistorised bridge-type detector that has been developed for experimental work on burn-out over a period of several years and which has proved extremely reliable. For nonuniformly heated channels, Lee and Obertelli (L1, L5) have used a number of bridge-type detectors, each covering a short length of the channel. By observing which bridge was first to trip off the power, the approximate location of burn-out was obtained. Rod-bundle arrangements present special problems, and methods of detection which have been used have been described by Adnams et al. (A1).

Three distinct types of burn-out signal have been noted by experimenters, and these are typified in the actual instrument traces shown in Fig. 4, which are copied from Lee and Obertelli (L4); they used both thermocouples and a bridge detector on uniformly heated tubes. The traces show the point of initiation of burn-out and the point at which the power has been tripped out, with good agreement between the two methods. Figure 4 introduces the words

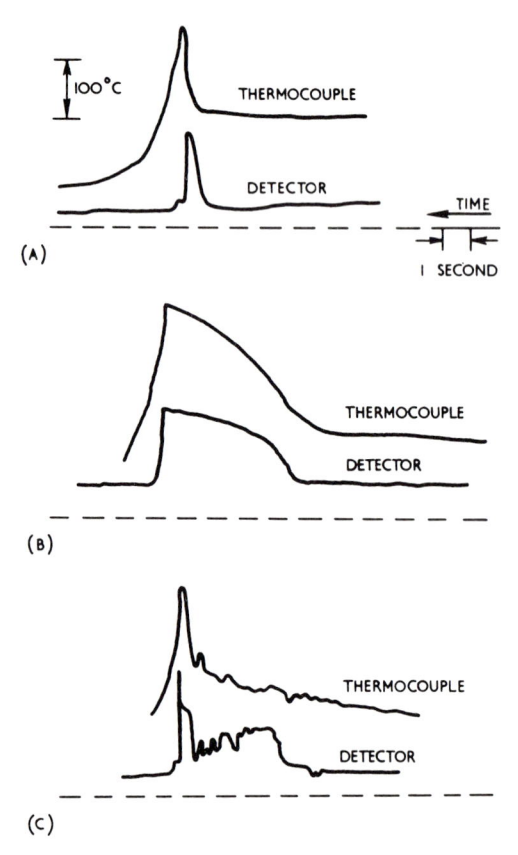

FIG. 4. High-speed recorder traces of burn-out in uniformly heated tubes with flowing water [from Lee and Obertelli (L4)]. (A) Fast burn-out: $L = 17$ in., $d = 0.22$ in., $G = 3.0 \times 10^6$ lb/hr-ft^2, $P = 955$ psia, $\Delta h = 10$ Btu/lb. (B) Slow burn-out: $L = 79$ in., $d = 0.22$ in., $G = 0.75 \times 10^6$ lb/hr-ft^2, $P = 990$ psia, $\Delta h = 151$ Btu/lb. (C) Slow burn-out with temperature oscillations: $L = 34$ in., $d = 0.424$ in., $G = 0.3 \times 10^6$ lb/hr-ft^2, $P = 565$ psia, $\Delta h = 16$ Btu/lb.

"fast" and "slow" burn-out, which are widely used in burn-out terminology, although they have no precise definition.

Fast burn-out, Fig. 4A, occurs when the temperature rise is very rapid, for example, less than one second elapsing between the initiation of burn-out and the time at which the metal temperature becomes dangerously high. Unless the channel power is quickly interrupted, a fast burn-out will usually result in physical burn-out. Lee and Obertelli (L4) report having examined a large number of instrument traces to see whether fast burn-out could be associated with any particular ranges of flow velocity, pressure, or quality at the burn-out point, but no generalization could be made. However, it does appear that in the case of water, fast burn-out is nearly always associated with subcooled or low-quality conditions at burn-out.

Slow burn-out (Figs. 4B and 4C) corresponds loosely with a temperature excursion taking several seconds. There are two types of slow burn-out: in the first, the temperature rise is smooth (Fig. 4B), and in the second, the temperature rise is accompanied by temperature oscillations (Fig. 4C). There is some indication that these temperature oscillations may depend on the wall thickness of the heated test channel, the possibility being that whereas oscillations may be inherent in the slow burn-out process, due perhaps to the vapor-blanketed area alternately contracting and enlarging, a thick wall has sufficient thermal inertia to damp out the effect and indicate smooth signals from a thermocouple or a Wheatstone bridge. Thus, Fig. 4B, for example, refers to a tube having a wall thickness of 0.082 in., whereas Fig. 4C is for a tube of 0.038-in. wall thickness. However, the cause of the oscillations is not certain, and some systematic experiments are needed to elucidate the point.

At the CISE Laboratories in Milan, where the phenomenon of fast and slow burn-out was first noted, the onset of random temperature oscillations has in itself been assumed to signify burn-out, the implication being that temperature oscillations always occur [Bertoletti *et al.* (B19) and Alessandrini *et al.* (A5)]. However, the CISE experiments have in the past been carried out with preformed mixtures of steam and water at entry to a heated test channel, and it is possible that this feature, which is known to produce flow disturbances (see Section III), may be the reason for the fact that temperature oscillations always occur.

Slow burn-out tends to be associated with high-quality burn-out conditions and to produce a not unduly excessive wall-temperature rise. In fact, there appears to be an extreme condition in which the temperature rise may hardly be noticeable, and it becomes difficult to say whether burn-out has occurred. These circumstances probably coincide with the jump discontinuity in Fig. 3 ceasing to exist for certain values of system parameters. The condition is effectively one in which, at the burn-out point, the heat-transfer coefficient is the same whether the surface is vapor-blanketed or liquid-wetted.

The customary experimental approach to obtaining a burn-out measurement for a given test-section geometry and test fluid is to fix the system pressure, mass velocity, and inlet temperature. A heat flux of about 80% of the expected burn-out value is then applied and the flux level raised by small increments until burn-out is signalled. Becker (B9), however, prefers to adopt a different approach, in which the system pressure, inlet temperature, and the heat flux are fixed, with the mass velocity set at a value high enough to prevent burn-out. The mass velocity is then slowly reduced until burn-out is signalled. Thus, Becker measures what perhaps more correctly may be called the burn-out mass-velocity. However, experiments first carried out by McEwen et al. (M5) and subsequently by Becker (B9) confirm that both the above methods will produce practically an identical set of values for the system variables at burn-out, and, therefore, either method can be assumed to give the burn-out flux. Yet another approach, described by Aladyev et al. (A3), is to fix the system pressure, mass velocity, and the heat flux, and then to slowly raise the liquid temperature at the heated channel inlet until burn-out is signaled. Aladyev et al. (A3) also refer to a very direct way of detecting burn-out by visually observing when a part of the test channel turns red in color. Under certain circumstances, the red spot was observed to be of no more than 0.1-in. diameter, which confirms how very localized burn-out can sometimes be.

D. Burn-Out in the Annular Flow Regime

As mentioned in the introduction, an important step forward has recently been made by a very convincing demonstration of a burn-out mechanism applicable to the annular flow regime. The demonstration has been made by measuring the flow rate of liquid on the wall of a heated tube under annular flow conditions with various heat flux levels almost up to burn-out. The first experiments were made by Hewitt et al. (H4) using a vertical heated rod on the outside of the bottom end of which a water film was formed, with steam being passed up the annular space between the rod and a glass shroud tube. This arrangement also permitted visual observation. Staniforth et al. (S8) repeated the experiments for Freon-12 flowing inside a heated tube, but using subcooled liquid at inlet (i.e., liquid below the saturation temperature) so that annular flow was produced higher up the tube in a more natural way. Subsequently, Hewitt et al. (H5) also used subcooled water at the inlet to a heated tube and confirmed their initial findings.

Figure 5 illustrates the experimental technique developed by Hewitt et al. for measuring the flow rate of liquid in the wall film as applied to flow inside a heated tube. As can be seen, a short length of porous sinter tube is positioned a few diameters beyond the outlet of the heated tube. The internal diameter of the sinter and the heated tube are made identical so as to avoid any flow

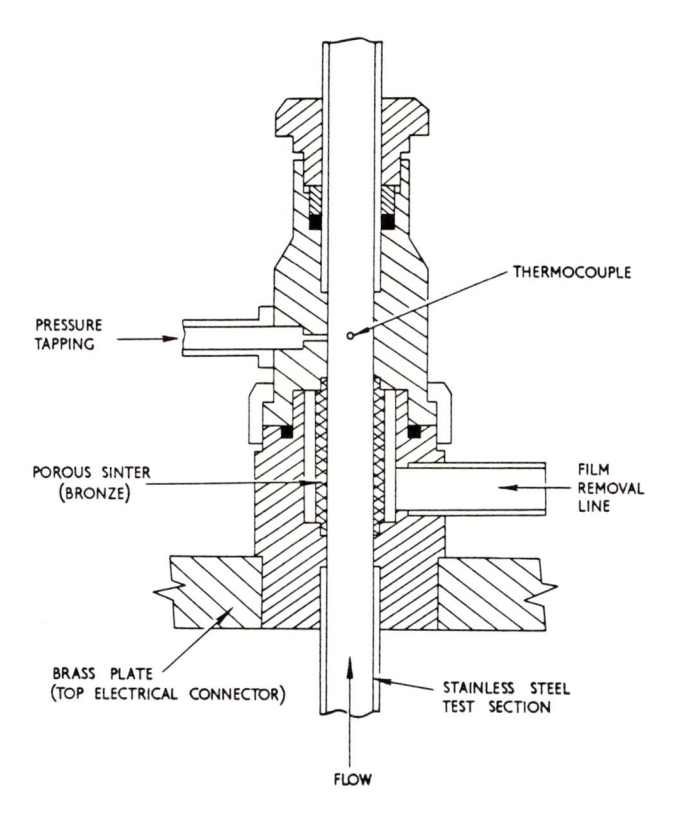

FIG. 5. Device for measuring the wall-film flow rate with annular flow in a tube [from Hewitt et al. (H5)].

disturbance. An annular chamber surrounds the sinter and communicates with a suction line leading to a condenser and extraction system. Figure 6 is a typical result given by Staniforth and Stevens (S7) showing the flow rate through the sinter as a function of the pressure difference across the sinter. Figure 6 shows the total flow rate of liquid plus vapor drawn through the sinter and also the flow rate of liquid alone as determined by a heat balance in the extraction system condenser. A characteristic of these plots in experiments where the exit quality was high (as in Fig. 6) is the plateau in the liquid flow rate curve. Now, in the earlier work of Hewitt et al. (H4), where visual studies were made at the same time as film flow measurements, it was seen that the appearance of the plateau coincided with the condition in which all the liquid flowing along the tube wall was drawn through the sinter. Thus, plots such as Fig. 6 provide a direct measurement of the liquid flow rate along the wall of the heated tube at the exit. Apparently the plateau is not so

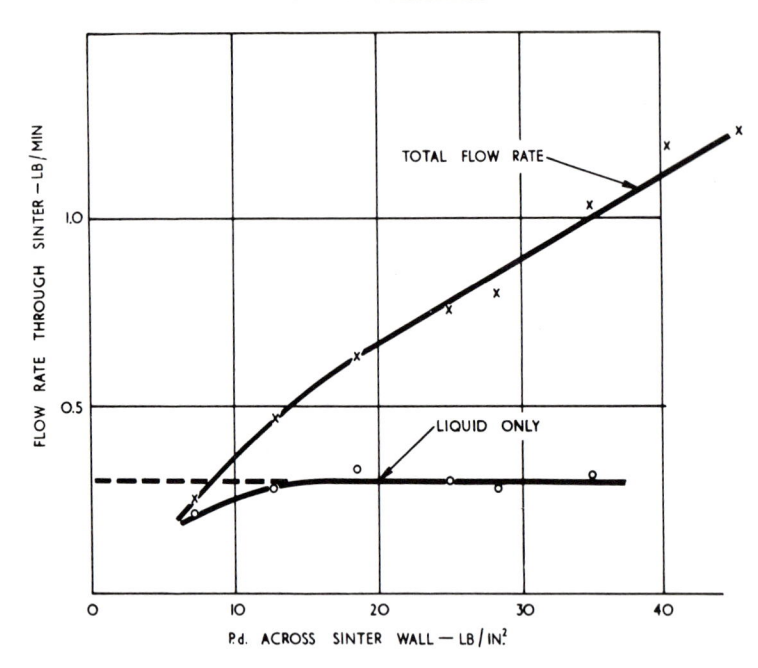

FIG. 6. Variation of flow rate through sinter element with pressure drop (P.d.) [from Staniforth and Stevens (S7)]. Based on tests with Freon 12 in a uniformly heated tube. $L = 102$ in., $d = 0.38$ in., $P = 155$ psia, $G = 0.162 \times 10^6$ lb/hr-ft^2, $\Delta h = 12.5$ Btu/lb, heat flux 8390 Btu/hr-ft^2.

clearly defined at low-exit qualities, since the vapor drawn in from the core now contains large amounts of entrained liquid.

The above experimenters have used the technique described to obtain flow rate measurements of the liquid wall-film at various mass velocities, tube dimensions, etc., and some typical results from Staniforth and Stevens (S7) are shown in Fig. 7. Also shown are the values of burn-out heat flux obtained at the four different mass velocities indicated. It can be seen that the liquid-film flow rate decreases steadily with increasing heat flux until at burn-out the flow rate becomes zero or very close to zero. We thus have confirmation of a burn-out mechanism in the annular flow regime which postulates a liquid film on the heated wall diminishing under the combined effects of evaporation, entrainment, and deposition until at burn-out, the film has become so thin that it breaks up into rivulets which cause dry spots and consequent overheating.

The annular flow regime is very extensive, and the above mechanism of burn-out is stated (S7) to be consistent with the film-flow measurement data over a range of exit qualities from 10 to 100% for uniformly heated round tubes. A summary of experimental observations on flow patterns produced

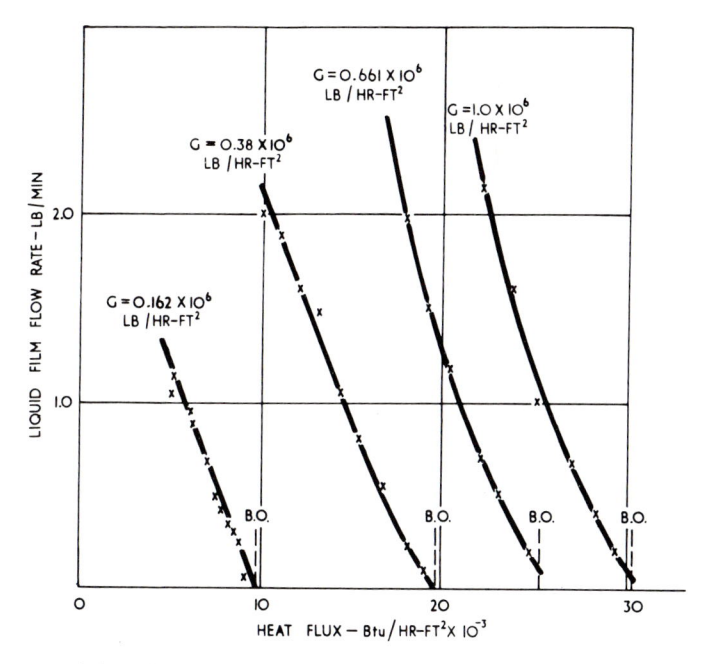

Fig. 7. Variation of wall-film flow rate with heat flux and mass velocity [from Staniforth and Stevens (S7)]. Based on tests with Freon 12 in a uniformly heated tube, $L = 102$ in., $d = 0.38$ in., $P = 155$ psia, $\Delta h = 12$ Btu/lb.

in two-phase flow is given by Bennett *et al.* (B15) including recent work on the application of X-ray photography to flow inside pipes. Flow-pattern diagrams that have been produced for steam and water mixtures suggest that annular flow exists over practically the whole quality range with mass velocities above 10^6 lb/hr-ft^2. The range tends to diminish with lower mass velocities, and at 10^5 lb/hr-ft^2, it is restricted to roughly 30–100% quality. Pressure appears to have comparatively little effect on the flow-pattern diagrams.

In order to develop the above burn-out mechanism further, it will be necessary to know more about the entrainment and deposition processes occurring. Experimentally, it is likely that these processes will be very difficult to measure separately and under conditions comparable to those prevailing in a boiling channel. From analysis of their film flow-rate data, Staniforth *et al.* (S8) have deduced that under burn-out conditions, the deposition of liquid droplets from the vapor core plays an important part in reinforcing the liquid film, particularly at high mass velocities. At low mass velocities, they conclude that deposition and entrainment rates must be nearly equal, and, therefore, since a thin liquid film can be expected to be tenacious and give rise to very little entrainment, they argue that both deposition and entrainment tend to zero near the burn-out location with low mass velocities.

E. The Connection between Burn-Out and Various Other Flow Patterns

The annular flow pattern discussed above shows a definite connection with burn-out, and enables a simple burn-out mechanism to be set forth. There are many other flow patterns referred to in the literature, however, and we will consider here what can be said about any connection they may have with burn-out. It does not follow that there must be a connection, as a flow pattern is essentially a description of the *bulk* conditions in a channel and depends upon the none-too-reliable results of visual observation, which is often impeded by optical distortion. Thus, although gross conditions may appear to change and one pattern give way to another, the hydrodynamic state prevailing close to the heated surface may remain practically unaffected and the burn-out mechanisms unaltered.

More than a score of adjectives can be found in the literature to describe supposedly different flow patterns, although only five of these appear to have a distinctive character. They are bubble, slug, froth, annular, and mist flow; in the case of a horizontal channel, there is also stratified flow. The subject is adequately covered by Scott in Volume 4 of the present series[1] and by Bennett *et al.* (B15). It is to be noted, however, that flow patterns have mainly been observed and studied in unheated systems using air and water or saturated water and steam at the channel inlet. There is increasing evidence that, while the same general patterns exist in a boiling system, they may have important and often subtle differences. An interesting example is that noted by Kirby (K2) in photographic studies of forced-convection boiling burn-out on a vertical, heated metal strip mounted along one of the walls of a test channel of square cross section. Glass windows on the adjacent walls gave visual access. Kirby confirmed the observation made earlier by Jiji (J6) that once a bubble boundary layer is well established in subcooled boiling (i.e., where the bulk liquid temperature is below the saturation temperature), the number of small bubbles present becomes negligible. Mainly, very large bubbles (or rather, lumps of steam) are to be seen, and these appear to slide over the heated surface in the direction of flow, apparently mopping up steam from any nucleating sites and preventing new bubbles from growing. The important thing Kirby noticed was that on cutting off the power supply to his heated strip, the lumps of steam immediately drifted away from the surface into the mainstream. Clearly this evidence shows that heat transfer introduces some mechanism which pushes steam toward the heated surface, thereby affecting the flow pattern.

The question at issue is of course whether different burn-out mechanisms may operate for different prevailing flow patterns. To the writer's knowledge,

---

[1] Scott, D. S., *Advan. Chem. Eng.* **4**, 200 (1963).

there is no direct evidence on this question, and the position at present is that only one burn-out mechanism, that described above for annular flow, has been identified with reasonable certainty. This is not to say that burn-out occurs only with annular flow; on the contrary, it can occur with any of the flow patterns, but only in the case of annular flow is the associated burn-out mechanism known.

An important piece of indirect evidence is considered to be the fact that no reliable data have ever been found in which the burn-out heat flux for a channel shows any sudden or significantly sharp change in value as a result of varying the system parameters. Figure 17, for example, shows some burn-out results for a uniformly heated round tube, and although the condition at burn-out varies in these results from highly subcooled to fairly high quality, it is apparent that the change in burn-out flux is smooth and continuous.

Many investigators have postulated that two physical burn-out regimes exist: one when net quality conditions prevail, and the other when conditions are subcooled. Separate burn-out correlations are to be found for these two regimes, the division of data having been made by using the heat-balance equation and assuming that thermal equilibrium exists. This latter assumption can be grossly untrue, however, as was well shown by Jiji (J6) in tests with water at pressures from 200 to 1000 psia. He observed lumps of steam (of more than $\frac{1}{4}$-diameter at 1000 psia) sliding along a vertical heated surface, and simultaneously measured temperatures in the main liquid stream as much as 200°F below the corresponding saturation temperature. These experiments show that when the heat-balance equation indicates zero net quality, the actual condition in the mainstream may be highly subcooled, and thus, division into the two regimes mentioned above is entirely artificial. What has in fact most likely happened is that investigators have found better correlation accuracy by dividing burn-out data into two parts, so that they are really introducing regimes of convenience without any evidence that there are physical differences.

F. The Post-Burn-Out Condition Giving Tolerable Wall Temperatures

To complete the broad picture of what is meant by burn-out, it is useful to consider further the particular burn-out condition which produces tolerable wall temperatures. Such a condition must occur, for example, in any practical, once-through boiler system, where a change from, for example, liquid water at inlet to superheated steam at outlet takes place in a single heated channel. Normal operation of nuclear reactors beyond burn-out has also been contemplated, and Collier *et al.* (C5) have described successful experiments with irradiated Zircaloy-clad rods operating continuously beyond burn-out through the use of a high-quality steam and water mixture giving what is called "fog cooling." Sorlie *et al.* (S6) have also given results of experiments carried out in the boiling-water loop of the General Electric Test Reactor in which UO$_2$ fuel rods were operated with boiling water at 1000 psia well beyond burn-out for about 5 min. The maximum fuel-rod surface temperature recorded was 2000°F.

The important information required in the design of the above systems is the channel-wall temperatures attained after burn-out. Unfortunately, very little experimental information of this kind has been published, the most important sources of data being Schmidt (S3), Polomik *et al.* (P3), Bagley (B1), Konkov and Modnikova (K5), and Kearsey (K1), all of which refer to water. Figure 8 shows an example given by Schmidt of wall temperature plotted against enthalpy for water flowing in an electrically heated tube 5-mm in diameter and 2.5-m long. Wall-temperature curves corresponding

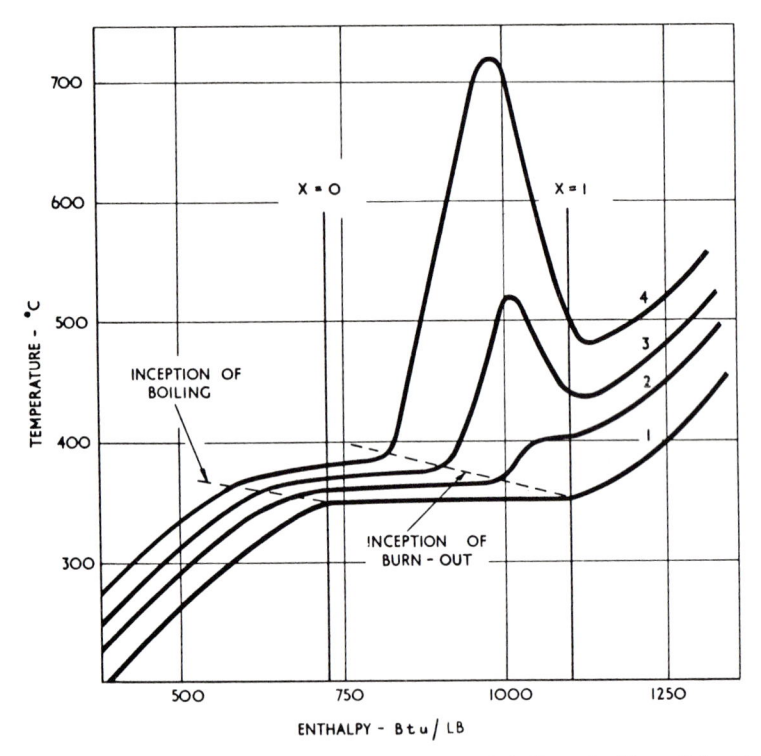

FIG. 8. Temperature distribution in a once-through boiler tube [from Schmidt (S3)]. $L = 98.5$ in., $d = 0.197$ in., $P = 2415$ psia, $\Delta h = 365$ Btu/lb, $G = 0.44 \times 10^6$ lb/hr-ft^2. Curve 1: water temperature. Curve 2: tube-wall temperature with heat flux of 92×10^3 Btu/hr-ft^2. Curve 3: tube-wall temperature with heat flux of 148×10^3 Btu/hr-ft^2. Curve 4: tube-wall temperature with heat flux of 221×10^3 Btu/hr-ft^2.

to three different heat flux levels are shown, the water mass-velocity being 0.44×10^6 lb/hr-ft^2 and the system pressure 2415 psia. On increasing the heat flux, the point of inception of burn-out moves back along the tube, and the inception points can be clearly seen in Fig. 8 for the three heat flux levels shown. From each inception point, the curves rise to produce a peak in wall temperature, the actual tube length involved in this rise being about 13 in. for the conditions indicated in Fig. 8. The question may be asked, why does the wall temperature not jump immediately to a higher value at the inception of burn-out? The answer is that burn-out does not start simultaneously all around the perimeter of a heated channel, as confirmed by Hewitt et al. (H4), who watched the liquid film on the outside of a heated rod, seen through a glass annulus, become very thin at the burn-out point and break up into revulets. Presumably, these thin and narrowing rivulets of liquid may in some way continue along a heated tube until all wetting finally ceases and the maximum wall temperatures as seen in Fig. 8 reached. Clearly, wall thickness will play an important part in determining the shape of the temperature profile between the point of inception of burn-out and the point of maximum temperature.

Wall temperatures drop after reaching the maximum in the case of the two highest heat flux levels in Fig. 8, and this is due to increasing convective heat transfer through the steam film, which now completely blankets the surface. The improved heat transfer is caused by the higher flow velocities in the tube as more entrained liquid is evaporated. Finally, at about 100% quality, based on the assumption of thermal equilibrium, only steam is present, and wall temperatures rise once more due to decreasing heat-transfer coefficients as the steam becomes superheated.

Figure 8 shows that increasing the heat flux at constant mass velocity causes the peak in wall temperature to increase and to move towards lower enthalpy or steam quality values. The increase in peak temperature is thus due not only to a higher heat flux, which demands a higher temperature difference across the vapor film at the wall, but to a lower flow velocity in the tube as the peaks move into regions of reduced quality. The latter effect of lower flow velocity is probably the dominant factor in giving "fast" burn-out its characteristically rapid and often destructive temperature rise, for, as stated earlier, fast burn-out is usually observed at conditions of subcooled or low quality boiling.

III. The Important System-Describing Parameters

The first logical step in analyzing any physical process is to try to ascertain what system parameters are important. The burn-out process involves a large number of important system parameters, and, while some of these can be named almost intuitively (such as heated channel length, channel cross-section geometry, system pressure, mass velocity, and inlet temperature, or enthalpy and the heat flux profile, all of which have been confirmed by experiment to be important), others are not so obvious.

A. Distinction between a Hard and a Soft Inlet

In burn-out experiments, a test section is part of a loop which may be open or closed, and the question arises as to whether or not any of the loop equipment, such as condensers, heaters, pumps, or pipe fittings, have any significant effect on the burn-out flux. This issue came to prominence at the Boulder Heat Transfer Conference in 1961 with a Russian paper by Aladyev (A4) describing burn-out experiments in which a branch pipe, connecting to a small vessel, was fitted close to the test section inlet. The test section itself was a uniformly heated tube 8 mm in diameter and 16 cm long. The results are reproduced in Fig. 9, and show burn-out flux plotted against exit steam quality. Curve (A) was obtained with the branch vessel filled with cold water,

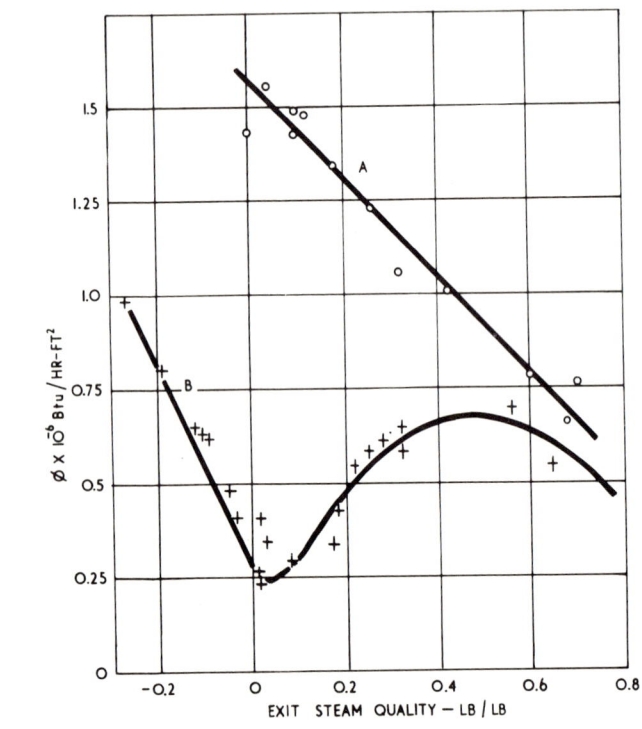

FIG. 9. Burn-out flux for water in a uniformly heated tube with (A) stable flow and (B) unstable flow produced by "soft" inlet conditions [from Aladyev (A4)]. $L = 6.28$ in., $d = 0.314$ in., $P = 1420$ psia, $G = 0.3 \times 10^6$ lb/hr-ft^2.

and curve (B) with the vessel containing a compressible medium such as steam or nitrogen. It was observed that curve (B) was associated with strong pressure and flow pulsations of about ½-sec period occurring in the test section, and resulted in greatly reduced burn-out fluxes, depending on the exit quality.

The Russian paper attracted much attention, although the phenomenon described by Aladyev had in fact been reported in the United States much earlier by Lowdermilk *et al.* (L10), who also showed that the larger the compressible volume at the test section inlet, the greater were the flow fluctuations and the lower the burn-out heat flux. The open test loop used by Lowdermilk had an unusual means of pumping water; it consisted of a rubber bag inside a gas cylinder. The bag was filled with water, which was compressed and delivered to the test section at the required rate by applying gas pressure to the outside of the bag. Even without the compressible volume at the branch connection near the test-section inlet, it was found that severe flow oscillations occurred in the test section, and these must be attributed to the elastic nature of the pumping device. Lowdermilk investigated the effect of throttling by a control valve near the test-section inlet and observed that throttling reduced the degree of oscillation and finally eliminated it, as shown by the changing value of the burn-out flux in Fig. 10. It would appear that once the resistance across the inlet control valve exceeds a certain amount, the test section and the pumping device are effectively decoupled and oscillations cease. Very similar results have been reported by Becker *et al.* (B14), who, using a closed loop and natural circulation, observed that closing the test-section inlet valve by a certain amount achieved burn-out results identical to those for forced convection with steady flow. Otherwise, the natural circulation loop gave lower burn-out heat fluxes.

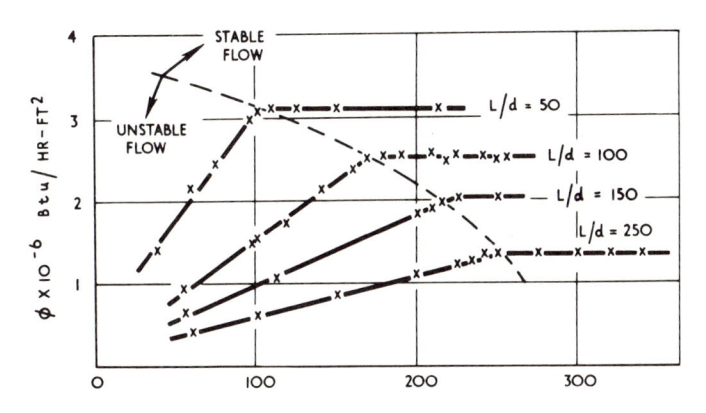

FIG. 10. Effect of an inlet flow restriction on burn-out flux in uniformly heated tubes [from Lowdermilk *et al.* (L10)]. $d = 0.076$ in., $G = 1.78 \times 10^6$ lb/hr-ft².

Some interesting results on the effect of inlet throttling have been given by Styrikovitch *et al.* (S17) using an open test loop as shown in Fig. 11. Steam and water at high pressure were delivered to the loop by way of a mixer and passed through a cooler/condenser and then through about 30 m of coiled pipe before entering the test section, which comprised a uniformly heated tube 16 cm long and 0.8 cm in diameter. The test-section pressure was maintained at 100 atm by throttling on one of the four control valves shown.

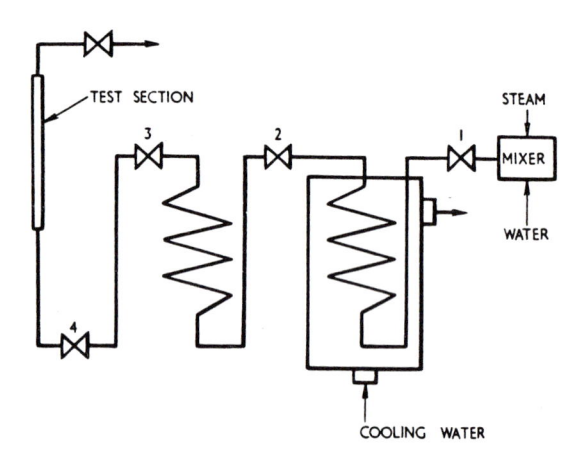

FIG. 11. Test loop for investigating the effect of inlet throttling on burn-out [from Styrikovitch *et al.* (S17)].

The mass velocity to the test section was kept constant at 0.295×10^6 lb/hr-ft^2, and the inlet condition was varied from subcooled water to quality conditions. Figure 12 shows the burn-out flux plotted against the inlet condition with throttling on each of the four control valves in turn. Only the curves are reproduced in Fig. 12, but the actual data points deviate very little from these curves. It can be seen that throttling with valves 2, 3, or 4 has little effect on burn-out with subcooled water inlet (shown as a negative quality in Fig. 12). For valve 1 however, with subcooled inlet, the burn-out flux is much reduced, and this is because of the presence of steam in the cooler/condenser between valve 1 and the test-section inlet, causing flow oscillations to develop. With quality inlet conditions to the test section, steam is always present in the inlet supply line, and, consequently, the burn-out flux is affected regardless of the valve used for throttling. Using valve 4, however, the compressible volume of steam in the supply line is not large, and, consequently, as can be seen in Fig. 12, the sharp drop in burn-out flux near zero inlet quality, which is caused by unstable flow, is almost eliminated. Other Russian work on the influence of inlet conditions to a test channel is described by Miropolski *et al.* (M13) and Doroshtchuk and Frid (D4).

Many experimenters have adopted the practice of feeding a preformed mixture of steam and water to their test sections, either out of interest in this type of system or else to avoid the power demanded by long channels. The CISE Laboratories in Italy have produced a considerable amount of data of this kind (S4), and a typical example of their results is shown in Fig. 13. The curves have a characteristic "swan-neck" shape similar to the Russian data for unstable flow conditions shown in Fig. 9, and the burn-out flux values are generally below those for normal steady-flow conditions.

The conclusion to be drawn from the above examples and many others is that "softness" in a boiling system, preceding the boiling channel inlet, may cause flow oscillations of low frequency. It is probably the pressure perturbations arising from the explosive nature of nucleate boiling that initiates the oscillation, and the reduced burn-out flux which follows probably corresponds to the trough of the flow oscillation, as a reduction in flow rate always drops the burn-out flux in forced-convection boiling.

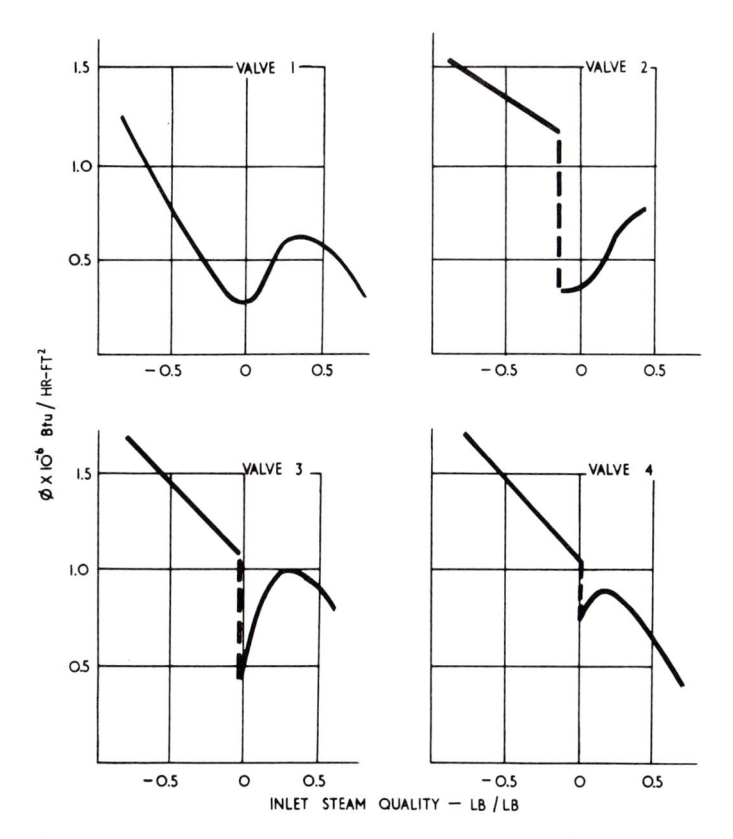

FIG. 12. Burn-out flux variation resulting from using the different throttling valves shown in Fig. 11 [from Styrikovitch (S17)].

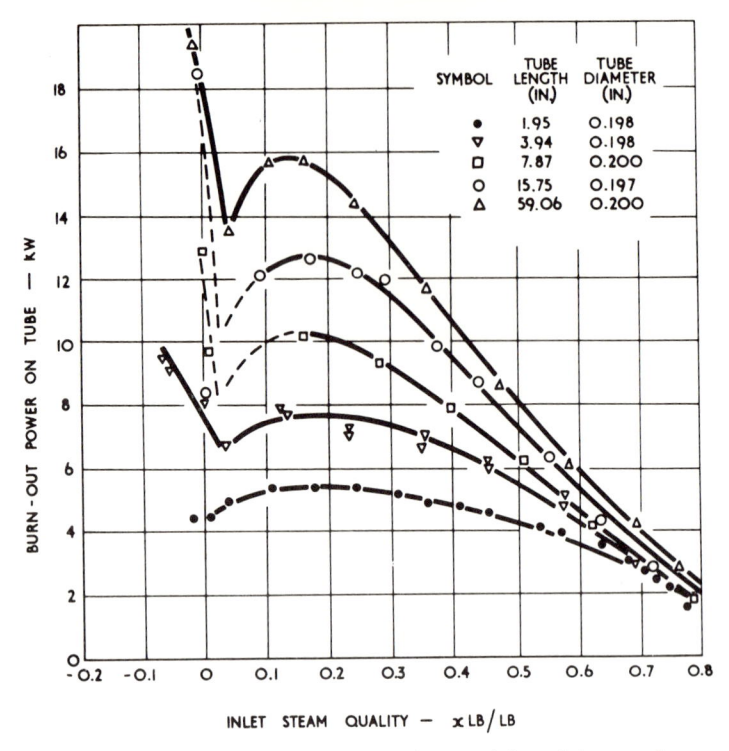

FIG. 13. Typical result of using a steam and water inlet mixture on burn-out in a uniformly heated tube [from Silvestri (S4)]. $P = 1018$ psia, $G = 0.8 \times 10^6$ lb/hr-ft^2.

The simple rule for ensuring stability is to have a "hard" inlet condition, which means designing pipework so as to avoid stagnant pockets that may collect gas or vapor, using a feed of liquid phase only with sufficient subcooling to avoid flashing in the feed lines, operating centrifugal pumps well away from any flat part of their characteristic curve, and positioning the pumps as close as possible to the boiling-channel inlet. If oscillations do occur, they can be stopped by a restriction in the supply line; in the case of boiling channels in parallel such restrictions are probably essential, although there is very little experimental evidence on the behavior of parallel channels. Casterline and Lee (C2) describe some results obtained at Columbia University using three ½-in. i.d. tubes in parallel, each having a 76-in. heated length, operating at water pressures of 500, 1000, and 1500 psia. It was observed that there were operating conditions which could cause pronounced hydraulic interaction between the heated channels. In many of these cases, the flow disturbances were of such magnitude that the flow to the most affected channel fell to zero at the minimum point of the oscillation. The tendency toward flow instability was greatest at low pressures, low mass velocities, and with flow restrictions in the exit piping. These flow instabilities can be expected to reduce the burn-out flux, and, in fact, significant reductions were observed.

The presence of softness downstream from the exit of a boiling channel does not generally affect burn-out; this can be inferred from the consistency of experimental results in which hard inlet conditions prevail, irrespective of the wide variety of downstream equipment such as steam separators, condensers and pressurizers, etc., used by different experimenters. The only trouble that may arise may come from having too severe a restriction in the outlet line, which may set up flow oscillations, as first noted by Lowdermilk and Weiland (L9). The same thing has been observed by Lee (L1) in burn-out experiments on long tubes, and was caused by a length of small-diameter pipe being used as an expansion bend at the boiling-channel outlet. Becker *et al.* (B14) has also reported that in his natural circulation loop, instability was worsened by throttling at the boiling-channel outlet. Lowdermilk and Weiland (L9) found, however, that a simple way of stopping the oscillations due to an outlet restriction was by fitting a branch line, connected to a compressible volume, between the restriction and the exit from the boiling channel.

From the point of view of systematic data analysis, it has been found that consistent results can be obtained only with burn-out data produced under stable conditions. The unstable condition leads to considerable scatter, depending on the particular setting of a valve, the size of a compressible volume, the method of preforming a steam and water mixture, and so on. These latter quantities have either been recorded with very low accuracy, or have not been recorded at all. Therefore, the unstable-conditions data appear to be of little value, except for qualitative purposes. In any case, one is usually not interested in instabilities apart from knowing how to avoid them, which is by having a hard inlet.

B. THE INFLUENCE OF GRAVITY ON BURN-OUT

Gravity can have a significant influence on pool-boiling burn-out, as shown by the centrifuge experiments of Costello and Adams (C8) and Ivey (I3). In a forced-convection system, however, where a gravity effect may influence burn-out as a consequence of the flow direction, or by the inclination of a channel, there is evidence that the effect can be negligible, although a really thorough examination has not been made. Zenkevich and Subbotin (Z2) used both upflow and downflow of water in a vertical channel with subcooled boiling and found no effect on the burn-out flux for mass velocities above 0.2×10^6 lb/hr-ft^2. Jacket *et al.* (J1) carried out tests on rectangular heated channels and round tubes with water at 2000 psia and reported no effect on burn-out as a result of inclining the test section by 45°. Barnett (B3) has made an appraisal of what data there is, and a typical example of one of his comparisons is shown in Fig. 14, which is based on data produced by Lee and Obertelli (L4). Barnett's overall conclusion was that the existing data showed no indication of any gravity effect on burn-out with forced convection.

The above conclusion must certainly be taken with a measure of reserve as regards the mass velocity, for at very low velocities it appears reasonable to expect that the relative motion between vapor and liquid in a boiling channel will be affected sufficiently to influence the burn-out flux. Barnett's conclusion also applies to simple channels, whereas Fig. 35 discussed in Section VIII,C shows that a rod-bundle system placed in a horizontal position is likely to incur a reduction in the burn-out flux at mass velocities less than 0.5×10^6 lb/hr-ft^2, presumably on account of flow stratification. Furthermore, gravitational effects induced in a boiling channel by such means as swirlers placed inside a round tube can certainly increase the burn-out flux as shown by Bundy et al. (B23), Howard (H10), and Moeck et al. (M15).

C. The Influence of Surface Conditions and Liquid Purity

Forced convection burn-out appears to be generally unaffected by surface conditions and liquid purity. The most convincing demonstration of this

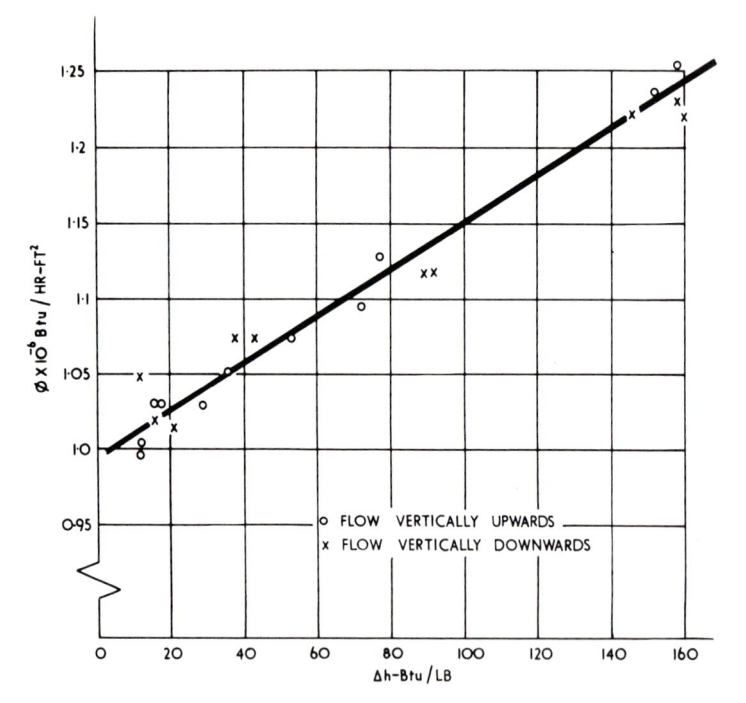

FIG. 14. Example showing that gravity has no significant effect on burn-out in a uniformly heated tube using water [from Barnett (B3)]. $L = 34$ in., $d = 0.424$ in., $P = 580$ psia, $G = 0.74 \times 10^6$ lb/hr-ft^2.

comes from the consistency that has been found to exist in all burn-out data for stable flow conditions. These data refer to a wide range of different surface materials with different finishes, degrees of aging, etc. Also, fluids in various states of purity have been used, including fluids saturated with different gases, but this factor appears to have no significant effect on burn-out. The consistency of the data, which supports the above argument, will be made evident later in Section VIII on data correlation.

A number of specific tests have been reported. Aladyev *et al.* (A4), who tested eleven tubes with different surfaces, viz., (a) polished tubes, (b) tubes with natural roughness, (c) tubes pickled with hydrochloric acid, (d) tubes with deep longitudional scores, and (e) tubes with many cross scores 0.12 to 0.15 mm deep. These experiments showed that the surface condition had practically no effect on the burn-out flux. De Bortoli *et al.* (D2), from tests at 2000 psia with water in rectangular channels concluded that the material or surface roughness up to 120 μin. did not affect the burn-out flux. The materials investigated were nickel, zirconium-2, and stainless steel type 304. Jens and Lottes (J5) and De Bortoli and Masnovi (D1) found no effect on burn-out of dissolved gas in water flowing in heated channels in the pressure range 500–2000 psia.

An exception to the above evidence is that of Durant and Mirshak (D5), who state that tests on roughened tubes showed burn-out heat flux values 50 to 100% above those for smooth tubes. The tests were done at about 50 psia using electrically heated tubes of $\frac{1}{2}$-in. o.d. and 24-in. length with water flowing through the annulus formed by a glass housing and the outer tube surface. Roughening was produced by cutting threads up to 0.009 in. deep and by a diamond knurl up to 0.013 in. deep. The results of Durant and Mirshak are misleading, however, as examination shows that the roughened tubes have been compared with smooth tube performance under significantly different operating conditions. In particular, the inlet water temperatures were generally lower in the case of the roughened tubes, which always produces higher burn-out flux values with stable flow. Also, the annular spacing between the roughened tubes and the circular glass housing was up to twice that used in the control experiments on a smooth tube, a factor which again is known to increase burn-out flux. Furthermore, the roughened-tube burn-out flux results quoted as giving 50 to 100% higher values were based on the smooth-tube surface area, whereas the actual area was as much as 67% higher. Surface increases of this magnitude hardly belong to the category of surface roughening as it is normally understood.

D. The Influence of Test-Section Wall Thickness

The wall thickness of a heated test section depends on the resistivity of the material used and on the electrical supply available. Since these factors vary considerably, the experimental data inevitably refer to a fairly wide range of wall thicknesses, and the question arises as to whether or not burn-out is significantly affected. We can invoke the same argument used previously with regards to surface conditions and liquid purity to conclude that any effect must be small, at least within the range of the data, since otherwise the accuracy achieved in correlating data from different sources would not be possible. This same argument must also apply to the question of whether differences exist between test sections heated by ac or dc, with the inference again being that there is no appreciable difference.

Aladyev et al. (A4) refer to specific tests using two tubes with wall thickness of 0.016 and 0.079 in., and they report no noticeable effect on the burn-out flux, but details of the tests are not given. Lee (L1) examined the question of wall thickness using two uniformly heated tube lengths with water at 1000 psia, and his results for a mass velocity of 1.5×10^6 lb/hr-ft² are shown in Fig. 15. It can be seen that with the 68-in. tube there is no difference between a wall thickness of 0.034 in. and a thickness of 0.082 in. With the 34-in. tube, however, the thicker wall gives about 7% higher burn-out flux values at low

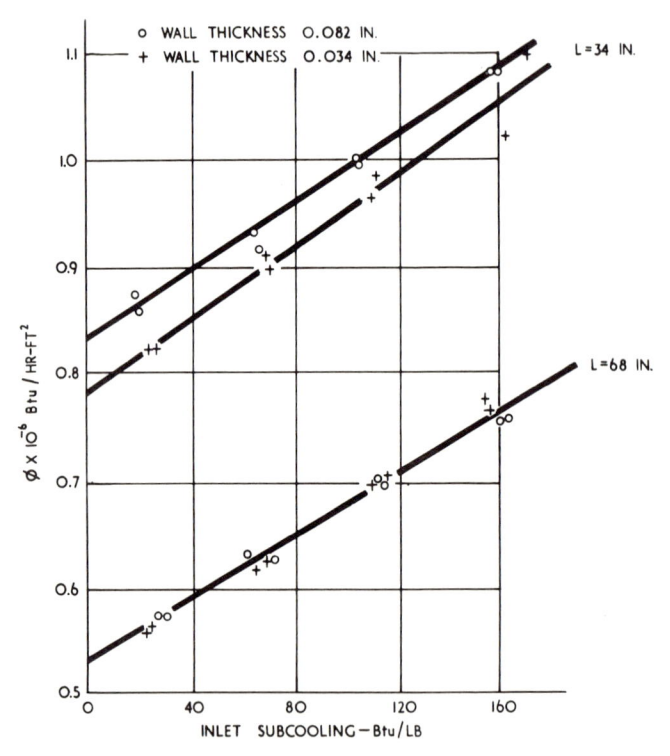

FIG. 15. Effect of wall thickness on burn-out in a uniformly heated tube using water [from Lee (L1)]. $d = 0.364$ in., $P = 1000$ psia, $G = 1.5 \times 10^6$ lb/hr-ft².

inlet subcooling. Lee repeated the tests shown in Fig. 15 at a higher mass velocity of 3×10^6 lb/hr-ft^2 and observed exactly the same wall-thickness effect. The maximum discrepancy of 7% is not inconsistent with the earlier argument about data correlation, as the corresponding mean variation is $\pm 3\frac{1}{2}\%$, and this is well within any reasonably acceptable limits of a data correlation.

Confirmation that a thicker tube wall may slightly increase the burn-out flux has been given by Hewitt et al. (H6). These authors have even gone to the extent of showing that not only the burn-out flux, but the wall-film flow rate prior to burn-out as well, is affected by the wall thickness, as shown in Fig. 16. The technique of measuring the film flow-rate has been described in Section II,D. The reason a thicker-walled tube should maintain a higher wall-film flow rate is not clearly understood.

IV. The Linear Relation between Burn-Out Flux and Inlet Subcooling

With the important system-describing parameters identified, the next step is to find out if any of them are related in a simple way to the burn-out flux. A simple relationship, provided it is universally valid, helps considerably in

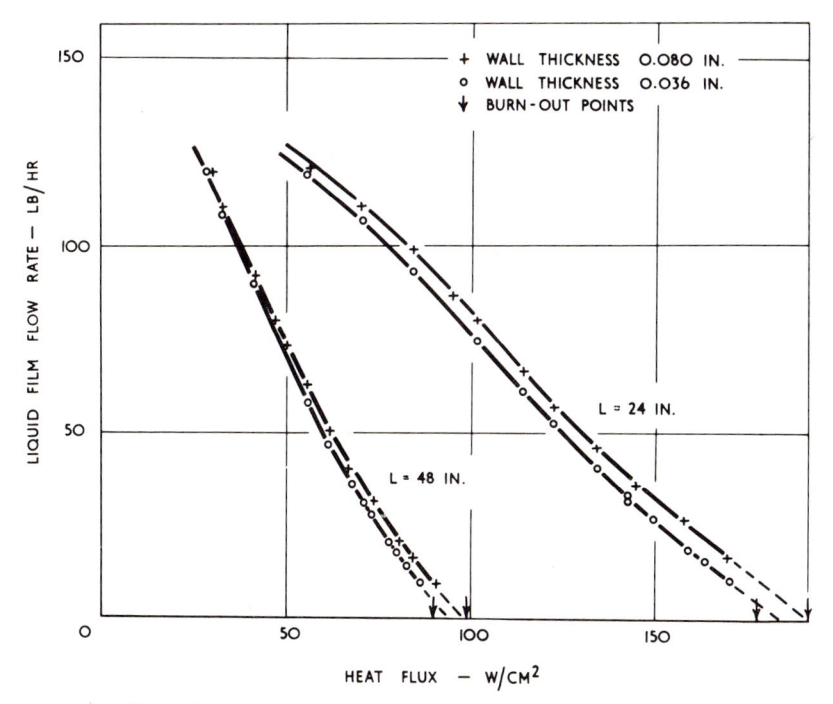

FIG. 16. Effect of tube-wall thickness on the film flow rate [from Hewitt et al. (H6)]. Based on tests with water in uniformly heated stainless steel tubes. $d = 0.366$ in., $P = 55$ psia, $G = 0.219 \times 10^6$ lb/hr-ft^2.

formulating an empirical equation to represent experimental data. Examination has shown that only the inlet subcooling comes close to satisfying the requirements, the feature observed being a linear relationship illustrated by the example in Fig. 17, which is based on Columbia University data (M8). This linear relationship was recognized in one of the first burn-out correlations produced for a forced-convection system in 1949 by McAdams et al. (M1). It has also been shown to exist in pool-boiling systems and Fig. 18 is an early example given by Farber (F1) in 1951. Boiling was at atmospheric pressure and the degrees of subcooling shown in Fig. 18 refer to the amount by which the bulk water temperature was below saturation.

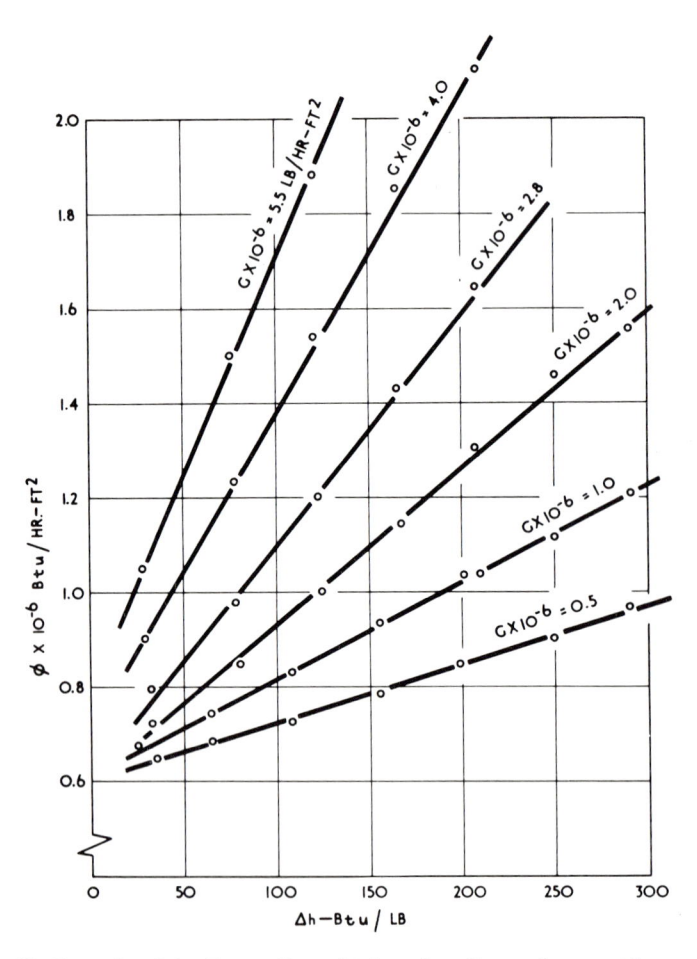

FIG. 17. Example of the linear effect of inlet subcooling on burn-out in a uniformly heated tube using water [from Matzner (M8)]. $L = 77$ in., $d = 1.475$ in., $P = 1000$ psia.

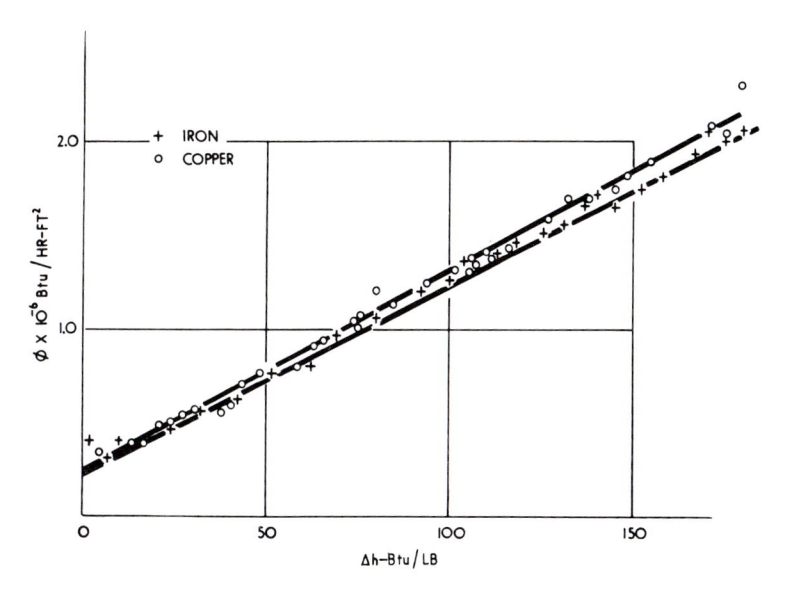

FIG. 18. Example of the linear effect of bulk fluid subcooling on burn-out for copper and iron wires with pool boiling of water at atmospheric pressure [from Farber (F1)].

The linear relationship is remarkably precise over a wide range of conditions including all types of test-section geometries, irrespective of whether subcooled or bulk boiling occurs. Slight deviations from linearity do exist, however, and an examination of these cases show that they are always associated with burn-out qualities in the region of 0 to 10%. The deviation is such that a best-fit straight line through experimental data rarely deviates from a best-fit experimental curve by more than 5%. Figure 19, based on data from Lee and Obertelli (L6), illustrates the character of the deviation by an upward curvature in one of the two graphs shown. The heat-balance equation indicates that burn-out qualities along the affected graph, which is for a mass velocity of 3×10^6 lb/hr-ft^2, are almost entirely within the range 0–10%. On the other hand, for the graph with the lower mass velocity (1.5×10^6 lb/hr-ft^2), the burn-out qualities are all greater than 10% and the graph is linear. It would appear that the deviation is somehow connected with the bubbly or slug-flow type of flow pattern in a boiling channel, as these patterns are confined roughly within the quality range 0–10%.

An adequate explanation of the linearity has not yet been found, although it would seem to have something simple yet basic to say about the burn-out mechanism. It will be seen in the following sections that it enables a very effective burn-out correlation to be derived. It is also considered that a plot of burn-out flux against inlet subcooling, as in Figs. 17 and 19, is the best

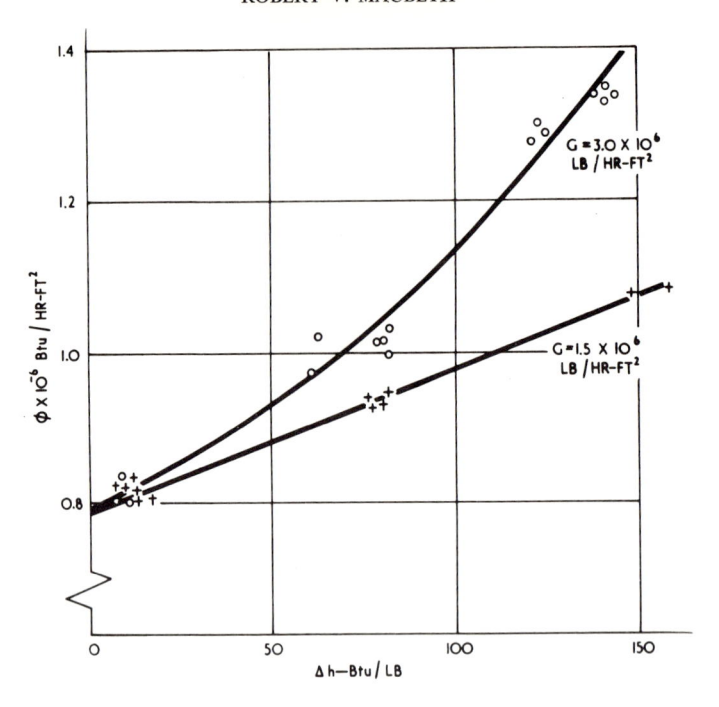

FIG. 19. Example of deviation from linearity between burn-out heat flux and inlet subcooling [from Lee and Obertelli (L6)]. Based on tests with water in a uniformly heated tube, $L = 34$ in., $d = 0.426$ in., $P = 1000$ psia.

way of graphically presenting burn-out data. These plots can reduce experimental work to a minimum by taking advantage of the linear characteristic where applicable, and they can be extremely useful in revealing faulty experimental work and in showing when soft inlet conditions exist, as illustrated by the example in Fig. 20. This example is taken from data of Bertoletti et al. (B19) for a uniformly heated tube which was supplied with water by a reciprocating pump. Although the experimenters used pulsation dampers on both the suction and delivery sides of their pump, Fig. 20 shows all the symptoms of a soft inlet condition, including a maximum in one of the experimental curves, and it seems certain that, probably unknown to the experimenters, severe flow oscillations were being set up in the boiling channel. It is not advisable to use a reciprocating pump in burn-out work.

V. Possible Contractions among the System-Describing Parameters

In addition to looking for simple relationships that may exist between the burn-out flux and the system-describing parameters, the possibility also exists of trying to find a simplification in the form of a contraction in the number

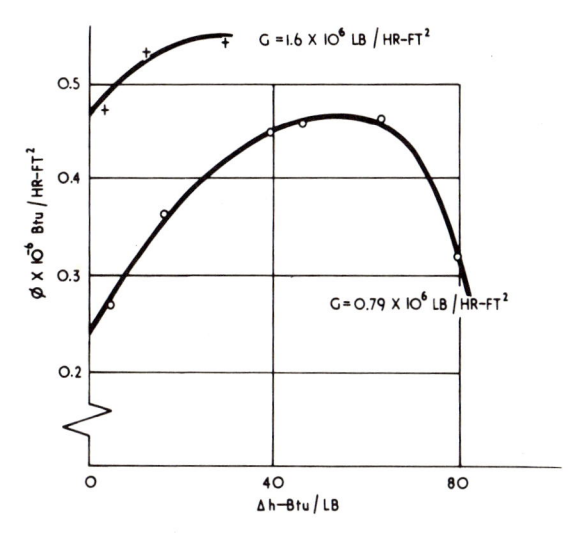

FIG. 20. Typical symptom of a "soft" inlet condition [from Bertoletti et al. (B19)]. Based on tests with water in a uniformly heated tube. $L = 43.3$ in., $d = 0.199$ in., $P = 1000$ psia.

of parameters. For example, in a uniformly heated round tube with a hard inlet, the burn-out flux ϕ can be represented by the equation

$$\phi = f(P, d, G, L, \Delta h) \qquad (1)$$

which is merely a statement of the fact that for a given fluid, burn-out flux is a function of five important system-describing parameters, pressure P, tube internal diameter d, mass velocity G, heated channel length L, and the degree of subcooling at the channel inlet, Δh. Now, a simplifying assumption often found in the literature is that the heated length and diameter of a tube affect burn-out only through their ratio, L/d. This hypothesis allows a contraction of two of the system parameters, and with it Eq. (1) reduces to

$$\phi = f(P, L/d, G, \Delta h) \qquad (2)$$

Hence, for a given P, G and Δh, the hypothesis will be correct if two tubes of different length, but with the same L/d ratio, give the same burn-out flux. Barnett (B3) has tested Eq. (2), and Fig. 21 gives one of his test results, showing that the hypothesis is not generally valid. Thus, any burn-out theory or correlation which combines L and d only as a ratio cannot generally be correct.

The above method of testing an hypothesis has proved a very effective tool in burn-out analysis, and has been a great help in appraising the many burn-out theories and correlations that have appeared in the literature.

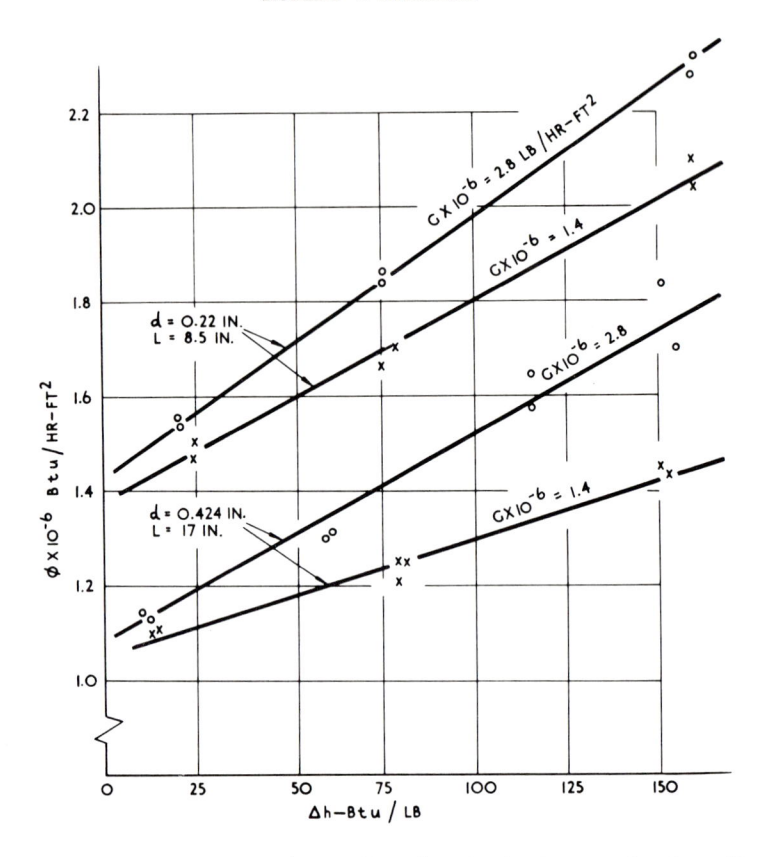

FIG. 21. Test of the hypothesis $\phi = f(L/d, G, \Delta h, P)$ using burn-out data for uniformly heated round tubes with water [from Barnett (B3)]. $P = 1000$ psia, $L/d = 40$ (approx.).

Equation (2), for instance, represents several such correlations, including the well-known Bettis correlation (D2) which is probably better identified as

$$\phi = f(P, L/d, G, H_{BO}) \tag{3}$$

where H_{BO} is the fluid enthalpy at the burn-out position. Equations (2) and (3) are equivalent, however, since H_{BO} and Δh are interchangeable by the heat-balance equation, which, for a uniformly heated round tube, is given by

$$H_{BO} = 4\phi L/dG + h_f - \Delta h \tag{4}$$

where h_f, an enthalpy, is of course a function of the pressure.

Another possible contraction considered by Barnett (B3) was suggested by the idea that the Reynolds number may be a dominant factor in determining burn-out, in which case the parameters d and G would combine as a product.

Barnett examined three likely combinations for a uniformly heated round tube:

$$\phi = f(dG, L, \Delta h, P) \tag{5}$$

$$\phi = f(dG, L/d, \Delta h, P) \tag{6}$$

$$\phi d = f(dG, L/d, \Delta h, P) \tag{7}$$

but none of these was found to be valid on being tested with suitable experimental data. It must be noted that these tests do not entirely invalidate a dG hypothesis, since many other combinations are possible; it can only be affirmed that the three particular ones examined are not consistent with the experimental data.

Equation (7) is an interesting example of the versatility of the hypothesis-testing technique that Barnett has exploited. Thus, suppose it is assumed that the important fluid properties which control the burn-out phenomenon are density ρ_L, specific heat, C_{PL}, and thermal conductivity k_L of the liquid phase, together with density of the vapor phase ρ_V and the latent heat λ. Then, for burn-out in a uniformly heated round tube we have

$$\phi = f(L, d, G, \Delta h, \rho_L, \lambda, C_{pL}, k_L, \rho_V) \tag{8}$$

Application of dimensional analysis to Equation (8) using a set of fundamental units containing heat, mass, length, time, and temperature, yields

$$\frac{\phi \, dC_{pL}}{k_L \lambda} = f\left[\frac{dGC_{pL}}{k_L}, \frac{L}{d}, \frac{\Delta h}{\lambda}, \frac{\rho_L}{\rho_V}\right] \tag{9}$$

which, for a given fluid, reduces to Eq. (7). Thus, in testing Eq. (7), not only is a particular dG hypothesis examined, but the hypothesis that the above set of fluid properties are the important ones is examined as well. The actual test result given by Barnett (B3) for Eq. (7) is reproduced in Fig. 22. As can be seen, this equation proves unacceptable.

To date, no valid contraction among the system-describing parameters has been found.

VI. The Local-Conditions Concept of Burn-Out

Thus far, the burn-out phenomenon has been discussed mainly in terms of the important system-describing parameters. This approach is preferable, since the system parameters are, in fact, the independent variables and they must uniquely and unequivocally determine the heat flux required to produce burn-out. It can be argued, however, that burn-out, being a local phenomenon, may be described entirely by local parameters; of this there can be no dispute. The problem is to find a description of these local parameters that works. Our knowledge of the hydrodynamic behavior of boiling two-phase flow, as well as analytical techniques, appear too limited at present to give any valid detailed description of local conditions. The only likely approach open is to consider whether or not gross quantities such as average quality or void fraction are of any relevance. At present, only quality would appear to be worth considering, since it is readily calculated (i.e., the mean channel flow quality) from the heat-balance equation, whereas the void fraction presents a problem as difficult as burn-out itself.

A. THE BARNETT LOCAL-CONDITIONS HYPOTHESIS

Barnett (B3) considered the relevance of quality to burn-out and made the important observation that for uniformly heated round tubes, the burn-out

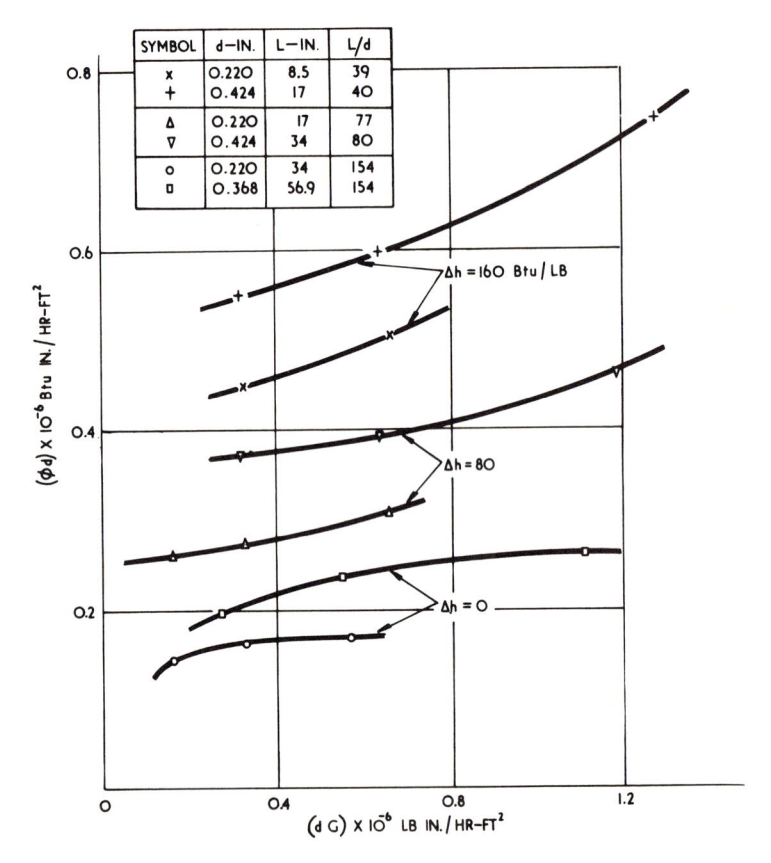

SYMBOL	d—IN.	L—IN.	L/d
x	0.220	8.5	39
+	0.424	17	40
▲	0.220	17	77
▽	0.424	34	80
○	0.220	34	154
□	0.368	56.9	154

FIG. 22. Test of the hypothesis $\phi d = f(dG, L/d, \Delta h, P)$ using burn-out data for uniformly heated round tubes with water at 1000 psia [from Barnett (B3)].

flux can be represented fairly accurately by four local conditions as follows:

$$\phi = f(P, G, d, \kappa) \tag{10}$$

where κ represents the quality at the burn-out position. It must be noted that κ may assume an equivalent negative value in the case of subcooled burn-out [see Eq. (12)].

It can be seen that κ in Eq. (10) replaces the system-describing parameters L and Δh in Eq. (1). A direct test of the hypothesis is therefore to plot ϕ against κ for fixed values of P, G, and d, with L and Δh varying. For the hypothesis to be correct, the data points must all lie on a smooth curve. Experience shows, however, that plotting ϕ against κ often produces an undue amount of scatter which may obscure and distort any true relationship existing. This enhanced scatter is caused by the cumulative effect of experimental errors in the various terms in the heat-balance equation from which the quality κ is derived.

The above difficulty is particularly true of burn-out data for water, and Barnett therefore used a different approach for testing Eq. (10). His method was first to change the equation into a form containing only system-describing parameters, which can be done by substitution among the following three expressions:

(1) A mathematical statement of the observation that a linear relationship effectively exists between ϕ and Δh:

$$\phi = f_1(P, d_c, G, L) + f_2(P, d_c, G, L)\,\Delta h \tag{11}$$

(In order to make the derivation completely general, the term d_c is used to represent the cross-sectional dimensions of any channel. In the case of a round tube, d_c equals d.)

(2) The heat-balance equation in its general form applicable to any uniformly heated channel:

$$\kappa \lambda + \Delta h = \phi P_h L / A G \tag{12}$$

where P_h is the total heated parameter, and A is the cross-sectional flow area of a channel.

(3) A general statement of the Barnett local-conditions hypothesis:

$$\phi = f(P, G, d_c, \kappa) \tag{13}$$

Eliminating Δh between Eqs. (11) and (12) and writing f_1 for $f_1(P, d_c, G, L)$ and f_2 for $f_2(P, d_c, G, L)$, we get

$$\phi = (f_1 - f_2 \kappa \lambda) \div (1 - f_2 P_h L / A G) \tag{14}$$

Eq. (14) shows a linear relation between ϕ and κ, and implies that Eq. (13) may be written:

$$\phi = f_3(P, G, d_c) + f_4(P, G, d_c)\kappa \tag{15}$$

From the similarity between Eqs. (14) and (15)

$$f_2 = \frac{-f_4}{\lambda - f_4 P_h L/AG} \tag{16}$$

and

$$f_1 = \frac{f_3}{1 - f_4 P_h L/AG\lambda} \tag{17}$$

Substituting these expressions for f_1 and f_2 in Eq. (11) gives

$$\phi = \frac{f_3 - f_4 \Delta h/\lambda}{1 - f_4 P_h L/AG\lambda}$$

and, finally, writing $A' = f_3$ C' and $C' = -AG\lambda/P_h f_4$ gives the general burn-out equation

$$\phi = \frac{A' + (A/P_h)G\,\Delta h}{C' + L} \tag{18}$$

where A' and C' are functions of P, G, and d_c only.

Effectively what Barnett (B3) did was to confirm the validity of Eq. (18) using burn-out data for water in round tubes. Subsequently, Eq. (18) was applied to more extensive round-tube data (M3, T1), to annuli (B6), rectangular channels (M3), and rod-bundle arrangements (M4), using simple mathematical expressions for A' and C', and a consistently high accuracy was achieved. Details of the respective correlations obtained are given in Section VIII.

A more detailed appraisal of the Barnett local-conditions hypothesis has been made possible by recent burn-out data for Freon. This is due to the high experimental accuracy of the Freon data, which derives from the nature of the fluid itself, since it allows a very precise determination of the burn-out point. The high accuracy means that a plot of ϕ against κ is not subject to the usual limitations, and an example from Staniforth and Stevens (S7) is shown in Fig. 23. It can be seen that the Barnett local-conditions hypothesis is not strictly valid because of a length effect appearing at low qualities. The discrepancy appears to be associated with tubes of short length and with high mass velocities. However, its effect on the general burn-out equation for uniformly heated channels [Eq. (18)] is to introduce systematic errors of only a few percent, which are too small to be of any practical significance.

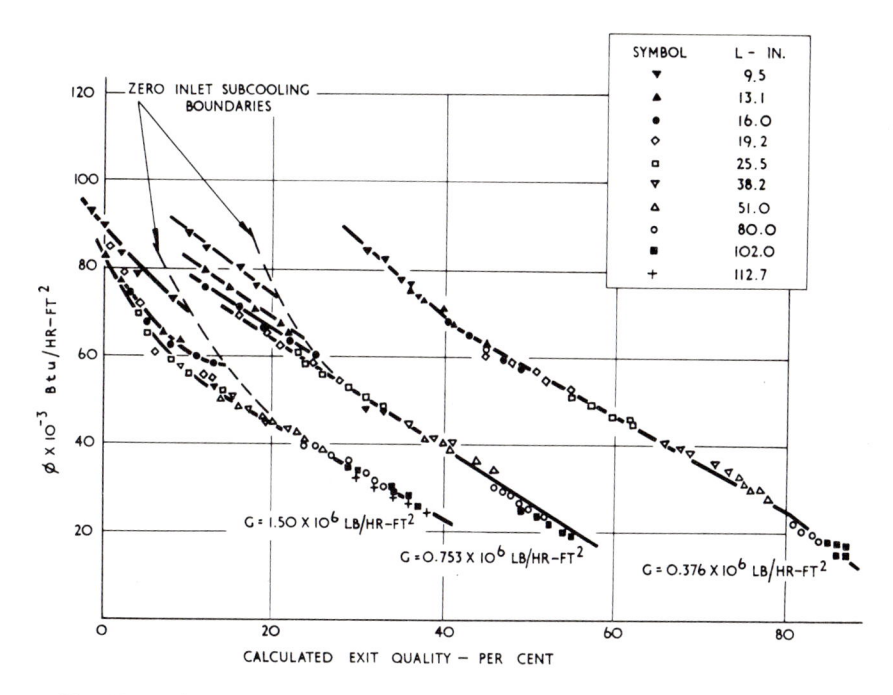

FIG. 23. Variation of burn-out heat flux with exit quality for uniformly heated tubes using Freon 12 [from Staniforth and Stevens (S7)]. $d = 0.344$ in., $P = 155$ psia.

There is a slight indication in Fig. 23 that the above discrepancy may disappear in the subcooled boiling region; Staniforth and Stevens were unable to investigate this point because of apparatus limitations. Further evidence is available, however, from earlier work by Styrikovich et al. (S17) using water in a uniformly heated round tube. They obtained a plot of ϕ against κ showing little of the usual obscuration due to enhanced scatter, and the plot is reproduced in Fig. 24 (with the data points omitted for clarity). A hard-inlet condition was mostly used, but some of the results are for a soft inlet obtained by using a preformed steam and water mixture. The transition point between single-phase and two-phase inlet has been marked on the curves in Fig. 24. It would seem that the discrepancy in the Barnett local-conditions hypothesis disappears in the subcooled boiling region.

B. The Kirby Local-Conditions Hypothesis

Kirby (K3) has found an interesting hypothesis that works in the net quality region. It is a contraction of Barnett's local-conditions hypothesis [Eq. (10)] and is therefore not strictly applicable at low qualities where slight

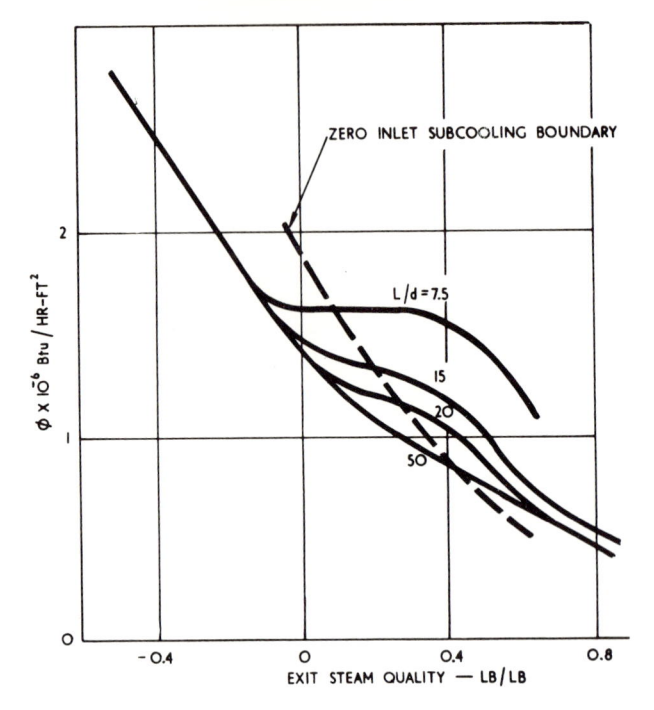

FIG. 24. Variation of burn-out heat flux with exit quality for uniformly heated tubes using water [from Styrikovich et al. (S17)]. $d = 0.315$ in., $P = 1420$ psia, $G = 0.55 \times 10^6$ lb/hr-ft².

deviations from the Barnett hypothesis occur. The contraction is in the parameters d and G and takes the form:

$$\phi = f(P, d^{1/2}G, \kappa) \tag{19}$$

A test of the hypothesis made by Kirby using Freon data is shown in Fig. 25; further confirmation using water data is also given by Kirby (K3). Whether the hypothesis is valid in the subcooled boiling region has not been examined, but it is considered unlikely since Eq. (19) represents a high degree of simplification, and can hardly be expected to apply to both subcooled and annular-flow boiling, which have markedly different physical characters.

VII. The Low-Velocity Burn-Out Regime

The characteristic that has been called the low-velocity regime (M2) would have been observed at a much earlier stage if burn-out data had existed for experiments covering a very wide range of mass velocities with the other system parameters fixed. Even today there is no really good direct example of

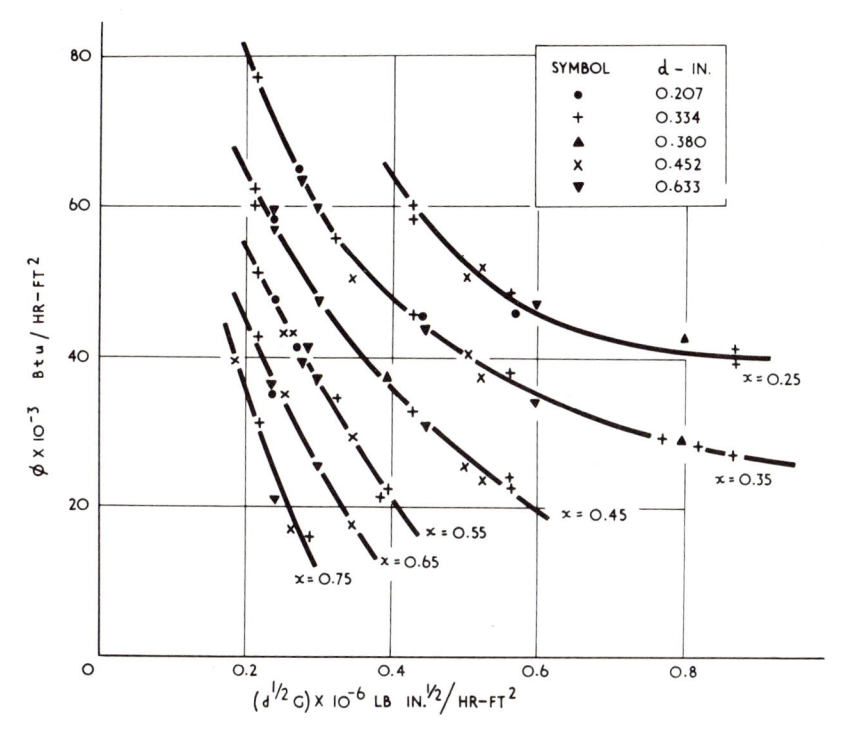

FIG. 25. Test of the hypothesis $\phi = f(P, d^{1/2}G, \kappa)$ using burn-out data for uniformly heated round tubes with Freon 12 at 155 psia [from Kirby (K3)].

the regime, for although data are available over a wide range of mass velocities, the other parameters keep changing. One of the best direct examples, from recent Freon data by Stevens et al. (S13), is shown in Fig. 26. The characteristic referred to is the almost direct linear relationship between ϕ and G that exists at low values of mass velocity, and which quickly changes to another relationship at higher velocities. The effect is more marked for low L/d ratios and produces a distinct knuckle in the curve.

In Fig. 26, the dotted lines correspond to 100% quality at the tube exit and represent the ϕ/G ratios obtained from the following heat-balance equation for a uniformly heated tube (with $\kappa = 1$):

$$\kappa\lambda + \Delta h = 4\phi L/dG \tag{20}$$

The dotted lines thus represent a limiting condition, and it can be seen that the low-velocity regime corresponds to the range of G values which give very high exit qualities at burn-out. The extent of the low-velocity regime for boiling water in uniformly heated round tubes has been estimated by plotting and cross-plotting data, and the approximate boundary limits obtained are

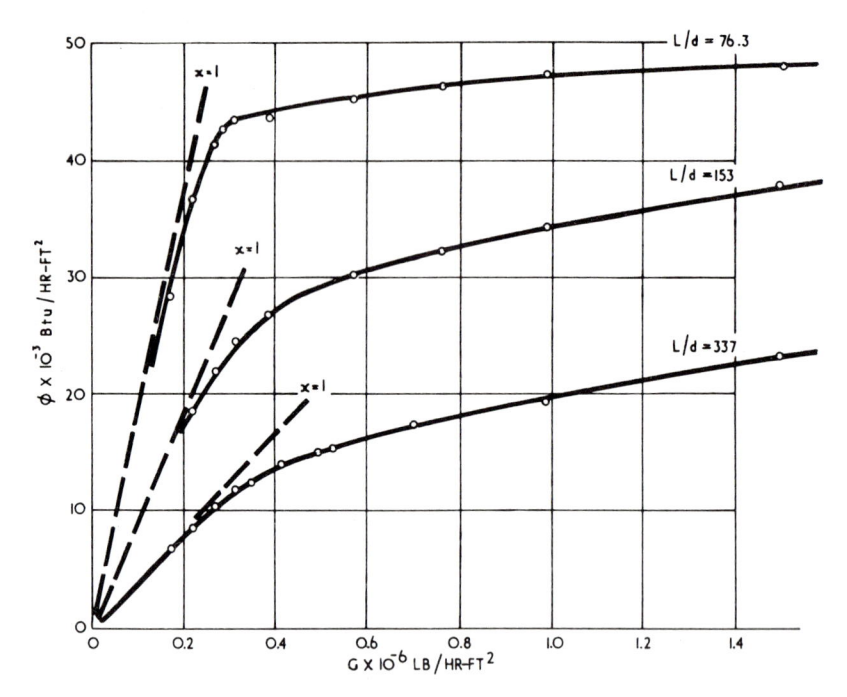

FIG. 26. Burn-out curves for Freon-12 in uniformly heated round tubes [from Stevens et al. (S13)]. $d = 0.334$ in., $P = 155$ psia, $\Delta h = 0$.

shown in Fig. 27. The regime would seem to disappear at zero pressure and towards the critical pressure, while at about 800 psia, it extends to fairly high mass velocities, depending on the L/d ratio; the regime is extended by increasing the L/d ratio. For channels other than round tubes it has been observed that Fig. 27 can be used by substituting L/d_h for L/d, where d_h is the heated equivalent diameter of the channel (4 × flow area/heated perimeter).

Because of the closeness of the low-velocity regime to the 100% quality line, it is to be expected that latent heat will be the dominant fluid property controlling burn-out in this regime. In fact, it has been found (M3) that latent heat alone can adequately represent all the low-velocity regime data, which includes pressures ranging from 15 to 2000 psia. For uniformly heated round tubes, for example, the appropriate burn-out equation obtained, based on Eq. (18), is:

$$\phi \times 10^{-6} = \frac{(G \times 10^{-6})(\lambda + \Delta h)}{158 d^{0.1}(G \times 10^{-6})^{0.49} + 4L/d} \qquad (21)$$

Equation (21) fits the published data belonging to the low-velocity regime (Fig. 27), with an rms error of 5.5%, and is applicable to any system pressure.

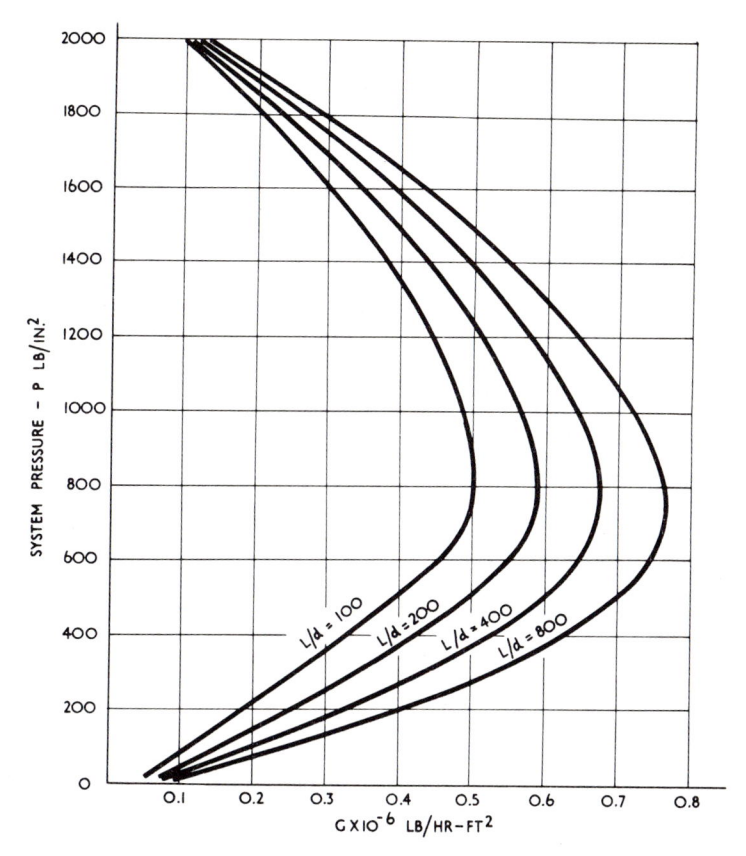

FIG. 27. Approximate boundary of the low-velocity burn-out regime for water flowing in round tubes. The low-velocity regime lies to the left of any given curve [from Macbeth (M2)].

VIII. The Correlation of Water Data for Uniformly Heated Channels

Producing burn-out correlations would appear to be almost a pastime; Milioti (M12), for example, was able to compile a total of 59 different burn-out correlations, and the number still grows. Most of these correlations are based on very restricted ranges of system parameters, however, and although they work well within the restrictions, they usually deviate markedly on extrapolation. Some of the earlier correlations are also readily seen to be inconsistent with now well-established experimental facts, even simple though important facts such as the linear or nearly linear relationship between ϕ and Δh. As mentioned earlier, the hypothesis-testing technique exploited by Barnett is a very effective tool for showing up defects, and the method has been used to show where less obvious faults, such as combining L and d only as a ratio, exist in some of the correlations.

The above processes of elimination led to many correlations being discarded, and in a recent catalog by Clerici et al. (C4), who examined "the most important burn-out correlations," the number listed is down to 20. Clerici et al. show graphically the results of some spot checks using the selected correlations to predict the burn-out heat flux for uniformly heated round tubes. The results indicate wide disparity, with calculated burn-out flux values differing in some cases by a factor of more than 2, so that a great deal more sorting out of burn-out correlations is still needed.

To make a proper appraisal is laborious, and it is not proposed to carry one out here. There are several criteria involved, for example, a correlation must be checked against all the appropriate and reliable data available. Experience indicates that a good correlation will achieve an rms error of about 6%, where error is defined by the ratio $(\phi_e - \phi_c)/\phi_e$. Examination of individual errors is also necessary to ensure that they are not in any way systematic, as this would indicate a fault in the correlation. A difficulty arises from the fact that when new data appear, extending the range of the system parameters, it nearly always transpires that a previously good correlation becomes unsatisfactory. Consequently, frequent changes in numerical constants or even alterations to the form of a correlation have to be made. In fact, several of those listed by Clerici et al. (C4) are in their second or even third modification, which is tiresome but necessary in maintaining good correlations.

For the purpose of the present text the burn-out data bank maintained by the Atomic Energy Establishment at Winfrith has been brought up to date by compiling on computer cards as much reliable published data referring to uniformly heated tubes, rectangular channels, rod-bundle arrangements, and annuli as could be found (up to May 1966). These data have then been used to optimize, by computer, the general form of burn-out correlation derived earlier and given by Eq. (18). The correlations obtained for the different categories of channel geometry are presented below, together with some discussion of their effectiveness and the parametric relationships they indicate. It is to be noted that these correlations refer to axially, uniformly heated channels with flowing water; the effect on burn-out of nonuniform heating, which can be considerable, is discussed in Section X, while in Section XI proposed methods of dealing with fluids other than water are considered.

While Eq. (18) is taken as the basis for a set of burn-out correlations which may be of some general value, it is appreciated that there are other correlations that may be equally effective or even more effective in representing the data; however, as indicated above, the job of resolving this question is extremely laborious and has yet to be done by anyone. As will be seen in the results given below, Eq. (18) achieves an rms error of the order of 6% in all cases, while the examination of individual errors shows that there is no sign of any serious systematic fault. It is considered that Eq. (18) gives as close a representation of the existing data as is likely to be achieved. For information about other correlations that have been developed using extensive ranges of data (a factor which is one of the most important requirements in developing correlations), the reader is referred to Bertoletti et al. (B18), Becker (B9, B10), and Tong (T3).

A. Round Tubes

The problem with the general formula, Eq. (18), which, for convenience, is written as follows for round tubes,

$$\phi \times 10^{-6} = \frac{A' + \frac{1}{4}d(G \times 10^{-6})\Delta h}{C' + L} \qquad (22)$$

is to find an adequate representation of A' and C'. The derivation given in Section VI,A shows that A' and C' are functions of P, d, and G. Now, apart from the low-velocity regime, the effect of pressure on burn-out is extremely complex, and since most of the data fall into 8 distinct pressure groups (15, 250, 560, 1000, 1250, 1550, 1800, and 2000 psia) it has been found convenient (L2, T1) to consider each of these separately. Thus, for a given pressure, the problem reduces to one of obtaining functions for d and G only.

The use of polynomial expressions is an obvious choice, but it was observed that they need a very large number of terms to represent the data adequately. The difficulty appears to lie in the marked change in the effect of mass velocity (occurring at low L/d ratios) when moving from the low to the high mass-velocity regime; this effect has been illustrated in Fig. 26. Therefore, for greater simplicity, the data were separated in accordance with Fig. 27 so that only for the high-velocity regime, to which the majority of data belong, did expressions for A' and C' have to be found. The low-velocity regime is already correlated, irrespective of pressure, by Eq. (21).

The polynomials finally chosen as having the smallest number of terms needed to represent the high-velocity regime data are as follows:

$$A' = y_0 d^{y_1}(G \times 10^{-6})^{y_2}[1 + y_3 d + y_4(G \times 10^{-6}) + y_5 d(G \times 10^{-6})] \qquad (23)$$

$$C' = y_6 d^{y_7}(G \times 10^{-6})^{y_8}[1 + y_9 d + y_{10}(G \times 10^{-6}) + y_{11}d(G \times 10^{-6})] \qquad (24)$$

The "y" values in these expressions were optimized by computer, an operation which is readily done for any given set of experimental data by substituting the expressions in Eq. (22) and using for an optimization criterion the requirement that the rms value of $(\phi_e - \phi_c)/\phi_e$ should be as small as possible.

The range of the system parameters in the data compilation for uniformly heated round tubes are given in Table I. Seven of the eight main pressure

TABLE I
Range of System Parameters for Burn-Out in Round Tubes Uniformly Heated with Liquid Water Inlet[a]

Actual or nominal pressure (psia)	Actual pressure range (psia)	Range of diameters (in.)	Range of lengths (in.)	Range of mass velocities (lb/hr-ft^2) $\times 10^{-6}$	No. of Experiments
15	—	0.04–0.94	1–33.84	0.0073–4.24	107
100	—	0.18	9.4	0.0096–0.068	20
285	250–310	0.045 & 0.21	4.5 & 6.0	0.04–11.6	71
560	500–650	0.12–0.426	3–68	0.028–7.82	534
720	680–760	0.12–0.22	6–68	0.028–5.99	37
1000	960–1065	0.12–1.475	3–144	0.02–13.7	934
1250	1152–1320	0.226–0.862	24.6–137	0.175–3.0	100
1550	1436–1655	0.075–1.76	6–79.7	0.02–7.28	527
1800	1750–1825	0.075–1.114	6–60	0.058–3.03	195
2000	—	0.075–0.436	3–72	0.023–7.79	649
2250	—	0.075	6 & 27.4	1.43–2.70	30
2500	—	0.075	6 & 27.4	1.5–2.86	30
2650	—	0.118	1.38 & 5.91	0.635–2.14	10
2750	—	0.075	6 & 27.4	1.36–2.79	30
Miscellaneous:					
Swedish	34–594	0.155–0.395	23.6–122.8	0.074–1.85	1576
Others	20–3000	0.12–0.226	3–24.6	0.037–5.05	10

[a] Derived mainly from Thompson and Macbeth (T1) and Lee (L2).

groups referred to above can clearly be seen; the eighth is a nominal pressure of 250 psia (actual pressure range 200–300 psia) which was selected so as to make use of the large amount of Swedish data.

The optimized y values and the corresponding rms errors achieved for each of the eight pressure groups are shown in Table II. It will be noted that the full polynomial expressions, Eqs. (23) and (24), were found unnecessary in some cases, and it is always advisable to keep the number of optimized constants to a minimum, since this tends to make a correlation more useful for extrapolation purposes. It has been found that doubling the rms errors gives a good indication of the maximum amount of scatter about a correlation line, and as a general rule it is advised that $\pm 15\%$ should be taken as the limit of correlation accuracy. Examination shows that this amount is compounded of roughly $\pm 6\%$ experimental scatter in data from any one source, a deviation of $\pm 6\%$ between data from different sources, and $\pm 3\%$ caused by defects in the correlation itself. These figures show that no correlation can be expected to represent the data with an accuracy better than about $\pm 12\%$.

A word of caution should be added with regard to the calculation of the burn-out flux for a pressure intermediate to the main pressure groups that have been correlated: this calculation must not be done by taking intermediate y values from Table II. The recommended procedure is to estimate the burn-out flux for the required conditions for the main pressure groups above and below the required pressure, and then to interpolate linearly. It must be also emphasized that while the above correlations can be used with confidence within the experimental ranges of the data, extrapolation outside these ranges should not be taken very far without allowing for a possible reduction in the accuracy obtained.

The influence of the system parameters P, d, and G on the burn-out flux are not easily seen from the y values in Table II, and it is therefore helpful to illustrate the general trends by considering some specific examples, using the correlation to obtain burn-out flux values.

1. *The Effect of Mass Velocity on Burn-Out*

The main characteristics of the effect of mass velocity are shown in Fig. 28. One interesting feature, curve (2), is a rapid rise of burn-out flux in the low-velocity regime to a value which thereafter remains practically independent of mass velocity. The primary condition which tends to induce this effect is a low value of Δh, but the pressure and L/d ratio are also important. At 1000 psia, for example, the L/d ratio must be less than about 100. The influence of the L/d ratio was shown in Fig. 26.

Curve (1) in Fig. 28 illustrates that at another pressure and with a higher L/d ratio the burn-out flux remains strongly dependent on mass velocity,

TABLE II

OPTIMIZED y VALUES AND rms ERRORS FOR EQS. (23) AND (24) APPLIED TO HIGH-VELOCITY REGIME BURN-OUT DATA FROM TABLE I

System pressure (psia)	15	250	560	1000	1250	1550	1800	2000
y_0	1120	106	237	114	93.3	58.0	194	65.5
y_1	1.19	0.847	1.20	0.811	1.10	0.834	2.09	1.19
y_2	1.37	0.677	0.425	0.221	0.575	0.224	0.593	0.376
y_3	0	0	−0.940	−0.128	0	−0.0336	−0.597	−0.577
y_4	0	0	−0.0324	0.0274	0	0.0755	−0.131	0.220
y_5	0	0	0.111	−0.0667	0	−0.296	−0.0482	−0.373
y_6	1000	60.3	19.3	127	88.5	48.3	231	17.1
y_7	1.4	1.4	0.959	1.32	1.46	0.823	1.93	1.18
y_8	1.05	0.937	0.831	0.411	1.00	0.121	0.612	−0.456
y_9	0	0	2.61	−0.274	0	0	−0.575	1.53
y_{10}	0	0	−0.0578	−0.0397	0	0	−0.255	2.75
y_{11}	0	0	0.124	−0.0221	0	0	0.110	2.24
rms error (%)	8.8	4.7	7.27	6.48	5.12	6.02	3.93	7.56
No. of experiments	88	237	225	802	100	527	195	615

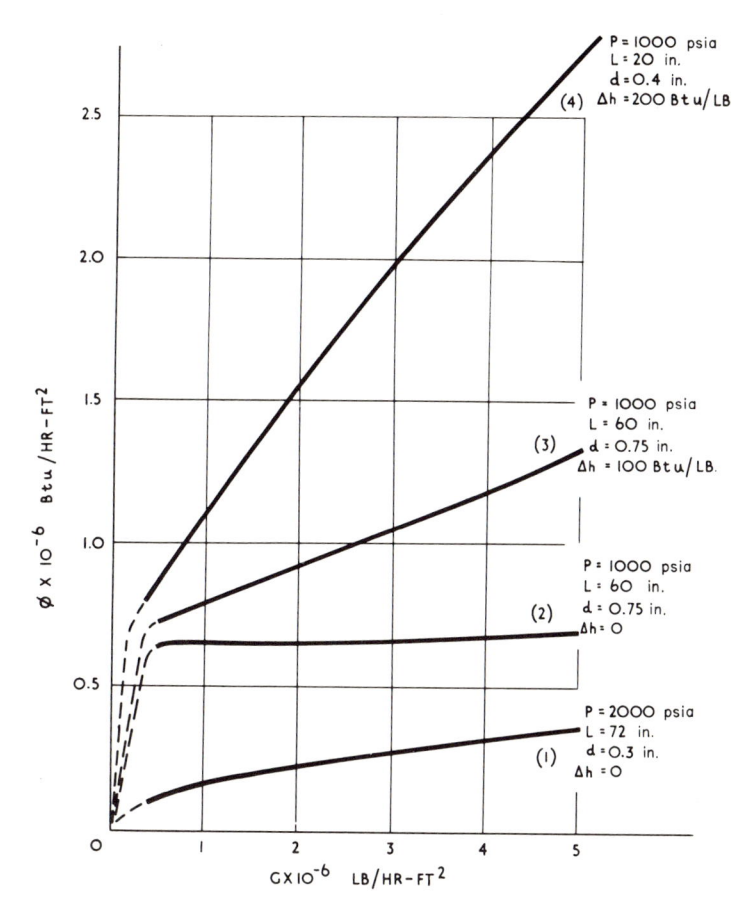

FIG. 28. Main characteristics of the effect of mass velocity on burn-out for uniformly heated round tubes using water.

even with zero inlet subcooling. Curve (3) shows the significant rise in burn-out flux that can be obtained by adding inlet subcooling to curve (2), and curve (4) shows that with a high Δh value and a small L/d ratio, the effect of mass velocity may become very pronounced.

Matzner (M6), in burn-out tests on a 0.930 in.-i.d. uniformly heated tube 77.6 in long and water at 1000 psia, has reported finding a region of operation where burn-out was reached by increasing the flow rate while maintaining both the heat flux and the inlet subcooling constant. In other words, a fall in burn-out flux with increasing mass velocity was indicated. Matzner observed that the effect occurred only at low values of inlet subcooling and at certain given flow rates.

Calculations show that the correlation complies with the above unexpected behavior, and the effect is present, very slightly, for the tube considered in curve (2) of Fig. 28.

2. *The Effect of Tube Diameter on Burn-Out*

Tube diameter shows a more regular pattern than mass velocity, as the examples in Fig. 29 indicate. The tendency is for the burn-out flux to rise smoothly as the diameter increases, and to approach a limiting value. A slight drop in burn-out flux is shown by the curves (1) and (2) in Fig. 29, and this feature has been confirmed by experiment. Lee (L1) has even noted that at very large diameters the slight drop may revert to a rise in burn-out flux.

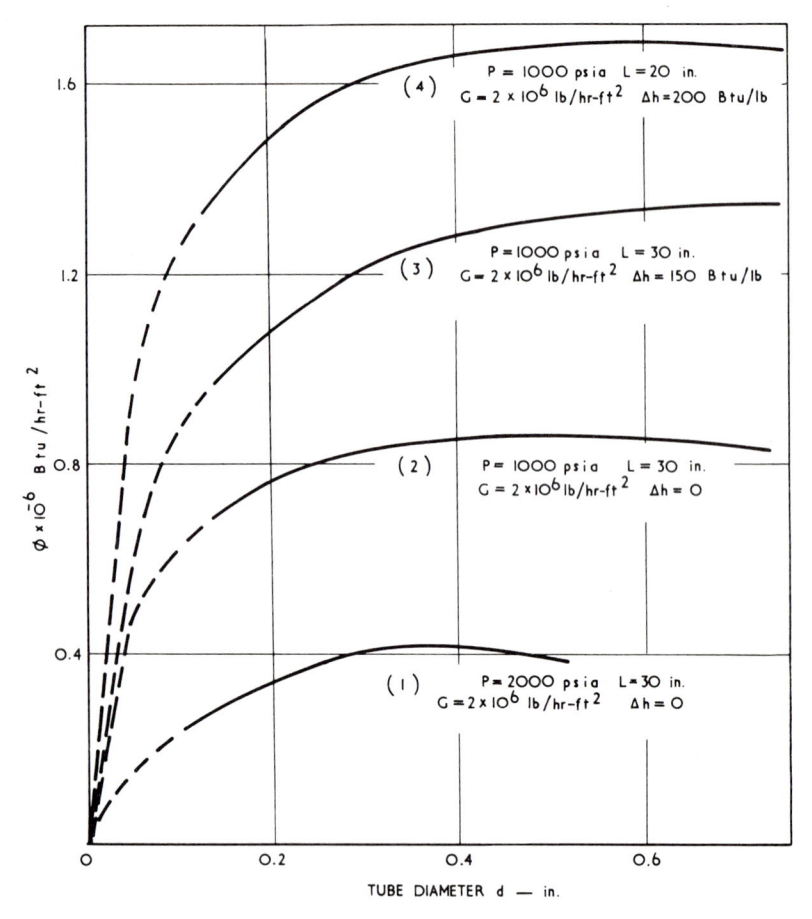

FIG. 29. Main characteristics of the effect of tube diameter on burn-out for uniformly heated round tubes using water.

3. The Effect of Pressure on Burn-Out

Figure 30 shows the main characteristics of the effect of pressure. It can be seen that a maximum heat flux occurs generally below 500 psia. The most interesting feature of Fig. 30, however, is a secondary maximum in the region 1500–2000 psia which tends to occur with moderate L/d ratios, large inlet subcooling, and high mass velocity.

A different view of the effect of pressure is obtained by considering a constant inlet temperature, instead of constant inlet subcooled enthalpy. This view is probably a more practical view, and some of its features are shown in Fig. 31. It can be seen that, compared with Fig. 30, the secondary maximum around 1500–2000 psia is much more pronounced, and may even exceed the low-pressure maximum.

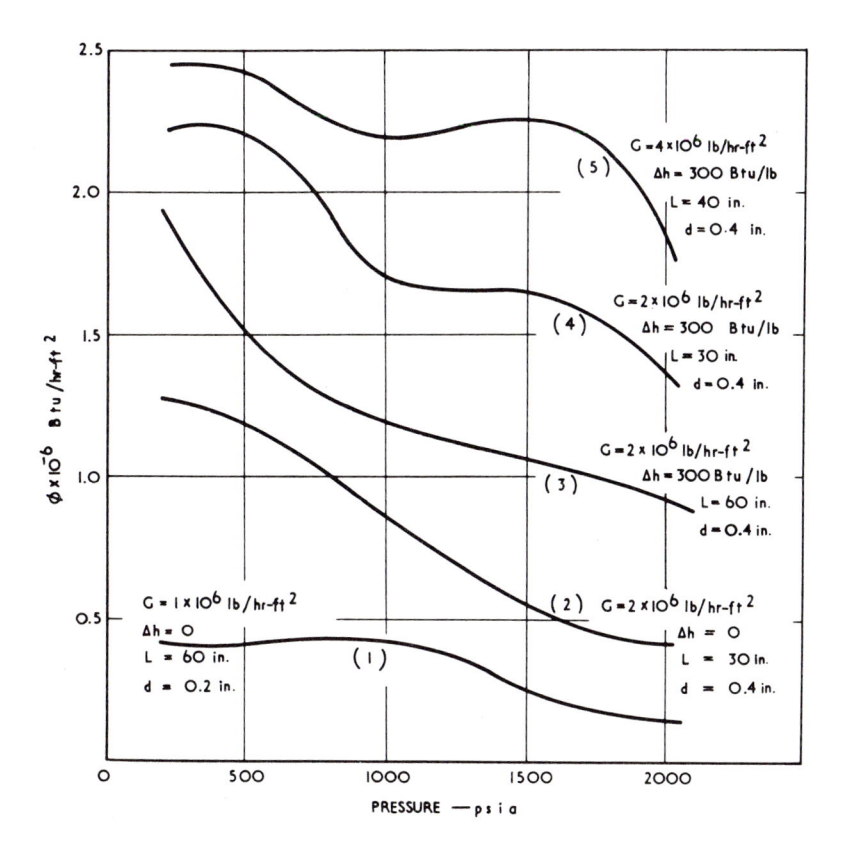

FIG. 30. Main characteristics of the effect of pressure on burn-out for uniformly heated round tubes using water. Curves based on constant inlet subcoolings (Δh).

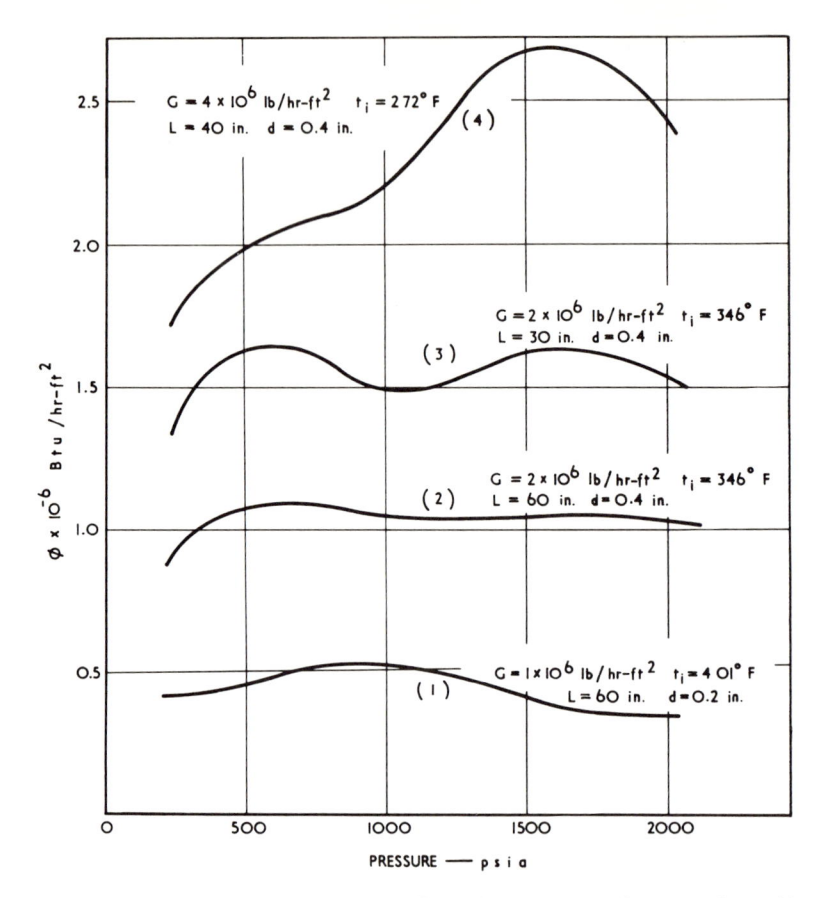

FIG. 31. Main characteristics of the effect of pressure on burn-out for uniformly heated round tubes using water. Curves based on constant inlet temperatures (t_i).

B. Rectangular Channels

Comparatively few burn-out data have been published for rectangular channel geometries; most of these were produced before 1958, and have been compiled by De Bortoli (D2). The limited range of the system parameters are shown in Table III, indicating that they mostly refer to a pressure of 2000 psia.

The rectangular channels tested have three characteristic dimensions: the flow width (which measures 1 in. for all the data shown in Table III), an equivalent heated width (which measures approximately 0.9 in.), and the internal spacing S between the flat heating surfaces. The heated width and the flow width are not the same, due to the way the channels are constructed,

TABLE III

RANGE OF SYSTEM PARAMETERS FOR BURN-OUT IN RECTANGULAR CHANNELS UNIFORMLY HEATED (LENGTHWISE) WITH LIQUID WATER INLET[a]

Pressure (psia)	Range of spacing or gap (in.)	Range of lengths (in.)	Range of mass velocities $\times 10^{-6}$ (lb/hr-ft^2)	No. of Experiments
600	0.050–0.097	12.1–27.0	0.77–2.57	22
800	0.059–0.101	6.0–27.0	0.18–3.62	41
1200	0.059–0.101	6.0–27.0	0.17–4.00	54
2000	0.050–0.101	6.0–27.0	0.02–4.78	394

[a] Compiled in De Bortoli et al. (D2).

as described in De Bortoli et al. (D2). The general burn-out correlation, Eq. (18), thus becomes

$$\phi \times 10^{-6} = \frac{A' + 0.555 S (G \times 10^{-6}) \Delta h}{C' + L} \tag{25}$$

As previously with round tubes, the problem becomes one of representing A' and C'. Simple power functions have been found adequate (M3), and the appropriate expressions are

$$A' = y_0 S^{y_1} (G \times 10^{-6})^{y_2} \tag{26}$$

$$C' = y_3 S^{y_4} (G \times 10^{-6})^{y_5} \tag{27}$$

The optimized y values and the corresponding rms errors achieved for each of the four pressure groups are given in Table IV. The errors are generally higher than usual, but examination shows they are attributable to the experimental scatter which is often found to be large in early burn-out data.

TABLE IV

OPTIMIZED y VALUES AND rms ERRORS FOR EQS. (26) AND 27) APPLIED TO BURN-OUT DATA IN TABLE III

Pressure (psia)	y_0	y_1	y_2	y_3	y_4	y_5	rms error (%)
600	190	0.928	0.640	8.1	1.4	3.93	6.1
800	46.5	0.390	0.391	104	1.4	0.007	12.9
1200	537	1.32	0.764	286	1.4	1.29	4.9
2000	202	1.08	0.669	370	1.4	0.725	9.4

C. Rod Bundles

Burn-out data and descriptive details of 24 different rod-bundle geometries, representing all known published work up to 1965, have been compiled and analyzed by Macbeth (M4). Data that have subsequently appeared are given by Matzner (M10), Janssen (J2), Edwards and Obertelli (E1), Becker *et al.* (B11), Moeck (M14), and Hesson (H3). All these data refer to water, and in most of the bundles the direction of water flow is vertically upwards, parallel to the heated rods; however, a few tests have also been made with the bundles horizontal, also using parallel flow. Nearly all the experiments have been performed at around 1000 psia, so that the correlation of rod-bundle data must be restricted to this pressure alone.

The main problem with rod-bundle data is the large number and complexity of the geometry parameters involved. There are two approaches to the problem: (1) a bundle can be considered as comprising a number of subchannels, or (2) a bundle can be considered as a whole. In method (1), the object is to attempt to describe the mass and enthalpy distribution among the various subchannels, and then by applying some single-channel burn-out correlation, which entails using an appropriate equivalent diameter, predict the subchannel in which burn-out will occur and the corresponding heat flux. The method is being studied at various places, but the difficulties are considerable and published information is limited. Some general comments on the subject have been given by Tong *et al.* (T4), and by Bowring (B22). Method (2), i.e., considering a bundle as a whole, was the one explored by Macbeth (M4), and the main findings are summarized below, since they establish a basis for applying the general burn-out correlation [Eq. (18)].

The data examination given in Macbeth (M4) refers to a total of 18 different bundles tested at around 1000 psia covering a wide range of geometries, including arrangement of 3, 4, 7, 12, and 19 rods per bundle. Lengths varied from 17 to 76 in., and in all cases the heat flux was uniformly applied to individual rods, although in some of the bundles the flux varied between rods to simulate conditions in a nuclear reactor. On an account of this flux variation, the ϕ value used in the analysis, was taken as the *average* flux on the *heated* rods in a bundle. Examination showed a close linear relationship existing between ϕ and Δh for all the rod-bundle data, and Fig. 32 is an example of data obtained at Columbia University (M9) for a 19-rod bundle. In this example, the heat-flux variation on the rods relative to the flux on the rods in the outer ring (see inset diagram in Fig. 32), was 1/0.806/0.766.

The first step in the analysis was to assess the effect of mass velocity for the 1000 psia data. By plotting and cross-plotting and finally by computer optimization it was found that 15 of the different geometries conformed very

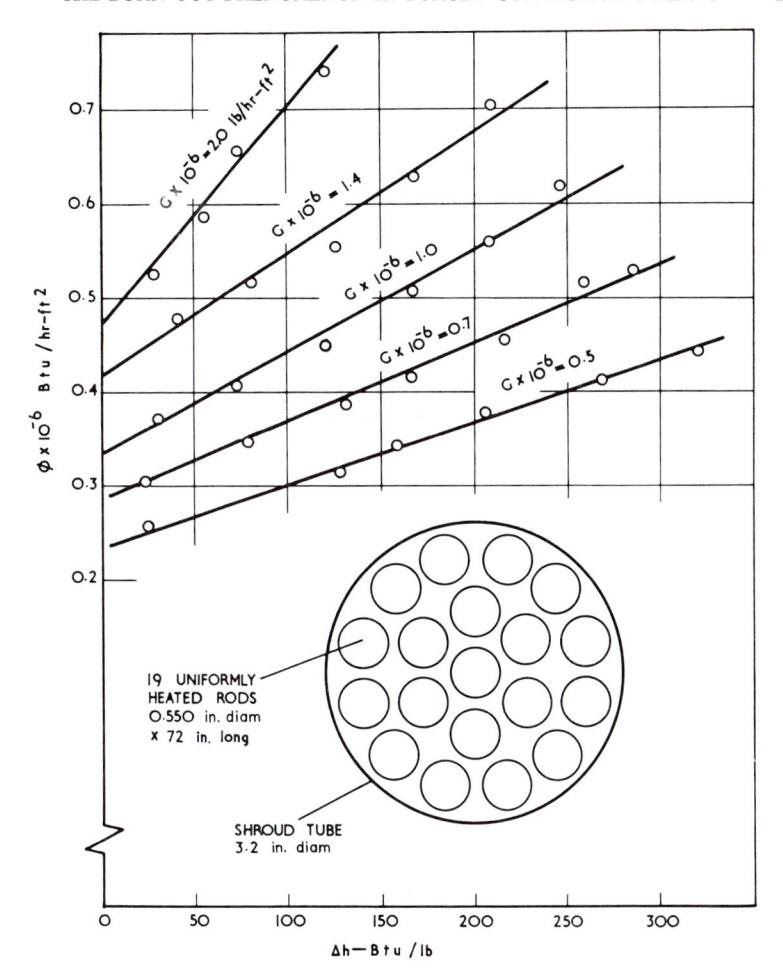

FIG. 32. Example of the linear effect of inlet subcooling on burn-out in a rod bundle using water at 1015 psia [from Matzner (M9)].

accurately with the following equation:

$$\phi \times 10^{-6} = K_1(G \times 10^{-6})^{1/2} + K_2(G \times 10^{-6}) \Delta h \qquad (28)$$

where K_1 and K_2 are numerical coefficients which vary with each bundle and are therefore geometry-dependent. Some examples of how well Eq. (28) fitted the data are shown in Fig. 33.

The remaining 3 of the 18 bundles at 1000 psia indicated a distinctly weaker G dependence, and were found to conform accurately with the equation

$$\phi \times 10^{-6} = K_1(G \times 10^{-6})^{1/4} + K_2(G \times 10^{-6}) \Delta h \qquad (29)$$

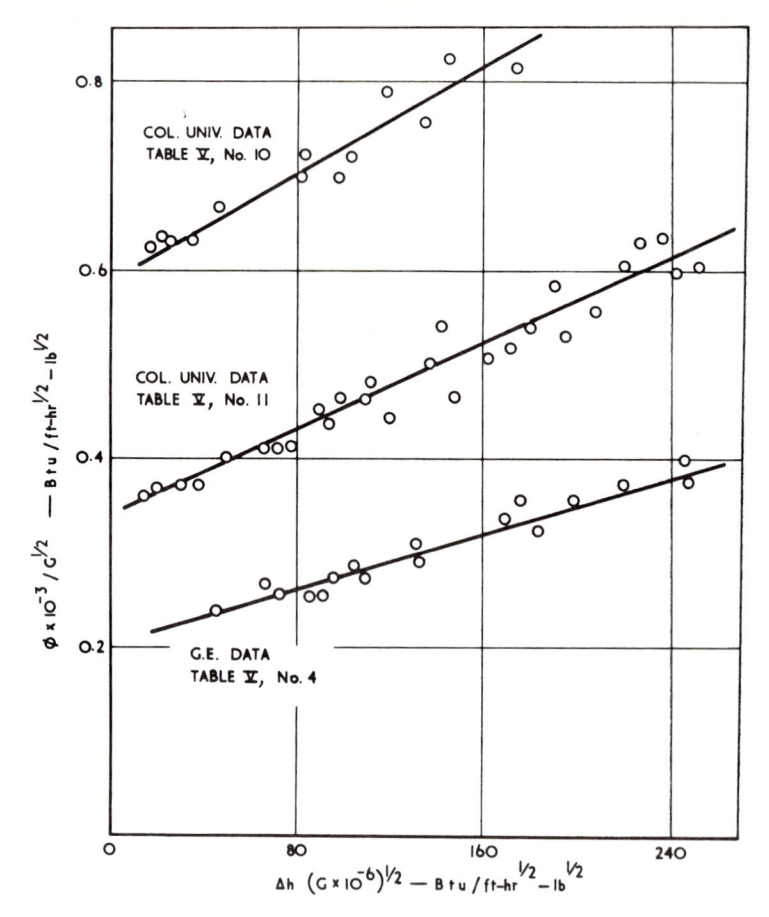

Fig. 33. Examples of correlating rod bundle burn-out data by the equation $\phi \times 10^{-6} = K_1(G \times 10^{-6})^{1/2} + K_2(G \times 10^{-6}) \Delta h$.

The cause of the weaker G dependence must be ascribed to some particular feature of geometry, although exactly what it is has not been found. All of the three bundles involved had their rods supported and correctly positioned by wires wrapped helically around certain of the rods, and it is possible that the wires caused an unfavorable distribution of steam and water. However, it is doubtful that the wire wraps were themselves responsible, since several of the bundles conforming to Eq. (28) were also wire wrapped. (Other devices used for rod supports are suitably spaced grids and ferrules.) The explanation most probably lies in a combination of the effects of the wire wraps with the effects of given rod diameters and rod spacings. For ease of identification, the data that conform with Eq. (28) are hereafter called "normal" data.

It can be seen that the above equations achieve the desirable objective of concentrating the geometry effects, which present the main difficulty in analyzing rod-bundle data, into two numerical coefficients. Their values, obtained by computer optimization, are tabulated in Table V, together with the important geometry parameters; the problem becomes one of finding which of these parameters may possibly give a smooth relationship with the K_1 and K_2 values. Examination showed that considerable simplification does appear to exist in a rod-bundle system; as an example, Fig. 34 shows a relationship between K_1 for all the "normal" data in Table V and the geometry term L/d_h. It can be seen that Fig. 34 separates the bundles with vertical upflow from those with horizontal flow, with the implication that horizontal flow produces a distinctly lower burn-out flux. Some doubt is cast on the truth of this implication, however, by more recent data (H2)

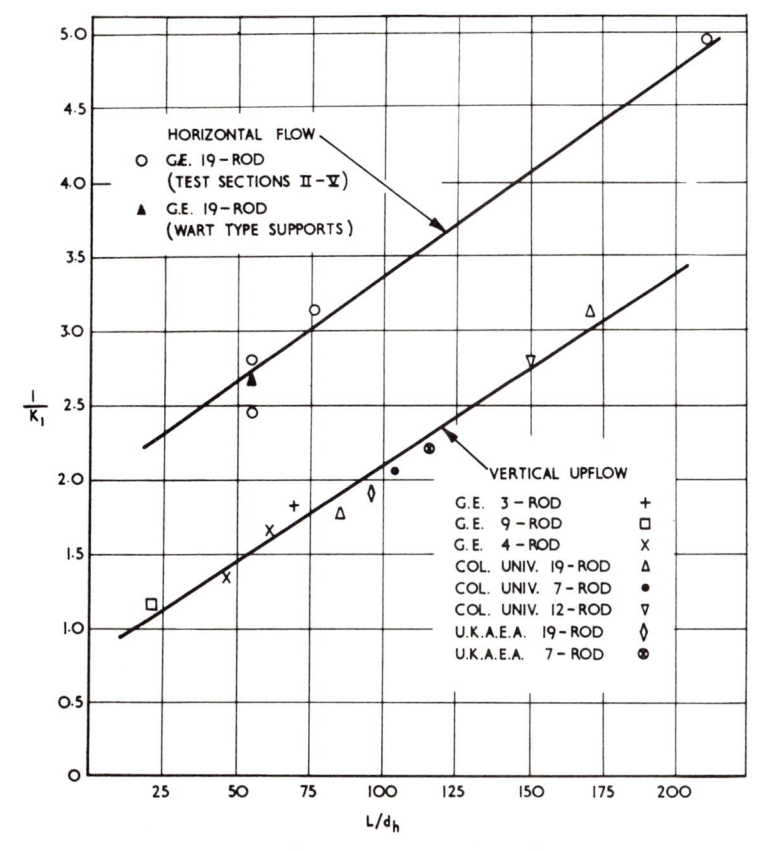

FIG. 34. A simple geometry relationship with the basic burn-out flux for rod bundles. $\phi_0 \times 10^{-6} = K_1(G \times 10^{-6})^{1/2}$. Details of the rod bundles are given in Table V.

TABLE V. Geometry Parameters and K_1 and K_2 Values for 1000 psia (Nominal) Rod-Bundle Data[a]

No.	Description of bundle[b]	L	K_1	$K_2 \times 10^3$	d_r	S_r	S_s	A	P_w	P_h	d_w	d_h	Data source ref. no.
1	GE 19-rod, test section II	19.5	0.320	0.40	0.629	0.015	0.060	2.40	47.7	37.5	0.201	0.256	W1
2	GE 19-rod, test section III	19.5	0.359	1.09	0.587	0.050	0.101	3.16	45.3	35.1	0.279	0.360	W1
3	GE 19-rod, test section IV	19.5	0.410	1.06	0.587	0.050	0.101	3.16	45.3	35.1	0.279	0.360	W1
4	GE 19-rod, test section V	76	0.201	0.73	0.587	0.050	0.101	3.16	45.3	35.1	0.279	0.360	W1
5	GE 19-rod	19.5	0.375	1.20	0.587	0.050	0.101	3.16	45.3	35.1	0.279	0.360	B8
6	GE 3-rod	54	0.549	1.82	0.250	0.062	0.132	0.454	5.11	2.35	0.356	0.772	J3
7	GE 9-rod	18	0.86	3.1	0.375	0.170	0.190	2.39	17.6	10.6	0.543	0.902	P2
8	GE 4-rod, test section I	36	0.733	3.47	0.437	0.187	0.135	1.07	10.2	5.50	0.418	0.778	H1
9	GE 4-rod, test section II	48	0.604	2.84	0.437	0.187	0.135	1.07	10.2	5.50	0.418	0.778	H1
10	CU 19-rod, test section I	36	0.563	1.01	0.550	0.083	0.083	3.49	42.8	32.8	0.326	0.425	M7
11	CU 19-rod, test section II	72	0.321	1.05	0.550	0.083	0.083	3.49	42.8	32.8	0.326	0.425	M9
12	CU 7-rod	37	0.487	1.13	0.550	0.083	0.089	1.08	19.0	12.1	0.228	0.359	M6
13	CU 12-rod	17	0.359	0.70	0.440	0.022	0.022	0.470	23.8	16.6	0.079	0.113	M11
14	UKAEA 7-rod	72	0.455	1.61	0.500	0.100	0.100	1.46	17.0	9.42	0.345	0.620	M4
15	UKAEA 19-rod	48	0.520	1.30	0.625	0.112	0.067	4.37	48.6	35.4	0.360	0.494	M4
16	GE 19-rod, test section I	18.5	0.631	1.72	0.564	0.074	0.112	3.50	43.9	33.6	0.319	0.417	W1
17	UKAEA 19-rod	48	0.372	1.07	0.625	0.085	0.085	4.30	48.6	37.3	0.354	0.461	M4
18	UKAEA 19-rod	48	0.341	1.10	0.625	0.085	0.085	4.30	48.6	37.3	0.354	0.461	M4

[a] Compiled in Macbeth (M4).

Notes: Nos. 1–5 were tested with horizontal flow and Nos. 6–18 with vertical upflow.

Nos. 2, 3, and 5 have identical rod geometry, but differ in the method used for supporting the rods.

Nos. 17 and 18 have identical rod geometry and rod supports, but have a different heat-flux variation between rods.

Nos. 1–15 are the "normal" data and conform to the equation: $\phi \times 10^{-6} = K_1(G \times 10^{-6})^{1/2} + K_2(G \times 10^{-6}) \Delta h$.

Nos. 16–18 are the nonnormal data and conform to the equation: $\phi \times 10^{-6} = K_1(G \times 10^{-6})^{1/4} + K_2(G \times 10^{-6}) \Delta h$.

[b] GE—General Electric; CU—Columbia University; UKAEA—United Kingdom Atomic Energy Authority.

which allows a direct comparison of the performance of a bundle in the two positions. These new data are for vertical upflow in a rod-bundle identical to one previously tested in the horizontal position—bundle No. 4 in Table V. The comparative results are given in Fig. 35, and it is clear that for a mass velocity greater than 10^6 lb/hr-ft^2 there is no difference between them, although the results for a G value of 0.5×10^6 lb/hr-ft^2 do show a distinctly lower burn-out flux in the horizontal position.

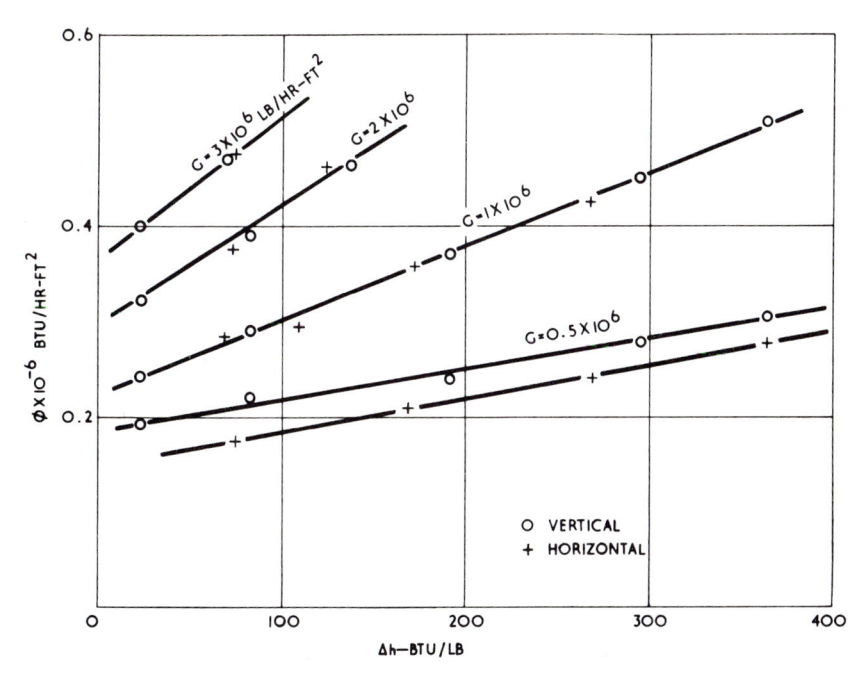

Fig. 35. Effect on burn-out of rod-bundle orientation. Bundle geometry as described for bundle No. 4, Table V. Test pressure 1215 psia.

There are several possible explanations of the apparent conflict between Figs. 34 and 35. One possible explanation, for example, would be that the vertical upflow curve drawn in Fig. 34 may not be a straight line, but should perhaps curve upwards (the data itself shows some signs of this) toward the uppermost point of the horizontal flow line, which corresponds to the rod bundle in question. It will be seen later, however, that this would not appear to be the explanation. In addition, there is the fact that all the horizontal flow data in Fig. 34, as well as the new data in Fig. 35, are for a test pressure of 1215 psia, whereas the vertical upflow data in Fig. 34 refer to 1000 psia. Although there is no evidence to indicate the effect of pressure in the case of rod-bundle systems with round tubes, it is found that increasing pressure from

1000 to 1215 psia may reduce the burn-out flux by as much as 20%, depending on geometry. The main difficulty in this matter is obviously insufficient experimental evidence.

The analysis given in Macbeth (M4) continues by assuming that the term d_h effectively represents the cross-sectional geometry of the normal rod-bundle data. If this assumption is correct, then the general correlation [Eq. (18)], may be applied by representing A' and C' as functions of G and d_h for a given pressure. It was found that simple power functions were adequate, and a correlation was obtained by computer optimization which predicted 97% of the vertical-upflow normal data (Nos. 6–15 inclusive in Table V) to within ±12%.

For the purpose of the present paper, additional vertical-upflow normal data that have become available since Ref. (M4) was written have been added to the computer optimization program, which brings the total number of different rod bundles involved up to 22 and the total number of experimental results to 637. The most important of these additions are the data given by Edwards and Obertelli (E1) from a series of burn-out tests on 37-rod bundles 144-in. long; these data represent a considerable extension to the range of system parameters. It is of interest to note that these 37-rod bundles, which have an L/d_h ratio of 271, all indicate a value for K_1 in Eq. (28) of approximately 0.21. The reciprical of K_1 is thus 4.8, and when plotted in Fig. 34 it lies almost exactly on an extrapolation of the straight line for vertical upflow. The other additional data have also been found to lie very close to this line, and it therefore seems to be a reliable means of estimating the basic burn-out flux at 1000 psia for rod bundles with vertical-upflow and normal characteristics. (Note that the flux obtained will be the average on all the heated rods taken together.)

The result of the computer optimization on the 1000 psia normal data with vertical-upflow is as follows:

$$\phi \times 10^{-6} = \frac{55.9 d_h^{0.81}(G \times 10^{-6})^{0.51} + 0.25 d_h(G \times 10^{-6})\Delta h}{35.8 d_h^{0.60}(G \times 10^{-6})^{0.32} + L} \tag{30}$$

where ϕ is the *average* flux on the *heated* rods in a bundle. Equation (30) predicts 95% of the total of 637 experimental results to within ±16%, and all but 12 of the results are within 20% limits. Detailed examination suggests that some of the scatter about the correlation line is caused by differences in the way the rods are supported in the various bundles. There is evidence, for example, that certain types of grid arrangements used in the 19-rod bundle results reported by Matzner (M10) give significantly higher burn-out flux values than Eq. (30) predicts. On the whole, however, it is surprising that the wide variety of rod supports used appear to have so little effect on burn-out.

Among the additional data examined, the only non-normal characteristics found were for some of the results in (M10), which again, as in the earlier investigation (M4), apply to 19-rod bundles with wire-wrap supports. The consistent pattern that now seems to be emerging is that it is only the use of wires wrapped around the rods that causes nonnormal characteristics. It seems that the addition of a wrap around the complete bundle, as was the case for test sections No. 10 and 11 in Table V, prevents the nonnormal characteristics from appearing.

D. ANNULAR TEST SECTIONS

An annular test section consisting of a heated rod inside an unheated shroud tube is of particular interest, since it raises the question of whether or not, from the point of view of burn-out behavior, it belongs to the family of rod bundles. Barnett (B6) has examined the question and has found clear evidence in support of the idea.

The first step was to correlate the annulus data, which, fortunately, are mostly for a pressure of 1000 psia, the same as is the case for rod bundles. A compilation of the 1000 psia data is included in Barnett's report (B6), most of the experimental work being due to Janssen and Kervinen (J3), a summary of the range of system parameters involved in the total of 744 experimental results listed is given below (in all cases, the central rod is uniformly heated, and there is liquid water inlet with vertical upflow):

Range of pressure	990–1024 psia
Range of inside diameter of shroud tube	0.551–4.006 in.
Range of heated-rod diameter	0.375–3.798 in.
Range of mass velocity	0.14×10^6–6.20×10^6 lb/hr-ft^2
Range of inlet subcooling........	0–412 BTU/lb
Range of heated length	24–108 in

Barnett used a more relaxed form of the general burn-out formula [Eq. (18)] by writing it as follows:

$$\phi \times 10^{-6} = \frac{A' + B' \Delta h}{C' + L} \tag{31}$$

where B', a function of G, d_o (the outside flow diameter), and d_i (the inside flow diameter), does not necessarily have the value $0.25 d_h (G \times 10^{-6})$ [where $d_h = (d_o^2 - d_i^2)/d_i$], as would be strictly required by Eq. (18). It has been shown in Section VI, that the Barnett local-conditions hypothesis is not rigorously valid, so that by introducing the term B' a higher correlation accuracy is to be expected.

Barnett confirmed that d_h does not fully describe the cross-sectional geometry for burn-out in an annulus, and he decided to introduce as an extra term—the wetted equivalent diameter, i.e., the hydraulic diameter $d_w = d_o - d_i$. The final expressions found suitable for A', B', and C' are

$$A' = y_0 d_h^{y_1}(G \times 10^{-6})^{y_2}\{1 + y_3 \exp[y_4 d_w(G \times 10^{-6})]\} \qquad (32)$$

$$B' = y_5 d_h^{y_6}(G \times 10^{-6})^{y_7} \qquad (33)$$

$$C' = y_8 d_w^{y_9}(G \times 10^{-6})^{y_{10}} \qquad (34)$$

The computer-optimized y values obtained for a number of conditions are given in Table VI. It can be seen that the first condition assumes simple power functions only and a value for B' strictly in compliance with Eq. (18). The rms error achieved is good, but marked improvements are obtained by relaxing the equations for A' and B' in stages, as shown, the final result giving a much better rms error. It was not necessary in the analysis to separate the data into low- and high-velocity regimes, as was the case for round-tube data, since the lowest mass velocity is not so low as to cause difficulty.

Having obtained an accurate correlation for the annulus data at 1000 psia, Barnett applied it, for the same pressure, to the rod-bundle data, and Figs. 36 and 37 are two very convincing demonstrations of the connection that clearly exists between an annulus and a rod bundle. The method used in applying the annulus correlation was to assume a d_i value equal to the diameter of the rods in the bundle considered, and a d_o value such that both annulus and bundle had the same heated equivalent diameter d_h. All the normal rod-bundle data with vertical upflow were found to be well represented by the annulus correlation, but the nonnormal data showed the same disparity that was found in the rod-bundle analysis. Thus, we have a further indication that the non-normal rod bundles are showing markedly different burn-out behavior.

The fact that d_h alone was sufficient to describe the cross section of a rod bundle in the rod-bundle correlation [Eq. (30), whereas an extra term d_w was found necessary for an annulus, is probably due to the fact that the rod-bundle geometries were very restricted in with regard to the range of rod diameters and rod pitches. If the d_h values for the rod bundles are plotted against corresponding d_w values, the result is a fairly smooth relationship, which is entirely fortuitous. The annulus data, however, show much more random scatter in a similar plot, indicating more variation in the geometry parameters. Application of the rod-bundle correlation [Eq. (30)] to the annulus data results in fairly close predictions, so that the need to introduce the term d_w in an annulus correlation is only slight.

Levy et al. (L8), have reported experiments with water at 1000 psia using a 102-in. long annulus comprising a 0.540-in. diameter uniformly heated rod

TABLE VI

Optimized y Values for Eqs. (32), (33), and (34) Applied to the 1000-psia Annular Burn-Out Data Compiled by Barnett (B6)

y_0	y_1	y_2	y_3	y_4	y_5	y_6	y_7	y_8	y_9	y_{10}	rms error
60.3	0.904	0.523	0	0	0.250	1.00	1.00	166	1.30	0.525	7.88
69.4	0.751	0.226	−0.672	−6.09	0.250	1.00	1.00	166	1.25	0.329	7.30
60.3	0.873	0.484	0	0	0.267	1.27	0.926	206	1.49	0.427	6.89
67.5	0.680	0.192	−0.744	−6.51	0.259	1.26	0.817	185	1.42	0.212	5.91

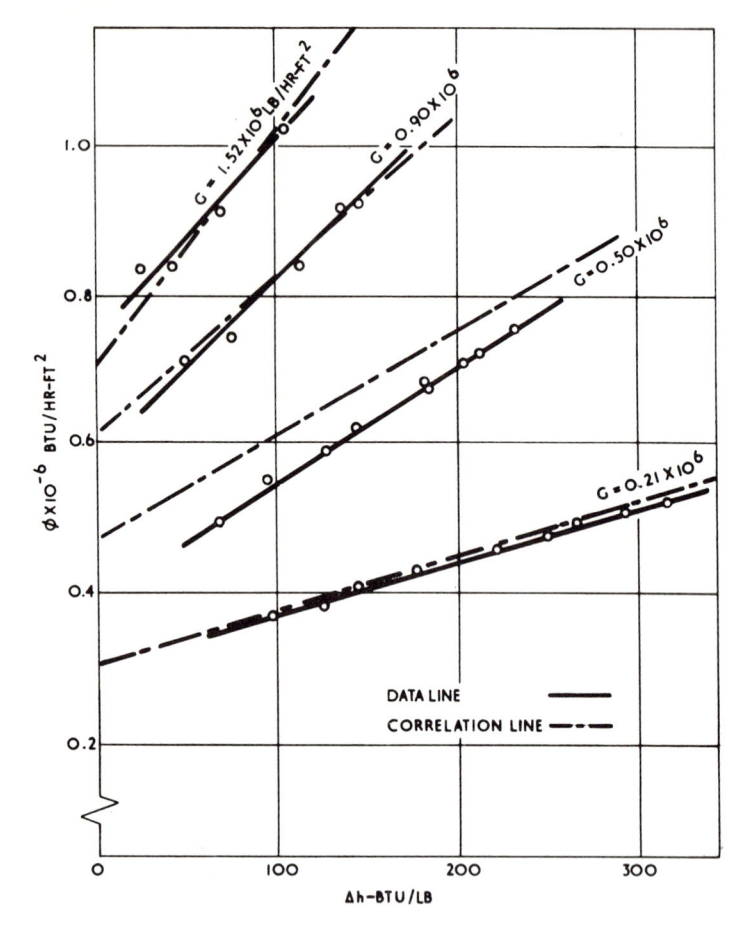

FIG. 36. Comparison of burn-out data for a 4-rod bundle (No. 9, Table V) with Barnett's annulus correlation [from Barnett (B6)].

inside an 0.875-in.-diameter shroud tube. Burn-out tests were performed with the heated rod made eccentric so as to give minimum gaps of 0.033, 0.061, 0.096 and 0.1675 in., the last figure being the concentric case. The experimental results, according to Levy et al., show a decrease in burn-out flux of from 15 to 36% by reducing the gap from 0.1675 to 0.033 in. An examination of these data made by Barnett (B4), however, showed that there were some discrepancies when the heat-balance equation was applied to the burn-out heat flux and quality values quoted. Barnett (B6) has obtained the correct values from a private communication with Levy et al., and on plotting the revised data on a ϕ versus Δh graph, taking the effect of mass velocity into account, has shown that the decrease in burn-out flux on reducing the

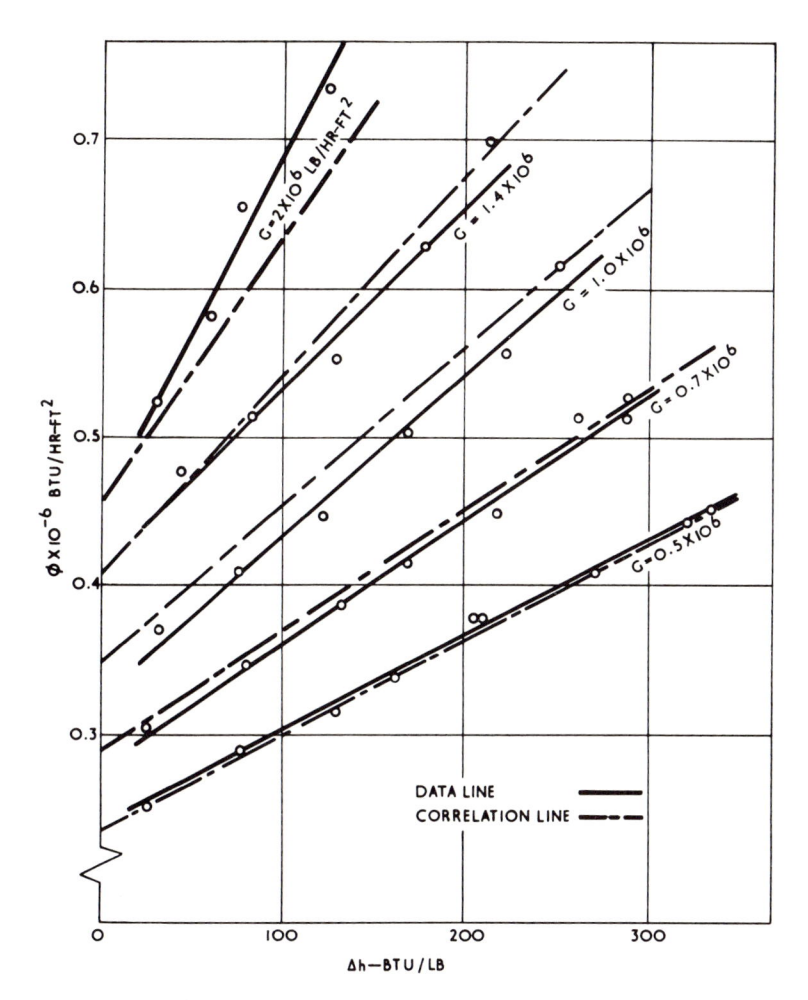

FIG. 37. Comparison of burn-out data for a 19-rod bundle (No. 11, Table V) with Barnett's annulus correlation [from Barnett (B6)].

gap from 0.1675 to 0.033 in. is, in fact, less than 15%. It would appear that the gap between a heated rod and the unheated shroud tube in a rod-bundle arrangement should not influence burn-out very much, except perhaps if the gap becomes extremely small. This view is supported by the rod-bundle correlation [Eq. (30)], which takes no account of the gap in question, yet represents with reasonable accuracy the wide range of normal bundle geometries, for which gaps vary from 0.022 to 0.190 in. The possibility that a very small gap may have a significant effect is suggested from experiments reported by Howard (H9) using a heated rod of 0.540-in. diameter and 73-in. length inside a shroud tube of 0.875-in. diameter. The heated rod was intentionally bowed over a short length so that it actually touched the shroud wall 5 in. below the outlet end of the heated length. The tests were with water at 1000 psia, and they indicate burn-out flux values 30–40% lower than the same annulus without bowing, the mass velocity and inlet subcooling being the same.

Experimental data have been produced for annuli in which both the inner rod and the shroud have been heated. Adorni et al. (A2), for example, refer to an annulus tested at 1000 psia, but using a preformed steam and water mixture which produced soft inlet conditions; these data are consequently loop-dependent. Becker et al. (B13) have also reported burn-out results for an annulus with dual heating, using liquid water at inlet with pressures varying from about 120 to 540 psia. The varying pressures make analysis of these data difficult, but two interesting conclusions appear to be that, for a given pressure and given inlet conditions, (1) putting a heat flux on the shroud tube has the effect of decreasing the burn-out flux on the inner rod, and (2) putting a heat flux on the rod has practically no effect on the burn-out flux on the shroud. Both of these conclusions apply within the limited range of the experiments, and it is not known how general they are.

The annulus test section has also been used to examine the influence on burn-out of various appendages that may be put in a channel. For example, Janssen and Kervinen (J3) attached circular ribs of 0.080-in. square section at about 1-in. pitch onto the inside of a 0.875-in. diameter shroud tube. The uniformly heated rod was of 0.375-in. diameter and 72-in. length; tests were done with water at 1000 psia. For a given inlet subcooling and mass velocity, the burn-out heat flux was found to be higher than that obtained without the circular ribs being present. At a mass velocity of 10^6 lb/hr-ft^2 for instance, the increase was about 15%. Lee (L3) has carried out tests with an annulus 31.5 in. long having a uniformly heated rod of 0.625-in. diameter and a shroud tube of 0.834-in. diameter. Tests were with water at 1000 psia. The effect on burn-out of a variety of appendages was studied, including:

(1) Wire wraps wound helically around the heated rod on a 9-in. pitch (single start) and on an 18-in. pitch (both single and two start).

(2) Short, longitudional bearing pads attached to the heated rod at a number of positions.

(3) Three longitudional strings of ceramic beads held against the heated rod over the full length by wires.

Compared with the performance of the annulus test section without any attachments, the above produced the following effects:

(1) The 9-in. wrap caused the burn-out flux to fall, the more so as the mass velocity increased; a 25% fall was recorded at $G = 2 \times 10^6$ lb/hr-ft^2.

(2) With the 18-in. wrap, and with the bearing pads, there was a tendency for the burn-out flux to increase slightly at low mass velocities; otherwise there was no significant effect.

(3) The beaded rod gave an increase in burn-out flux of about 30% at $G = 0.75 \times 10^6$ lb/hr-ft^2, reducing to a 10-15% improvement at higher mass velocities.

The above result for the 9-in. wrap is interesting, since the drop in burn-out flux it causes is similar to the effect of using 9-in. wraps in rod bundles. In the previous section, these bundles were identified as having nonnormal characteristics.

IX. The Equivalent-Diameter Hypothesis

The connection that has been shown in Section VIII to exist between burn-out in a rod bundle and in an annulus leads to the question of whether or not a link may also exist between, for example, a round tube and an annulus. Now, a round tube has its cross section defined uniquely by one dimension—its diameter; therefore if a link exists between a round tube and an annulus section, it must be by way of some suitably defined equivalent diameter. Two possibilities that immediately appear are the hydraulic diameter, $d_w = d_o - d_i$, and the heated equivalent diameter, $d_h = (d_o^2 - d_i^2)/d_i$; however, there are other possible definitions. To resolve the issue, Barnett (B4) devised a simple test, which is illustrated by Figs. 38 and 39. These show a plot of reliable burn-out data for annulus test sections using water at 1000 psia. Superimposed are the corresponding burn-out lines for round tubes of different diameters based on the correlation given in Section VIII. It is clearly evident that the hydraulic and the heated equivalent diameters are unsuitable, as the discrepancies are far larger than can be explained by any inaccuracies in the data or in the correlation used.

By examining many plots similar to Figs. 38 and 39, Barnett concluded that there can be no satisfactory definition for an equivalent diameter, the inaccuracies involved being too large to be acceptable. For example, Fig. 39 may suggest that $d = 1.5$ in. is an appropriate equivalent diameter for the annular test section considered, but as a general rule it is found that for another value of the mass velocity, or a different channel length, the data points will no longer coincide with the $d = 1.5$-in. line. Barnett's tests included rod-bundle data as well as annuli, and he showed that the same conclusion as above applies to a rod bundle, which is to be expected, since an annulus belongs to the family of rod bundles, as shown earlier. The implications of this finding are important because in the past it has been customary to use correlations for single tubes in the assessment of burn-out in bundles, using an equivalent diameter to connect the two cases.

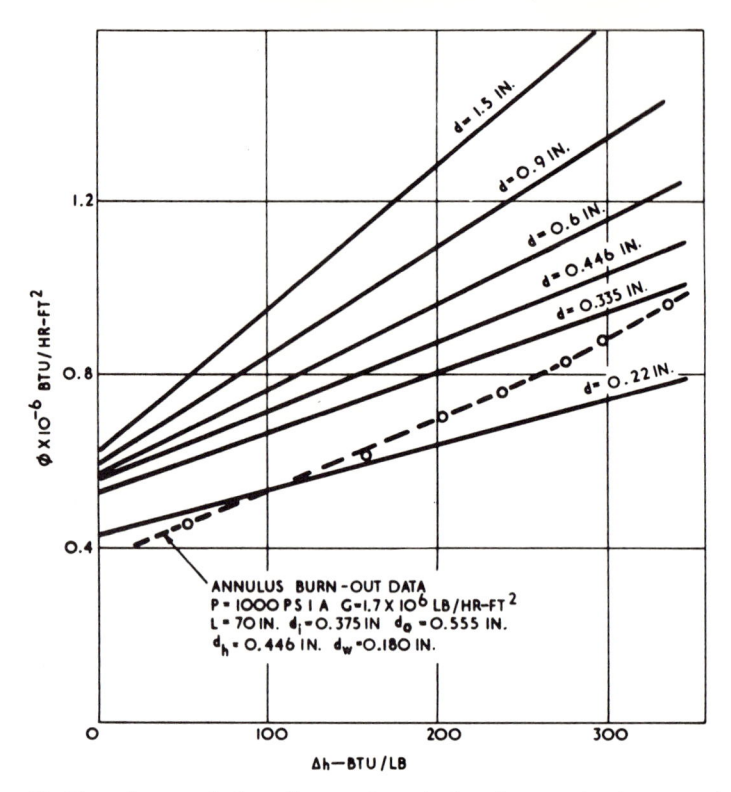

FIG. 38. Test of an equivalent-diameter hypothesis using annulus burn-out data for water [from Barnett (B4)]. The continuous lines are evaluated from the correlation of round tube data given in Section VIII, A for the same pressure, mass velocity, and length as the annulus data, and for the diameters shown.

X. The Effect on Burn-Out of Nonuniform Heating

All the references to burn-out have thus far been concerned with uniformly heated channels, apart from some of the rod bundles where the heat flux varies from one rod to another, but which respond to analysis in terms of the average heat flux. In a nuclear-reactor situation, however, the heat flux varies along the length of a channel, and to find what effect this may have, some burn-out experiments on round tubes and annuli have been done using, for example, symmetrical or skewed-cosine axial heat-flux profiles. Tests with axial non-uniform heating in a rod bundle have not yet been reported.

One result of nonuniform heating is that burn-out does not always occur at the channel outlet as it does with uniform heating, provided instabilities are avoided and the flow is vertically upwards. Consequently, the problem of analysis is made that much more difficult. The obvious first step is to see if

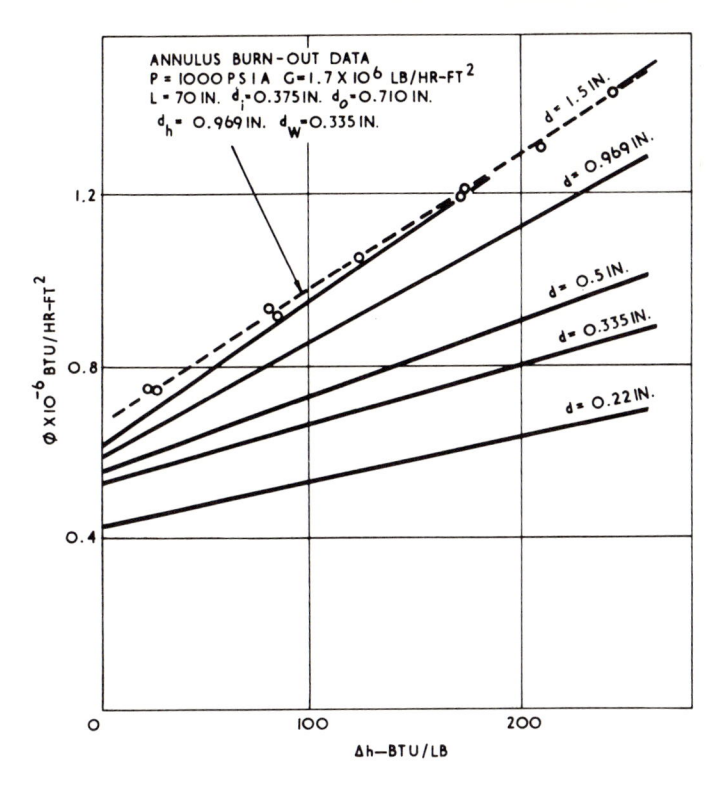

FIG. 39. Test of an equivalent diameter hypothesis using annulus burn-out data for water [from Barnett (B4)]. The continuous lines are evaluated from the correlation of round tube data given in Section VIII, A for the same pressure, mass velocity, and length as the annulus data, and for the diameters shown.

any connection can be found between burn-out in uniformly and nonuniformly heated channels. The Barnett local-conditions hypothesis of Eq. (10), which works extremely well with uniform heating, provides, for example, a ready means of testing one connection that may exist. Tests show, however, that this hypothesis cannot provide the necessary link; Fig. 40 is an example given by Barnett (B4) using the data of Swenson *et al.* (S18) for experiments with water at 2000 psia on a 0.422-in. diameter tube 72 in. long having a skewed-cosine heat-flux profile with a peak flux about 18 in. from the outlet end. Figure 40 shows the value of the peak heat flux when burn-out is signalled somewhere in the tube plotted against inlet subcooling for three different mass velocities. With cosine heat-flux profiles it is usually found that a number of adjoining burn-out detectors record burn-out signals almost simultaneously, so that a precise burn-out position cannot generally be specified. Lee (L1) for example, in tests on a cosine-heated tube 144-in. long, reports burn-out occurring over a 32-in. length starting 4 in. from the outlet end. For the data in Fig. 40, the spread recorded was mostly of the order of 3 in.

The Barnett local-conditions hypothesis has been applied to the above example by rewriting the general burn-out correlation for uniformly heated round tubes [Eq. (22)] as follows:

$$\phi \times 10^{-6} = \frac{1}{C'} \{A' - \tfrac{1}{4}d(G \times 10^{-6})\kappa\lambda\} \tag{35}$$

This is derived by substituting from the heat-balance expression, Eq. (20). Now, the skewed-cosine heat-flux profile being considered gives a known functional relationship between the flux and the quality at any position along the channel. By equating this relationship with Eq. (35), a solution can be obtained giving the local values of ϕ and κ at the predicted burn-out position. The corresponding peak flux can then be evaluated, and in this way the predicted burn-out lines for the three mass velocities in Fig. 40 can be drawn.

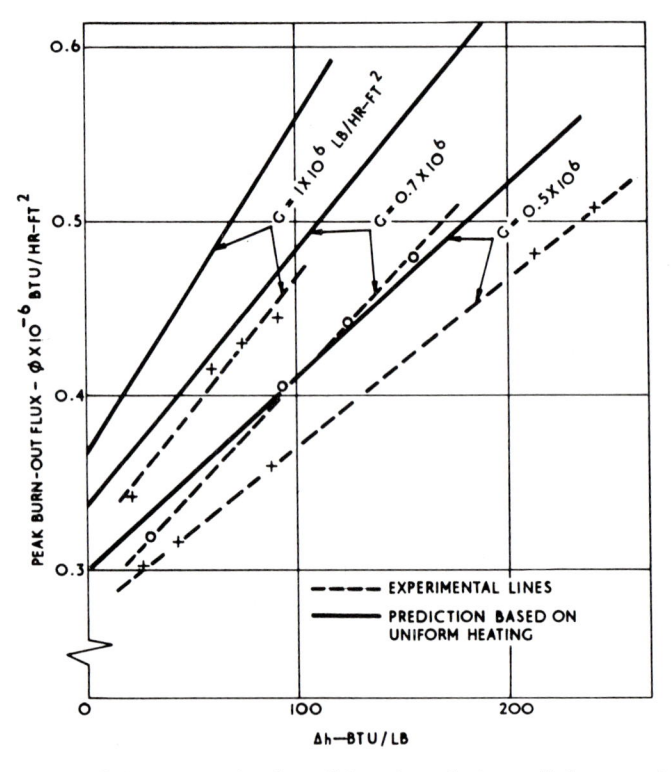

Fig. 40. Test of the Barnett local-conditions hypothesis applied to a tube with a skewed-cosine heat-flux profile [from Barnett (B4)]. Fluid: water, $d = 0.422$ in., $L = 72$ in., $P = 2000$ psia.

The A' and C' terms are given by Eqs. (23) and (24), and appropriate y values for a pressure of 2000 psia are listed in Table II.

It can be seen from Fig. 40 that the discrepancy between predicted and experimental lines is too large to be acceptable, and that, therefore, the Barnett local-conditions hypothesis must be rejected as unable to link together a uniformly and nonuniformly heated channel. General confirmation of this finding has been shown by Barnett (B4), Lee (L1), Lee and Obertelli (L5) and Stevens et al. (S14) for a wide range of cosine and other heat-flux profiles. An example of a hot-patch test made by Stevens et al. (S14) is shown in Fig. 41 for a round tube of 0.380-in. diameter with Freon-12 at 155 psia. In these tests, the tube was uniformly heated to give the control results, and for the hot-patch experiments, the heat flux was doubled over a 4-in. length at the tube outlet. Figure 41 is a plot of ϕ against κ, measured at the tube outlet, where burn-out occurred, and it therefore gives a direct test of the local-conditions hypothesis. It can be seen once again that the hypothesis is unable to link together a uniformly and a nonuniformly heated channel.

There is an exception to the above generalization which is shown in Russian data reported by Smolin et al. (S5) for a round tube of 0.291-in. diameter and 150-in. long with water at pressures of 1420 and 2560 psia. In

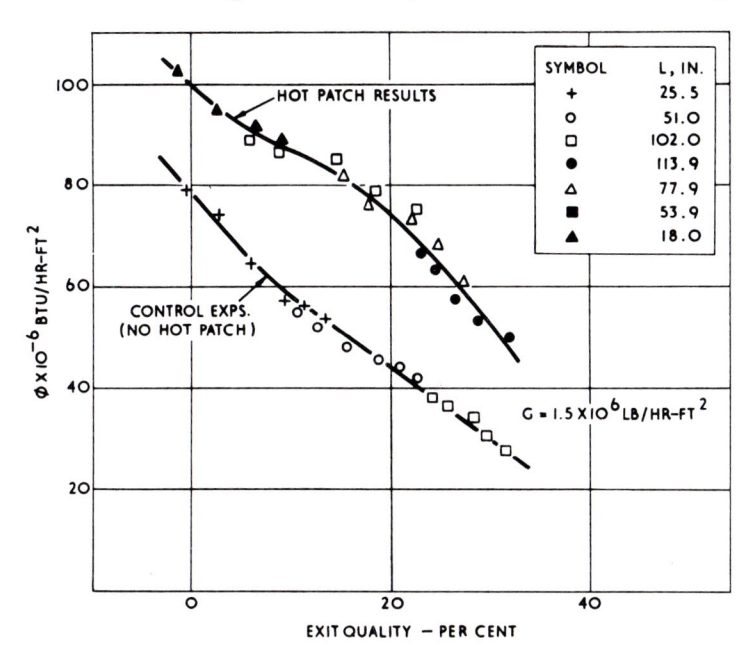

FIG. 41. Test of the Barnett local-conditions hypothesis applied to uniformly heated tubes of various length, with the heat flux doubled over the last 4 in. at the outlet end [from Stevens et al. (S14)]. Fluid: Freon-12, $d = 0.38$ in., $P = 155$ psia.

one case, the heat flux on the tube was uniform, and in another case it was arranged so that the heat flux on the upper 59 in. of the tube was $2\frac{1}{2}$ times that on the lower 91 in. The heat flux on the uniformly heated tube and on the upper part of the other tube was fixed at 0.368×10^6 Btu/hr-ft^2, and the burn-out mass velocity was determined for different inlet subcoolings. The results, given in Fig. 42, show the quality at burn-out, which occurred at the tube outlet in both cases, plotted against the burn-out velocity. It can be seen that the local-conditions hypothesis is effective in linking together the two sets of burn-out data. The result is an interesting one, since it suggests that if, in a nonuniformly heated channel, there is a fairly long length of uniformly heated portion before the burn-out position, then a "memory" of the earlier heat-flux profile is lost and burn-out is dependent only on the local conditions.

No really effective way has yet been found for predicting burn-out in nonuniformly heated channels, but there is one method that does appear consistently to get within about $\pm 20\%$. The method was first demonstrated

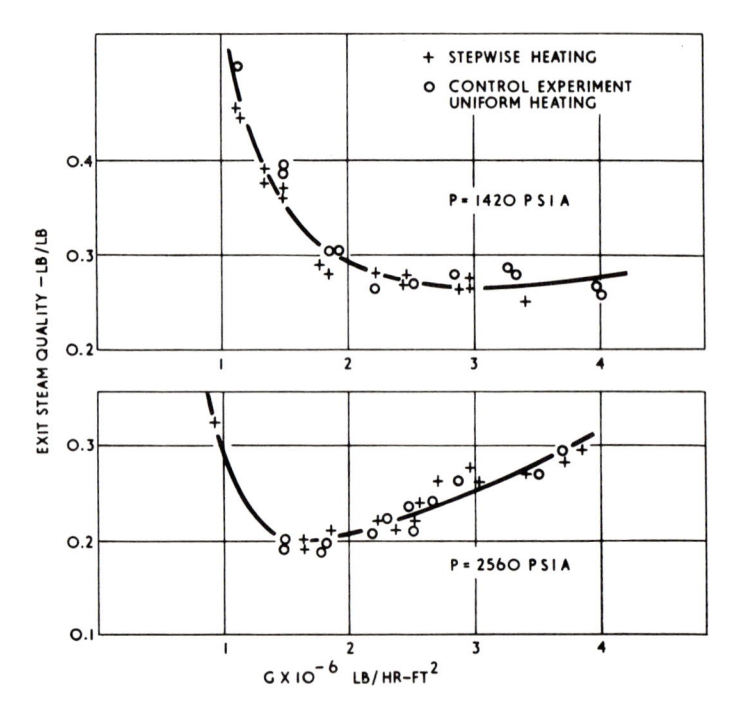

FIG. 42. Test of the Barnett local-conditions hypothesis applied to a vertical tube with heat flux on the upper 59 in. of the tube (the outlet end) $2\frac{1}{2}$ times that on the lower 91 in. [from Smolin et al. (S5)]. Fluid: water, $d = 0.291$ in., $L = 150$ in., Fixed heat flux at outlet end: 0.37×10^6 Btu/hr-ft^2.

by Cook (C6), and the example he gave is reproduced in Fig. 43. The data points shown are for a symmetrical-cosine heated rod of 0.540-in. diameter and 108-in. length contained within a 0.875-in.-diameter shroud tube using water at 1000 psia. The plot is of total burn-out power on the rod against inlet subcooling for two mass velocities, and the lines drawn are based on calculations for a uniform axial heat flux. (These lines check almost exactly with the burn-out correlation given in Section VIII,D for an annulus.) The hypothesis proposed by Cook, one which is supported by the evidence in Fig. 43, is that burn-out with nonuniform power distribution may be reliably predicted from uniform power-distribution data. This total-power hypothesis has been checked for a wide variety of heat-flux profiles by Lee and Obertelli (L5), Lee (L1), Biancone *et al.* (B20), Bertoletti *et al.* (B17), and Barnett (B4), and confirmed to within about $\pm 20\%$.

That the total-power hypothesis cannot be correct, however, is well shown from an experiment reported by Lee and Obertelli (L5). In this experiment, a symmetrical-cosine heated tube of 0.383-in. diameter and 72-in. length was

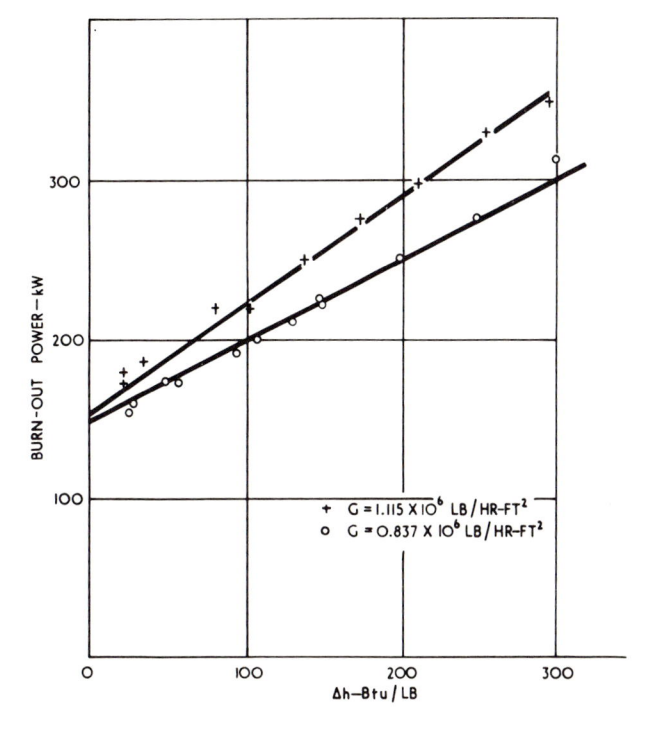

FIG. 43. Test of the total-power hypothesis applied to an annulus with a symmetrical-cosine heated rod [from Cook (C6)]. Fluid: water, $d_i = 0.540$ in., $d_o = 0.875$ in., $L = 108$ in., $P = 1000$ psia.

tested with water at 1000 and 1600 psia, and the burn-out power was obtained for a range of mass velocities and inlet subcooling. It was noted that burn-out signals were always recorded over an area of the tube which extended to distances less than 60 in. from the tube inlet. Consequently, it was decided to repeat the experiments with the heat flux made zero over the last 12 in., the heat-flux profile over the first 60 in. remaining as before. It was found that the peak fluxes at burn-out were hardly altered, and it was concluded that the heat input downstream of any part of the burn-put area has no affect on burn-out. The implication of this conclusion with regard to the total-power hypothesis is obvious.

XI. The Technique of Modeling Forced-Convection Burn-Out

Modeling is a valuable technique in almost any sphere of engineering, as it allows the behavior of full-size equipment or systems to be studied at a fraction of the cost with greater speed and with the opportunity of exploring the effect of modifications easily. In large, complex boiling systems, the need for modeling techniques is very real, and considerable progress has recently been made in establishing successful techniques for dealing with the problem of forced-convection burn-out. There are two approaches, and the present state of these is summarized below.

A. The Scaling-Law Approach

The scaling-law approach has been developed by Barnett (B5) based on experimental work of Lee and Obertelli (L4) for water and of Stevens et al. (S13) for Freon. To establish a set of scaling laws, the key operation is to find what fluid properties are important in the behavior of the phenomenon being studied. Three properties can be named right away, the latent heat (λ) and the densities of the liquid and vapor phases (ρ_L and ρ_V, respectively), since these are essential for correctly scaling such average conditions in a boiling channel as vapor quality and voidage. With regard to other important properties however, it is not possible to present any convincing theoretical arguments. Hence, any method for determining the significance of a particular additional property must involve the use of experimental data. An ideal technique for doing this would be to select a fluid, vary each property individually, and examine the effect of these variations upon the behavior of the system. However, since each property is dependent on pressure and temperature, individual variation cannot be achieved, and therefore this method is not feasible. The need for experimental data for two or more fluids is thus apparent, and the method adopted by Barnett is as follows:

A variety of possible sets of important properties are chosen for testing, such as those shown in Table VII. Two properties among the sets listed will be unfamiliar. These are $\beta = d\theta_s/dP$ and $\gamma = -d(\rho_L/\rho_V)_s/dP$, where θ_s is the saturation temperature corresponding to the pressure P.

The family of uniformly heated round tubes has the simplest heat flux distribution and test-section geometry for which burn-out may occur, and it was selected for the preliminary investigation to test the different sets of important properties. Now, for a uniformly heated tube, the burn-out flux for a given fluid can be represented by the equation

$$\phi = f(L, d, G, P, \Delta h) \tag{36}$$

Since, however, at any given pressure there is for all substances a unique value of the density ratio ρ_L/ρ_V evaluated at the saturation temperature, Eq. (36) may be rewritten as

$$\phi = f(L, d, G, \rho_L/\rho_V, \Delta h) \tag{37}$$

This allows the pressure, which has significance only when the fluid experiencing it is named, to be replaced by a dimensionless group of properties which has significance for all fluids. Thus, the first requirement in constructing a model of a system is that the inlet pressures should be selected so that the density ratio of the two phases in the model is the same as that in the system

TABLE VII

Possible Sets of Scaling Laws for Forced Convection Burn-Out and the Implied Scaling Factors for Water and Freon-12

Set no.	Properties assumed important in addition to λ, ρ_L, and ρ_V	L and d	G	Δh	ϕ
1	C_{pL}, k_L, β	0.589	2.33	11.7	27.3
2	C_{pL}, k_L	0.669	2.05	11.7	24.0
3	C_{pL}, k_L, σ	1.16	1.19	11.7	13.9
4	C_{pL}, k_L, γ	0.717	1.91	11.7	22.4
5	β, k_L, γ	0.483	1.91	11.7	22.4
6	μ_L, σ	0.102	4.01	11.7	47.0
7	μ_V, σ	1.49	1.05	11.7	12.3
8	μ_L, γ	0.213	1.91	11.7	22.4
9	μ_L, k_L, β	0.320	1.27	11.7	14.9
10	μ_V, γ	0.814	1.91	11.7	22.4
11	σ, γ	0.446	1.91	11.7	22.4
12	C_{pV}, k_V, β	0.371	1.97	11.7	23.1
13	C_{pV}, k_V, γ	0.381	1.91	11.7	22.4
14	C_{pV}, k_V, σ	0.627	2.49	11.7	29.1

of interest. The density ratio is a dimensionless group of properties only, as distinct from dimensionless groups that may incorporate both properties and system parameters, and by making the density ratio the same in the model as in the system of interest we rule out the possibility of also being able to make other dimensionless groups of properties the same, except possibly by chance. Herein lies an obstacle to scaling, for if another dimensionless group of properties, e.g., the Prandtl number, was really important to the burn-out phenomenon, then the practical application of the scaling-law technique would be generally impossible. The sets of important fluid properties that have been examined by Barnett, therefore, as listed in Table VII, are such that only one dimensionless group of properties, ρ_L/ρ_V, can be formed from amongst their number.

If the properties in set 1, for example, are assumed to be the important ones, the implication is that Eq. (37) may be generalized by writing

$$\phi = f(L, d, G, \rho_L/\rho_V, \Delta h, \lambda, \rho_L, \rho_V, C_{pL}, k_L, \beta) \tag{38}$$

The application of dimensional analysis to Eq. (38) yields, under the assumption that heat and macroscopic kinetic energy are fundamentally different physical quantities (so that five units are required—heat, mass, length, time, and temperature), the expression

$$\frac{\phi \beta^{1/2} C_{PL}^{1/2}}{\lambda^{3/2} \rho_L^{1/2}} = f\left[\frac{L}{d}, \frac{dC_{PL}^{1/2}\lambda^{1/2}\rho_L^{1/2}}{k_L \beta^{1/2}}, \frac{G\beta^{1/2}C_{PL}^{1/2}}{\lambda^{1/2}\rho_L^{1/2}}, \frac{\rho_L}{\rho_V}, \frac{\Delta h}{\lambda}\right] \tag{39}$$

The validity of set 1 can be tested by ascertaining whether Eq. (39) is itself valid; similar procedures are followed for the other sets of properties.

Most of the tests made so far have used water and Freon-12 (CCl$_2$F$_2$), and the scaling factors implied by the various possible sets of scaling laws may be calculated from the physical properties for these two fluids. The appropriate scaling factors based on water at 1000 psia, for which $\rho_L/\rho_V = 20.63$, are listed in Table VII. As an example of how the scaling factors are calculated, the group $\Delta h/\lambda$ in Eq. (39) will have the same value for water and Freon-12 if

$$\frac{(\Delta h)_{\text{water}}}{(\Delta h)_{\text{Freon 12}}} = \frac{(\lambda)_{\text{water}}}{(\lambda)_{\text{Freon 12}}} = 11.7$$

The precise method of making a test is best seen by referring to Fig. 44, which is an actual test of Set 4 of the scaling laws and is reproduced from Barnett (B5). The figure shows experimental data for water in a tube at three different mass velocities. Experiments were carried out on Freon-12 using parametric values for L, d, G, and Δh in accordance with the implied scaling factors shown in Table VII. The burn-out-flux values obtained were then

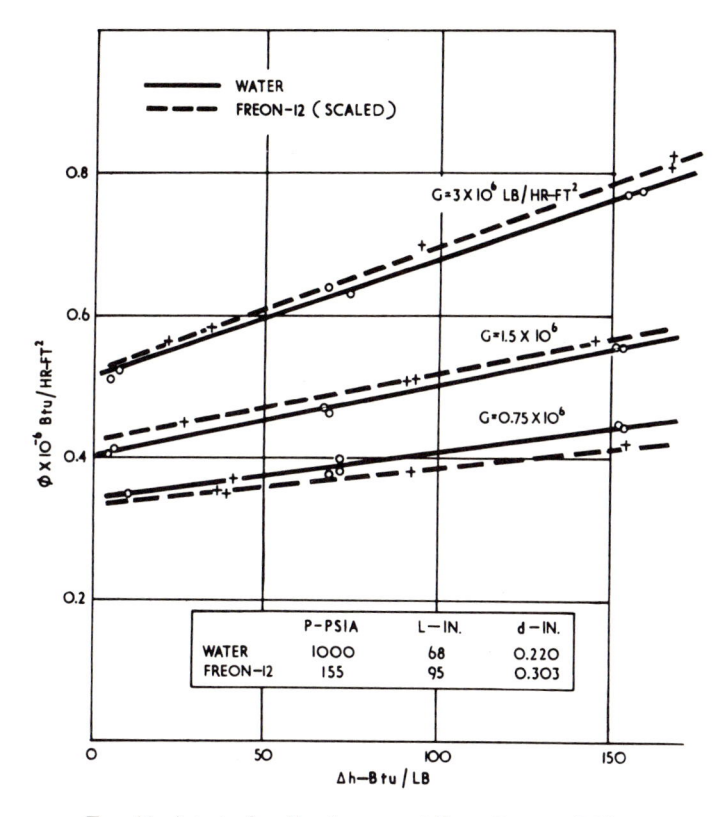

FIG. 44. A test of scaling laws, set 4 [from Barnett (B5)].

multiplied by the scaling factor for ϕ to give the water equivalents of Freon-12, and it is these values that are shown plotted in Fig. 44, together with the original water data. The closeness of the results for the two fluids indicates that set 4 of the scaling laws is a valid set within the range of the parameters shown, and further tests at other pressures and for other tube sizes tend to confirm this finding. An example of a test of set 3 of the scaling laws is given in Fig. 45, and shows that this particular set is definitely not valid.

Barnett's conclusions on the scaling-law investigations thus far carried out are:

(1) Of the possible sets of scaling laws containing only the dimensionless group of properties ρ_L/ρ_V, set 4, which includes λ, ρ_L, ρ_V, γ, C_{pL}, and k_L, appears closest to consistency with the available experimental evidence for burn-out in uniformly heated round tubes.

(2) The available evidence suggests that another property must be added to set 4 in order to reduced errors below $\pm 7\%$. This property is probably

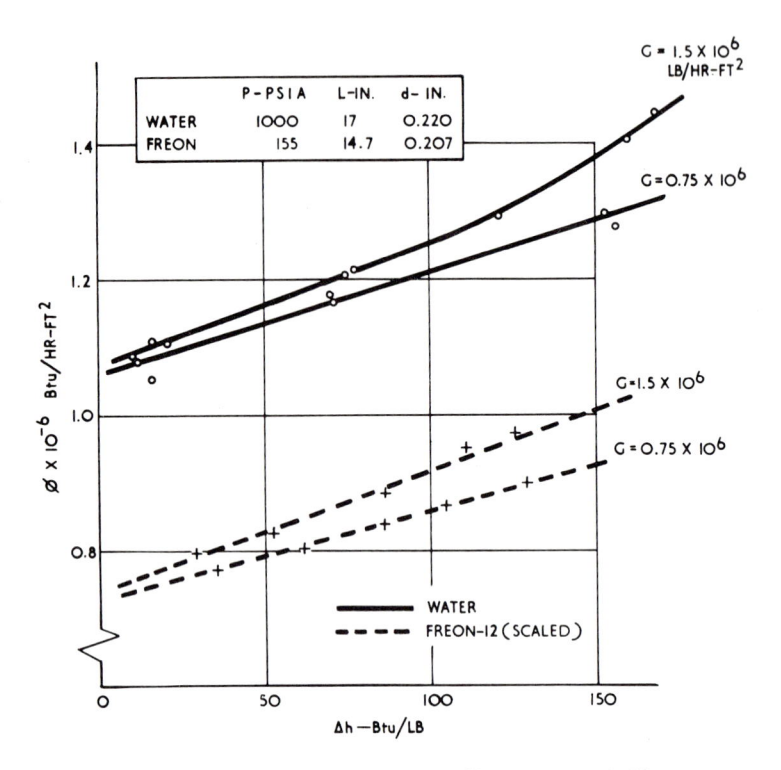

FIG. 45. A test of scaling laws, set 3 [from Barnett (B5)].

β, which unfortunately gives rise to a second dimensionless group of properties, viz., $\lambda\gamma/\beta C_{pL}$.

(3) If the second conclusion is correct, the Freons cannot be used to model burn-out in water-cooled systems to better than $\pm 7\%$, based on the available evidence for uniformly heated round tubes.

(4) Data for other geometries, e.g., annuli, must be obtained in order that set 4 may be tested more widely. The object of such tests would be to see if significantly larger errors occur with other geometries or whether errors never exceed $\pm 7\%$. If errors never exceed $\pm 7\%$, set 4 may present a useful method of obtaining assessments of the behavior of boiling channels which are sufficiently accurate for design purposes.

B. The Scaling-Factor Approach

A less rigorous approach to modeling than the one described above is an empirical procedure whereby water data for a given geometry may be linked with that of another fluid by numerical scaling factors without actually identifying the fluid properties which make up the factors. The procedure has been developed by Staniforth and Stevens (S7), and consists of the following logical sequence of steps:

(1) Note is first made of the fact that λ, ρ_L, and ρ_V must be important properties influencing burn-out.

(2) The ratio ρ_L/ρ_V is assumed to be the same for both the modeling fluid and the fluid of interest. This step establishes the corresponding pressures, but it must be noted that it is subject to the condition that there will not be another important dimensionless group of fluid properties, as was discussed earlier.

(3) A scaling factor for Δh is assumed to be given by making the ratio $\Delta h/\lambda$ the same for both the modeling fluid and the fluid of interest.

(4) The ratio L/d is taken to be the same, which is axiomatic in any modeling procedure.

(5) From the above, the burn-out equation for, say, a round tube can be written as follows:

$$\phi = f(L/d, d, G, \rho_L/\rho_V, \Delta h/\lambda) \tag{40}$$

which shows that scaling factors are still required for ϕ, d, and G.

(6) It is noted that $\phi/\lambda G$ is a dimensionless group, and that, consequently, it is possible to write:

$$\text{Scaling factor for } \phi = (\lambda/\lambda_m) \times \text{Scaling factor for } G$$

where λ and λ_m refer respectively to the fluid of interest and to the modeling fluid. This equation means that only two scaling factors have to be found.

For convenience, Staniforth and Stevens took a number of arbitrary scaling factors for d and in each case compared experimental curves of ϕ versus G for water against the corresponding curves for Freon-12. This means that the dimensionless groups in Eq. (40) were the same for each fluid, and an arbitrary scaling factor for d applied. By comparing two sets of curves, it was found possible to make them coincide by multiplying either coordinate by a numerical factor which is then the appropriate scaling factor. The method has been found to be consistent and accurate, and the results obtained from tests on uniformly heated round tubes are given in Fig. 46.

The scaling-factor approach has been applied by Stevens and Wood (S12) to 19-rod bundle test sections, using water at 1000 psia and Freon-12 at the corresponding pressure of 155 psia. The scaling factors shown in Fig. 46 were used in these rod-bundle experiments, the linear-scaling factor being taken as unity; in fact, the same rod bundles were used for both fluids. The results indicate excellent agreement, the discrepancies being no greater than might be expected between sets of water data from different sources.

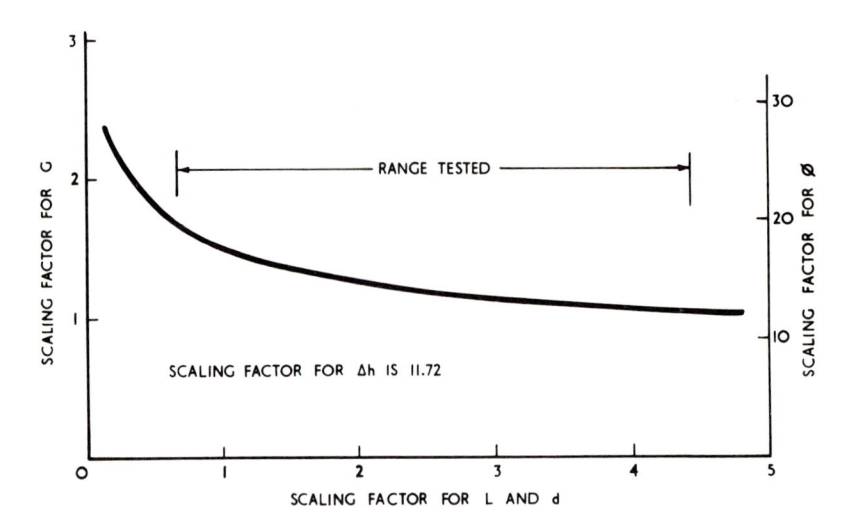

FIG. 46. Scaling factors which relate burn-out data for water at 1000 psia and Freon-12 at 155 psia [from Staniforth and Stevens (S7)].

Nomenclature

A	Cross-sectional flow area of a channel, in.^{2}	f	Denotes an unknown function
A'	An unknown function [see Eq. (18)]	G	Average mass velocity, lb/hr-ft^2
B'	An unknown function [see Eq. (31)]	H_{BO}	Average fluid enthalpy at position of burn-out, Btu/lb
C'	An unknown function [see Eq. (18)]	h_f	Enthalpy of saturated liquid, Btu/lb
C_p	Specific heat at constant pressure	K_1, K_2	Numerical coefficients defined by Eq. (28)
d	Tube internal diameter, in.	k	Thermal conductivity
d_c	Represents the terms that describe a channel cross-section	L	Heated channel length, in.
		P	System pressure, psia
d_h	Heated equivalent diameter $(4A/P_h)$, in.	P_h	Total heated perimeter, in.
		P_w	Total wetted perimeter, in.
d_i	Inside flow diameter of a concentric annulus, in.	S	Internal spacing between flat heating surfaces of a rectangular channel, in.
d_o	Outside flow diameter of a concentric annulus, in.	S_r	Rod to rod spacing or gap, in.
d_r	Rod diameter, in.	S_s	Rod to shroud spacing or gap, in.
d_w	Hydraulic diameter $(4A/P_w)$, in.	$y_0, \ldots, y_n$	Numerical values optimized by computer

GREEK LETTERS

$\beta = d\theta_s/dP$
$\gamma = -d(\rho_L/\rho_V)_s/dP$
Δ_h Degree of subcooling at channel inlet, expressed as an enthalpy deficit, Btu/lb
θ_s Saturation temperature corresponding to pressure P.
κ Average flowing steam fraction or quality, lb/lb

λ Latent heat, Btu/lb
μ Viscosity
ρ Density
σ Surface tension
ϕ General symbol for burn-out heat flux, Btu/hr-ft^2
ϕ_0 Basic burn-out heat flux ($\Delta h = 0$), Btu/hr-ft^2

SUBSCRIPTS

e Experimental value
C Calculated value

L Liquid phase
V Vapour phase

References

A1. Adnams, D. J., Salt, K. J., and Wintle, C. A., Methods of detecting burnout in rod clusters, AEEW-R.432 (1965).

A2. Adorni, N., Bertoletti, S., Lesage, J., Lombardi, C., Peterlongo, G., Soldaini, G., Weckermann, F. J., and Zavattarelli, R., Results of wet steam cooling experiments; pressure drop, heat transfer and burnout measurements in annular tubes with internal and bilateral heating, CISE-R.31 (1961).

A3. Aladyev, I. T., Dodonov, L. D., and Udalov, V. S., Critical heat fluxes for water flow in pipes, *At. Energ.* (*USSR*) **6**, 74 (1959).

A4. Aladyev, I. T., Miropolsky, Z. L., Doroshchuk, V. E., and Styrikovitch, M. A., Boiling crisis in tubes, *Intern. Heat Transfer Conf., Boulder, Colorado*, 1961, paper No. 28, 237.

A5. Alessandrini, A., Bertoletti, S., Gaspari, G. P., Lombardi, C., Soldaini, G., and Zavattarelli, R., Critical heat flux data for fully developed flow of steam and water mixtures in round vertical tubes with an intermediate nonheated section, CISE-R69 (1963).

B1. Bagley, R., The application of heat transfer data to the design of once-through boiler furnaces, *Proc. Inst. Mech. Engrs.* (*London*) **180**, Pt. 3C (1965–1966).

B2. Barnett, P. G., The scaling of forced convection boiling heat transfer, AEEW-R.134 (1963).

B3. Barnett, P. G., An investigation into the validity of certain hypotheses implied by various burn-out correlations, AEEW-R.214 (1963).

B4. Barnett, P. G., The prediction of burn-out in non-uniformly heated rod clusters from burn-out data for uniformly heated round tubes, AEEW-R.362 H.M. Stationery Office, London (1964).

B5. Barnett, P. G., Scaling of burn-out in forced convection boiling heat transfer, *Proc. Inst. Mech. Engrs.* (*London*) **180**, Pt. 3C (1965–1966).

B6. Barnett, P. G., A correlation of burn-out data for uniformly heated annuli and its use for predicting burn-out in uniformly heated rod bundles, AEEW-R.463 (1966).

B7. Barte, D. R., Bankoff, S. G., and Colahan, W. J., Summary of conference on bubble dynamics and boiling heat transfer held at the Jet Propulsion Laboratory, June 14th–15th, 1956, JPL Memo. No. 20-137, Calif. Inst. Tech., Pasadena.

B8. Batch, J. M., and Hesson, G. M., Comparison of boiling burn-out data for 19-rod bundle fuel elements with wires and warts, HW 80391, G. E. Hanford Lab., Richland. Washington (1964).

B9. Becker, K. M., An analytical and experimental study of burn-out conditions in vertical round ducts, AE-178, Stockholm (1965).

B10. Becker, K. M., A correlation for burn-out predictions in vertical rod bundles, paper presented at *Symp. Boiling and Two-Phase Flow, EURATOM, ISPRA, June* 1966.

B11. Becker, K. M., and Hernborg, G., Measurements of the effects of spacers on the burn-out conditions for flow of boiling water in a vertical annulus and a vertical 7-rod cluster, AE-165, Stockholm (1965).

B12. Becker, K. M., and Persson, P., An analysis of burn-out conditions for flow of boiling water in a vertical round duct, *J. Heat Transfer* **86**, 515 (1964).

B13. Becker, K. M., Hernborg, G., Bode, M., and Erikson, O., Burn-out data for flow of boiling water in vertical round ducts, annuli and rod clusters, AE-177, Stockholm (1965).

B14. Becker, K. M., Jahnberg, S., Haga, I., Hansson, P. T. and Mathisen, R. P., Hydrodynamic instability and dynamic burn-out in natural circulation two-phase flow. An experimental and theoretical study, *Nukleonik* **6**, 224 (1964).

B15. Bennett, A. W., Hewitt, G. F., Kearsey, H. A., Keeys, R. K. F., and Lacey, P. M. C., Flow visualisation studies of boiling at high pressure, *Proc. Inst. Mech. Engrs. (London)* **180**, Pt. 3C (1965–1966).

B16. Berenson, P. J., Transition boiling heat transfer, *Natl. Heat Transfer Conf., 4th, A.I.Ch.E/ASME, 1960*, Preprint No. 18.

B17. Bertoletti, S., Gaspari, G. P., Lombardi, C., and Zavattarelli, R., Critical heat flux data for fully developed flow of steam and water mixtures in round vertical tubes with non-uniform axial power distribution, CISE-R.74 (1963).

B18. Bertoletti, S., Gaspari, G. P., Lombardi, C., Peterlongo, G., Silvestri, M., and Tacconi, F. A., Heat Transfer crisis with steam-water mixtures, *Energia Nucl. (Milan)* **12**, 3 (1965).

B19. Bertoletti, S., Gaspari, G. P., Lombardi, C., Peterlongo, G., Silvestri, M. Soldaini, G., and Zavatterelli, R., Critical heat flux data for fully developed flow of steam and water mixtures in round vertical tubes, CISE-R.63 (1962).

B20. Biancone, F., Companile, A., Galimi, G., and Goffi, M., Forced convection burn-out and hydrodynamic instability experiments for water at high pressure. 1. Presentation of data for round tubes with uniform and nonuniform power distribution, EUR 2490e (1965).

B21. Birdseye, D. E., Experimental investigation of heat transfer characteristics of liquid nitrogen tetroxide, JPL Tech. Rep. No. 32-37 (1960).

B22. Bowring, R. W., Burn-out prediction in multi-rod clusters by subchannel analysis, paper presented at *Sympo. Two-Phase Flow, Grenoble, May 1965*.

B23. Bundy, R. D., Gambill, W. R., and Wansborough, R. W., Heat transfer, burn-out and pressure drop for water in swirl flow through tubes with internal twisted tapes, *Natl. Heat Transfer Conf., 4th, A.I.Ch.E., August 1960*.

C1. Carr, J. B., Burn-out in net quality by liquid film destruction, CONF-595-1 (1964).

C2. Casterline, J. E., and Lee, D. M., Flow instability and critical heat flux in heated parallel channels, TID-21403, Columbia Univ. Eng. Res. Lab. (Oct. 27, 1964).

C3. Cathro, K. J., and Tait, R. W. F., Heat transfer to liquids boiling inside tubes. I. The climbing-film evaporator, *Australian J. Appl. Sci.* **8**, 279 (1957).

C4. Clerici, G. C., Garriba, S., Sala, R., and Tozzi, A., A catalogue of the commonest burn-out correlations for forced convection in the quality region, paper presented at *Symp. Boiling and Two-Phase Flow, EURATOM, ISPRA, June, 1966*.

C5. Collier, J. G., Lane, A. D., Pace, R. D., and Winter, E. E., The first experimental irradiation of fog cooled fuel, AECL Rept. CRFD-1164 (1963).

C6. Cook, W. H., Fuel cycle program. A boiling water reactor research and development program, 1st Quart. rept., Aug. 1960–Sept. 1960, GEAP-3558 (1960).
C7. Core, T. C., and Sato, K., Determination of burnout limits of Polyphenyl coolants, IDO-28007 (1958).
C8. Costello, C. P., and Adams, J. M., Burnout heat fluxes in pool boiling at high accelerations, *Intern. Develop. Heat Transfer* Pt. II, Paper No. 30. ASME (1961).
C9. Coulson, J. M., and McNelly, M. J., Heat transfer in a climbing film evaporator, *Trans. Inst. Chem. Engrs. (London)* **34**, 247 (1956).
D1. De Bortoli, R. A., and Masnovi, R., Effect of dissolved hydrogen on burnout for water flowing vertically upwards in round tubes at 2000 psia, WAPD-TH-318 (1957).
D2. De Bortoli, R. A., Green, S. J., Le Tourneau, B. W., Troy, M., and Weiss, A., Forced convection heat transfer burnout studies for water in rectangular channels and round tubes at pressure above 500 psia, WAPD-188 (1958).
D3. Dimmock, T. H., Heat transfer properties of anhydrous ammonia, Rept. No. RMI-124-S1, Reaction Motors, Denville (1957).
D4. Doroshtchuk, V. E., and Frid, F. P., On the problem of the influence of inlet throttling and of the heated length on the critical heat flux for round tubes, *Teploenerg.* (Sept. 1959).
D5. Durant, W. S., and Mirshak, S., Roughening of heat transfer surfaces as a method of increasing heat flux at burnout, DP-380 (July 1959).
E1. Edwards, P. A., and Obertelli, J. D., Burnout and pressure drop data on 37-rod clusters in high pressure water, AEEW-R488 (1966).
E2. Ellion, M. E., A study of the mechanism of boiling heat transfer, JPL Memo. No. 20-88, Calif. Inst. Technol., Pasadena, California (1954).
F1. Farber, E. A., Free convection heat transfer from electrically heated wires, *J. Appl. Phys.* **22**, 1437 (1951).
F2. Farber, E. A., and Scorah, R. L., Heat transfer to water under pressure, *Trans. ASME (Am. Soc. Mech. Engrs.)* **70**, 369 (1948).
F3. Firstenberg, H., Goldmann, K., Lo Bianco, L., Preiser, S., and Rabinowitz, G., Compilation of experimental forced convection quality burnout data with calculated Reynolds numbers, NDA-2131-16 (1960).
G1. Gambil, W. R., and Bundy, R. D., High flux heat transfer characteristics of pure ethylene glycol, *A.I.Ch.E. (Am. Inst. Chem. Eng.) J.* **9**, 55 (1963).
G2. Goldmann, K., Firstenberg, H., and Lombardi, C., Burnout in turbulent flow—a droplet diffusion model, *J. Heat Transfer* **83**, 158 (1961).
G3. Grace, T. M., The mechanism of burnout in initially subcooled forced convection systems, TID-19845 (1963).
G4. Grace, T. M., and Isbin, H. S., Comment on paper by Becker, *J. Heat Transfer* **86**, 524 (1964).
G5. Griffith, P., The correlation of nucleate boiling burnout data, ASME Paper No. 57-HT-21 (April 1957).
H1. Hench, J. E., Multirod (four rod) critical heat flux at 1000 psia, GEAP-4358 (1963).
H2. Hesson, G. M., Fitzsimmons, D. E., and Batch, J. M., Comparison of boiling burnout data for 19-rod bundles in horizontal and vertical positions, HW-83443 Rev. 1, G. E. Hanford Lab., Richland, Washington (1964).
H3. Hesson, G. M., Fitzsimmons, D. E., and Batch, J. M., Experimental boiling burnout heat fluxes with an electrically heated 19-rod bundle test section, BNWL-206 (1965).
H4. Hewitt, G. F., Kearsey, H. A., Lacey, P. M. C., and Pulling, D. J. Burnout and nucleation in climbing film flow, AERE-R4374 (1963).

H5. Hewitt, G. F., Kearsey, H. A., Lacey, P. M. C., and Pulling, D. J., Burnout and film flow in the evaporation of water in tubes, AERE-R4864 (1965).

H6. Hewitt, G. F., Kearsey, H. A., Lacey, P. M. C., and Pulling, D. J. Burnout and film flow in the evaporation of water in tubes, *Proc. Inst. Mech. Engrs.* (*London*) **180**, Pt. 3C (1965–1966).

H7. Hines, W. S., Forced convection and peak nucleate boiling heat transfer characteristics for hydrazine flowing turbulently in a round tube at pressures to 1000 psia, Rept. No. 2059, Rocketdyne, Canoga Park, California (1959).

H8. Hoffman, H. W., and Krakoviak, A. I., Convective boiling with liquid potassium, *Proc. Heat Transfer and Fluid Mech. Inst., Stanford, California, 1964*.

H9. Howard, C. L., Fuel cycle program. A boiling water reactor research and development program. Eleventh Quart. Progr. Rept., Jan–Mar. 1963, GEAP-4215.

H10. Howard, C. L., Methods for improving the critical heat flux for BWR's, GEAP-4203 (1963).

I1. Isbin, H. S., Vanderwater, R., Fauske, H., and Singh, S., A model for correlating two phase, steam water, burnout heat transfer fluxes, *J. Heat Transfer* **83**, 149 (1961).

I2. Ivashkevich, A. A., Critical heat fluxes in the forced flows of liquids in channels, *At. Energ.* (*USSR*) **8**, 44 (1960) [English transl.: Longo, J., Jr., KAPL-1744(1957)].

I3. Ivey, H. J., Acceleration and the critical heat flux in pool boiling heat transfer, *Proc. Inst. Mech. Engrs.* (*London*) (1962).

J1. Jacket, H. S., Roarty, J. D., and Zerbe, J. E., Investigation of burnout heat flux in rectangular channels at 2000 psia, *Trans. ASME* (*Am. Soc. Mech. Engrs.*) **80**, 391 (1958).

J2. Janssen, E., Two-phase flow and heat transfer in multirod geometries, GEAP-4798 (1965).

J3. Janssen, E., and Kervinen, J. A., Burnout conditions for single rod in annular geometry, GEAP-3899 (1963).

J4. Jens, W. H., and Leppert, G., Recent developments in boiling research, *J. Am. Soc. Naval Engrs.* **67**, 137 (1955).

J5. Jens, W. H., and Lottes, P. A., Analysis of heat transfer, burnout pressure drop and density data for high pressure water, ANL-4627 (1951).

J6. Jiji, L. M., Incipient boiling and the bubble boundary layer formation over a heated plate for forced convection flow in a pressurized rectangular channel, Ph.D. Thesis, Univ. of Michigan, Ann Arbor, 1962.

K1. Kearsey, H. A., Steam water heat transfer—post burnout conditions, *Chem. Process Eng.* p. 455 (Aug. 1965).

K2. Kirby, G. J., A visual study of forced convection boiling, AEEW-R281, H.M. Stationery Office, London (1965).

K3. Kirby, G. J., Burnout in climbing film two-phase flow. A review of theories of the mechanism, AEEW-R470, H.M. Stationery Office, London (1966).

K4. Kirby, G. J., private communication.

K5. Konkov, A. S., and Modnikova, V. V., Experimental Investigation of the conditions of deterioration of heat transfer during boiling in tubes, *Teploenerg.* **9**, No. 8, 77 (1962).

L1. Lee, D. H., An experimental investigation of forced convection burnout in high pressure water. 3. Long tubes with uniform and nonuniform axial heating, AEEW-R355 (1965).

L2. Lee, D. H., An experimental investigation of forced convection burnout in high pressure water. 4. Large diameter tubes at about 1600 psia, AEEW-R 479 (1966).

L3. Lee, D. H., private communication.

L4. Lee, D. H., and Obertelli, J. D., An experimental investigation of forced convection burnout in high pressure water. 1. Round tubes with uniform flux distribution, AEEW-R213 (1963).

L5. Lee, D. H., and Obertelli, J. D., An experimental investigation of forced convection burn-out in high pressure water. 2. Preliminary results for round tubes with non-uniform axial heat flux distribution, AEEW-R309 (1963).

L6. Lee, D. H., and Obertelli, J. D., An experimental investigation of burnout with forced convection high-pressure water, *Proc. Inst. Mech. Engrs. (London)* **180**, Pt. 3C (1965–1966).

L7. Levy, S., Prediction of the critical heat flux in forced convection flow, GEAP-3961 (1962).

L8. Levy, S., Polomik, E. E., Swan, C. L., and McKinney, A. W., Eccentric rod burnout at 1000 psia with net steam generation, *Intern. J. Heat Mass Trans.* **5** (July 1962).

L9. Lowdermilk, W. H., and Weiland, W. F., Some measurements of boiling burnout, NACA Res. Memo. E54K10 (1955).

L10. Lowdermilk, W. H., Lanzo, C. D., and Siegel, B. L., Investigation of boiling burn-out and flow stability for water flowing in tubes, NACA Tech. Note 4382 (1958).

M1. McAdams, W. H., Kennel, W. E., Minden, C. S., Carl, R., Picornell, P., and Dew, J. E., Heat transfer at high rates to water with surface boiling, *Ind. Eng. Chem.* **41** (1949).

M2. Macbeth, R. V., Burnout analysis. 3. The low-velocity burnout regime, AEEW-R.222 (1963).

M3. Macbeth, R. V., Burnout analysis. 4. Application of a local conditions hypothesis to world data for uniformly heated round tubes and rectangular channels, AEEW-R.267 (1963).

M4. Macbeth, R. V., Burn-out analysis. 5. Examination of published world data for rod bundles, AEEW-R.358 (1964).

M5. McEwen, L. H., Batch, J. M., Foley, D. J., and Kreiter, M. R., Heat transfer beyond burnout for forced convection bulk boiling, ASME Paper No. 57-SA-49 (1957).

M6. Matzner, B., Basic experimental studies of boiling fluid flow and heat transfer at elevated pressures, TID 12574, Columbia Univ. (1961).

M7. Matzner, B., Basic experimental studies of boiling fluid flow and heat transfer at elevated pressures, TID 15637, Columbia Univ. (1962); TID 14439, (1961).

M8. Matzner, B., Basic experimental studies of boiling fluid flow and heat transfer at elevated pressures, TID-18978, Columbia Univ. (1963).

M9. Matzner, B., Basic experimental studies of boiling fluid flow and heat transfer at elevated pressures, TID 18296, Columbia Univ. (1963).

M10. Matzner, B., Experimental performance evaluation of proposed fuel elements for the steam generating heavy water reactor, NOR-1643, Columbia Univ. (1964).

M11. Matzner, B., and Biderman, R., Heat transfer and hydraulic studies for SNAP-4 fuel element geometries, Topical Rept. 1 and 2 Task XV of contract AT (30-3)-187, Columbia Univ. (1963).

M12. Milioti, S., A survey of burnout correlations as applied to water-cooled nuclear reactors, M.Sc. Thesis, Dept. of Nucl. Eng., Penn. State Univ., University Park, Pennsylvania, 1964.

M13. Miropolski, Z. L., Shitsman, M. E., Mostinski, I. L., and Stavrovski, A. A., Influence of inlet conditions on the critical heat flux for water boiling in tubes, *Teploenerg.* (Jan. 1959).

M14. Moeck, E. O., Dryout in a 19-rod bundle cooled by steam/water fog at 515 psia, ASME Paper 65-HT-50 (1965).

M15. Moeck, E. O., Wikhammer, G. A., Macdonald, I. P. L., and Collier, J. G., Two methods of improving the dryout heat flux for high pressure steam/water flow, AECL-2109 (1964).

N1. Noel, M. B., Experimental investigation of heat transfer characteristics of hydrazine and a mixture of 90 per cent hydrazine and 10 per cent ethylenediamine, JPL Tech. Rept. No. 32-109 (June 1961).

N2. Noel, M. B., Experimental investigation of the forced convection and nucleate boiling heat transfer characteristics of liquid ammonia, JPL Tech. Rept. No. 32-125 (July 1961).

N3. Nukiyama, S., Experiments on the determination of the maximum and minimum values of the heat transferred between a metal surface and boiling water, *J. Soc. Mech. Engrs.* (*Japan*) **37**, 367 (1934) [English transl. by Brickley, S. G., AERE-Transl. 854 (1960)].

P1. Perroud, P., Rebiere, J., and Weil, L., Burnout flux for liquid hydrogen and liquid nitrogen in forced convection, CENG/ASP-64-01 (1964).

P2. Polomik, E., and Quinn, E. P., Multirod burnout at high pressure, GEAP-3940 (1962).

P3. Polomik, E. E., Levy, S., and Sawochka, S. G., Heat transfer coefficients with annular flow during once-through boiling of water to 100 per cent quality at 800, 1100, and 1400 psia, GEAP-3703 (1961).

R1. Randles, J., A theory of burnout in heated channels at low mass velocities, AEEW-R279, H.M. Stationery Office, London (1963).

R2. Reynolds, J. M., Burnout in forced convection nucleate boiling of water, Mass. Inst. Inst. Technol. Tech. Rept. No. 10, NONR-1848 (39) (July 1957).

R3. Robinson, J. M., and Lurie, H., Critical heat flux of some polyphenyl coolants, *Symp. Boiling Phenomena in Nucl. Heat Transfer 1962*, Am. Inst. Chem. Eng., Preprint No. 156.

S1. Sachs, P., and Long, R. A. K., A correlation for heat transfer in stratified two-phase flow with vaporization, *Intern. J. Heat Mass Transfer* **2**, 222 (1961).

S2. Salt, K. J., and Wintle, C. A., Design and operation of a transistorized bridge-type detector for burnout in boiling heat transfer experiments, AEEW-R.330, H.M. Stationery Office, London (1964).

S3. Schmidt, K. R., Thermodynamic investigations of highly loaded boiler heating surfaces, *Mitt. Ver. Grosskesselbesitzer* **63**, 391 (1959) (AEC-tr-4033).

S4. Silvestri, M., A research program in two-phase flow, CISE Rept. (1963).

S5. Smolin, V. N., Polyakov, V. K., and Esikov, V. I., An experimental investigation of heat transfer crisis, *J. Nucl. Energy* **19**, 209 (1965).

S6. Sorlie, T., Levy, S., Lyons, M. F., and Boyden, J. E., Experience with BWR fuel rods operating above critical heat flux, *Nucleonics* p. 62 (April 1965).

S7. Staniforth, R., and Stevens, G. F., Experimental studies of burnout using freon-12 at low pressure, with reference to burnout in water at high pressure, *Proc. Inst. Mech. Engrs.* (*London*) **180**, Pt. 3C (1965–1966).

S8. Staniforth, R., Stevens, G. F., and Wood, R. W., An experimental investigation into the relationship between burnout and film flow-rate in a uniformly heated round tube, AEEW-R430, H. M. Stationery Office, London (1965).

S9. Sterman, L. S., and Stiushin, N. G., An investigation into the influence of speed of circulation on the values of critical heat flows for liquid boiling in tubes, *J. Tech. Phys.* (*USSR*) **22**, 446 (1952).

S10. Sterman, L. S., Stiushin, N. G., and Morozov, V. G., An investigation of the dependence of critical heat flux on the rate of circulation, *J. Tech. Phys.* (*USSR*) **1**, 2250 (1956).

S11. Stevens, G. F., private communication.
S12. Stevens, G. F., and Wood, R. W., A comparison between burnout data for 19-rod cluster test sections cooled by freon-12 at 155 psia, and water at 1000 psia in vertical upflow, AEEW-R.468 (1966).
S13. Stevens, G. F., Elliott, D. F., and Wood, R. W., An experimental investigation into forced convection burnout in freon, with reference to burnout in water, AEEW-R.321 (1964).
S14. Stevens, G. F., Elliott, D. F., and Wood, R. W., An experimental comparison between forced convection burnout in freon-12 flowing vertically upwards through uniformly and nonuniformly heated round rubes, AEEW-R426, H. M. Stationery Office, London (1965).
S15. Stock, B. J., Observations on transition boiling heat transfer phenomena, ANL-6175 (1960).
S16. Stroebe, G., Baker, E., and Badger, W., Boiling film heat transfer coefficients in a long tube vertical eavporator, *Trans. Am. Inst. Chem. Engrs.* **35**, 17 (1939).
S17. Styrikovitch, M. A., Miropolski, Z. L., Shitsman, M. E., Mostinski, I. L., Stavrovski, A. A., and Faktorovitch, L. E., Effect of upstream elements on critical boiling in a vapor generating pipe, *Teploenerg.* (May 1960) (USAEC Transl. AEC-tr-4740).
S18. Swenson, H. S., Carver, J. R., and Kakarala, C. R., The influence of axial heat flux distribution on the departure from nucleate boiling in a water-cooled tube, ASME Paper No. 62-WA-297 (1962).
T1. Thompson, B., and Macbeth, R. V., Burnout in uniformly heated round tubes: A compilation of World data with accurate correlations, AEEW-R.356, H. M. Stationery Office, London (1964).
T2. Tippets, F. E., Analysis of the critical heat flux condition in high pressure boiling water flow, ASME Paper No. 62-WA-161 (1962).
T3. Tong, L. S., "Boiling Heat Transfer and Two-Phase Flow." Wiley, New York, 1965.
T4. Tong, L. S., Chelmer, H., and McCabe, E. A., Hot channel factors for flow distribution and mixing in core thermal design, WCAP-2211 (1963).
T5. Tong, L. S., Currin, H. B., and Engel, F. C., DNB (burnout) studies in an open lattice core, WCAP-3736 (1964).
T6. Tong, L. S., Currin, H. B., and Thorp, A. G., New DNB (burnout) correlations, WCAP-1997 (1962).
V1. Van Gasselt, M. L. G., and Van Meel, D. A., Research on the influence of the conditions of flow, subcooling and composition, on the burnout heat flux of polyphenyl reactor cooling agents, EUR 2299e (1965).
W1. Waters, E. D., Hesson, G. M., Fitzsimmons, D. E., and Batch, J. M., Boiling burnout experiments with 19-rod bundles in axial flow, HW 77303., G. E. Hanford Lab., Richland, Washington (1963).
Z1. Zenkevitch, B. A., Similitude relations for critical heat loading in forced liquid flow *At. Energ., USSR*, **4**, 74 (1958) [English transl.: Merte, H., WAPD-AD-Th-539 (1959)].
Z2. Zenkevich, B. A., and Subbotin, B. I., Critical heat fluxes in subcooled water with forced circulation, *J. Nucl. Energy, Pt. B* **1**, No. 2, 134 (1959).

GAS–LIQUID DISPERSIONS

William Resnick* and Benjamin Gal-Or[†]

Technion-Israel Institute of Technology
Haifa, Israel

I. Introduction	296
A. Variables Affecting Operating Efficiency	296
B. Analytical and Empirical Approaches	299
II. Experimental Studies	300
A. Experimental Evaluation of Contacting Efficiency	300
B. Volumetric Mass-Transfer Coefficients	303
C. Average Bubble Diameter and Bubble-Size Distribution	307
D. Gas Holdup and Average Residence Time	312
E. Gas Residence-Time Distribution	314
F. Flow Patterns and Bubble Velocities	316
G. Optimum Design and Operating Conditions	319
III. Effect of Surface-Active Agents	327
A. Experimental Studies	327
B. Theoretical Studies	329
IV. Mathematical Models	333
A. Absorption in Agitated Liquid with Simultaneous Chemical Reaction—Film Model	335
B. Absorption in Agitated Liquid with Simultaneous Chemical Reaction—Penetration Model	336
C. Film-Penetration Model	339
D. Film-Penetration Model with Simultaneous Chemical Reaction	341
E. Mass Transfer with Chemical Reaction from a Moving Gas Bubble	344
F. Gas Absorption with Chemical Reaction—Time-Dependent Bulk Concentration	346
G. Steady-State Convective Diffusion	347
H. Steady-State Convective Diffusion with Simultaneous First-Order Irreversible Chemical Reaction	350

* Department of Chemical Engineering.

† Department of Aeronautical Engineering. Also at the Department of Chemical Engineering, University of Pittsburgh, Pittsburgh, Pennsylvania.

I. Residence-Time Model for Total Mass Transfer with and without Chemical Reaction 353
J. Effect of Bubble-Size Distribution and Holdup on Mass- or Heat-Transfer Rates 361
K. Effect of Holdup on Convective Mass Transfer......................... 371
L. Coupled Heat Transfer and Multicomponent Mass Transfer, with Residence-Time and Bubble-Size Distributions................................... 374
M. Additional Mathematical Studies.. 386
Nomenclature 388
References 390

I. Introduction

Many operations in chemical engineering require the dispersion of gases in a liquid phase. One simple way to achieve this is to bubble the gas through the liquid by using sparge pipes at the bottom of the contactor. Mixing impellers of the turbine type are frequently used to increase the rate of mass- or heat-transfer over that obtained without the use of the mixer. Consequently, the contacting of gas and liquid in a mixing vessel equipped with a rotating impeller is generally used for the absorption of sparingly soluble gases. In many cases, the physical diffusion is accompanied by a simultaneous chemical reaction in the liquid phase. Some of the more common applications are in biochemical fermentation, hydrogenation, hydrocarbon oxidation and various other oxidation processes.

A contactor equipped with an impeller is generally considered to be the most flexible and most generally effective gas disperser known. A porous sparger has no advantage except at very low power inputs.

Efficient contact is produced between the phases in agitated gas–liquid contactors and, therefore, this type of equipment can also be useful for those absorption and stripping operations for which conventional plate or packed towers may not be suited. It may also be useful where the operation involves the contact of three phases—say, gas, liquid, and suspended solids. The latter application could be represented by the low-pressure polymerization of ethylene with solid catalysts (F5).

In this chapter we shall refer mainly to mechanically agitated gas–liquid dispersions. However, most of the theoretical and experimental conclusions also apply to any type of gas–liquid dispersion.

A. Variables Affecting Operating Efficiency

Recent studies on heat- and mass-transfer to and from bubbles in liquid media have primarily been limited to studies of the transfer mechanism for single moving bubbles. Transfer to or from swarms of bubbles moving in an arbitrary liquid field is very complex and has been analyzed theoretically in certain simple cases only (G3, G5, G6, G8, M3, R9, W1).

Let us first examine the main variables that affect the absorption rate in agitated gas–liquid contacting. Some of the more important ones are:

System Variables:
    1. Viscosity of the continuous phase
    2. Density of the continuous phase
    3. Interfacial surface tension
    4. Partial pressure of the component being absorbed
    5. Diffusion coefficients
    6. Chemical reaction rate constants
    7. Concentration of adventitious surfactants

Operating Variables:
    8. Rotational speed of the impeller
    9. Gas flow rate
    10. Liquid volume and liquid flow rate

Equipment Variables:
    11. Impeller type and diameter
    12. Geometric ratios of the agitation equipment

No exact theoretical analysis has as yet been possible because of the large number of variables involved and the complex mechanisms governing the transfer mechanism in a gas–liquid dispersion. The following section analyzes in a qualitative manner some of the effects produced by the mixing impeller in the disperser. It will serve to show some of the interrelationships involved as well as to illustrate the difficulties in the path of arriving at an exact mechanism.

The mixing impeller is used primarily to subdivide the incoming gas into bubbles and to disperse these bubbles throughout the agitated liquid phase. The shear produced by the impeller blades on both liquid and gas causes the incoming gas to be subdivided into numerous bubbles which have relatively small diameters compared to the diameter obtained by free bubbling. In general the results are (G2):

Increase in interfacial area. The total surface area for diffusion is increased because the bubble diameter is smaller than for the free-bubbling case at the same gas flow rate; hence there is a resultant increase in the overall absorption rate. The overall absorption rate will also increase when the diffusion is accompanied by simultaneous chemical reaction in the liquid phase, but the increase in surface area only has an appreciable effect when the chemical reaction rate is high; the absorption rate for this case is then controlled by physical diffusion rather than by the chemical reaction rate (G6).

Increase in bubble contact time. The bubbles are swept by the turbulent flow currents in the liquid and the result is an increased contact time between the bubbles and the agitated phase; that is, the time from the moment of bubble formation until its escape from the liquid is increased. Consequently the gas holdup in the vessel, the total surface area for diffusion, and the overall absorption rate increase. However, the instantaneous mass transfer flux is *decreased* with an increase in contact time (D10, G5). Recent work by the authors on bubble residence time in the liquid phase, coupled with conclusion based on the literature, indicate that bubble contact time is an important variable which controls the rate of mass transfer in the vessel (G6, G8).

Increase in mass-transfer rate per unit area. As stated above, agitated gas–liquid contactors are used, in general, when it is necessary to deal with sparingly soluble gases. According to the terminology of the film theory, absorption is then controlled by the liquid resistance, and agitation of the liquid phase could increase the mass-transfer rate per unit area. As will be

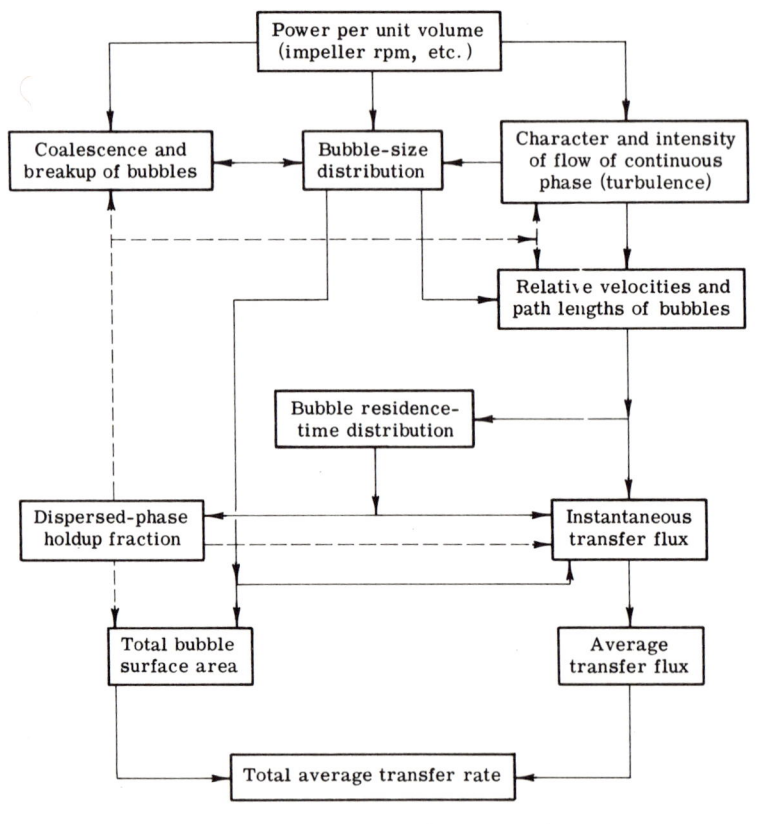

FIG. 1. Relationships between agitation intensity and transfer rates at constant gas flow [after Gal-Or and Walatka (G9)]. At increased gas flow, the holdup fraction Φ is increased mainly by an increase in the number of bubbles produced.

explained later, an increase in agitation intensity may produce only a small or even negligible effect on the mass-transfer coefficient per unit area.

The main relationships between the agitation intensity of the dispersion and the total mass-transfer rate are summarized qualitatively for constant gas flow rate by Fig. 1 (G9) wherein interaction effects among the bubbles are indicated by dashed lines. Intermediate phenomena not shown, such as the direct and feedback effects between coalescence and mass transfer (G5, G9), should also be considered.

This brief discussion of some of the many effects and interrelations involved in changing only one of the operating variables points up quite clearly the reasons why no exact analysis of the dispersion of gases in a liquid phase has been possible. However, some of the interrelationships can be estimated by using mathematical models; for example, the effects of bubble-size distribution, gas holdup, and contact times on the instantaneous and average mass-transfer fluxes have recently been reported elsewhere (G5, G9).

B. Analytical and Empirical Approaches

One approach to the analysis of gas–liquid dispersion has been to develop simplified theoretical models that provide an approximation to the real processes and mechanisms operating in the system. Typically, their application to the design problem requires a knowledge of factors such as bubble velocity relative to the agitated liquid and rate of surface renewal or film thickness as functions of mixing conditions. Even to the extent that these variables are real in the sense of being defined in terms of measurable quantities, they are difficult to determine; those factors that are merely conceptual are impossible to evaluate. Even with the aid of indeterminate variables, the models have commonly permitted the development of equations only for the rate of absorption per unit area of the interface between phases. It is not possible, therefore, to calculate by their aid the total rate of absorption for equipment design without the necessity of extensive laboratory or pilot-plant tests. However, some recently proposed simplified models (G5, G6) do overcome some of these difficulties and may be used to predict the total rate of diffusion with and without chemical reaction. A summary of the mathematical models is given in Section IV.

Another approach to the design problem is to determine empirical correlations based on experimental work and to adopt these correlations for scale-up. In many of the published works the latter approach is investigated. The correlations are such that the volumetric mass-transfer coefficient is generally reported as a function of one or more of the equipment, system, or operating variables cited above. Empirical correlations can be used confidently for scale-up only for equipment that has complete geometrical similarity to the equipment for which the correlation was derived, and operates in the same ranges and with the same chemical reaction system.

II. Experimental Studies

A. Experimental Evaluation of Contacting Efficiency

Mass-transfer rates have been determined by measuring the absorption rate of a pure gas or of a component of a gas mixture as a function of the several operating variables involved. The basic requirement of the evaluation method is that the rate step for the physical absorption should be controlling, not the chemical reaction rate. The experimental method that has gained the widest acceptance involves the oxidation of sodium sulfite, although in some of the more recent work, the rate of carbon dioxide absorption in various media has been used to determine mass-transfer rates and interfacial areas.

1. *Sulfite-Oxidation Method—Kinetic Survey*

This method involves measurement of the oxidation rate of an aqueous sodium sulfite solution catalyzed by cupric or cobaltous ions. The oxygen absorbed reacts with the sulfite according to the equation:

$$O_2 + 2SO_3^= \rightarrow 2SO_4^=$$

In spite of its wide application, the mechanisms of this reaction remain obscure. Many diverse arguments have been published since the reaction was first investigated in 1897 (B1, C5, C9, F7, J6, M5, P9, R2, S5, W2, W4, Y1, Y4). Cooper *et al.* (C9) introduced this method as a yardstick for the measurement of volumetric mass-transfer coefficients in gas–liquid contacting. Karow *et al.* (K1) later concluded that the sulfite oxidation is suitable for fermentation process scale-up studies. Cooper *et al.* established that the reaction proceeds at a rate independent of sulfite ion concentration over wide concentration ranges. In their work they considered the sulfite oxidation to be of zero order with respect to both sulfite and sulfate concentration.

Yagi and Inoue (Y1) and Westerterp *et al.* (W2, W4) have recently presented the results of extensive research concerning this subject. Yagi and Inoue investigated this reaction by measuring the oxygen concentration in the aqueous solution by a polarographic method. The materials used in their experiments were scrupulously purified. They determined that the reaction is first-order with respect to both oxygen and sulfite concentration. When the sulfite concentration is very high compared to the oxygen concentration, the process may be regarded as absorption accompanied by first-order chemical reaction with respect to the dissolved oxygen. The rate equation for the reaction in the presence of cobaltous catalyst at 20°C was expressed as follows:

$$\frac{d[O_2]}{dt} = -1.4[1 + 1.46 \times 10^7[CoSO_4]][Na_2SO_3][O_2] \quad \frac{\text{g-mole}}{\text{liter sec}} \quad (1)$$

Their measured rate constant was less than that estimated from wetted-wall column absorption experiments (S2).

Westerterp et al. reported the first-order reaction rate constant with respect to oxygen concentration in a solution at 30°C containing 100 g of sodium sulfite per liter. The catalyst concentration was 0.001 g-mole/liter. They found that k is 37,000 sec^{-1} for the CoSO$_4$ catalyst and 9800 sec^{-1} for CuSO$_4$ catalyst. For the same sodium sulfite concentration but with copper sulfate concentration greater than 0.005 g-mole/liter, the reaction rate constant as a function of temperature is approximated by:

$$k \simeq 7.7 \times 10^{12} \exp(-12,300/RT) \quad \text{sec}^{-1} \quad (2)$$

Some of the most important characteristics of this reaction were summarized as (W4, Y1):

(a) The reaction is very sensitive to catalyst traces. Cu, Co, Fe, Ce, Mn, and ozone increase, whereas ethyl alcohol, glycerol, and mannitol decrease, the reaction rate.

(b) At sulfite concentration below 0.01 molar, the conversion rate is first-order with respect to sulfite and independent of the oxygen concentration.

(c) At sulfite concentration above about 0.02 molar, the reaction rate is first-order with respect to oxygen and independent of the sulfite concentration.

(d) At copper concentration above 0.001 molar, the reaction is independent of the copper concentration.

(e) For the reaction with copper as a catalyst, a chain mechanism is assumed: $Cu^{\cdot\cdot} + SO_3^{=} \rightarrow (\text{complex}) \xrightarrow{O_2} Cu^{\cdot\cdot} + SO_4^{=}$.

(f) Cobalt is a more active catalyst than copper and is less sensitive to impurities.

2. Other Methods

Measurement of the absorption rate of carbon dioxide in aqueous solutions of sodium hydroxide has been used in some of the more recent work on mass-transfer rate in gas–liquid dispersions (D6, N3, R4, R5, V5, W2, W4, Y3). Although this absorption has a disadvantage because of the high solubility of CO_2 as compared to O_2, it has several advantages over the sulfite-oxidation method. For example, it is relatively insensitive to impurities, and the physical properties of the liquid can be altered by the addition of other liquids without appreciably affecting the chemical kinetics. Yoshida and

Miura (Y3) have studied the effect of the addition of glycerol to water on the reaction rate constant. The reaction in the liquid phase is second order. Their values for k'' at 28°C, which indicate a slight increase with increasing viscosity, are given in Table I.

TABLE I

EFFECT OF THE ADDITION OF GLYCEROL TO WATER ON THE REACTION RATE CONSTANT[a]

Glycerol concentration (Weight %)	Reaction Rate constant (M^3/kg-mole sec.)
0.0	8400
12.3	11,000
26.0	14,800
32.5	15,000

[a] From the data of Yoshida and Miura (Y3).

Vassilatos et al. (V5) have reviewed this reaction. The mechanism they propose is:

$$CO_2 + OH^- \rightarrow HCO_3^- \quad \text{(moderately rapid)}$$
$$HCO_3^- + OH^- \rightarrow CO_3^= + H_2O \quad \text{(very fast)}$$

and the second-order rate constant in m^3/kg-mole sec was given by:

$$\log_{10} k'' = 13.635 - (2895/T) \tag{3}$$

This equation is proposed for the range 0–40°C and for NaOH concentrations from 0.005 to 0.05 molar. These authors also reviewed and used the reaction between CO_2 and aqueous ammonia. They propose mechanisms and report kinetic data.

Danckwerts et al. (D6, R4, R5) recently used the absorption of CO_2 in carbonate–bicarbonate buffer solutions containing arsenate as a catalyst in the study of absorption in a packed column. The CO_2 undergoes a pseudo first-order reaction and the reaction rate constant is well defined. Consequently this reaction could prove to be a useful method for determining mass-transfer rates and evaluating the reliability of analytical approaches proposed for the prediction of mass transfer with simultaneous chemical reaction in gas–liquid dispersions.

Westerterp et al. (W4) and Yoshida and Miura (Y3) utilized the CO_2–NaOH system for obtaining the interfacial area of dispersions with turbine and vaned-disk impellers. Vassilatos et al. (V5) used CO_2 absorption in NaOH solutions to evaluate absorption rates in systems with known interfacial area and compared the rates with those predicted by theoretical models.

Johnson *et al.* (J5) have used the hydrogenation of α-methylstyrene catalyzed by palladium–alumina in powder form in agitated vessels. The physical diffusion of hydrogen through the liquid is the rate-controlling step. The total resistance of this transfer consisted of two separate resistances, one in the liquid adjoining the bubbles and another in the liquid adjoining the suspended solid particles.

Emmert and Pigford (E2) have studied the reaction between carbon dioxide and aqueous solutions of monoethanolamine (MEA) and report that the reaction rate constant is 5400 liter/mole sec at 25°C. If it is assumed that MEA is present in excess, the reaction may be treated as pseudo first-order. This pseudo first-order reaction has been recently used by Johnson *et al.* (J4) to study the rate of absorption from single carbon dioxide bubbles under forced convection conditions, and the results were compared with their theoretical model.

Gal-Or and Hoelscher (G5) have recently developed a fast and simple transient-response method for the measurement of concentration and volumetric mass-transfer coefficients in gas–liquid dispersions. The method involves the use of a transient response to a step change in the composition of the feed gas. The resulting change in the composition of the liquid phase of the dispersion is measured by means of a Clark electrode, which permits the rapid and accurate analysis of oxygen or carbon dioxide concentrations in a gas, in blood, or in any liquid mixture.

B. Volumetric Mass-Transfer Coefficients

Cooper *et al.* (C9) were the first to determine mass-transfer coefficients by measuring the oxidation rate of sodium sulfite in an aqueous solution catalyzed by cupric ions. Their data were taken for a vaned-disk agitator with 16 blades and for a flat paddle. The ratio of agitator-to-tank diameter was 0.4, and the ratio of paddle to tank diameter was 0.25. The tank was equipped with four baffles, with baffle-width to tank diameter ratio of 0.1.

In evaluating their results they assumed the film theory, and, because the oxygen is sparingly soluble and the chemical reaction rate high, they also assumed that the liquid film is the controlling resistance. The results were calculated as a volumetric mass-transfer coefficient based, however, on the gas film. They found that the volumetric mass-transfer coefficient increased with power input and superficial gas velocity. Their results can be expressed as follows:

$$Ks \propto P_v^{0.95} \tag{4}$$

$$Ks \propto V_s^{0.67} \tag{5}$$

TABLE II
EMPIRICAL CORRELATIONS OF VOLUMETRIC MASS-TRANSFER COEFFICIENT Ks AND SPECIFIC INTERFACIAL AREA s[a]

| Experimental conditions ||||| Quantity correlated | Correlations ||| Reference |
| Liquid volume (liter) | D_T (m) | $\frac{L}{D_T}$ | Impeller Type | Blades || Value of exponent of: |||
						V_s	N	P_v	
2.7; 67	0.15; 0.44	0.4	Vaned-disk	16	↑	0.67	—	0.95	C9
11; 11,000	0.24; 2.44	0.25	Paddle	2		0.67	—	0.57	C9
1.9	0.13	0.7	Turbine	2		0.75	1.67	—	J5
0.8	0.10	0.5	Turbine	6		0.21	2.04	—	H22
0.8	0.10	0.5	Turbine	6		0.13	2.40	—	H22
0.2	0.06	0.47	Vaned-disk	4		0.14	2.17	—	H22
1.4	0.12	0.30	Paddle	2		0.05	1.79	—	H22
1.4	0.12	0.30	Paddle	4		0	1.26	—	H22
2.7	0.15	0.20; 0.5	Paddle	4	Ks	0	3.00	—	F6
1.5	0.13	0.5	Turbine	6		0.68	2.00	—	S9
2.7; 41	0.15; 0.38	0.40	Turbine	12		0.67	2.00	—	Y4
2.7; 41	0.15; 0.38	0.40	Vaned-disk	16		0.40–0.84	1.29–2.05	—	Y4
1.5	?	?	Propeller	?		0.40	1.70	—	M5
20	0.33	0.30; 0.46	Turbine	4		?	2.4	—	E1
26,600	3.05	0.20	Turbine	8		0.4	—	0.53	O3
1.9	0.13	0.7	Turbine	8	↓	1.0	3.00	—	P10
3.95	0.16; 0.50	0.24–0.39	Various			?	—	0.43–0.95	K2
22.3	0.305	0.33	Turbine	6		—	—	0.78	R12
6.3–98	0.2–0.5	0.1–0.63	Turbine	?		0.33–0.75	<1.55	—	P4

						$\leftarrow$ s $\rightarrow$		
12.3; 104	0.25; 0.51	0.6	Paddle	4	—	1.5	—	V8
6.3; 104	0.20; 0.51	0.33	Turbine	6	0.50	—	0.40	C1
6.3; 104	0.21; 0.51	0.33	Turbine	6	0.50	—	0.40	C2
12.3; 155	0.25; 0.58	0.40	Turbine	12	0.75	1.1	—	Y3
12.3; 155	0.25; 0.58	0.40	Vaned-disk	16	—	0.7–0.9	—	Y3
44	0.38	0.33; 0.40	Paddle	5	0.60	—	0.35	P11
2.2–570	0.14–0.90	0.2–0.7	Turbine	6	0[b]	1.0	—	W2, W4
2.7–170		0.2–0.7	Turbine	4	0[b]	1.0	—	W2, W4
5.5	0.191	0.5–0.7	Paddle	2	0[b]	1.0	—	W2, W4
5.5	0.191	0.4–0.7	Propeller	3	0[b]	1.0	—	W2, W4

[a] See Eqs. (6)–(9).
[b] At agitation rates above some minimum agitation rate.

where P_v is the power input per unit volume, V_s is the superficial linear gas velocity, and Ks is the volumetric mass-transfer coefficient—the product of the mass transfer coefficient per unit area and the interfacial area per unit volume.

In 1960, Yoshida *et al.* (Y4), working with a geometrically similar system and with the sulfite-oxidation method, confirmed the results reported by Cooper *et al.* They also showed that the gas film does not offer any resistance to the mass transfer of oxygen from air to the sodium sulfite solution. In addition, they found that the mass-transfer coefficient per unit area was equal for water and for aqueous sodium sulfite.

A summary of a number of correlations proposed for volumetric mass-transfer coefficients and specific interfacial area is presented in Table II, which includes data additional to those of Westerterp *et al.* (W4). It is apparent that disagreement exists as to the numerical values for the exponents. This is due, in part, to the lack of geometric similarity in the equipment used. In addition, variation in operating factors such as the purity of the system (surfactants), kind of chemical system, temperature, etc., also contribute to the discrepancies. To summarize Table II:

(a) For the volumetric mass-transfer coefficient,

$$Ks \propto V_s^m N^n \tag{6}$$

$$Ks \propto V_s^m P_v^\omega \tag{7}$$

where the ranges of the exponents are: $0 \leqslant m \leqslant 1.0$; $1.26 \leqslant n \leqslant 3.0$; $0.43 \leqslant \omega \leqslant 0.95$.

(b) For the specific interfacial area,

$$s \propto V_s^\kappa N^M \tag{8}$$

$$s \propto V_s^\kappa P_v^\Delta \tag{9}$$

where the ranges of the exponents are: $0 \leqslant \kappa \leqslant 0.75$; $0.7 \leqslant M \leqslant 1.5$; $0.35 \leqslant \Delta \leqslant 0.40$.

It should be noted that these correlations merely demonstrate the results obtained by different investigators and cannot be used directly for equipment design nor to evaluate the exact manner in which Ks, s, or K depend on the operating variables cited. From the divergent results it is apparent that other, possibly unrecognized, operating factors are also involved. In general, however, the correlations do indicate a strong dependence of the volumetric mass-transfer coefficient on the specific interfacial area. This also is in agreement with the experimental finding of Calderbank and Moo-Young (C4) that power dissipation has but little effect on mass-transfer coefficients per unit area for gas–liquid dispersions. They found that for bubbles less than

2.5 mm in average diameter, the data can be correlated by

$$k_L = 0.31 \left[\frac{(\rho_c - \rho_d)\mu_c g}{\rho_c^2} \right]^{1/3} [N_{Sc}]^{-2/3} \qquad (10)$$

whereas for bubbles greater than 2.5 mm in diameter,

$$k_L = 0.42 \left[\frac{(\rho_c - \rho_d)\mu_c g}{\rho_c^2} \right]^{1/3} [N_{Sc}]^{-1/2} \qquad (11)$$

In addition, it was concluded that the liquid-phase diffusion coefficient is the major factor influencing the value of the mass-transfer coefficient per unit area. Inasmuch as agitators operate poorly in gas–liquid dispersions, it is impractical to induce turbulence by mechanical means that exceeds gravitational forces. They conclude, therefore, that heat- and mass-transfer coefficients *per unit area* in gas dispersions are almost completely unaffected by the mechanical power dissipated in the system. Consequently, the total mass-transfer rate in agitated gas–liquid contacting is changed almost entirely in accordance with the interfacial area—a function of the power input.

C. Average Bubble Diameter and Bubble-Size Distribution

1. *Average Bubble Size**

Yoshida and Miura (Y3) reported empirical correlations for average bubble diameter, interfacial area, gas holdup, and mass-transfer coefficients. The bubble diameter was calculated as

$$d_{32} = \frac{\sum_{i=1}^{n} d_i^3}{\sum_{i=1}^{n} d_i^2} = \frac{6\Phi}{s} \qquad (12)$$

where Φ is the fractional gas holdup. Their results can be summarized as follows:

$$d_{32} \propto N^{-0.1} L^{0.1} \qquad (V_s = 20 \text{ ft/hr}) \qquad (13)$$
$$d_{32} \propto N^{-0.1} L^{0.2} \qquad (V_s = 40 \text{ ft/hr}) \qquad (14)$$
$$d_{32} \propto N^{-0.1} L^{0.3} \qquad (V_s = 90 \text{ ft/hr}) \qquad (15)$$

where N is the rotational speed of the impeller and L is the impeller diameter. The measurements were made for air–CO_2 mixtures in sodium hydroxide solutions at temperatures from 10 to 30°C. A vaned-disk and a turbine agitator were used and the ratio of agitator to contactor diameter was 0.4. Rotational speeds in the range of 60–400 rpm were investigated.

In their work they found that the average bubble diameter varied from 1.5 to 4.5 mm. It is apparent from the correlations that altering the impeller

---
* For a general formulation of mean bubble sizes see Eqs. (261)–(264).

rotational speed cannot result in drastic changes in the average bubble diameter.

The findings of Yoshida and Miura are apparently at variance with those reported by Vermeulen *et al.* (V8), who found that the average bubble diameter at constant holdup varied according to:

$$d_{32} \propto N^{-1.5} \tag{16}$$

In Vermeulen's work, a paddle impeller stirred fixed amounts of gas and liquid in a closed vessel. When the impeller was brought to the proper speed (240–360 rpm), the liquid and the gas that had been above it were dispersed together and completely filled the vessel. It is impossible to extrapolate from this experimental set-up to the usual type of gas–liquid contacting operation.

Calderbank *et al.* (C1–C4), who worked with systems quite similar geometrically to that of Yoshida and Miura, found that the average bubble diameter for air in water at 15°C ranged from 3 to 5 mm. Westerterp *et al.* (W2–W4) found the range to be 1–5 mm for air in sodium sulfite solution at 30°C. In addition, they noted that any increase in interfacial area between the bubbles and the liquid was due primarily to the increase in gas holdup, and the average bubble diameter was essentially unaffected by the impeller speed and was approximately 4.5 mm (W3).

Calderbank (C1) reports that smaller bubble diameters are obtainable when a solute is present that will inhibit coalescence of the bubbles. Explanation of some of the interrelations between coalescence, breakup, and bubble size in various gas–liquid systems is given in an extensive study by Vermeulen (V7).

2. *Bubble-Size Distribution*

Gal-Or (G4) has recently reported bubble-size distribution data in air–water dispersions. The equipment used to evaluate the bubble-size distribution is a new type of multistage gas–liquid contactor without pressure drop in each stage, in which the gas is drawn in from the bottom of the vessel. Typical bubble-size—cumulative-volume data are given in Fig. 2.† The data show that for 99% of the bubbles, $0.1 \leqslant a \leqslant 1.4$ mm. The "surface mean radius" a_{32} for these data was found to be 0.07 cm. Other experimental results with this type of equipment show that this average bubble size is lower than the average size obtained with the common types of gas–liquid contactors (C1, W3, Y4), resulting in a higher total surface area for diffusion. This is mainly due to the high shearing rates produced between the impeller and stator. Another reason is probably contributed by the lower pressure inside

† Further experimental data on cumulative number density are reported in Ref. G3.

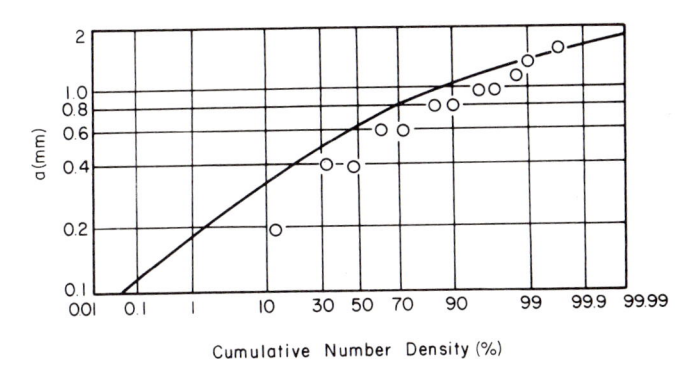

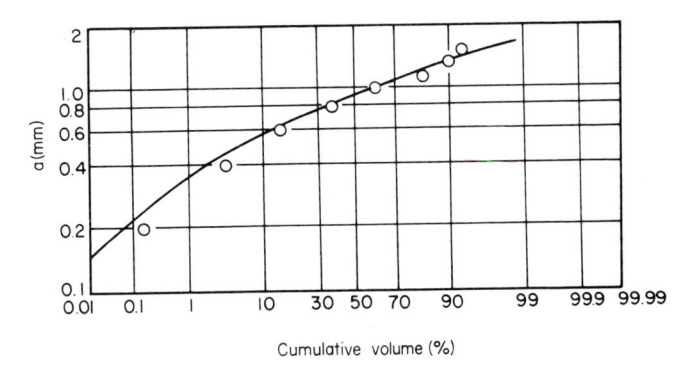

FIG. 2. Typical data for the bubble-size distribution in a gas–liquid dispersion produced in a new type of contactor without a pressure drop per stage (G4). Dispersed phase: air. Continuous phase: water. The solid lines were calculated from Eq. (17) and Eq. (258) or (260). [after Gal-Or and Hoelscher (G5)].

the bubble at the moment of its formation, resulting in a slight compression due to hydrostatic pressure in the liquid bulk.

The data in Fig. 2 are compared with a proposed particle-size distribution function $f^*(a, \bar{a}_v)$ (B4, G5):

$$f^*(a, \bar{a}_v) = \frac{16}{\pi \bar{a}_v^3} a^2 \exp\left[-\left(\frac{4}{\bar{a}_v^3 \sqrt{\pi}}\right)^{2/3} a^2\right] \quad (17)$$

in which a_v is the "mean volume radius" defined by

$$\bar{a}_v = \left[\frac{\sum n_i a_i^3}{\sum n_i}\right]^{1/3} \quad (18)$$

Equation (17) indicates that the entire distribution may be determined if one parameter, $\bar{a}_v$, is known as a function of the physical properties of the system and the operating variables. It is constant for a particular system under constant operating conditions. This equation has been checked in a batch system of hydrosols coagulating in Brownian motion, where $\bar{a}_v$ changes with time due to coalescence and breakup of particles, and in a liquid–liquid dispersion, in which $\bar{a}_v$ is not a function of time (B4, G5). The agreement in both cases is good. The deviation in Fig. 2 probably results from the distortion of the bubbles from spherical shape and a departure from random collisions, coalescence, and breakup of bubbles.

Equation (17) is a type of normalized gamma distribution function,

$$\int_0^\infty f^*(a, \bar{a}_v) \, da = 1 \tag{19}$$

and is of the same type as the Maxwell–Boltzmann velocity distribution of gaseous atoms. The cumulative density function [cf. Eq. (258)] is obtained from

$$\text{Cumulative density } (\%) = 100 \left\{ 1 - 4 \left(\frac{\alpha^3}{\pi} \right)^{1/2} \int_0^\infty a^2 \exp(-\alpha a^2) \, da \right\}$$

$$= 100 \left\{ \text{erf}[a\sqrt{\alpha}] - 2 \left(\frac{\alpha}{\mu} \right)^{1/2} a \exp[-\alpha a^2] \right\} \tag{20}$$

in which

$$\alpha = \left(\frac{16\sqrt{\pi N_v}}{3\Phi} \right)^{2/3} = \left(\frac{4}{\sqrt{\pi \bar{a}_v^3}} \right)^{2/3} \tag{21}$$

There is a maximum and minimum particle size in any particle distribution, and the limits of the integral would preferably be these limits. However, in examining Eq. (20), we obtain for $\bar{a}_v = 0.12$ cm, for example, that the cumulative density percent of particles having $0.02 \leqslant a \leqslant 0.2$ cm ranges from 0.6% to 95.4%, and for particles having $0.01 \leqslant a \leqslant 0.3$ cm, from 0.06% to 99.956%. Thus the error introduced in evaluating this integral is usually small. Using this particle-size distribution, the relationship between a_{32} and $\bar{a}_v$ in the dispersion was evaluated by Gal-Or and Hoelscher (G5):

$$a_{32} = \frac{\int_0^\infty a^3 f^*(a, \bar{a}_v) \, da}{\int_0^\infty a^2 f^*(a, \bar{a}_v) \, da} = \frac{\Gamma(3.0)}{\Gamma(2.5)\sqrt{\alpha}} = 1.148 \bar{a}_v \tag{22}$$

Experimental evaluation of this result was made by Gal-Or (G3) who found that $a_{32}/\bar{a}_v = 1.133$ for air-distilled water dispersion and $a_{32}/\bar{a}_v = 1.27$ for air-tap-water dispersions. The dispersed-phase holdup fraction, Φ, is thus given by (G5)

$$\Phi = \frac{4}{3}\pi N_v \int_0^\infty a^3 f^*(a, \bar{a}_v)\, da \qquad (23)$$

in which N_v, the number of bubbles per unit volume of dispersion, is defined by:

$$N_v = \frac{\Phi}{\frac{4}{3}\pi \bar{a}_v^3} \qquad (24)$$

Consequently, the total interfacial area of the bubbles per unit dispersion volume, s, is given by (G5) [cf. Eq. (259)]

$$S = \frac{\int_0^A dA}{V_v} = \frac{3\Phi}{\bar{a}_v^3}\int_0^\infty a^2 \frac{16}{\pi \bar{a}_v^3} a^2 \exp\left[-\left(\frac{4}{\sqrt{\pi \bar{a}_v^3}}\right)^{2/3} a^2\right] da$$

$$= 0.872\frac{3\Phi}{\bar{a}_v} = \frac{3\Phi}{a_{32}} \qquad (25)$$

The final result is very well known in experimental studies and is usually used to evaluate a_{32} from known Φ and s values. Other studies in this field include the work by Shinnar and Church (S7) who used the Kolmogoroff theory of local isotropy to predict particle size in agitated dispersions, and an analysis by Levich (L3) on the breakup of bubbles. Levich derived an expression for the critical bubble radius (the radius at which breakup begins):

$$a_{cr} \cong \frac{\sigma}{u^2}\left(\frac{3}{K_f \rho_d \rho_c^2}\right)^{1/3} \qquad (26)$$

where K_f is the drag coefficient. In the case of water, this gives $a_{cr} \cong 1.8$ cm, which agrees well with experimental data (L3).

Bubble stability and breakup were reviewed by Hinze (H16). Early stages in the motion and breakup of two-dimensional air bubbles in water have been followed by Rowe and Partridge (R8), using high-speed cinephotography. The initial diameter of their circular bubble was about 4 in.

Anderson (A2) has derived a formula relating the bubble-radius probability density function (B3) to the contact-time density function on the assumption that the bubble-rise velocity is independent of position. Bankoff (B3) has developed bubble-radius distribution functions that relate the contact-time density function to the radial and axial positions of bubbles as obtained from resistivity-probe measurements. Soo (S10) has recently considered a particle-size distribution function for solid particles in a free stream:

$$f(a, a_0, \Delta a) = \frac{\pi^{-1/2} a}{\Delta a}\exp[-(a - a_0)^2/(\Delta a)^2] \qquad (27)$$

where a_0 is a characteristic, or nearly mean, particle size and Δa is the range of size variation.

A general statistical-mechanical formulation proposed by Hulburt and Katz (H20) analyses the main problems in particle technology. For the purposes of engineering analysis, it takes as known the dependence of particle nucleation and growth on the particle environments and incorporates this dependence into differential equations showing how such quantities as average particle size vary with position in the processing equipment. These differential equations can then be placed alongside corresponding differential equations for the environmental factors—composition, temperature, bulk velocity—and the overall set of equations can be viewed as a mathematical characterization of the processing system. Some simplified analyses for applications were presented, but the applicability of this theory to gas–liquid dispersions still remains to be pursued and extended.

D. Gas Holdup and Average Residence Time

In 1944, Foust et al. (F2) studied air holdup in water in baffled vessels agitated with a special impeller developed for gas dispersion. The impeller consisted of arrowhead-shaped blades mounted on a flat disk. The gas holdup was determined by measuring the liquid level before the air was introduced and while the air was fed at a point underneath the impeller. They found that the gas holdup ranged from 2% to 10% of the air–free-liquid volume.

They defined an "average contact time" as the average residence time of a bubble per unit length of dispersion. Their experimental data were correlated by:

$$\bar{\theta}_d = \frac{Hu}{YQ} = C\left(\frac{P_v}{V_s}\right)^{0.47} \qquad (28)$$

where $\bar{\theta}_d$ is the average contact time per unit height of dispersion, Hu is the volume of gas holdup, Q is the volumetric gas flow rate, and Y is the height of dispersion. The dimensional constant C is a function of liquid depth and geometric ratios, and varies from 1.26 to 1.65 when P_v is expressed as horsepower per cubic foot and V_s as feet per second.

According to their measurements, the gas holdup increases with the gas velocity but the "average contact time" drops. This is not surprising, as will be shown. The volumetric gas flow rate is

$$Q = V_s A_T \qquad (29)$$

where A_T is the cross-sectional area of the tank. By substituting for Q in the definition of $\bar{\theta}_d$ [Eq. (28)], it becomes apparent that the "average contact time" is the fractional gas holdup Φ divided by the superficial gas velocity, or

$$\bar{\theta}_d = \frac{\Phi}{V_s} \propto \left(\frac{P_v}{V_s}\right)^{0.47} \tag{30}$$

and

$$\Phi \propto P_v^{0.42} V_s^{0.53} \tag{31}$$

Calderbank (C1) obtained similar results for the air–water system with a six-blade turbine. His data for gas holdup could be expressed as

$$\Phi \propto P_v^{0.4} V_s^{0.5} L^{1.2} \tag{32}$$

and for the vaned-disk impeller

$$\begin{aligned}
\Phi &\propto N^{0.8} L & (V_s &= 20 \text{ ft/hr}) \\
\Phi &\propto N^{0.7} L & (V_s &= 40 \text{ ft/hr}) \\
\Phi &\propto N^{0.6} L & (V_s &= 90 \text{ ft/hr})
\end{aligned} \tag{33}$$

Clark and Vermeulen (C8) measured gas holdup in three different liquids —isopropyl alcohol, ethylene glycol, and water. They measured the increase in holdup with agitation as compared to no agitation, and correlated their results as a function of the volumetric gas velocity, Weber number, P/P_0, and a geometric factor. Typical volumetric gas holdup values reported in the literature vary from about 2% to 40% of the total dispersion volume (C1, C2, C8, F2, G10).

Gal-Or and Resnick (G8) measured average residence time in a system that was geometrically similar to those used by Cooper *et al.* (C9) and Yoshida *et al.* (Y4) with air–distilled water and air–sodium sulfite solutions of the same concentration as used by these investigators. The ratio of impeller to tank diameter was 0.4 in one series (as in the work of Cooper and Yoshida) and 0.3 in a second series. Gal-Or and Resnick reported their results as an average residence time in seconds per foot of gas-free liquid, $\bar{\theta}_h$. The average residence time was calculated from the equation

$$\bar{\theta} = \frac{\int_0^\infty \theta c(\theta) \, d\theta}{\int_0^\infty c(\theta) \, d\theta} \tag{34}$$

where $c(\theta)$ is the system response to a pulse of helium tracer injected into the incoming air. For both impeller-to-tank diameter ratios and the two chemical systems, they correlated their data by:

$$\bar{\theta}_h = 0.153 (P_v/V_s)^{0.45} \tag{35}$$

where the ratio P_v/V_s is expressed in units of pounds per cubic foot. The data reported by Foust *et al.* (F2), recalculated on this same basis, resulted in

$$\bar{\theta}_h = 0.1(P_v/V_s)^{0.45} \tag{36}$$

The same exponent was obtained in both works, although in one case a vaned-disk impeller was used, and in the other, arrowhead-shaped blades. Although the nominal residence time can be determined from a knowledge of gas holdup and volumetric gas flow rate, the nominal residence time (Eq. 36) will equal the statistical average residence time (Eq. 34) for certain cases only (L2).

Experimental data on gas holdup in a sparged contactor have been reported in the literature (F1, Y2).

E. Gas Residence-Time Distribution

The residence time or "contact time" discussed in the preceding section is a simple average. Hyman (H21) pointed out that the residence time of any one gas molecule could vary widely from the mean because of the tortuous paths followed by the gas bubbles from the gas inlet to the surface. Knowledge of gas residence time is important for design purposes and is necessary for an understanding of the behavior of gas–liquid dispersions (W5). Relatively little experimental effort has been devoted to this area.

The first experimental study appears to have been made by Hanhart *et al.* (H8). Their experimental system consisted of a cylindrical vessel 60 cm in diameter which was terminated by a conical head for gas collection. The impeller shaft and gas inlet entered through the tank bottom. Agitation was provided by six-blade turbine impellers. In all runs, the dispersion height was maintained at 90 cm. The residence time was determined experimentally by determining the system response to a "step" forcing function. The step change was effected by switching from air to a mixture containing 90% air and 10% hydrogen. The exit gas was analyzed continuously by a thermal conductivity cell.

To analyze their data, they assumed that the flow in the conical top was characterized as a perfectly-mixed regime (the conical volume was agitated separately by a fan) and that a plug flow regime characterized the flow through the piping system.

Their conclusions are that the gas residence-time distribution in their mixing vessel is intermediate between that to be expected from one perfectly-mixed vessel and that from two perfectly-mixed vessels of equal size in cascade. The cascade behavior of two equal-sized mixers is approached with a relatively large impeller located half-way between the bottom and top surfaces. The response curve becomes similar to that of one perfectly-mixed vessel when small impellers are used or if the impeller is located below the half-way point.

However, their experimental system was not equivalent to the system normally used in laboratory or in industrial gas–liquid contacting.

These investigators did not report the average residence time as a function of the impeller speed and gas velocity. They did report, nevertheless, that for the air–water and air–one-normal sodium sulfate solution the average gas residence time ranged from 4 to 7 sec. They report that the maximum possible average residence time of the gas in the dispersion is 7 sec for a dispersion height of 90 cm. This average residence-time range was observed over a range of volumetric air rates of 1.1 to 3.2 liter/sec and impeller speeds of 1.67 to 22.5 rps.

Gal-Or and Resnick (G2, G8) studied gas residence-time distribution by measuring the system response to a pulse of helium tracer injected into the air. The experimental system consisted of a cylindrical tank 39 cm in diameter with an air-free liquid depth of 39 cm. Two different vaned-disk impellers with 16 blades were used, one with a ratio of impeller to tank diameter of 0.3 and the other with a ratio of 0.4. The system was equivalent to those used by Cooper *et al.* (C9) and Yoshida *et al.* (Y4). The exit gas was sampled for analysis by a collection funnel at the dispersion surface. It was noted that the radial location of the collection funnel had no effect on the measurements. This is in accord with the "perfectly-mixed" stage reported by Hanhart *et al.* (H8). Special precautions were taken to minimize lags and to ensure that the unavoidable lags were equal and reproducible for all the runs. The lags in the piping were measured and appropriate corrections for them were made in the interpretation of the experimental results. The dependence of the average residence-time results on power consumption and gas velocity was given in Section II,D.

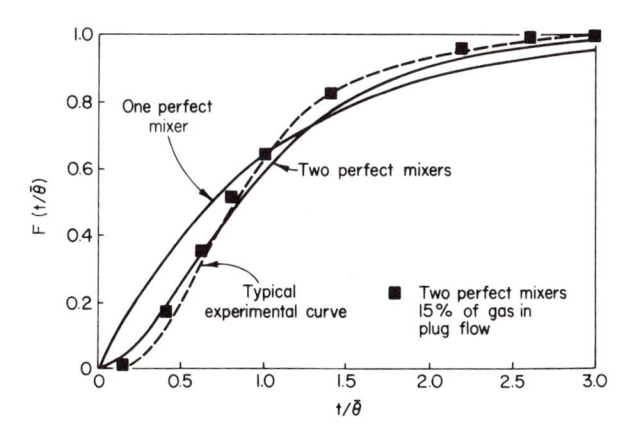

Fig. 3. Typical residence-time distribution curves in a gas–liquid dispersion [after Gal-Or and Resnick (G8)].

A typical residence-time distribution curve obtained by Gal-Or and Resnick is shown in Fig. 3. It approaches those that would be obtained for two equal-sized perfectly-mixed vessels in series with part of the gas moving through the vessels in plug flow, rather than for one-perfectly mixed vessel. This finding was confirmed by visual and photographic observations that two gas mixing sections were present in the gas–liquid contactor, one above and one below the impeller (G7). When the gas bubble leaves the impeller it may enter the upper zone immediately or may be drawn down into the lower zone by the liquid. The bubbles circulate in both of these zones. As time progresses, some bubbles leave the bottom section and enter the top zone where they may again circulate before they leave the system.

F. Flow Patterns and Bubble Velocities

1. *Experimental Studies*

The flow patterns of agitated liquid have been studied extensively (A1, B11, F6, K5, M6, N2, R12, V5), usually by photographic methods. Apparently no work has been reported on bubble-flow patterns and relative velocities in agitated gas–liquid dispersions. Some simple pictures have been presented that only show the same details that may be seen with the unaided eye (B11, F6, Y4).

A somewhat qualitative study of bubble paths and their average velocities relative to that of the agitated liquid in a dispersion is reported in a recent work by Gal-Or and Resnick (G7). The technique employed was based on photographing tracer particles in horizontal and vertical light planes, and was applied to investigate the velocity of gas bubbles by photographing the bubbles themselves. Solid spherical tracer particles were also introduced into the liquid and their streaks were used to determine the liquid velocity. The difference between these two velocities as measured under the same stirring conditions then gives the relative velocity between the bubble and the stirred liquid. From the experimental results it is apparent that two zones of mixing exist; one above the impeller and the other beneath it. It was also found that the bubble-flow patterns produced by a vaned-disk impeller are approximately helical and that in each mixing zone the helix progresses around a circular axis in the direction of impeller rotation. The flow pattern of the bubbles was found to be similar to that of the agitated medium. The average velocity of the bubbles relative to the agitated medium rises slightly with an increase in the impeller rotational speed. The results also showed that over a large range of rotational speed, the local dispersion density of the bubbles throughout the whole vessel does not differ appreciably from the average, and no fluid or bubble dead spaces exist.

In Section I, a qualitative schematic description of the main connection between increased agitation intensity and increased total mass-transfer rate was given. It can readily be seen from this description that further research in gas and liquid flow patterns and in the area of relative bubble velocities in dispersions will contribute to the basic knowledge necessary for understanding the real mechanisms occurring in these systems.

2. *Estimation of Bubble Velocity**

In general, the velocity of a bubble relative to the continuous phase in the dispersion can be expressed as (G9);

$$U_{dis} = \psi(\mu_c, \rho_c, a, g, t, \Gamma, \Phi, P_v, \sigma) \tag{37}$$

where the viscosity and the density of the gas, as well as other weak effects such as that of diffusion, are neglected. Theoretical studies on bubble velocity have normally been limited to studies of a steady motion of single bubbles under the influence of gravity. It is clear, however, that the suspended bubbles entrained by the turbulent eddies (bubbles which describe complicated trajectories in the fluid) are moving with a relative velocity different from that derived for the case of single bubble rise. Experimental observations (G7) indicate that these velocities may be of the same order of magnitude as those predicted for the steady rise of single bubbles. Thus, a short discussion of some of the analyses available for the case of a single rising bubble will be of value in estimating the effect of dispersed phase holdup on bubble velocities in a swarm as well as in evaluating the effect of surface-active agents on bubble velocity and on convective mass transfer.

Studying the steady motion of a single medium-size bubble rising in a liquid medium under the influence of gravity, Levich (L3, L4) solved the continuity equation simultaneously with the equations of motion by introducing the concept of a boundary layer for the case of a bubble. This boundary layer accounts for the zero, or extremely low, shear stress at the interface. Despite some errors in deriving the equations, his result was later confirmed with minor improvements (A4, M3, M10).

Assuming that the velocity distribution for flow past a gas bubble differs relatively little from the velocity distribution in an ideal liquid, and neglecting the curvature of the boundary layer, Levich finds that

$$U = \frac{\rho_c g a^2}{9\mu_c} \tag{38}$$

---

* See also Sections IIIA and IIIB for the effect of surface-active agents on bubble velocity.

is applicable for clean interfaces and bubble diameters less than 0.2 cm where $50 < N_{Re} < 800$. For higher Reynolds numbers and larger bubbles, pulsations appear within the bubble; then the bubble is deformed and no longer rises in a straight line. Assuming that the distribution of liquid velocities around the bubble remains the same, in principle, as for moderate Reynolds numbers, Levich estimates that for high Reynolds numbers and large bubbles,

$$U \simeq \left[\frac{2}{15}\frac{\sigma^2 g}{\rho_c \mu_c}\frac{\rho_d}{\rho_c}\right]^{1/5} \tag{39}$$

For $N_{Re} \ll 1$, the problem of bubble motion is closely related to that of the motion of a liquid drop in a liquid medium, and can consequently be derived from the Rybczynski–Hadamard formula (H2, R13);

$$U = \frac{2(\rho_c - \rho_d)(\mu_c + \mu_d)ga^2}{3\mu_c(2\mu_c + 3\mu_d)} \tag{40}$$

that was derived from the Navier–Stokes equations for the steady rise of drops with clean interfaces. When ρ_d and μ_d are neglected in comparison to ρ_c and μ_c, Eq. (40) gives

$$U = \tfrac{1}{3}\rho_c g a^2/\mu_c \tag{41}$$

which is applicable when

$$\frac{\rho_c^2 g a^3}{3\mu_c^2} \ll 1 \tag{42}$$

To evaluate the effect of holdup on bubble velocity, Marrucci (M3) used a spherical cell model of radius b such that

$$\Phi = (a/b)^3 \tag{43}$$

and considered an irrotational flow between two concentric spheres exhibiting a relative velocity U, as treated by Lamb (L1). Marrucci (M3) showed that the effect of holdup can thus be expressed as

$$U_{\text{swarm}} = U_{\text{single}}(1 - \Phi)^2/(1 - \Phi^{5/3}) \tag{44}$$

for $1 \ll N_{Re} < 300$ [cf. Eq. (69)]. Recent experimental work (M4) shows that U_{swarm} may be predicted approximately by Eq. (44). However, an increase in bubble velocity may be observed (M11) with increased Φ, due mainly to "chimney" effects in which massive continuous upward gas currents may appear in the dispersion and lead to increased "bubble" velocities. This field is still open to extensive theoretical and experimental research.

Astarita and Apuzzo (A3, A4) have recently reviewed the available theoretical knowledge of the motion of gas bubbles in Newtonian liquids and the possibilities of extension to non-Newtonian liquids both purely viscous and viscoelastic. They also presented experimental data on the rising velocities of gas bubbles in a variety of non-Newtonian liquids. Using direct dimensional analysis, Astarita (A3) shows that the terminal velocity of gas bubbles through a viscoelastic liquid is given by an equation of the form:

$$N_{Le} = f(N_{Re}, N_{As}) \tag{45}$$

where N_{Le} is the Levich number ($\mu_c U / \rho_c g a^2$), the ratio of viscous to buoyant forces, and N_{As} is the Astarita number ($\rho_c g a / G$), the ratio of the stress to the elastic modulus G.

According to Boussinesq (B12) the velocity of drop fall was expressed as

$$U = \frac{2}{3} \frac{(\rho_c - \rho_d) g a^2}{\mu_c} \cdot \frac{\mu_c + \mu_d + (2e/3a)}{2\mu_c + 3\mu_d + (2e/a)} \tag{46}$$

where e is the "surface viscosity coefficient". The quantity e does not represent ordinary viscosity, but a second viscosity for the surface motion that can be related (L3) to the retarding effects that are produced by the presence of adventitious surface-active impurities. A further discussion of the effects of surface-active agents on bubble velocity is given in Section III.

G. Optimum Design and Operating Conditions

In gas–liquid dispersions, the bubbles are suspended in a liquid which is either stirred artificially by mechanical agitation or by the bubbles themselves as they pass through the fluid. In general, mechanical agitation is favored if the amount of liquid is large compared with the gas flow rate, and if any solids are present in the system. It is also favored when a reaction is conducted in a batch operation with respect to the liquid (C8). A porous sparger has no advantage except at very low power inputs. A summary of the use of different types of spargers is given by Gaden (G1). An experimental study of the effects of superficial gas velocity and gas holdup on heat transfer in a sparged contactor is reported by Fair et al. (F1). The performance of this type of equipment has been studied rather extensively (D12, F1, O4, T1, V5). In view of the numerous applications of agitated gas–liquid contacting, a short summary on the optimum use of single as well as multiple impellers is given below.

1. *Use of Single Impeller*

The main functions of the impeller are to subdivide the entering gas into small bubbles, to disperse them throughout the liquid, and to maximize the gas residence time. Several impeller designs have proved to promote optimum gas–liquid transfer. Flat-blade turbines and vaned-disk impellers are used extensively. The center-disk turbine with flat blades is particularly satisfactory if the gas is introduced below the impeller (P3, P6). In the vaned-disk type, the blades are located below the plate only. The radial discharge and recirculation patterns of these impellers promote a long gas path in the liquid, and consequently increase gas residence time. Propellers are of little value in promoting gas–liquid contacting. The gas introduced into the axial flow pattern produced by the propeller may rapidly escape from the liquid, and the possibility that an individual bubble will be recaptured and recirculated is slight. As a result, the gas residence time is not maximized.

Rushton (R11) in 1954 presented a graph showing contacting efficiency as a function of impeller diameter at constant power input. He found that the rate of mass transfer between phases increased to a maximum and then decreased as the impeller diameter increased. The optimum occurred at a ratio of impeller to tank diameter of about 0.25, a ratio which is much smaller than that found for liquid blending.

More recently, Westerterp (W3) studied the optimum design for an agitated gas–liquid disperser. He also suggested, as a general rule, that small-diameter impellers operating at high rotational speed be used to obtain efficient utilization of the power input. He found that an optimum ratio of agitator to vessel diameter $(L/D_T)_{OPT}$ exists when the total interfacial area is a maximum at a given power input. This conclusion is important because, as stated earlier, the capacity of a gas–liquid contactor is, under most circumstances, determined mainly by the total interfacial area. Westerterp reported empirical correlations for six-blade turbines to determine this optimum. He concluded that if the energy supplied were to increase, then the diameter of the optimum agitator would decrease and shaft rotational speed increase. In general, his optimum ratios of impeller to tank diameter are less than the value of 0.25 recommended by Rushton (R11). For very high gas loads (>0.10 m/sec) and with low viscosity liquids, he recommends no agitation.

2. *Use of Multiple Impellers*

Mixers for gas–liquid contacting are often equipped with multiple impellers on a single shaft with the gas injected below the lowest impeller. Large commercial systems may use relatively deep cylindrical vessels. The usual ratio of liquid-depth to vessel-diameter lies between 1 and 3 (R3), and the maximum recommended is 4 (P3). In such deep systems, the question arises as to whether the use of multiple impellers would provide more effective contact than a single impeller. It must be kept in mind that if the impellers are spaced too widely the result may be ineffective agitation between the fields of action of the impellers. On the other hand, if they are too close, interference may occur between the flow streams from the adjacent impellers.

Several investigators have dealt with this problem (K1, O2, P3, R3, R12). Richards (R3) recommended that the impellers should be 1.5 impeller diameters apart. He added, however, that the correct spacing is a function of the medium being agitated. In general, he recommends that impellers should not be fitted less than 1 impeller diameter apart, and that something between 1 and 2 diameters is satisfactory. This criterion may take into account a possible transition from one flow pattern to another as the characteristics of the medium change during the course of the operation.

Karow *et al.* (K1), Rushton *et al.* (R12), and Oldshue (O2) studied rates of absorption with multiple impellers, and their results indicate that incorrect spacing and operating conditions result in decreased capacities. Rushton *et al.* (R12) found that for ratios of liquid depth to tank diameter less than 2.5 it was not possible to alter the absorption coefficients more than $\pm 10\%$ by the use of multiple impellers. They achieved a 25% increase in the absorption coefficient by the use of multiple impellers for a system with liquid depth equal to 4 tank diameters, at a power input of about 3.5 hp/1000 gal. In the same system, the coefficient could be decreased by 50% if the impellers were not properly positioned. Advantages of multiple turbines accrue only at high air flows or at high power levels (1.8–3.6 hp/1000 gal). Disadvantages of multiple impellers may come about at low air flows, low power inputs, and with improper spacing of impellers.

In designing multiple impeller configurations, an advantage may be gained by using a self-induction impeller located near the free surface of the dispersion, and thus recycling unreacted gas from the vapor space (P3).

Gal-Or (G4) has recently developed a multistage gas–liquid contactor that has no pressure drops. The unit consists of contacting compartments and a common shaft with multiple impellers. As a result of the impeller action, the gas is sucked from stage to stage through openings provided in the separating trays. The impeller acts as an agitator for the dispersion and provides pumping for the gas. Thus, the gas is self-induced from the bottom of the vessel and is subdivided in each compartment into small-sized bubbles that are dispersed throughout the liquid phase. Consequently, the pressure drop across the liquid phase is eliminated and a "negative" pressure drop is established instead. By applying vacuum at the top, the lower pressure at the bottom of the unit allows heat-sensitive materials to be handled under lower temperature conditions. Use of stators with radial vanes causes a very high shear on the liquid and incoming gas, resulting in a very small and relatively uniform diameter of bubbles with a high total surface area for diffusion. When entrainment problems are encountered and a compact unit is desirable, additional impellers may be installed between the stages to remove tiny drops from the gas phase, and consequently the tray spacing may be significantly decreased.

The experimental results with a one-stage unit (Fig. 4) show that the vacuum produced in the lower compartment under operating condition is 6–10 cm of water.

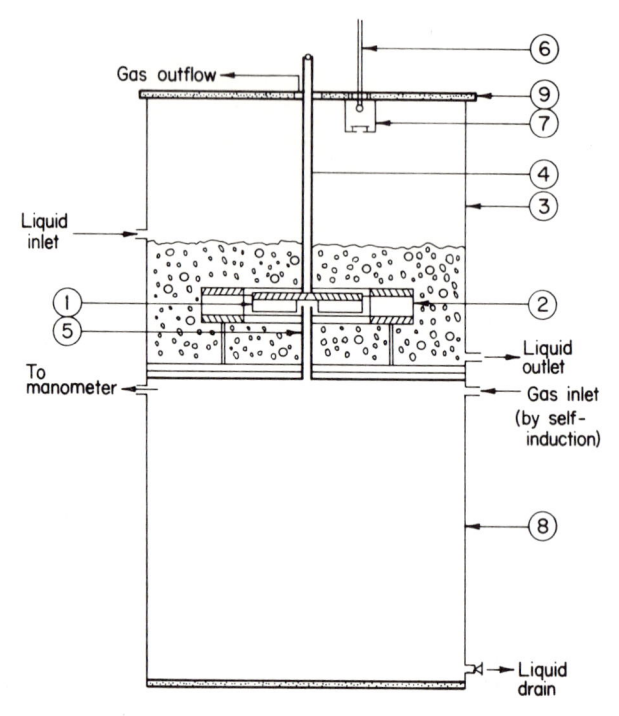

FIG. 4. Sectional view of one stage of a new type of gas–liquid contactor without pressure drop. (1) Vaned-disk impeller; (2) stator; (3) contacting tank; (4) impeller shaft; (5) gas inlet pipe (gas is self-induced through this pipe into the dispersion); (6) gas outlet thermometer; (7) thermometer pocket; (8) drainage tank; (9) lid [after Gal-Or (G4)].

3. Power Requirements

Hyman (H21) reviewed the literature published up to about 1960 on power requirements for gas–liquid agitated systems. This section will report work published subsequent to his review.

Yoshida et al. (Y4) obtained data on power consumption in a 25-cm-diameter vessel. Four different types of vaned-disk impellers and one turbine were used, but for all impellers the ratio of impeller to tank diameter was maintained at 0.4. Their results, as would be expected, showed that the power dissipation increased rapidly with agitator speed and decreased with increasing gas rate for vaned-disk impellers. At certain agitator speeds, the amount of gas held below the vaned disk reaches a saturation value that apparently varies with the gas rate. For the turbine impeller, the gas rate had a smaller effect on power than that obtained with vaned-disk impellers. At a superficial gas velocity of 400 ft/hr, the turbine, for example, required 60% of the no-gas power at the same speed, whereas the power requirements for a six-vaned disk impeller was less than 20% of the equivalent no-gas power.

Michel and Miller (M7) developed a completely empirical dimensional correlation for power requirements:

$$P \propto \left(\frac{P_0^2 NL^3}{Q^{0.56}}\right)^{0.45} \tag{47}$$

where P_0 is the power requirement for the no-gas condition and P is the requirement under gassing conditions. Similar results were obtained by Oyama and Endoh (O5) and Gal-Or and Resnick (G2, G8). However, work of Clark (C8) and others indicates that the predominant effect of this relation is to make P essentially proportional to P_0, with the possibility that the influence of Q is inadequately correlated.

The power necessary for an impeller operating in a liquid can be readily calculated from published plots of the power number as a function of the Reynolds number, but aeration results in a decrease in the power requirement for a given impeller speed. A drive that is designed on a no-gas basis will therefore be oversized for the gassed condition, whereas a drive designed for the gassed condition would be undersized if the gas rate were lower than the design condition or if the agitator must also operate under a no-gas condition —for example, during startup or shutdown.

Although the Michel–Miller correlation does not have any theoretical basis, it should be useful for making estimates. A word of caution is nevertheless in order. At extremely high gassing rates (>0.5–1 volume/volume of liquid/min), Hamer and Blakebrough (H3) found that a flooding or channeling condition was reached and the gassing rate had no further effect on the power requirements.

Clark and Vermeulen (C8) later reported an extensive experimental study of power requirements in agitated gas–liquid systems. They correlated their data in dimensionless form as a function of fractional gas holdup, Weber number, and a geometrical factor. Their correlation is shown in Fig. 5.

Unfortunately, relatively little information has been made available for industrial gas–liquid contactors. Further data from industry could permit significant tests of the reliability of the present correlations and their applicability to scale-up. Steel and Maxon (S11) reported on the power requirements during novobiacin fermentation in 20- and 250-liter pilot-plant vessels and in 12,000- and 24,000-gal vessels. The comparative data are difficult to evaluate because of changes that occurred in viscosity and gas retention during the course of the fermentation. In addition, geometric similarity did not prevail

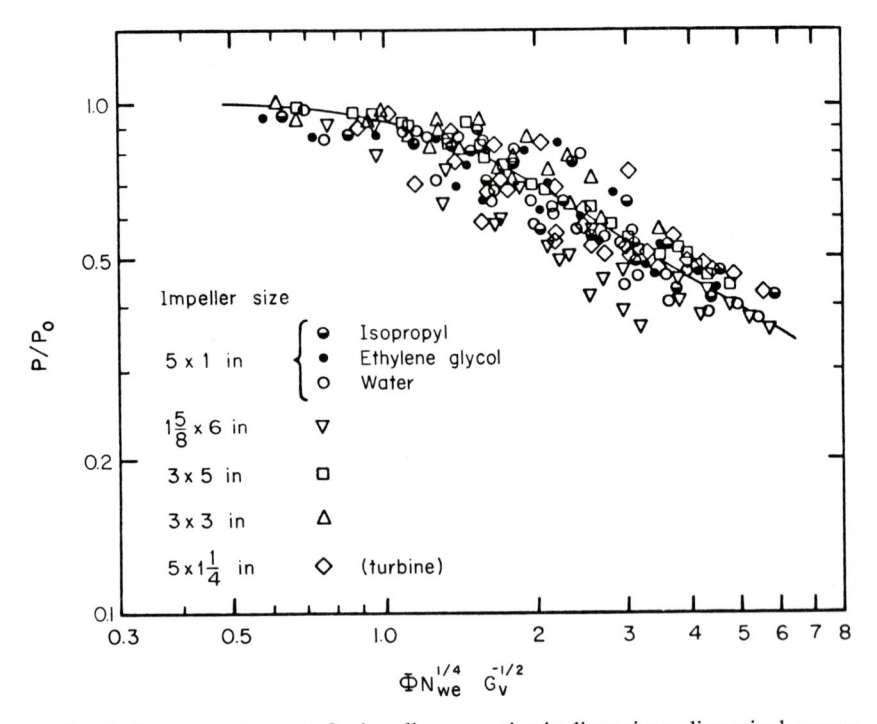

FIG. 5. Power requirements for impellers operating in dispersions, dimensionless parameters [after Clark and Vermeulen (C8)].

among the different vessels. Their results do confirm, in a qualitative manner, the behavior already reported. They observe that the power consumption dropped, for example, when the gas holdup increased. In the 24,000-gal vessel dissipation dropped from 1.0 to about 0.8 hp/100 gal while the gas holdup increased from about 5 to 30%. Also for the 24,000-gal fermenter, at constant agitation speed the power requirements at the end of fermentation dropped from about 0.75 to 0.6 hp/100 gal when the air rate was increased from 900 to 1500 CFM. Steel and Maxon also emphasize some of the difficulties involved in obtaining industrial data on complex non-Newtonian systems (S11).

4. *Scale-up by Dimensional Analysis*

Although the absorption of a gas in a gas–liquid disperser is governed by basic mass-transfer phenomena, our knowledge of bubble dynamics and of the fluid dynamic conditions in the vessel are insufficient to permit the calculation of mass-transfer rates from first principles. One approach that is sometimes fruitful under conditions where our knowledge is insufficient to completely define the system is that of dimensional analysis.

Pavlushenko et al. (P4) in their dimensional analysis considered Ks, the volumetric mass transfer coefficient, to be a function of μ_c, ρ_c, L, D_T, N, V_s, and g. They determined the following relationship for the dimensionless groupings:

$$\frac{Ks}{N} = B\left(\frac{\rho_c N L^2}{\mu_c}\right)^{-a'} \left(\frac{N^2 L}{g}\right)^{-k'} \left(\frac{D_T}{L}\right)^{j'} \left(\frac{V_s}{N}\right)^{i'} \quad (48)$$

From their experimental work, they obtained the following values for the exponents for a turbine impeller:

For	$-a'$	$-k'$	$-j'$	i'
$N_{Re} < 1.8 \times 10^5$	1.44	0.22	0.67	0.33
$N_{Re} > 1.8 \times 10^5$	0.09	0.165	0.67	0.75

Other exponents were obtained for the other types of impellers they investigated.

In their analysis, however, they neglected the surface tension and the diffusivity. As has already been pointed out, the volumetric mass-transfer coefficient is a function of the interfacial area, which will be strongly affected by the surface tension. The mass-transfer coefficient per unit area will be a function of the diffusivity. The omission of these two important factors, surface tension and diffusivity, even though they were held constant in Pavlushenko's work, can result in changes in the values of the exponents in Eq. (48). For example, the omission of the surface tension would eliminate the Weber number, and the omission of the diffusivity eliminates the Schmidt number. Since these numbers include variables that already appear in Eq. (48), the groups in this equation that also contain these same variables could end up with different values for the exponents.

Gal-Or (G2) has attempted to separate the problem into three parts—the factors that affect the interfacial area, those that affect the mass-transfer coefficient per unit area, and those involving the equipment geometrical ratios.

Experimental observations showed that the interfacial area can be represented as:

$$s = \psi(L, \rho_c, \mu_c, N, V_s, \sigma, g) \quad (49)$$

in which σ includes the effects of surface-active impurities and the viscosity and density of the gas can be neglected.

Equation (49) can be cast into the following dimensionless groupings:

$$sL \propto \left(\frac{L^2 N \rho_c}{\mu_c}\right)^{B_1} \left(\frac{V_s}{LN}\right)^{B_2} \left(\frac{L^3 N^2 \rho_c}{\sigma}\right)^{B_3} \left(\frac{LN^2}{g}\right)^{B_4} \quad (50)$$

The first dimensionless group on the right is the Reynolds number, the second represents the ratio of the gas velocity to the impeller tip speed, the third is the Weber number, and the fourth is the Froude number.

The mass-transfer coefficient per unit area for clean interfaces can be written as:

$$K_L = \psi'(L, \mu_c, \rho_c, N, D) \tag{51}$$

in which the effects of bubble radius and dispersed-phase holdup fraction on K_L are not included because these are dependent variables that are a function of operating and physical variables.

Equation (51) can be put into the dimensionless groupings:

$$\frac{K_L L}{D} \propto \left(\frac{L^2 N \rho_c}{\mu_c}\right)^{B_5} \left(\frac{\mu_c}{\rho_c D}\right)^{B_6} \tag{52}$$

The first group is the Sherwood number, the second is the Reynolds number, and the third the Schmidt number. If surfactants are present, their effect on μ should also be included [cf. Eq. (66) or (68)].

Some of the exponents in Eqs. (50) and (52) can be evaluated from experimental data. For example, Calderbank and Moo-Young (C4) investigated several chemical systems and found that the mass-transfer coefficient per unit area was a function of the Schmidt number to the power of from 0.50 to 0.67; this would also be the value of B_6. In addition, they found that agitation had no effect on K_L; therefore, B_5 is equal to zero.

Calderbank (C1) found that the interfacial area could be expressed as

$$s = \frac{P_v^{0.4} \rho_c^{0.2} V_s^{0.5}}{\sigma^{0.6}} \tag{53}$$

From his results, B_2 equals 0.5 and B_3 equals 0.6. Multiplying Eq. (50) by (52) and substituting the appropriate values for the known exponents, the result is:

$$\frac{K_L s L^2}{D} \propto (N_{\text{Re}})^{B_1} \left(\frac{V_s}{LN}\right)^{0.5} (N_{\text{We}})^{0.6} (N_{\text{Fr}})^{B_4} (N_{\text{Sc}})^{B_6} \tag{54}$$

where $0.5 \leq B_6 \leq 0.67$. In order to be able to apply Eq. (54) to varying geometric systems it is necessary to include the effects of the various geometric ratios of the equipment. The general dimensionless equation would then appear as

$$\frac{K_L s L^2}{D} \propto (N_{\text{Re}})^{B_1} \left(\frac{V_s}{LN}\right)^{0.5} (N_{\text{We}})^{0.6} (N_{\text{Fr}})^{B_4} (N_{\text{Sc}})^{B_6}$$

$$\times \left(\frac{H}{D_T}\right)^{B_7} \left(\frac{L}{D_T}\right)^{B_8} \left(\frac{W}{L}\right)^{B_9} \left(\frac{L_i}{L}\right)^{B_{10}} \cdots \tag{55}$$

where $0.5 \leq B_6 \leq 0.67$. As long as fundamental general methods for calculating the total mass-transfer rate in gas–liquid dispersions are not available, it may be necessary to rely on correlations that are based on dimensional analysis and to improve them. The separation of the problem into its more basic parts (mass-transfer coefficient per unit area, interfacial area, and geometric factors, plus the inclusion of diffusivity and surface tension) has resulted in a general dimensionless equation that permits experimental data from one system to be applied to another and those for one set of operating conditions to be applied to another. A great deal of experimental work will be necessary, however, before the values for the various exponents are known.

III. Effect of Surface-Active Agents

A. Experimental Studies

Experiments with fluids having clean interfaces are relatively rare. There is ample evidence of the presence of minute amounts of surface-active agents in all systems but those purified with the most extraordinary care, and these agents can have a profound effect on the hydrodynamics and rates of mass or heat transfer in dispersions. Many unsuspected impurities in a system can be surface-active, and even the equipment itself can supply enough such contamination to affect the results. The amount of surface-active agents present may be so small that no noticeable change in any ordinary physical property—including interfacial tension—is detected (K4). Consequently, much of the experimental data reported in the literature must be used with caution, due to the absence of information regarding the purity of the fluids used.

In studying interfacial area in gas–liquid and liquid–liquid dispersions, Vermeulen *et al.* (V8) found large deviations between different mixtures having the same nominal composition. In air–kerosene mixtures, for example, they found that measured drop diameters were approximately one-tenth of those obtained in other gas–liquid systems, and were completely inconsistent with them in terms of the effects of measured physical variables. Bottle-shaking tests on multiple samples of a large number of organic liquids showed that the ability to produce fine dispersions of air was a random one and was associated with impurities. They concluded that these phenomena could be attributed to dynamic effects involving concentration of trace impurities at the interface. Rodger *et al.* (R7), who studied the interfacial area of liquid–liquid dispersions, claimed that the settling time for a dispersion is a sensitive measure of the effect of surface-active contamination.

In studying gas absorption from single bubbles, Hammerton and Garner (H7) clearly demonstrated that contamination of the surface of a rising bubble by surface-active material initially present in trace amount can convert a mobile liquid surface into a rigid structure, and that such a conversion is accompanied by a marked reduction in absorption rate.

Talc particles sprinkled on the surface may become virtually immobile if the surface is even slightly contaminated, indicating that the surface elements become stagnant and are setting up a considerable resistance to the "clearing" of the surface by eddies of liquid approaching obliquely (D9).

The mechanisms by which this interaction occurs may be divided into two distinct groups (S4): first, the hydrodynamic behavior of a multiphase system can be changed by the addition of surface-active agents, and, as a result, the rate of mass transfer is altered; secondly, surface contaminants can interfere directly with the transport of matter across a phase boundary by some mechanism of molecular blocking.

In general, the contents of any bubble moving with respect to the external liquid (as is the case in gas–liquid dispersions) tend to circulate because of external viscous shear. However, circulation may be greatly retarded if surface-active agents are present, since molecules of surfactants are swept toward the rear of the bubble and set up a surface tension gradient opposing the external shear. Circulation within the bubble is reduced or even stopped if the interfacial tension gradient and the external viscous drag are of the same order of magnitude. Circulation may also be retarded or stopped by rigid films of insoluble surfactants or solid particles.

Experimental observations (S3) indicate that a stagnant cap is formed over the rear of the droplet as surface-active agents are added, and that this cap tends to enlarge with increasing concentrations until the entire droplet is enveloped. Thus, circulation may occur only in the front portion of the bubble. In contrast to this mechanism, Thorsen and Terjesen (T3) and Garner (G11) concluded that most of the mass transfer takes place at the rear of the bubble.

Garner and Hale (G12) concluded that the reduced mass-transfer rates are primarily due not to the area occupied by the surface-active agents, but to the absorption of the solute on the surface film; in addition, they suggested a third resistance at the interface for which equilibrium cannot exist. However, the interfacial resistances that were found in some cases may be due to the approximate nature of the calculations involved. Scriven and Pigford (S6) have made accurate measurements on phase equilibrium at the gas–liquid interface during absorption, and concluded from these results and those of other investigators that equilibrium prevails at a freshly formed, relatively clean CO_2–water interface (distilled water and 99.9% pure CO_2). The same statement, they added, applies to the absorption of other slightly soluble gases in water. Experiments show that the addition of 15 ppm of pentapropylbenzenesulphonate to water decreases the absorption of oxygen from bubbles by as much as 67% under the low shear conditions that obtain between two concentric cylinders (T4). Davis (D7, D9) found that the mass-transfer coefficient per unit surface area for absorption of various gases into stirred water can be reduced by a factor of from 1.8 to 2.2 by film-covered interfaces. Protein concentrations as low as 1 mg/m² of interface were enough to cause this marked reduction. These experimental results demonstrate the magnitude of the possible errors in theoretical treatments that do not take into account adventitious surfactants.

Adventitious surfactants also have a marked effect on the mechanism of coalescence. In studying the coalescence of curved water surfaces, Lindblad (L8) used aged distilled water that was stored for about 30 h in a polyethylene bottle opened to the air through a narrow polyethylene tube inserted in the water. He found that if fresh distilled water (water exposed not longer than 1 h to the air) was used, the delay time in coalescence was approximately half as long. Consequently, he concluded that this difference is due to some form of contamination which settled into the water or onto the water surface.

B. Theoretical Studies

An evaluation of the retardation effects of surfactants on the steady velocity of a single drop (or bubble) under the influence of gravity has been made by Levich (L3) and extended recently by Newman (N1). A further generalization to the domain of flow around an ensemble of many drops or bubbles in the presence of surfactants has been completed most recently by Waslo and Gal-Or (W1). The terminal velocity of the ensemble is expressed in terms of the dispersed-phase holdup fraction and reduces to Levich's solution for a single particle when Φ approaches zero. The basic theoretical principles governing these retardation effects will be demonstrated here for the case of a single drop or bubble. Thermodynamically, this is a case where coupling effects between the diffusion of surfactants (first-order tensorial transfer) and viscous flow (second-order tensorial transfer) takes place. Subject to the Curie principle, it demonstrates that this retardation effect occurs on a nonisotropic interface. Therefore, it is necessary to express the concentration of surfactants Γ, as it varies from point to point on the interface, in terms of the coordinates of the interface, i.e.,

$$\Gamma = \Gamma(x_1, x_2, x_3) \tag{56}$$

where x_1, x_2, and x_3 are related to each other by the equation of the surface. A change in Γ results in a variation in surface tension along the surface, which leads in turn to the appearance of surface forces that change the boundary conditions as compared to those for a clean interface. In the presence of surfactants, they become

$$\tau_t = \operatorname{grad} \sigma = \frac{\partial \sigma}{\partial \Gamma} \operatorname{grad} \Gamma = (\tau_d)_{tt} - (\tau_c)_{tt} \tag{57}$$

$$\tau_\sigma = (\tau_d)_{nn} - (\tau_c)_{nn} \tag{58}$$

where τ_σ is the capillary pressure $= 2\sigma(\Gamma)/a$, τ_t the tangential force per unit of liquid surface directed from points of greater to points of lesser surface tension, $(\tau_d)_{tt}$ and $(\tau_c)_{tt}$ the tangential components of viscous stress tensor for the dispersed and continuous phases, respectively (assumed equal in magnitude for a clean interface), and $(\tau_d)_{nn}$ and $(\tau_c)_{nn}$ the normal components of the viscous stress tensor for the dispersed and continuous phase, respectively. The symmetrical effect of viscous flow on the diffusional flux of surfactants at the interface is discussed by Waslo and Gal-Or (W1).

In order to solve the complete problem of liquid flow around a single spherical drop or bubble in the presence of surface-active agents it is necessary to solve the equations for viscous liquid motion simultaneously with a conservation equation for dissolved surface-active agents;

$$\frac{\partial \Gamma}{\partial t} + \text{div}_s(\Gamma \mathbf{v}_t - D_s^{\,s}\,\text{grad}_s\,\Gamma) + J_n = 0 \tag{59}$$

where J_n acts as a source (or sink) of surfactants (note that the domain of Eq. (59) is only the interface and J_n takes into account the flux of surfactants from another set of space coordinates). Equation (59) allows one to express τ_t and τ_σ in boundary conditions (57) and (58) in terms of the gradient of Γ. In using this equation the assumption is made that the surface-active agents are soluble only in the continuous phase. Restricting his solution to a steady viscous fall regime of a drop where $N_{Re} \ll 1$, Levich (L3) has solved the equations for viscous liquid motion with boundary conditions (57), (58), and

$$(v_c)_r = -U\cos\theta \quad \text{as} \quad r \to \infty \tag{60}$$

$$(v_c)_\theta = +U\sin\theta \quad \text{as} \quad r \to \infty \tag{61}$$

$$(v_d)_r = (V_c)_r = 0 \quad \text{at} \quad r = a \tag{62}$$

$$(v_c)_\theta = (V_d)_\theta \quad \text{at} \quad r = a \tag{63}$$

$$(v_d)_r \text{ and } (v_d)_\theta \text{ are finite at } r = 0 \tag{64}$$

The flux of surface-active agents from the surface into the bulk of the liquid may be controlled by the slower of the following processes: 1) adsorption or desorption of surfactants at the surface or 2) diffusion of surfactants from the liquid bulk to the surface. Consequently, Levich evaluated the solution for a single drop

$$U = \frac{2}{3}\left(\frac{(\rho_c - \rho_d)ga^2}{\mu_c}\right)\left(\frac{\mu_c + \mu_d + \gamma}{2\mu_c + 3\mu_d + 3\gamma}\right) \tag{65}$$

for two cases:

(a) *Adsorption as the controlling rate:* Here the retardation coefficient is

$$\gamma = \frac{2\Gamma_0}{2\alpha a}\left|\frac{\partial \sigma}{\partial \Gamma}\right| \qquad (66)$$

in which

$$\alpha = \frac{\partial P}{\partial \Gamma} - \frac{\partial Q}{\partial \Gamma} \qquad (67)$$

and Γ_0 is the equilibrium concentration of surfactants at the liquid surface.

(b) *Diffusion of surfactants from the liquid bulk to the surface as the controlling rate:* Here, the retardation coefficient is

$$\gamma = \frac{2RT\Gamma_0^2 \bar{\delta}}{3D_s a C^0} \qquad (68)$$

where $\bar{\delta}$ is the average thickness of the diffusion boundary layer and c^0 is the concentration of surfactants in the liquid bulk that corresponds to the equilibrium concentration Γ_0 on the surface (C^0 can be related to Γ_0 by a Langmuir adsorption isotherm, for example).

Waslo and Gal-Or (W1) recently generalized the Levich solution [Eq. (65)] by evaluating the effect of Φ and γ on the terminal velocity of an ensemble of spherical drops of bubbles. Their solution is

$$U_{\text{ensemble}} = \frac{2}{9}\left(\frac{(\rho_c - \rho_d)ga^2}{\mu_c}\right)$$

$$\times \left(\frac{3\mu_c(1 - \Phi^{1/3})(1 - \Phi^{5/3}) + \left(3 - \frac{9}{2}\Phi^{1/3} + \frac{9}{2}\Phi^{5/3} - 3\Phi^2\right)(\mu_d + \gamma)}{2\mu_c(1 - \Phi^{5/3}) + (3 + 2\Phi^{5/3})(\mu_d + \gamma)}\right) \qquad (69)$$

It should be noted that these equations were derived for the case of falling drops. Nevertheless, the following general conclusions should also apply for bubbles in dispersions:

(a) The liquid motion carries the molecules of surfactants to the surface of the bubble, where they are swept toward the rear portion of the bubble. The resulting accumulation of surfactants lowers the surface tension in the rear part of the bubble, giving rise to a force that is exerted along the surface of the bubble. This force retards the motion of the surface elements and reduces the bubble velocity. Consequently, the contact time of the liquid surface elements with the gas phase is increased, causing the internal circulation as well as mass- and heat-transfer fluxes to decrease. At sufficiently high values of the retardation coefficient, motion on the surface of the bubble becomes fully damped. Consequently, the contact time of the liquid surface elements may be as long as the residence time of the bubble in the dispersion; in other words, from the moment of bubble formation until it bursts on the free surface of the dispersion. At that moment, the surface elements remain, and are mixed with the liquid phase (G5). This conclusion may prove very helpful in evaluating mathematical models for gas–liquid and liquid–liquid dispersions (G5, G9).

(b) On examining Eqs. (65), (66), and (68), it appears that the Rybczynski–Hadamard equation [Eq. (40)] is obtained when the sum $(\mu_c + \mu_d)$ is much larger than γ. In other words, when the viscosity of the liquid in a gas–liquid dispersion is high, we may expect the effect of surface-active agents to decrease or to vanish. Thus, if the effect of molecular blocking (which may reduce the transfer through the interface due to the presence of surfactants) is very small, we should expect experimental data for high-viscosity liquids containing surfactants to agree with the theories of convective diffusion to or from clean interfaces. (For further discussion of this point see Sections IV, J and IV, K).

(c) Equations (66) and (68) predict that, as the particle size is reduced, the effect of surface-active agents is increased. For insoluble surfactants, however, Levich (L3) suggests the expression

$$\gamma = \frac{2\Gamma_0 a}{3D_s^s} \left| \frac{\partial \sigma}{\partial \Gamma} \right| \tag{70}$$

which predicts the opposite effect for the particle size. In this case, the surfactants remain on the surface of the particle during its entire contact time with the continuous phase.

Davies *et al.* (D9) have recently measured the rates of absorption of various gases into turbulently stirred water both with carefully cleaned surfaces and with surfaces covered with varying amounts of surfactants. That hydrodynamic resistances, rather than monolayer resistances, are predominant in their work is consistent with the high sensitivity of k_L to very small amounts of surface contamination and also with the observation that a limit to the reduction in k_L is found (D7, D9). This is in agreement with the results of Lindland and Terjesen (L9), who found that after a small concentration of surfactant had been used further additions caused but little change in terminal velocity (L9).

Consequently, for obtaining good reproducibility (especially for comparison of the results of various investigators), experimental work should be designed to be of such "contamination degree" that adventitious fluctuations in surfactant impurities will not alter the transfer coefficient. For example, a protein "contamination degree" of 1 mg/m^2 of interface or higher is enough to obtain constant k_L values (D9). Many of the contradictory results reported in the literature are probably due to surface-active impurities that are present in all but the most extraordinarily well purified systems.

The reader interested in more details is referred to the work by Lochiel (L10), Davies (D7), and Kintner (K4) and to the numerous works which they review. In general, what must be kept in mind is that fluids in most laboratory, and in all industrial, operations cannot be treated theoretically as though their surfaces were those of pure fluids.

IV. Mathematical Models

The overall set of partial differential equations that can be considered as a mathematical characterization of the processing system of gas–liquid dispersions should include such environmental parameters as composition, temperature, and velocity, in addition to the equations of bubble-size and residence-time distributions that describe the dependence of bubble nucleation and growth on the bubble environmental factors. A simultaneous solution of this set of differential equations with the appropriate initial and boundary conditions is needed to evaluate the behavior of the system. Subject to the Curie principle, this set of equations should include the possibilities of coupling effects among the various fluxes involved. In dispersions, the possibilities of couplings between fluxes that differ from each other by an odd tensorial rank exist. (An example is the coupling effect between diffusion of surfactants and the hydrodynamics of bubble velocity as treated in Section III.) As yet no analytical solution of the complete set of equations has been found because of the mathematical difficulties involved. To simplify matters, the pertinent transfer equation is usually solved independently, with some simplifying assumptions.

Most theoretical studies of heat or mass transfer in dispersions have been limited to studies of a single spherical bubble moving steadily under the influence of gravity in a clean system. It is clear, however, that swarms of suspended bubbles, usually entrained by turbulent eddies, have local relative velocities with respect to the continuous phase different from that derived for the case of a steady rise of a single bubble. This is mainly due to the fact that in an ensemble of bubbles the distributions of velocities, temperatures, and concentrations in the vicinity of one bubble are influenced by its neighbors. It is therefore logical to assume that in the case of dispersions the relative velocities and transfer rates depend on quantities characterizing an ensemble of bubbles. For the case of uniformly distributed bubbles, the dispersed-phase volume fraction Φ, particle-size distribution, and residence-time distribution are such quantities.

The dispersed-phase holdup fraction is, for example, responsible for many important interactions. These are indicated by the dashed lines of Fig. 1, which show the main interrelationships that govern the capacity of a given dispersion. Some of these interrelationships, such as the effects of residence-time distribution, bubble-size distribution, dispersed-phase holdup, and contact time on the average concentrations and instnataneous and total average transfer rates have recently been estimated by using simplified mathematical models (see Sections IV, I, J, K, L). Another important variable is the presence of surfactants in all but the most extraordinarily well-purified systems. These can have a profound effect on the hydrodynamics and transfer rates in the dispersion [see Sections III, IV, G, and IV, K and Gal-Or *et al.* (G9, W1)].

The general case of coupled heat-transfer and multicomponent mass-transfer in swarms of bubbles with residence-time and size distributions is treated in Section IV, L. In previous sections simplified cases of uncoupled mass transfer are considered starting from the most simplified models available for gas–liquid dispersions.

The simplified equation (for the general equations, see Section IV, L) in the case of unsteady-state diffusion with a simultaneous chemical reaction in isothermal, incompressible dilute binary solutions with constant ρ and D and with coupled phenomena neglected is

$$D^\alpha \nabla^2 c^\alpha = \mathbf{v} \cdot \nabla c^\alpha + R^{*\alpha} + \partial c^\alpha / \partial t \tag{71}$$

where the superscript α refers to each of the corresponding phases involved. To evaluate the mass (or heat) transfer rate in the case of gas–liquid dispersions, equations of the type of (71) should be evaluated separately for each phase and solved simultaneously. However, in the case of dispersions, where swarms of bubbles are moving with different local velocity components relative to an arbitrary moving fluid, the question arises as to the evaluation of $\mathbf{v}$—the velocity vector relative to an arbitrary moving continuous phase—a criterion which is very difficult to evaluate (see Sections II, F and III for further discussion). Thus, the mathematical analysis involved in arriving at an analytical solution is still very complicated, and, consequently, further simplifying assumptions are made. Usually, these assumptions are that the interface is clean and that the resistance to transfer in the gas phase is negligible. Consequently only one differential equation is solved for the component that dissolves in the continuous phase. The last assumption is justified for sparingly soluble gases being absorbed in a liquid.

Another assumption, which is commonly made when the diffusion is accompanied by chemical reaction, is that the effect of the heat of reaction is negligible.

Equations describing the transfer rate in gas–liquid dispersions have been derived and solved, based on the film-, penetration-, film-penetration-, and more advanced models for the cases of absorption with and without simultaneous chemical reaction. Some of the models reviewed in the following paragraphs were derived specifically for gas–liquid dispersion, whereas others were derived for more general cases of two-phase contact.

A. Absorption in Agitated Liquid with Simultaneous Chemical Reaction—Film Model

In this model, developed by Lightfoot (B9, L5, L6) for the case of sparingly soluble gas being absorbed in agitated liquid with simultaneous chemical reaction, the following assumptions were made:

(1) Each bubble is surrounded by a hypothetical stagnant film of liquid of thickness δ. The thickness is negligible in comparison with the bubble diameter.
(2) The concentration at the interface is constant.
(3) Steady state is established immediately.
(4) All the resistance to mass transfer is in the liquid film.
(5) The curvature of the bubble is negligible.
(6) The dissolved gas is removed from the liquid by an irreversible first-order reaction.

The one-dimensional equation for the steady state becomes, for this case:

$$D\, \partial^2 c/\partial y^2 = kc \tag{72}$$

where D is the diffusivity of dissolved gas through the liquid, assumed constant; k is the first-order reaction rate constant for the removal of dissolved gas; and c is the concentration of dissolved gas. The boundary conditions are:

$$c = c_i \quad \text{at} \quad y = 0 \tag{73}$$

$$c = c_L \quad \text{at} \quad y = \delta \tag{74}$$

where c_L is the concentration of the dissolved gas in the bulk liquid. The solution is:

$$\frac{c}{c_i} = \frac{(c_L/c_i)\sinh[y(k/D)^{1/2}] + \sinh[(\delta - y)(k/D)^{1/2}]}{\sinh[\delta(k/D)^{1/2}]} \tag{75}$$

From this Lightfoot obtained:

$$J = \frac{\delta(k/D)^{1/2}}{\sinh[\delta(k/D)^{1/2}]} \Bigg[\cosh[\delta(k/D)^{1/2}] - \frac{1}{\cosh[\delta(k/D)^{1/2}] + (k/D)^{1/2}[(V - A\delta)/A]\sinh[\delta(k/D)^{1/2}]}\Bigg] \tag{76}$$

where J is the ratio of the rate of absorption to rate of physical absorption in tank containing no dissolved gas, V is the volume of liquid in the tank, and A is the total bubble surface area. The qualitative behavior of this film model is shown in Fig. 6.

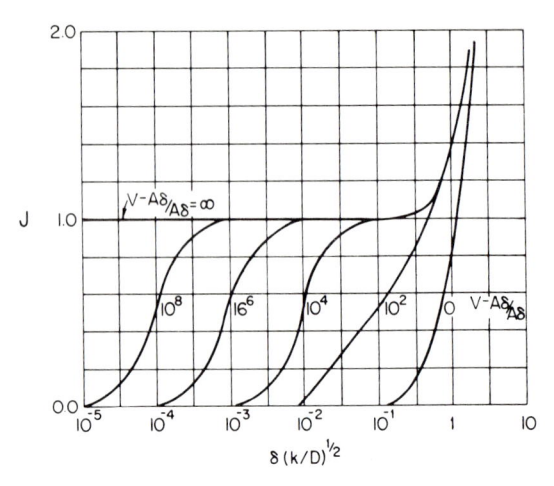

FIG. 6. Film model for diffusion with simultaneous irreversible first-order chemical reaction [after Lightfoot (L5)].

This approach to the problem is purely theoretical, since this model is based on a characteristic stagnant film thickness which is difficult to estimate or to measure. In addition, this model does not give any information as to the value of A, which must be determined separately by some other method. As a result, it is impossible to estimate the total mass-transfer rate in the disperser with the aid of this model only.

B. Absorption in Agitated Liquid with Simultaneous Chemical Reaction—Penetration Model

Again for the case of the sparingly soluble gas whose absorption is accompanied by a simultaneous irreversible first-order reaction, Lightfoot (L5, L6) made the following assumptions:

(1) The gas bubbles are of uniform size and concentration.
(2) Internal circulation occurs inside the bubbles, and the resistance to diffusion in the gas phase is negligible compared to that in the liquid phase.
(3) The bubbles are distributed uniformly throughout the vessel.
(4) There are no gross variations in the relative velocities between the bubble and the liquid, nor are there gross variations in the concentration of the solution.
(5) The concentration at the interface is constant and equal to c_i.
(6) The bubble curvature is negligible.
(7) The dissolved gas is removed by a first-order irreversible reaction.
(8) The system is at a pseudosteady state, i.e., the overall concentration in the tank is constant, and the effect of the rate of change of this composition is negligible.

GAS-LIQUID DISPERSIONS

The absorption is assumed to occur into elements of liquid moving around the bubble from front to rear in accordance with the penetration theory (H13). These elements maintain their identity for a distance into the fluid greater than the effective penetration of dissolving gas during the time required for this journey. The differential equation and initial and boundary conditions for the rate of absorption are then

$$D \frac{\partial^2 c}{\partial y^2} = kc + \frac{\partial c}{\partial t} \tag{77}$$

$$c = c_0 \quad \text{at} \quad t = 0 \quad \text{for} \quad y > 0 \tag{78}$$

$$c = c_i \quad \text{at} \quad y = 0 \quad \text{for} \quad t \geq 0 \tag{79}$$

$$c = c_0 \exp(-kt) \quad \text{at} \quad y = \infty \quad \text{for} \quad t \geq 0 \tag{80}$$

where c_0 is the concentration of dissolved gas in a liquid element meeting the gas phase at the front of the bubble, i.e., at the site of its turning around the bubble. These equations can be solved to yield:

$$\frac{c}{c_i} = \frac{1}{2} \left\{ e^{-\psi} \operatorname{erfc}\left[\frac{\psi}{2(kt)^{1/2}} - (kt)^{1/2}\right] + e^{\psi} \operatorname{erfc}\left[\frac{\psi}{(2kt)^{1/2}} + (kt)^{1/2}\right] \right\}$$
$$+ \frac{c_0}{c_i} e^{-kt} \operatorname{erf}\left[\frac{\psi}{2(kt)^{1/2}}\right] \tag{81}$$

where

$$\psi = y(k/D)^{1/2} \tag{82}$$

which is a dimensionless group.

According to the penetration theory, which assumes equal contact for each element (H13), one can write that

$$t_0 \simeq d/U \tag{83}$$

where d is the bubble diameter, U is the velocity of the bubble relative to the fluid- and t_0 is the time required for the liquid element to move completely around the bubble. In order to find c_0 in Eq. (81), it was assumed that complete mixing occurred between bubbles after time t_0. The material balance then becomes:

$$\int_0^{V/A} c(t_0, y) \, dy = \frac{V}{A} c_0 \tag{84}$$

By combining Eqs. (81) and (84), the following expression is obtained for c_0:

$$\frac{c_0}{c_i} = \frac{\operatorname{erf}\sqrt{\tau_0} + \frac{1}{2}\left\{e^{\psi_0} \operatorname{erfc}\left[\frac{\psi_0}{2\sqrt{\tau_0}} + \sqrt{\tau_0}\right] - e^{-\psi_0} \operatorname{erfc}\left[\frac{\psi_0}{2\sqrt{\tau_0}} - \sqrt{\tau_0}\right]\right\}}{\psi_0\left[1 - e^{-\tau_0} \operatorname{erf}\left[\frac{\psi_0}{2\sqrt{\tau_0}}\right]\right] - \sqrt{\tau_0}\, e^{-\tau_0}\left(\frac{2}{\sqrt{\tau_0}}\right)\left[\exp\left(-\frac{\psi_0^2}{4\tau_0}\right) - 1\right]} \tag{85}$$

where

$$\tau_0 = kd/U = kt_0 \tag{86}$$

$$\psi_0 = \frac{V}{A}\left(\frac{k}{D}\right)^{1/2} \tag{87}$$

and the average absorption rate per unit area becomes

$$\bar{N}_A = \frac{1}{t_0}\int_0^{t_0}\left[-D\left(\frac{\partial c}{\partial y}\right)_{y=0}\right]dt$$

$$= \frac{c_i(k/D)^{1/2}}{\tau_0}\left\{\left(\tau_0 + \frac{1}{2} - \frac{c_0}{c_i}\right)\operatorname{erf}\sqrt{\tau_0} + \left(\frac{\tau_0}{\pi}\right)^{1/2}e^{-\tau_0}\right\} \tag{88}$$

or

$$N^* = \frac{\bar{N}_A}{c_i}\left(\frac{\pi\tau_0}{4D}\right)^{1/2} = \frac{1}{2}e^{-\tau_0} + \left(\tau_0 + \frac{1}{2} - \frac{c_0}{c_i}\right)\frac{\sqrt{\pi}}{4\tau_0}\operatorname{erf}\sqrt{\tau_0} \tag{89}$$

where N^* is the ratio of the absorption rate in the presence of chemical reaction to the rate of physical absorption when the tank contains no dissolved gas. Values for N^* as a function of τ_0 are shown graphically in Fig. 7. A comparison between Figs. 6 and 7 shows that the relationships for both the film model and the penetration model are qualitatively similar.

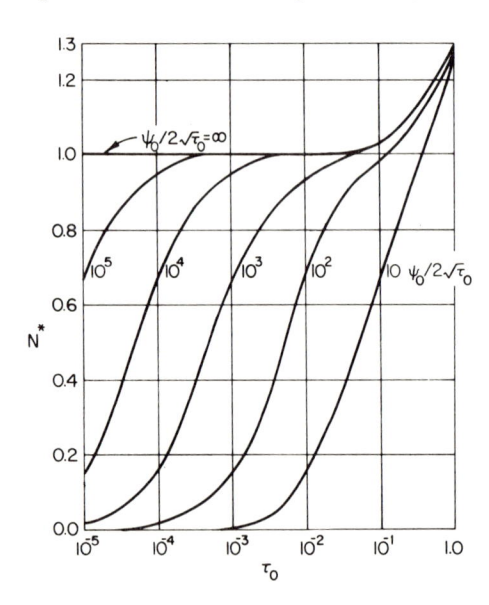

FIG. 7. Penetration model for diffusion with simultaneous irreversible first-order chemical reaction [after Lightfoot (L5)].

The model in its present form cannot be used for the design of gas–liquid contacting systems, for several reasons. The model requires a knowledge of the average bubble velocity relative to the fluid, U, a variable that is not available in most cases. This model only permits the calculation of the average rate per unit of area, and unless data are available from other sources on the total surface area available in the vessel, the model by itself does not permit the calculation of the overall absorption rate.

C. Film-Penetration Model

The film-penetration concept was introduced by Dobbins (D11) and Toor and Marchello (T5). Toor and Marchello (T5) showed that the film and penetration theories are not separate and unrelated concepts, but are actually complementary. In their work they solved the diffusion equations for the case where there is no chemical reaction. They considered the transfer between a gas and a stirred liquid which has its surface randomly replaced by eddies of fresh fluid from the bulk of the liquid. If the eddies remain on the surface a short period of time, each element may be assumed to absorb matter at the interface by unsteady-state diffusion. As the lifetime of the element increases, the penetration into the element increases, and, after a long enough time, a steady gradient is set up in the element, no more accumulation takes place, and material is then transferred through the element. Thus, the old elements obey the film theory, young ones the penetration theory, and the middle-aged ones have characteristics of both mechanisms. In the middle-aged case, the penetration has reached the inner side of the element but the steady-state gradient has not yet been established. If elements of all ages are present, all three types of transfer take place simultaneously, and this model, which includes all the cases, is called the film-penetration model. The mass-transfer equation in one dimensional form,

$$D\, \partial^2 c/\partial y^2 = \partial c/\partial t \tag{90}$$

is solved with the following initial and boundary conditions:

$$c = c_L \quad \text{at} \quad t = 0 \quad \text{for} \quad y > 0 \tag{91}$$

$$c = c_i \quad \text{at} \quad y = 0 \quad \text{for} \quad t \geqslant 0 \tag{92}$$

$$c = c_L \quad \text{at} \quad y = L \quad \text{for} \quad t \geqslant 0 \tag{93}$$

where L is a distance below the surface at which the concentration remains constant at c_L, the concentration of a freshly formed surface. It is assumed that c_i and c_L are independent of time. The difference between the film-penetration model and the penetration model is due to the final boundary condition, which lets transfer into an old element approach the steady-state value, whereas the penetration model, which makes L infinite, excludes this limit. The solutions for the instantaneous transfer rates are, for short times,

$$N_A = (c_i - c_L)\left(\frac{D}{\pi t}\right)^{1/2}\left[1 + \sum_{n=1}^{\infty} \exp - \frac{n^2 L^2}{Dt}\right] \tag{94}$$

and, for long times,

$$N_A = (c_i - c_L)\frac{D}{L}\left[1 + 2\sum_{n=1}^{\infty} \exp \frac{-n^2\pi^2 Dt}{L^2}\right] \tag{95}$$

These equations are equivalent, but the first converges rapidly for short times and the second for long times. The equations clearly show that for short times the penetration theory is approached:

$$N_A = (c_i - c_L)(D/\pi t)^{1/2} \tag{96}$$

and for long times the film theory is approached:

$$N_A = (c_i - c_L)D/L \tag{97}$$

The penetration theory holds for the region where t is much less than L^2/D, the film theory for the region where t is much greater than L^2/D. This comparison is shown in Fig. 8, which clearly shows that the film and penetration theories are asymptotes of the film-penetration model.

Although this approach permits a greater qualitative understanding of the mass-transfer mechanisms that govern the transfer between two phases, it still does not permit quantitative calculations to be made, because the thickness L and the concentration c_L are unknown and the total surface area of the bubbles is not included in the model.

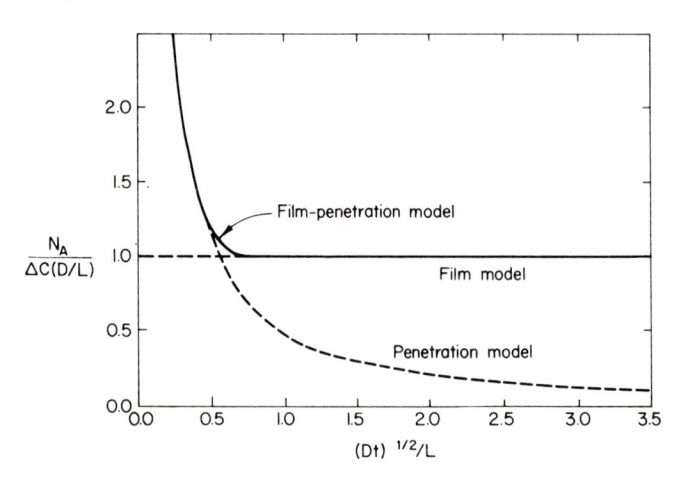

FIG. 8. Comparison of film-penetration model with film and penetration models [after Toor and Marchello (T5)].

D. Film-Penetration Model with Simultaneous Chemical Reaction

In 1963 and in 1965, Huang and Kuo (H18, H19) applied the film penetration model to the mechanism of simultaneous mass transfer and chemical reaction.

1. Mass Transfer Accompanied by Irreversible First-Order Chemical Reaction

For this case the partial differential equation in one dimension becomes (H18):

$$D \frac{\partial^2 c}{\partial y^2} = kc + \frac{\partial c}{\partial t} \tag{77}$$

Using the following initial and boundary conditions:

$$c = c_i \quad \text{for} \quad y = 0 \quad \text{at} \quad t \geqslant 0 \tag{98}$$
$$c = c_L \quad \text{for} \quad y = L \quad \text{at} \quad t \geqslant 0 \tag{99}$$
$$c = c_L \quad \text{for} \quad y > 0 \quad \text{at} \quad t = 0 \tag{100}$$

the following expression is obtained for the concentration gradient within the liquid element:

$$c = c_i - \frac{c_i - c_L}{L}$$
$$\times \left[y + \frac{2L}{\pi} \sum_{n=1}^{\infty} \frac{1 + n^2\pi^2\alpha^* \exp[-(1 + n^2\pi^2\alpha^*)kt \sin(n\pi y/L)]}{n(1 + n^2\pi^2\alpha^*)} \right] \tag{101}$$

where

$$\alpha^* = DL^2/k \tag{102}$$

The instantaneous mass-transfer rate becomes

$$N_A = \frac{D}{L}(C_i - C_L)\left[1 + 2\sum_{n=1}^{\infty} \frac{1 + n^2\pi^2\alpha^* \exp[-(1 + n^2\pi^2\alpha^*)kt}{(1 + n^2\pi^2\alpha^*)} \right] \tag{103}$$

With the aid of the surface-renewal concept, Huang and Kuo also derived the equation for the average mass-transfer rate:

$$\bar{N}_A = \frac{D}{L}(c_i - c_L)\left[1 + 2\sum_{n=1}^{\infty} \frac{1 + p/k}{1 + (p/k) + n^2\pi^2\alpha^*} \right] \tag{104}$$

where p is the rate of surface renewal.

2. Mass Transfer Accompanied by Reversible Chemical Reaction

Despite its frequent occurrence in chemical engineering systems, a theoretical model for mass transfer with a reversible reaction has not been completely analyzed, mainly because of the mathematical difficulties.

In the film-penetration model (H19), it is assumed that the reactant A penetrates through the surface element by one-dimensional unsteady-state molecular diffusion. Convective transport is assumed to be insignificant. The diffusing stream of the reactant A is depleted along the path of diffusion by its reversible reaction with the reactant B, which is an existing component of the liquid surface element. If such a reaction can be represented as

$$A + nB \underset{k_{-1}}{\overset{k_1'}{\rightleftarrows}} E \qquad (105)$$

and if the order of reaction is $(n+1)$th order, the material balances in a one-dimensional system give

$$\frac{\partial c_A}{\partial t} = D_A \frac{\partial^2 c_A}{\partial y^2} - k_1' c_A c_B^n + k_{-1} c_E \qquad (106)$$

$$\frac{1}{n}\frac{\partial c_B}{\partial t} = \left(\frac{D_B}{n}\right)\frac{\partial^2 c_B}{\partial y^2} - k_1' c_A c_B^n + k_{-1} c_E \qquad (107)$$

$$\frac{\partial c_E}{\partial t} = D_E \frac{\partial^2 c_E}{\partial y^2} + k_1' c_A c_B^n - k_{-1} c_E \qquad (108)$$

which are nonlinear partial differential equations. However, in many actual operations, the amount of B is large and remains nearly constant during the diffusional and reaction processes. Thus, the term $k_1' c_B^n$ is approximately constant and the above system with a high-order reaction may be treated as a pseudo first-order reversible reaction represented as

$$A \underset{k_{-1}}{\overset{k_1}{\rightleftarrows}} E \qquad (109)$$

where E is the product.

Under these conditions, Eqs. (106)–(108) are replaced by the following two linear partial differential equations:

$$\frac{\partial c_A}{\partial t} = D_A \frac{\partial^2 c_A}{\partial y^2} - k_1 c_A + k_{-1} c_E \qquad (110)$$

$$\frac{\partial c_E}{\partial t} = D_E \frac{\partial^2 c_E}{\partial y^2} + k_1 c_A - k_{-1} c_E \qquad (111)$$

Using the following initial and boundary conditions,

$$t = 0, \quad y > 0, \quad c_A = c_{AL}, \quad c_E = c_{EL} \qquad (112)$$

$$t > 0, \quad y = 0, \quad c_A = c_{Ai}, \quad \partial c_E/\partial y = 0 \qquad (113)$$

$$t > 0, \quad y = L, \quad c_A = c_{AL}, \quad c_E = c_{EL} \qquad (114)$$

and solving simultaneously by using the Laplace transformation, Huang and Kuo (H19) evaluated the Laplace transform $\bar{c}_A(y, p)$ of the concentration of reactant A.

Introducing Danckwerts' surface age distribution (D4)

$$\phi(t) = pe^{-pt} \tag{115}$$

the average mass-transfer rate per unit area was then expressed as

$$\bar{N}_A = \int_0^\infty -D_A\left(\frac{\partial c_A}{\partial y}\right)_{y=0} \phi(t)\, dt = -D_A p\left(\frac{\partial \bar{c}_A(y, p)}{\partial y}\right)_{y=0}$$

$$= \frac{\dfrac{D_A}{L}(U_0^2 - V_0^2)[c_{Ai} - f(c_{AL}, c_{EL})]}{(W_0 - V_0^2)\dfrac{\tanh U_0 L}{U_0 L} - (W_0 - U_0^2)\dfrac{\tanh V_0 L}{V_0 L}} \tag{116}$$

where

$$W_0 = \frac{k_1}{D_A}(1+\beta), \qquad \beta = p/k_1, \qquad K_1 = k_{-1}/k_1$$

from this

$$U_0 = \left[\frac{k_1}{2D_A}\left\{1 + \beta + \frac{K_1 + \beta}{D_{EA}}\right.\right.$$

$$\left.\left. + \left(\left[1 + \beta + \frac{K_1 + \beta}{D_{EA}}\right]^2 - \frac{4\beta}{D_{EA}}(1 + \beta + K_1)\right)^{1/2}\right\}\right]^{1/2} \tag{116a}$$

$$V_0 = \left[\frac{k_1}{2D_A}\left\{1 + \beta + \frac{K_1 + \beta}{D_{EA}}\right.\right.$$

$$\left.\left. - \left(\left[1 + \beta + \frac{K_1 + \beta}{D_{EA}}\right]^2 - \frac{4\beta}{D_{EA}}(1 + \beta + K_1)\right)^{1/2}\right\}\right]^{1/2} \tag{117}$$

$$f(c_{AL}, c_{EL}) = \frac{1}{1 + \beta + K_1}\left\{\left(\frac{c_{AL} - K_1 c_{EL}}{U_0^2 - V_0^2}\right)\left[\frac{k_1}{D_A}(1 + \beta + K_1)\right.\right.$$

$$\times \left(\frac{1}{\cosh U_0 L} - \frac{1}{\cosh V_0 L}\right)$$

$$\left.\left. + \left(\frac{U_0^2}{\cosh V_0 L} - \frac{V_0^2}{\cosh U_0 L}\right)\right]\right.$$

$$\left. + (K_1 + \beta)c_{AL} + K_1 c_{EL}\right\} \tag{118}$$

From a study of the behavior of the model, Huang and Kuo concluded that:

(1) For mass transfer with irreversible and reversible reactions, the film-penetration model is a more general concept than the film or surface renewal models which are its limiting cases.

(2) The backward reaction tends to increase the resistance to mass transfer. If the backward reaction rate is very small compared with the forward reaction rate, the transfer rate is at its highest value. Then, as the backward reaction rate is increased, the transfer rate begins to decline. When the backward reaction rate approaches infinity, the chemical reaction exerts no influence on the mass transfer and the system behaves as if no chemical reaction is involved.

However, in considering the applicability of this model, the question arises as to the magnitude of the unknown parameters p, L, c_{AL}, and c_{EL}. In addition, to use this model for gas–liquid dispersions requires evaluation of the total surface of the bubbles by some other method.

E. Mass Transfer with Chemical Reaction from a Moving Gas Bubble

A slightly different approach was taken by Gill (G15), who considered the case of a bubble moving through a stationary liquid with mass transfer accompanied by simultaneous first-order chemical reaction. His assumptions were as follows:

(1) The gas bubbles move with a constant velocity U with respect to a stationary fluid.

(2) The bubble diameter is constant.

(3) The bubble curvature can be neglected.

(4) The gas inside the bubble is in circulation and the movement through the fluid can be considered to be frictionless.

(5) The bubble is released into a pure liquid that does not contain any dissolved gas and immediately assumes its constant velocity.

(6) Equilibrium exists at the gas–liquid interface and the concentration is constant at c_i.

(7) The dissolved gas is removed by first-order irreversible chemical reaction.

The appropriate partial differential equation that results is

$$D\frac{\partial^2 c}{\partial y^2} = kc + U\frac{\partial c}{\partial S} + \frac{\partial c}{\partial t} \tag{119}$$

where S is the distance from the leading edge of the bubble along its surface. Equation (119) is converted to the following dimensionless form:

$$\frac{\partial^2 c_1}{\partial y_1^2} = \alpha_1 c_1 + \frac{\partial c_1}{\partial S_1} + \frac{\partial c_1}{\partial t'} \tag{120}$$

where

$$c_1 = c/c_i, \quad t' = tD/l^2, \quad S_1 = SD/Ul^2, \quad y_1 = y/l, \quad \alpha_1 = kl^2/D \tag{121}$$

and l is the unit of length that is large enough so that the concentration of the dissolved gas is zero at a distance l normal to the bubble surface. Equation (120) was solved with the following initial and boundary conditions:

When the bubble is placed into the liquid:

$$c_1(0, S_1, y) = 0 \tag{122}$$

The liquid at the beginning of the bubble path is free of dissolved gas:

$$c_1(t', 0, y) = 0 \tag{123}$$

Equilibrium is obtained at the interface:

$$c_1(t', S_1, 0) = 1 \tag{124}$$

The concentration of the dissolved gas at a distance l is zero:

$$C_1(t', S_1, 1) = 0 \tag{125}$$

The solution that is obtained for the instantaneous reaction rate per unit area is:

$$N_A = -\frac{Dc_i}{l}\left(\frac{\partial c_1}{\partial y_1}\right)_{y_1=0}$$

$$= \frac{Dc_i}{l}\left\{1 + 2\sum_{n=1}^{\infty} \frac{1}{\alpha_1 + n^2\pi^2}\left[\alpha_1 + n^2\pi^2\left(\begin{array}{c}\exp[-(\alpha_1 + n^2\pi^2)t'] \\ \exp[-(\alpha_1 + n^2\pi^2)S_1]\end{array}\right)\right]\right\} \tag{126}$$

in which the upper exponential is to be read if $t' < S_1$ and the lower if $t' \geqslant S_1$. If the initial period is neglected,

$$N_A = \frac{Dc_i}{l}\left[1 + 2\sum_{n=1}^{\infty} \frac{\alpha_1 + n^2\pi^2 \exp[-(\alpha_1 + n^2\pi^2)S_1]}{\alpha_1 + n^2\pi^2}\right] \tag{127}$$

and, if there is no chemical reaction, $\alpha_1 = 0$ and

$$N_A = \frac{Dc_i}{l}\left[1 + 2\sum_{n=1}^{\infty} \exp(-n^2\pi^2 S_1)\right] \tag{128}$$

which is equivalent to the film-penetration theory.

For this model, the basic criteria, l and t', are unknown.

F. Gas Absorption with Chemical Reaction—Time-Dependent Bulk Concentration

Gill and Nunge (G 16) solved the equation for diffusion accompanied by simultaneous chemical reaction with a changing concentration of the bulk liquid. They assume a film of thickness l in which the diffusion and simultaneous reaction take place, and suppose the liquid bulk outside this film to be completely mixed and to have a constant and uniform concentration. Their partial differential equation in one dimension is

$$\frac{\partial^2 c}{\partial y_1^2} = \alpha_1 c + \frac{\partial c}{\partial t'} \tag{129}$$

where y_1, α_1, and t' are the dimensionless variables of (121). With the initial and boundary conditions,

$$c(0, y_1) = c_0 \tag{130}$$

$$c(t', 0) = c_i \tag{131}$$

$$-\alpha_1 \frac{\partial c(t', 1)}{\partial y_1} = \frac{\partial c(t', 1)}{\partial t'} + \alpha_1 c(t', 1) \tag{132}$$

the solution is

$$\frac{c}{c_i} = 1 + 2 \sum_{n=1}^{\infty} \left\{ \left(\frac{c_0}{c_i \beta_n} - \frac{\beta_n}{\alpha_1 + \beta_n^2} \right) \exp[-(\alpha_1 + \beta_n^2)t'] + \frac{\alpha_1}{\beta_n(\beta_n^2 + \alpha_1)} \right\}$$

$$\times \left\{ \frac{\beta_n^2 + N_v^*}{\beta_n^2 + N_v^{*2} + N_v^*} \right\} \sin(\beta_n y_1) \tag{133}$$

where

$$N_v^* = Al/(V - Al) \tag{134}$$

and the β_n are the eigenvalues. The instantaneous mass-transfer rate per unit surface area is thus given by

$$N_A = \frac{2Dc_i}{l} \sum_{n=1}^{\infty} \left[\left(\frac{\beta_n}{\alpha_1 + \beta_n^2} - \frac{c_0/c_i}{\beta_n} \right) \exp[-(\alpha_1 + \beta_n^2)t'] + \frac{\alpha_1}{\beta_n(\beta_n^2 + \alpha_1)} \right]$$

$$\times \left[\frac{\beta_n(\beta_n^2 + N_v^{*2})}{\beta_n^2 + N_v^{*2} + N_v^*} \right] \tag{135}$$

The average reaction rate for a pseudosteady state is calculated according to

$$\bar{N}_A = \int_0^{\bar{\theta}} N_A \frac{dt}{\bar{\theta}} \tag{136}$$

where, in order to simplify the solution, it is assumed that all the bubbles spend the same amount of time $\bar{\theta}$ in the vessel. In support of this latter assumption they quote Hanratty (H 9), who shows that the use of different residence-time distributions does not have any great effect on the calculated results.

This model is, in essence, very similar to the one solved by Huang and Kuo (H 18), but differs from it in that it adds the dependence of the bulk liquid concentration on time.

G. STEADY-STATE CONVECTIVE DIFFUSION

The estimation of the diffusional flux to a clean surface of a single spherical bubble moving with a constant velocity relative to a liquid medium requires the solution of the equation for convective diffusion for the component that dissolves in the continuous phase. For steady-state incompressible axisymmetric flow, the equation for convective diffusion in spherical coordinates is approximated by

$$v_r \frac{\partial c}{\partial r} + \frac{v_\theta}{r} \frac{\partial c}{\partial \theta} = D \left[\frac{1}{r^2} \frac{\partial}{\partial r} \left(r^2 \frac{\partial c}{\partial r} \right) \right] \tag{137}$$

when $\partial^2 c/\partial \theta^2$ is small. The concentration distribution in the boundary layer can be approximated by (L3, R9)

$$v_r \frac{\partial c}{\partial r} + \frac{v_\theta}{r} \frac{\partial c}{\partial \theta} = D \frac{\partial^2 c}{\partial r^2} \tag{138}$$

if the thickness of the diffusion boundary layer is much smaller than the bubble diameter.

With the boundary conditions

$$c = c_i = \text{constant} \quad \text{for} \quad r = a \tag{139}$$

$$c = c_0 = \text{constant} \quad \text{for} \quad r \to \infty \tag{140}$$

Levich (L3, L4) has demonstrated that for $N_{\text{Re}} < 1$, the local diffusional flux to the surface of the bubble is given by

$$N_A^* = \left(\frac{3}{\pi} \right)^{1/2} \left[\frac{D \mu_c U}{2a(\mu_c + \mu_d)} \right]^{1/2} \frac{(1 + \cos \theta)}{(2 + \cos \theta)^{1/2}} (c_0 - c_i) \tag{141}$$

To evaluate the average diffusional flux, the total mass-transfer rate from the entire surface of the bubble must be divided by that entire surface:

$$\bar{N}_A = \frac{\int_0^{2\pi} \int_0^{\pi} N_A^* a^2 \sin \theta \, d\phi \, d\theta}{4\pi a^2}$$

$$= \frac{2}{(6\pi)^{1/2}} \left[\frac{D \mu_c U}{a(\mu_c + \mu_d)} \right]^{1/2} (c_0 - c_i) \tag{142}$$

By substituting the Rybczynski—Hadamard formula

$$U = \frac{2(\rho_c - \rho_d)(\mu_c + \mu_d)ga^2}{3\mu_c(2\mu_c + 3\mu_d)} \tag{40}$$

in Eq. (142), one obtains (G5):

$$\overline{N}_A = \frac{2}{3\sqrt{\pi}} \left[\frac{D(\rho_c - \rho_d)g}{2\mu_c + 3\mu_d} \right]^{1/2} (c_0 - c_i)\sqrt{a} \tag{143}$$

With the density and viscosity of the gas neglected in comparison to those of the liquid, the corresponding mass-transfer coefficient is

$$k_L = 0.379 \left[\frac{D\rho_c g}{2\mu_c} \right]^{1/2} \sqrt{a} \tag{144}$$

In their study of the effect of particle-size distribution on mass-transfer in dispersions, Gal-Or and Hoelscher (G5) show that when the variable particle size is replaced by the surface mean radius a_{32}, the error introduced is usually very small (see Section IV, J). Consequently if a in Eq. (144) is replaced by a_{32}, that equation can be compared with the experimental correlations [Eq. (10) and (11)] proposed by Calderbank and Moo-Young (C4) for mass transfer in dispersions (see Fig. 9).

Evaluation of Eq. (144) for the Calderbank and Moo-Young systems by using their values they give for D, ρ_c, μ_c, and bubble diameter (C2, C4) results in the solid lines shown in Fig. 9, indicating that the data for "large" and "small" bubbles, including their "transition region," are generally predicted by Eq. (144). This clearly demonstrates that k_L is proportional to the square root of the radius of the bubble, a result in contradiction to the conclusion of Calderbank and Moo-Young that k_L is not a function of the bubble diameter.

In studying mass-transfer rates in dispersions, Gal-Or and Resnick (G7, G8) measured contact times and average velocities of air bubbles relative to water–glycerol solutions in a mechanically agitated vessel. They found that these velocities lie in a range corresponding to small and moderate Reynolds numbers. Using these data, the result is that $N_{Re} < 1$ only in the case of very small bubbles and high concentrations of glycerol. However, Levich (L3) implies that Eq. (141) is valid at Reynolds numbers substantially greater than unity (but where the motion remains laminar). By solving the bubble-motion problem at moderate Reynolds numbers, Levich shows that under such conditions the flow past a bubble is basically without separation and the zone of turbulent motion covers only a very narrow area at the bubble's wake (at $N_{Re} = 625$, for example, the separation zone extends 2°

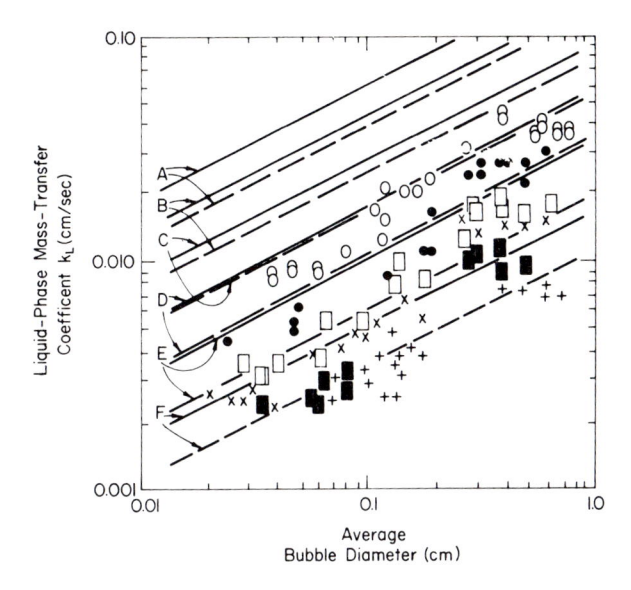

FIG. 9. Comparison of Eqs. (144) (solid line) and (145) (dashed line) with the experimental data of Calderbank and Moo-Young (C4) for the absorption of CO_2 at 25°C in water containing the following percentages by weight of glycerol [after Gal-Or and Walatka (G9)]

Theoretical curves: A) 14; B) 26; C) 38; D) 52.2; E) 63.0; F) 69.5.
Experimental points: ○ 14; ● 26; □ 38; × 52.2; ■ 63.0; + 69.5.

to either side of the line of symmetry). Levich's solution for this case is (see also Section II, F)

$$U = \frac{\rho_c g a^2}{9\mu_c} \tag{38}$$

which was found to be in complete agreement with experimental data for extraordinarily purified systems (L3). Substitution of Eq. (38) in Eq. (142) gives the result (G9)

$$k_L = 0.154 \left[\frac{D\rho_c g}{\mu_c}\right]^{1/2} \sqrt{a_{32}} \tag{145}$$

which is essentially the same as Eq. (144) with a evaluated as a_{32}, but gives lower mass-transfer rates, as shown by the dashed lines in Fig. 9.

Both Eq. (144) and Eq. (145) give values higher than the observed data, in particular for water–glycerol systems containing lower concentrations of glycerol. It should be mentioned that, in this comparison with experiment, the assumption is made that the suspended bubbles entrained by the fluid, which describe complicated trajectories in the fluid, are moving relative to the fluid at an average slip velocity U given by Eq. (40) or (38). This approximation is seemingly in agreement with experimental observations (G7), and means that the bubble motion is supposed quasisteady as if it were occurring in a steady acceleration field of gravity; it may not be exact. Disagreement between Eq. (144) and (145) and the experimental data can also arise from two effects that may decrease the relative velocity of the bubbles in dispersions (G9), as mentioned in Sections II and III:

(a) *Presence of Minute Amounts of Surface-Active Impurities.* Surface-active impurities form some kind of "skin" on the surface of the bubble that effectively retards internal circulation; consequently, U and k_L are reduced. The theoretical conclusion (see Section III) that adventitious surfactant impurities will not influence the transfer at high viscosities is confirmed by the improved agreement between Eq. (145) and experimental data at high viscosities of water–glycerol solutions. As noted above, the work of Calderbank and Moo-Young was probably influenced by very small traces of surfactants that are present in all but the most extrarodinarily purified systems.

(b) *Interaction between Bubbles in a Swarm.* According to Marrucci (M3) a bubble in a swarm with a holdup fraction Φ moves relative to the continuous phase at a velocity lower by the ratio $(1 - \Phi)^2/(1 - \Phi^{5/3})$ than that at which it would move in the absence of other bubbles (see Section II).

According to Levich (L3) Eq. (142) can approximate the convective diffusional flux even in the domain $1 < N_{\text{Re}} < 600$. Hence the application of the Marrucci correction [Eq. (44)] to Eq. (38) before combining it with Eq. (142) gives

$$(k_L)_{\text{swarm}} = 0.154 \left[\frac{D\rho_c g}{\mu_c}\right]^{1/2} \frac{(1 - \Phi)}{(1 - \Phi^{5/3})^{1/2}} \sqrt{a_{32}} \qquad (146)$$

as the mass-transfer coefficient for swarms of bubbles in very clean dispersions (or for dispersions in high-viscosity liquids even when containing surface-active impurities) (G9). As an example, for a 20% gas volume fraction, Eq. (146) gives a reduction of 16% in the mass-transfer rates predicted by Eq. (145) for single bubbles. Thus, introducing the effect of dispersed phase holdup may result in a better agreement between the theoretical and experimental data for gas–liquid dispersions in practical systems of high-viscosity liquids.

H. Steady-State Convective Diffusion with Simultaneous First-Order Irreversible Chemical Reaction

In deriving the equation for this case, Johnson et al. (J4) assume steady-state conditions with viscous, incompressible axisymmetric flow around single spherical gas bubbles. For this case, using spherical coordinates, the following equation is considered:

$$D\left[\frac{\partial^2 c}{\partial r^2} + \frac{2}{r}\frac{\partial c}{\partial r} + \frac{\cot\theta}{r^2}\frac{\partial c}{\partial \theta} + \frac{1}{r^2}\frac{\partial^2 c}{\partial \theta^2}\right] = v_r\frac{\partial c}{\partial r} + \frac{v_\theta}{r}\frac{\partial c}{\partial \theta} + kc \quad (147)$$

The boundary conditions used are:

$$\begin{aligned} c &= c_i & \text{at} \quad r &= a \\ c &= 0 & \text{at} \quad r &\to \infty \\ \partial c/\partial \theta &= 0 & \text{at} \quad \theta &= 0, \theta = \pi \end{aligned} \quad (148)$$

Equation (147) is made dimensionless by using the following definitions:

$$r' = r/a, \quad c' = c/c_i, \quad v_r' = v_r/U,$$
$$v_\theta' = v_\theta/U, \quad k' = a^2 k/D, \quad N_{pe} = 2aU/D \quad (149)$$

Substituting and dropping the primes converts equation (147) to

$$v_r\frac{\partial c}{\partial r} + \frac{v_\theta}{r}\frac{\partial c}{\partial \theta} = \frac{2}{N_{pe}}\left[\frac{\partial^2 c}{\partial r^2} + \frac{2}{r}\frac{\partial c}{\partial r} + \frac{\cot\theta}{r^2}\frac{\partial c}{\partial \theta} + \frac{1}{r^2}\frac{\partial^2 c}{\partial \theta^2} - kc\right] \quad (150)$$

and the boundary conditions become:

$$\left.\begin{aligned} c &= 1 & \text{at} \quad r &= 1 \\ c &= 0 & \text{at} \quad r &\to \infty \\ \partial c/\partial \theta &= 0 & \text{at} \quad \theta &= 0, \theta = \pi \end{aligned}\right\} \quad (151)$$

Numerical solutions were obtained in the form $c = f(r, \theta)$ using the velocity components as given by Hamielec et al. (H4, H5):

$$\begin{aligned} v_\theta &= \left\{1 - \frac{A_1}{r^3} - \frac{2A_2}{r^4} - \frac{3A_3}{r^5} - \frac{4A_4}{r^6}\right\}\sin\theta \\ &\quad + \left\{-\frac{B_1}{r^3} - \frac{2B_2}{r^4} - \frac{3B_3}{r^5} - \frac{4B_4}{r^6}\right\}\sin\theta\cos\theta \\ v_r &= -\left\{1 + \frac{2A_1}{r^3} + \frac{2A_2}{r^4} + \frac{2A_3}{r^5} + \frac{2A_4}{r^6}\right\}\cos\theta \\ &\quad - \left\{\frac{B_1}{r^3} + \frac{B_2}{r^4} + \frac{B_3}{r^5} + \frac{B_4}{r^6}\right\}\{2\cos^2\theta - \sin^2\theta\} \end{aligned} \quad (152)$$

where if μ^* is the ratio of the viscosity of the dispersed phase to that of the continuous phase,

$$\left.\begin{aligned} A_2 &= -\frac{125 + 120\mu^*}{60 + 29\mu^*} - A_1 \frac{140 + 75\mu^*}{60 + 29\mu^*} \\ A_3 &= \frac{135 + 153\mu^*}{60 + 29\mu^*} + A_1 \frac{108 + 63\mu^*}{60 + 29\mu^*} \\ A_4 &= -\frac{40 + 47.5\mu^*}{60 + 29\mu^*} - A_1 \frac{28 + 17\mu^*}{60 + 29\mu^*} \\ B_2 &= -B_1 \frac{140 + 69\mu^*}{60 + 27\mu^*} \\ B_3 &= B_1 \frac{108 + 57\mu^*}{60 + 27\mu^*} \\ B_4 &= -B_1 \frac{28 + 15\mu^*}{60 + 27\mu^*} \end{aligned}\right\} \quad (153)$$

From the c's, the local mass transfer was calculated by the formula

$$k_L = \frac{-D}{c_i}\left(\frac{\partial c}{\partial r}\right)_{r=a} \quad (154)$$

or its dimensionless equivalent,

$$N_{sh} = -2\left(\frac{\partial c'}{\partial r'}\right)_{r'=1} \quad (155)$$

and the results are shown in Fig. 10 as an average of the local Sherwood number up to the flow separation point.

An attempt has been made by Johnson and co-workers to relate such theoretical results with experimental data for the absorption of a single carbon dioxide bubble into aqueous solutions of monoethanolamine, determined under forced convection conditions over a Reynolds number range from 30 to 220. The numerical results were found to be much higher than the measured values for noncirculating bubbles. The numerical solutions indicate that the mass-transfer rate should be independent of Peclet number, whereas the experimentally measured rates increase gradually with increasing Peclet number. The discrepancy is attributed to the experimental technique, where-

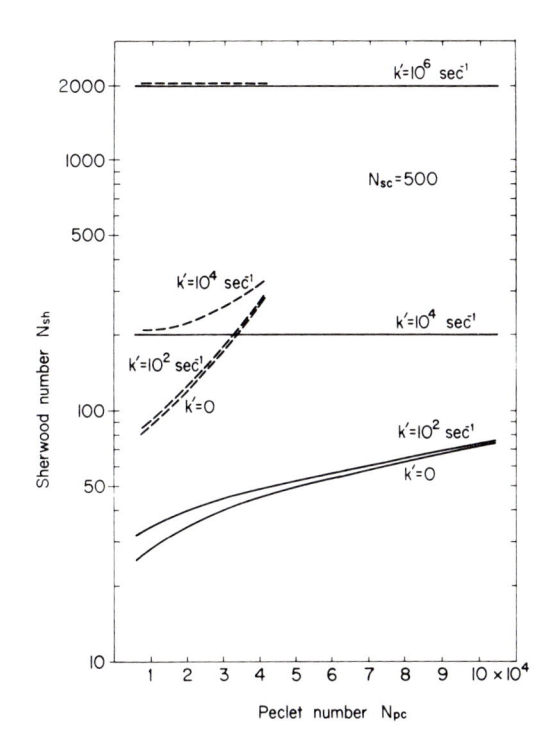

FIG. 10. Numerical solutions of the forced-convection mass-transfer equation for the case of irreversible first-order chemical reaction [after Johnson et al. (J4)]: (Solid lines—rigid spheres; dashed lines—circulating gas bubbles).

by the nozzle tip (which holds the suspended bubble) is too large in comparison with the bubble size being examined. In a more recent study (H4), Hamielec and co-workers used air bubbles to oxidize acetaldehyde to peracetic acid in a solution of ethylacetate with cobaltous acetate as catalyst. These experimental results agree well with the theory for $N_{Sc} = 89.5$, $k' = 10$, and a Reynolds number range between 816 and 1193.

This model considers a single bubble, and consequently its direct applicability to gas–liquid dispersions is still limited.

I. Residence-Time Model for Total Mass Transfer with and without Chemical Reaction

Gal-Or and Resnick (G2, G6, G8) recently proposed a theoretical model, based on the gas residence time, for total mass transfer in a gas–liquid agitated contactor. They assumed that the number of bubbles in the vessel at any instant is constant, and the total liquid volume is subdivided into a number of equal-volume elements equivalent to the number of bubbles in the vessel. Each bubble is then enveloped by a spherical shell whose volume is equal to the volume of the liquid element. Each bubble is introduced into each element for a retention time equal to the average residence time of the bubbles. At time equal to $\bar{\theta}$, the bubble was removed, the liquid element completely mixed, and another bubble introduced. Thus, the contact and mixing occurred alternately, as in the case of the penetration and film-penetration theories (see also Section IV,J).

By making use of the average bubble residence time, or the gas holdup, it was possible with this model to calculate not only the radius of the spherical shell but also the mass transfer per unit area, the total interfacial area, the average concentration of dissolved gas in the liquid, and the *total* average mass-transfer rate in the vessel. A change in agitation intensity or gas flow rate would affect $\bar{\theta}$ and would therefore change the model's radius, average dissolved gas concentration, total bubble surface, and total average absorption rate. Thus, the effects of mixing and flow patterns in the vessel were depicted by an indirect mechanism.

The following assumptions were made: (1) The gas bubbles are evenly distributed throughout the liquid phase and have constant radius and composition; (2) the concentration of the gas–liquid interface is constant and equal to c_i; (3) no gross variations occur in liquid composition throughout the vessel; and (4) the gas is sparingly soluble, and, in the case of a chemical reaction, it is removed by a first-order irreversible reaction with respect to the dissolving gas.

Under constant operating conditions, the number of bubbles in the vessel is constant, and their total volume or gas holdup is equal to $Q\bar{\theta}$, where $\bar{\theta}$ is the average residence-time of the bubbles and Q is the volumetric gas flow rate. The number of bubbles in the vessel at any instant is then given by

$$N_b = 3Q\bar{\theta}/4\pi a^3 \tag{156}$$

where a is the radius of the bubble. In accordance with the assumption that each bubble is enveloped by a spherical element, the total liquid volume V in the vessel is subdivided into equal-volume elements v_L equal in number to the number of bubbles in the vessel. Thus

$$v_L = \frac{4\pi V a^3}{3Q\bar{\theta}} = \frac{4}{3}\pi(b^3 - a^3) \tag{157}$$

where b is equal to the outer radius of the shell;

$$b = a\left(1 + \frac{V}{Q\bar{\theta}}\right)^{1/3} = a\Phi^{-1/3} \qquad (158)$$

The total surface available for mass transfer will be given by

$$A = 4N_b \pi a^2 = 3Q\bar{\theta}/a \qquad (159)$$

A change in the agitation intensity or gas flow rate will change $\bar{\theta}$. As a result, b, c_0, and the total mass-transfer rate are affected. Thus, the effect of mixing in the vessel is considered by indirect mechanisms.

The differential equation and boundary conditions for this model at constant ρ and D are

$$D\left[\frac{1}{r^2}\frac{\partial}{\partial r}\left(r^2 \frac{\partial c}{\partial r}\right)\right] = kc + \frac{\partial c}{\partial t} \qquad (160)$$

$$\partial c/\partial \theta = \partial c/\partial \phi = 0$$

$$\begin{array}{llll} c = c_0 & \text{at } t = 0 & \text{for } a < r \leqslant b & (a) \\ c = c_i & \text{at } r = a & \text{for } t \geqslant 0 & (b) \\ c = c_0 e^{-kt} & \text{at } r = b & \text{for } t \geqslant 0 & (c) \end{array} \qquad (161)$$

The last boundary condition results from the assumption that for the relatively short contact times occurring in real systems, the effect of diffusion at b is negligible, and, therefore, a change in concentration at this point results only from chemical reaction.

The solution is (G6):

$$c = \frac{c_i a}{r}\left[\frac{\sinh[(b-r)(k/D)^{1/2}]}{\sinh[(b-a)(k/D)^{1/2}]} + 2\pi e^{-kt}\sum_{n=1}^{\infty}\sin\left[\frac{n\pi(b-r)}{b-a}\right]\right.$$

$$\left. \times \frac{(-1)^n n \exp[-n^2\pi^2 Dt/(b-a)^2]}{n^2\pi^2 + (k/D)(b-a)^2}\right] - \frac{c_0 a}{r}\left[\frac{b-r}{b-a}e^{-kt} + \frac{2e^{-kt}}{\pi}\right.$$

$$\left. \times \sum_{n=1}^{\infty}\frac{(-1)^n}{n}\exp[-n^2\pi^2 Dt/(b-a)^2]\sin\left[\frac{n\pi(b-r)}{b-a}\right]\right] + c_0 e^{-kt}$$

$$(162)$$

This solution is shown graphically in Fig. 11.

In order to evaluate c_0, a mass balance in the liquid before and after mixing at time $t = \bar{\theta}$ gives

$$\int_a^b [4\pi r^2 c(r, \bar{\theta})] \, dr = \int_a^b [4\pi r^2 c_0] \, dr \qquad (163)$$

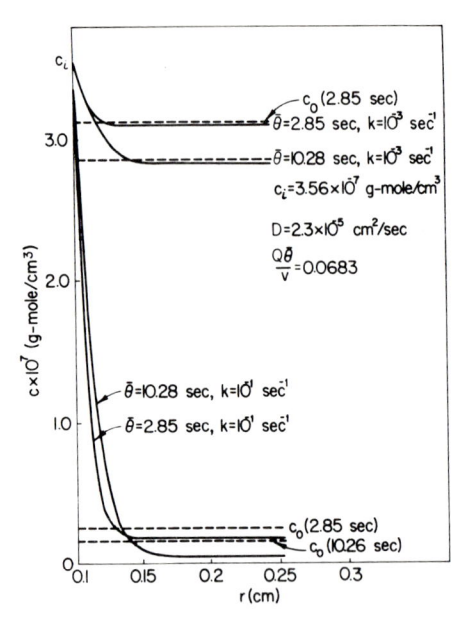

FIG. 11. Change in concentration with the radius of the residence-time model for constant gas holdup and varying contact time and reaction rate [after Gal-Or and Resnick (G2,G6)].

Evaluating the integrals and rearranging gives the final result:

$$c_0 = \frac{c_i \left[a \coth \xi + \dfrac{b-a}{\xi} - \dfrac{b}{\sinh \xi} + 2\xi e^{-\beta_0} \sum_{n=1}^{\infty} \dfrac{\exp\left[\dfrac{-n^2 \pi \beta_0}{\xi^2}\right][b(-1)^n - a]}{\xi^2 + n^2 \pi^2} \right]}{\sqrt{\dfrac{k\, a^2 V}{D\, 3Q\bar{\theta}}}\,(1 - e^{-\beta_0}) + \xi e^{-\beta_0}\left[\dfrac{b + 2a}{6} + \dfrac{2}{\pi^2} \sum_{n=1}^{\infty} \dfrac{\exp\left[-\dfrac{n^2 \pi^2 \beta_0}{\xi^2}\right][b(-1)^n - a]}{n^2}\right]} \quad (164)$$

This equation, which may be used directly to predict average concentration in the liquid phase under various operating conditions, is shown graphically in Fig. 12. Combining Eq. (164) with the expression for the total average rate of gas absorption in the vessel,

$$\bar{N}_T = \frac{A}{\bar{\theta}} \int_0^{\bar{\theta}} \left[-D\left(\frac{\partial c}{\partial r}\right)_{r=a} \right] dt \quad (165)$$

gives Eq. (166)

$$\bar{N}_T = \frac{3Q\bar{\theta}Dc_i}{a\beta_0(b-a)} \left[\frac{\beta_0(b-a)}{a} + \xi\beta_0 \coth \xi + 2\xi\pi^2 \sum_{n=1}^{\infty} \frac{n^2}{(\xi^2+n^2\pi^2)^2} - 2e^{-\beta_0}\xi^2\pi^2 \sum_{n=1}^{\infty} \frac{n^2 \exp[-n^2\pi^2\beta_0/\xi^2]}{(\xi^2+n^2\pi^2)^2} \right.$$

$$a \coth \xi + \frac{b-a}{\xi} - \frac{b}{\sinh \xi} + 2\xi e^{-\beta_0} \sum_{n=1}^{\infty} \frac{\exp\{-n^2\pi^2\beta_0/\xi^2\}\{b(-1)^n - a\}}{\xi^2 + n^2\pi^2}$$

$$+ \left(\frac{k}{D}\right)^{1/2} \frac{a^2 V}{3Q\bar{\theta}} (1 - e^{-\beta_0}) + \xi e^{-\beta_0} \left[\frac{b+2a}{6} + \frac{2}{\pi^2} \sum_{n=1}^{\infty} \frac{\exp[-n^2\pi^2\beta_0/\xi^2][b(-1)^n - a]}{n^2} \right]$$

$$\times \left\{ \frac{b}{a}(e^{-\beta_0} - 1) + 2\xi^2 \sum_{n=1}^{\infty} \frac{1}{\xi^2 + n^2\pi^2} + 2\xi^2 e^{-\beta_0} \sum_{n=1}^{\infty} \frac{\exp[-n^2\pi^2\beta_0/\xi^2]}{\xi^2 + n^2\pi^2} \right\} \right] \quad (166)$$

in which

$$\xi = a \left\{ \left(1 + \frac{V}{Q\bar{\theta}}\right)^{1/3} - 1 \right\} \left(\frac{k}{D}\right)^{1/2} \quad (167)$$

$$\beta_0 = k\bar{\theta} \quad (168)$$

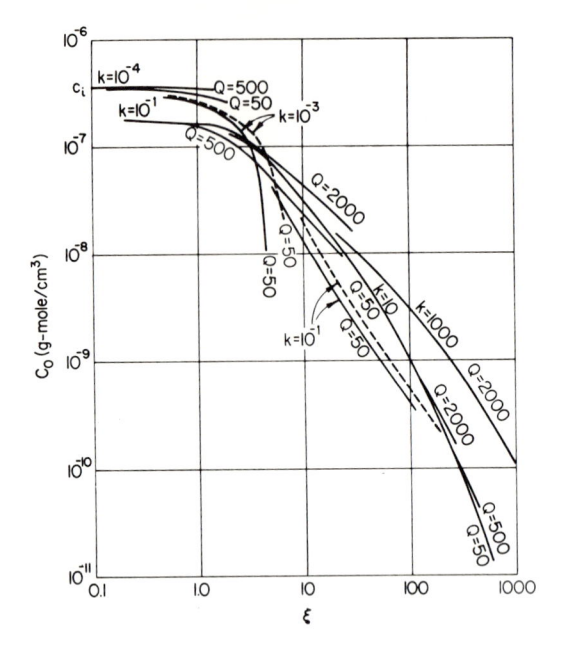

FIG. 12. Behaviour of c_0 [eq. (164)] as a function of $\xi = a(k/D)^{1/2}\{[1 + (V/Q\bar{\theta})]^{1/3} - 1\}$ [after Gal-Or and Resnick (G2,G6)]: Solid lines—$a = 0.1$ cm; dashed lines—$a = 0.2$ cm. (Values of k in sec^{-1}, values of Q in cm^3/sec.)

The behavior of $\bar{N}_T/V$ is given in Fig. 13 as a function of $Q\bar{\theta}/V$ and in Fig. 14 as a function of ξ for various values of k, Q and a. From these figures it can be deduced that:

(1) The total average rate of absorption in the vessel increases with the rate of chemical reaction.

(2) The average absorption rate increases with the gas holdup in the vessel. This may be caused by a change in the gas flow rate or the agitation intensity. The effect is augmented as k increases.

(3) An increase in bubble diameter will generally depress the total average absorption rate. This effect diminishes as the chemical reaction rate is decreased.

(4) For chemical reaction-rate constants greater than 10 sec^{-1}, $\bar{N}_T$ increases linearly with the total bubble surface area, i.e., linearly with the gas holdup. In other words, the agitation rate only affects the total bubble surface area and has almost no effect on the rate of absorption per unit area. This result is in accordance with the work of Calderbank and Moo-Young (C4), discussed in Section II.

(5) For low values of the chemical reaction-rate constant, the reaction

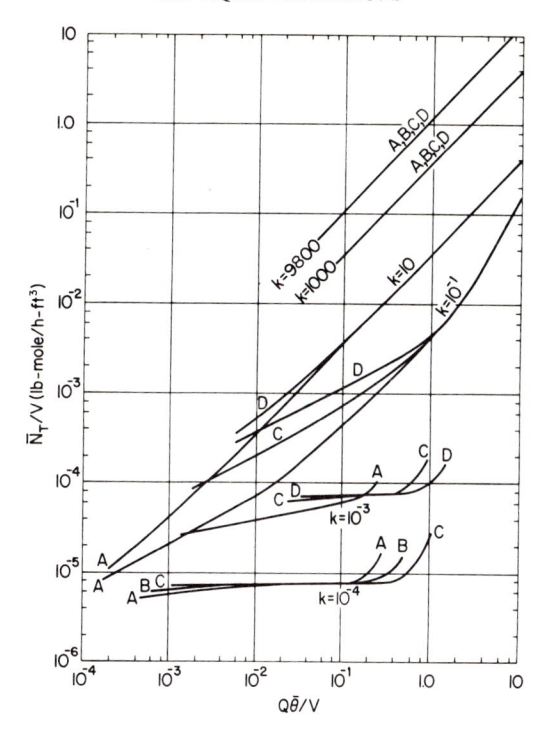

FIG. 13. Total average volumetric absorption rate in the dispersion, $\bar{N}_T/V$, as a function of gas holdup ratio $Q\bar{\theta}/V$, chemical reaction rate k (in sec^{-1}), and the following values of volumetric gas flow rate Q (in cm^3/sec): A) 50, B) 150, C) 500, D) 2000 [after Gal-Or and Resnick (G2,G6)] ($a = 0.1$ cm.)

rate controls the total average rate of absorption, and the contact time has little influence. An increase in the reaction rate constant increases the total average rate until, at high values of reaction rate, diffusion controls the total average rate of gas absorption.

Solutions for diffusion with and without chemical reaction in continuous systems have been reported elsewhere (G2, G6). In general, all the parameters in this model can be determined or estimated, and the theoretical expressions may assist in the interpretation of mass-transfer data and the prediction of equipment performance.

Experimental work was undertaken (G8) to provide the information necessary to permit a test of this theoretical model. The system used bore complete geometrical and chemical similarity to that used by Cooper *et al.* (C9) so that their mass-transfer rate measurements, along with the average residence-time and power-consumption results determined in the experimental work (see Section II,D), were used to compare the experimental values with the model.

The theoretical behavior of the model as calculated with the aid of a digital computer is shown graphically in Fig. (15) in the form of full or broken lines. The data points represent the experimental values obtained by Cooper et al. (C9), which correlate reasonably well with the predicted ones. As indicated by Eq. (166), the expression for $\bar{N}_T$ is directly proportional to c_i. Thus, by using a computer to calculate $\bar{N}_T$ for one value of c_i, it is possible to predict $\bar{N}_T$ for any desired partial pressure of oxygen. Consequently, a better agreement between the theory and experimental results is possible if the outlet partial pressure of oxygen is estimated (by using Fig. 15) and the average partial pressure of oxygen is used to recalculate the average c_i. However, reliable values for the parameters, such as c_i, D, and k, are still not known for this system.

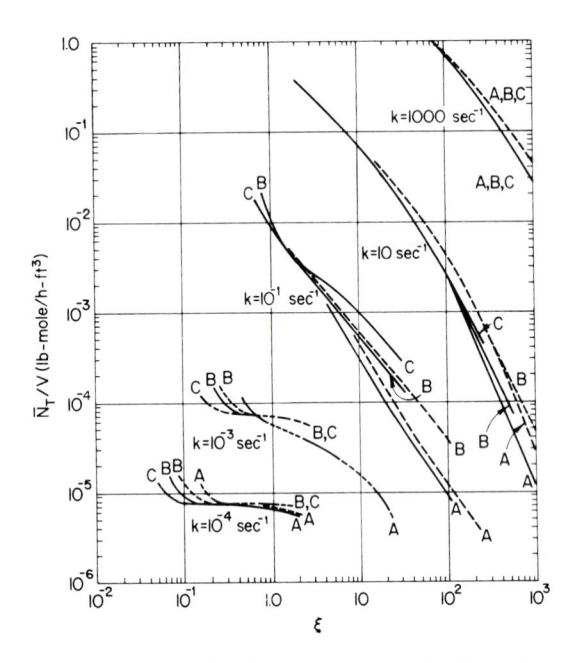

FIG. 14. Total average volumetric absorption rate in the dispersion as a function of $\xi = a(k/D)^{1/2}\{[1 + (V/Q)]^{1/3} - 1\}$ [after Gal-Or and Resnick (G2,G6)]: solid lines—$a = 0.1$ cm; dashed lines—$a = 0.2$ cm. A) $Q = 50\,\text{cm}^3/\text{sec}$; B) $Q = 500\,\text{cm}^3/\text{sec}$; C) $Q = 2000\,\text{cm}^3/\text{sec}$.

In conclusion, therefore, the model proposed by Gal-Or and Resnick predicts mass-transfer rates that correlate reasonably well with the experimental data now available. Further experimental work as well as accurate values for the reaction-rate constants, diffusivities, and solubilities for other systems will be needed to check this model over a broad range of operating and system variables. It would appear that experimental mass-transfer data could be correlated as $\bar{N}_T/V$ as a function either of ξ or of $Q\bar{\theta}/V$ (the gas holdup ratio).

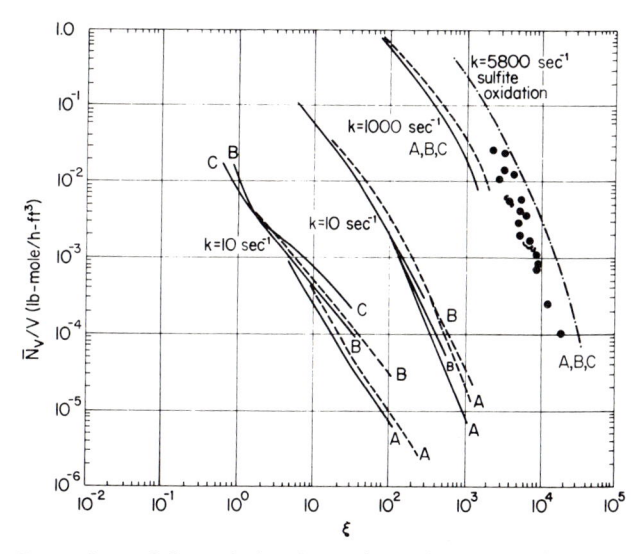

FIG. 15. Comparison of theoretical and experimental mass-transfer rates: N_T/V as a function of $\xi = a(k/D)^{1/2}\{[1 + (V/Q\bar{\theta})]^{1/3} - 1\}$. Solid lines: $a = 0.1$ cm, $c_i = 2.81 \times 10^{-7}$ g-mole/cm^3; dashed lines (— — —): $a = 0.2$ cm, $c_i = 2.81 \times 10^{-7}$ g-mole/cm^3; dashed line (·—·—): $a = 0.2$ cm, $c_i = 1.46 \times 10^{-7}$ g-mole/cm^3. A) $Q = 50$ cm^3/sec; B) $Q = 500$ cm^3/sec; C) $Q = 2000$ cm^3/sec. Experimental data ● (C9,G8). [After Gal-Or and Resnick (G2,G8)].

J. Effect of Bubble-Size Distribution and Holdup on Mass- or Heat-Transfer Rates

Most studies on heat- and mass-transfer to or from bubbles in continuous media have primarily been limited to the transfer mechanism for a single moving bubble. Transfer to or from swarms of bubbles moving in an arbitrary fluid field is complex and has only been analyzed theoretically for certain simple cases. To achieve a useful analysis, the assumption is commonly made that the bubbles are of uniform size. This permits calculation of the total interfacial area of the dispersion, the contact time of the bubble, and the transfer coefficient based on the average size. However, it is well known that the bubble-size distribution is not uniform, and the assumption of uniformity may lead to error. Of particular importance is the effect of the coalescence and breakup of bubbles and the effect of these phenomena on the bubble-size distribution. In addition, the interaction between adjacent bubbles in the dispersion should be taken into account in the estimation of the transfer rates in dispersions. This interaction increases as the dispersed-phase holdup fraction increases, causing a change in bubble velocity and affecting concentration profiles between the bubbles. As a result, the rate of mass- (or heat-) transfer is changed accordingly.

Gal-Or and Hoelscher (G5) have recently proposed a mathematical model that takes into account interaction between bubbles (or drops) in a swarm as well as the effect of bubble-size distribution. The analysis is presented for unsteady-state mass transfer with and without chemical reaction, and for steady-state diffusion to a family of moving bubbles.

1. *Unsteady-State Model for Transfer between Swarms of Bubbles and Agitated Liquid*

This model is proposed for the case of transfer from a swarm of bubbles (with bubble-size distribution) suspended in an agitated liquid with interaction between adjacent bubbles in the presence of surfactants.

Certain hydrodynamical problems, as well as mass-transfer problems in the presence of surface-active agents, have been investigated theoretically under steady-state conditions (L3, L4, L10, R9). However, if we take into account the fact that in gas–liquid dispersions, the nonstationary term must appear in the equation of mass- or heat-transfer, it becomes apparent that an exact analysis is possible if a mixing–contacting mechanism is adopted instead of a theoretical streamline flow around a single bubble sphere.

The surfactants that are present in practical agitated dispersions tend to form some kind of "skin" around the bubble which effectively retards the movement of the surface elements adjacent to the interface. Talc particles sprinkled on the surface become virtually immobile if the surface is even slightly contaminated, indicating that the surface elements are stagnant and are setting up a considerable resistance to the "clearing" of the surface by eddies of liquid approaching obliquely (D9). This may be the reason that mass- and heat-transfer coefficients are little affected by the amount of mechanical power dissipated in mixing of the dispersion (C4). According to Garner *et al.* (G12, G13) and Davies and Mayer (see D11), the control of mass transfer may change from surface renewal to diffusion through a stagnant film in the presence of surfactants. Consequently, the actual mechanism under these conditions may be in some agreement with the behavior of the "surface elements" that were proposed by Toor and Marchello in the "film-penetration" theory (H18, H19, T5). The contact time of these surface elements may be as long as the residence time the drop or bubble in the dispersion, i.e., from the moment of drop or bubble formation until it bursts on the free surface of the dispersion. When the bubble or drop bursts on the free surface of the dispersion, the surface elements stay and are mixed with the continuous phase. These assumptions, coupled with the inclusion of interaction effects between adjacent particles in a swarm, may give a better approximation to the actual mechanism in practical dispersions (especially for highly dispersed phase holdups) than the assumptions of a steady theoretical streamline flow around a single bubble sphere. Thus, for the purpose of this analysis, a type of film-penetration mixing–contacting mechanism is considered, coupled with the effects of bubble-size distribution and dispersed-phase holdup (G5).

In the proposed model, the total volume of the dispersion (or of the reactor) is subdivided into many subvessels (or subreactors). In each, Gal-Or and Hoelscher considered a fraction of the total number of bubbles having a radius between a and $a + da$. The distribution of bubbles in the subreactors

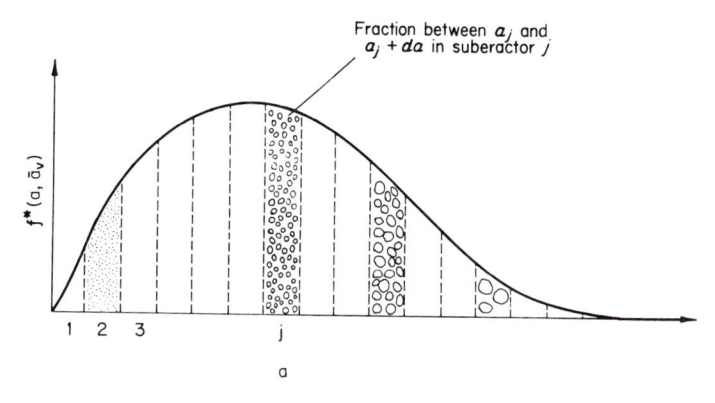

FIG. 16. Distribution of bubbles in the model of subreactors as a function of $f^*(a, \bar{a}_v)$ [after Gal-Or and Hoelscher (G5)].

is given by $f^*(a, a_v)$, as described qualitatively in Fig. 16. The dispersed-phase volume in any subreactor is thus given by

$$V_v \, d\Phi = \tfrac{4}{3}\pi a^3 N_v \, V_v f^*(a, a_v) \, da \tag{169}$$

in which

$$N_v = \frac{\Phi}{\tfrac{4}{3}\pi \bar{a}_v^{\,3}} \tag{24}$$

In each subreactor, the volume of the continuous phase V_j is again subdivided into a number of equal-volume elements $(v_L)_j$, equivalent to the number of bubbles in the subreactor. Each bubble is then considered to be enveloped by a spherical shell of volume $(v_L)_j$. Thus, for a sparingly soluble gas (G5),

$$(v_L)_j = \frac{4V_j \pi a^3}{3Q_j \bar{\theta}_j} = \tfrac{4}{3}\pi(b_j^3 - a^3) \tag{170}$$

or

$$b_j = a\left(1 + \frac{V_j}{Q_j \bar{\theta}_j}\right)^{1/3} \tag{171}$$

In each subreactor, the dispersed-phase holdup fraction should be the same as the overall fraction Φ; thus

$$b_j = a\Phi^{-1/3} = a/\Lambda \tag{172}$$

Consequently, for the same fractional holdup there are more bubbles per unit volume as the bubble size decreases, causing b_j (which is a yardstick for the distance between adjacent bubbles) to decrease, resulting in more and more interaction between adjacent bubbles. This is one type of interaction that is taken into account in this analysis.

Each bubble in each subreactor is then assumed to be introduced into each element for a retention time equal to the average residence time of the bubbles. At $t = \bar{\theta}$, the bubble is removed, the continuous phase in the shell is completely mixed, and another bubble is introduced. Thus, the total number of operating models is equal to the total number of bubbles in the disperser. Changing agitation intensity, for example, will change Φ, $\bar{\theta}$, and $\bar{a}_v$, and consequently the b_j distribution, rates of mass fluxes, and the total interfacial area are changed. Thus, the effects of operating and system variables are indirectly interacting in this semidynamic model.

On solving Eq. (160) with the boundary conditions (161a), (161b), and

$$\left(\frac{\partial c}{\partial r}\right)_{r=b_j} = 0, \quad t \geq 0 \tag{173}$$

one finds (B7)

$$c_j = \left(\frac{c_i a}{r}\right)\left\{\cosh\left[(b_j - r)\left(\frac{k}{D}\right)^{1/2}\right] - \frac{1}{b_j}\left(\frac{D}{k}\right)^{1/2}\sinh\left[(b_j - r)\left(\frac{D}{k}\right)^{1/2}\right]\right\}$$
$$+ \frac{1}{r}\sum_{n=1}^{\infty}\left(c_0 - \frac{c_i \lambda_n^2}{\lambda_n^2 + (k/D)}\right)\left\{\frac{2a \sin[\lambda_n(r-a)]\exp[-(\lambda_n^2 D + k)t]}{\lambda_n\{b_j \sin^2[\lambda_n(b_j - a)] - a\}}\right\} \tag{174}$$

in which λ_n are the positive roots of

$$\lambda_n b_j = \tan[\lambda_n(b_j - a)] \tag{175}$$

Boundary condition (173) does not imply that the profiles of the concentrations are unaffected by the distance between adjacent bubbles, as expressed in the equation for b_j. (P1)(W1).

The instantaneous flux at the surface of the bubbles in subreactor j will thus be given by (G7) Eq. (176).

$$N_{Aj} = -D\left(\frac{\partial c}{\partial r}\right)_{r=a}$$

$$= \frac{D}{a}\left[\frac{c_i}{\Omega}\left\{(1-\Phi^{1/3})\cosh\xi + \left(\sqrt{\frac{k}{D}}a - \frac{\Phi^{1/3}}{a}\sqrt{\frac{D}{k}}\right)\sinh\xi\right\}\right.$$

$$\left. - 2\sum_{n=1}^{\infty} c_i \frac{\left(\frac{D}{k}\left[\left(\sqrt{\frac{k}{D}}\bar{a}_v - \frac{\Phi^{1/3}}{\bar{a}_v}\sqrt{\frac{D}{k}}\right)\tanh\bar{\xi} + 1 - \Phi^{1/3}\right]\sqrt{1 - \frac{\Phi^{1/3}}{\bar{a}_v}\sqrt{\frac{D}{k}}\tanh\bar{\xi}}\right) - \Upsilon}{\frac{\bar{a}_v^2}{3}\left(\frac{1}{\Phi} - 1\right) - \Upsilon} - \frac{\lambda_n^2}{\lambda_n^2 + \frac{k}{D}}\right\}$$

$$\times \frac{\exp[-(\lambda_n^2 D + k)t]}{\Phi^{-1/3}\sin^2[\lambda_n a(\Phi^{-1/3} - 1)] - 1}\right] \quad (176)$$

$$\bar{\xi} = \bar{a}_v(\Phi^{-1/3} - 1)\sqrt{\frac{k}{D}}; \quad \xi = a(\Phi^{-1/3} - 1)\sqrt{\frac{k}{D}}; \quad \Omega = \cosh\xi - \frac{\Phi^{1/3}}{a}\sqrt{\frac{D}{k}}\sinh\xi$$

$$\Upsilon = 2\sum_{n=1}^{\infty} \frac{\exp[-(\lambda_n^2 D + k)\bar{\theta}]\left[1 - \Phi^{-1/3}\cos\bar{\eta} + \frac{\sin\bar{\eta}}{\lambda_n \bar{a}_v}\right]}{\left(\lambda_n^2 + \frac{k}{D}\right)(\Phi^{-1/3}\sin\bar{\eta} - 1)}; \quad \bar{\eta} = \lambda_n \bar{a}_v(\Phi^{-1/3} - 1) \quad (177)$$

in which

The dependence of N_{Aj} on bubble size for various chemical reaction rates and for $t = 0.001$, 0.01, and $t = \bar{\theta} = 2.85$ sec is shown in Fig. 17. For a fast chemical reaction with $k = 10^4$ sec^{-1}, N_{Aj} is not a function of time because the gradient is very steep and formed almost instantaneously. As the chemical reaction-rate decreases, N_{Aj} decreases too. However, for smaller chemical reaction rates, an influence of time is observed: as t increases, N_{Aj} decreases. For most of the practical region of particle sizes, N_{Aj} does not change as a changes; however, for very small values of a, N_{Aj} tends to increase slightly, reach a maximum, and then decrease. This phenomenon may be explained in terms of two opposing effects: The first is the curvature effect that increases the diffusional flux as a decreases; the second is the interaction between adjacent particles. For a given contact time and holdup, b_j (a measure of the distance between adjacent particles) decreases as a decreases, causing the diffusional flux to decrease with decreasing a. The last process begins to control the rate at sufficiently small values of a. As the chemical reaction-rate decreases, this maximum in N_{Aj} moves to higher values of a as a result of a deeper penetration of diffused matter as k decreases. In other words, for smaller chemical reaction-rates, the interaction between adjacent particles will start to control at relatively higher particle sizes.

To evaluate the total rate of mass transfer in the vessel, N_T, an integration over the entire surface of the population of bubbles having the distribution of sizes

$$N_T = \int_0^A N_{Aj} \, dA \tag{178}$$

is done, followed by integration with respect to time to evaluate the average rate of mass transfer under the pseudosteady-state conditions:

$$\bar{N}_T = \frac{\int_0^{\bar{\theta}} N_T \, dt}{\int_0^{\bar{\theta}} dt} \tag{179}$$

The first integration is carried out by using the expression

$$dA = \frac{3\Phi V_v}{\bar{a}_v^{\,3}} a^2 f^*(a, \bar{a}_v) \, da \tag{180}$$

On substituting Eq. (17) for $f^*(a, \bar{a}_v)$, the result of integration (178) is as shown in Eq. (181).

$$N_T = \frac{48\Phi V_v D c_i}{\pi \bar{a}_v^6} \int_0^\infty a^3 \exp\left[-\left(\frac{4}{\sqrt{\pi \bar{a}_v^3}}\right)^{2/3} a^2\right]$$

$$\times \left\{ \frac{(1 - \Phi^{1/3}) \cosh\left[a(\Phi^{-1/3} - 1)\sqrt{\frac{k}{D}}\right]}{\cosh\left[a(\Phi^{-1/3} - 1)\sqrt{\frac{k}{D}}\right] - \frac{\Phi^{1/3}}{a}\sqrt{\frac{D}{k}} \sinh\left[a(\Phi^{-1/3} - 1)\sqrt{\frac{k}{D}}\right]} \right.$$

$$+ \frac{\left(a\sqrt{\frac{k}{D}} - \frac{\Phi^{1/3}}{a}\sqrt{\frac{D}{k}}\right) \sinh\left[a(\Phi^{-1/3} - 1)\sqrt{\frac{k}{D}}\right] - \frac{\Phi^{1/3}}{a}\sqrt{\frac{D}{k}} \sinh\left[a(\Phi^{-1/3} - 1)\sqrt{\frac{k}{D}}\right]}{\cosh\left[a(\Phi^{-1/3} - 1)\sqrt{\frac{k}{D}}\right] - \frac{\Phi^{1/3}}{a}\sqrt{\frac{D}{k}} \sinh\left[a(\Phi^{-1/3} - 1)\sqrt{\frac{k}{D}}\right]}$$

$$\frac{\left[\sqrt{\frac{D}{k}}\left\{\left(\bar{a}_v\sqrt{\frac{k}{D}} - \frac{\Phi^{1/3}}{\bar{a}_v}\sqrt{\frac{D}{k}}\right)\tanh\bar{\xi} + 1 - \Phi^{1/3}\right\} \middle/ \left(1 - \frac{\Phi^{1/3}}{\bar{a}_v}\sqrt{\frac{D}{k}}\tanh\bar{\xi}\right)\right] - \Upsilon}{\frac{\bar{a}_v^2}{3}\left(\frac{1}{\Phi} - 1\right) - \Upsilon}$$

$$\left. - 2\sum_{n=1}^{\infty} \frac{\exp[-(\lambda_n^2 D + k)t]}{\Phi^{-1/3} \sin^2[\lambda_n a(\Phi^{-1/3} - 1)] - 1} \cdot \frac{\lambda_n^2}{\lambda_n^2 + \frac{k}{D}} \right\} da \quad (181)$$

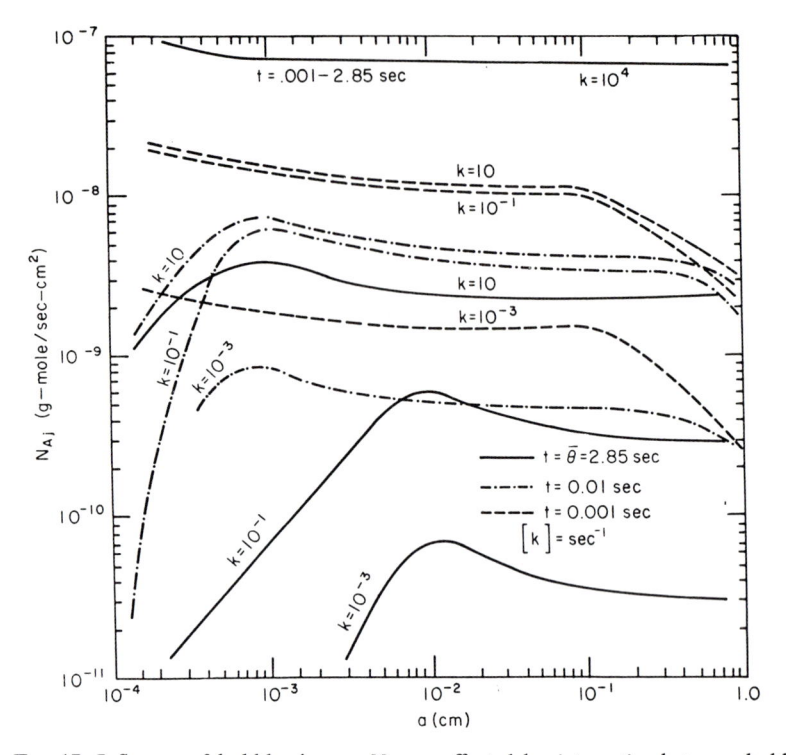

FIG. 17. Influence of bubble size on N_{Aj} as affected by *interaction* between bubbles (as a function of a in subreactors and for constant gas holdup), chemical reaction rates, and contact times. N_{Aj} was calculated from Eq. (176) with $c_i = 1.46 \times 10^{-7}$ gr-mole/cm³; $D = 2.3 \times 10^{-5}$ cm²/sec; $\Phi = 0.064$; $\bar{a}_v = 0.1$ cm; $\bar{\theta} = 2.85$ sec [after Gal-Or and Hoelscher (G5)].

For a fast chemical reaction $k \gg 1$ and $c_0 \to 0$, and Eq. (181) gives

$$N_T = \bar{N}_T = \frac{48\Phi V_v D c_i}{\pi \bar{a}_v^6} \left[\frac{1 - \Phi^{1/3}}{2\alpha^2} + \left(\frac{k}{D}\right)^{1/2} \frac{\Gamma(2.5)}{2\alpha^{2.5}} \right] \qquad (182)$$

For the particle-size distribution adopted here [Eq. (17)] it was found [Eqs. (22) and (25)] that $\bar{a}_v$ is $a_{32}/1.148$ and s, the interfacial area per unit volume of dispersion, is $(3\Phi/a_{32})$. Consequently,

$$k_L = \frac{\bar{N}_T}{sc_i} = 1.135 \frac{D}{a_{32}} \left[1 - \Phi^{1/3} + 0.885 a_{32} \left(\frac{k}{D}\right)^{1/2} \right] \qquad (183)$$

The effect of particle-size distribution on the mass-transfer coefficient is obtained for this case by using Eq. (176). Thus, when the variable particle size a is replaced by a_{32}, the mean

$$k_L = \frac{N_{Aj}}{c_i} = \frac{D}{a_{32}}\left[1 - \Phi^{1/3} + a_{32}\left(\frac{k}{D}\right)^{1/2}\right] \tag{184}$$

Comparison of Eq. (184) with Eq. (183) shows the effect of size distribution for the case of fast chemical reaction with simultaneous diffusion. This serves to emphasize the error that may arise when one applies uniform-drop-size assumptions to drop populations. Quantitatively the error is small, because $1 - \Phi^{1/3}$ is small in comparison with the second term in the brackets [i.e., $k_L \simeq (kD)^{1/2}$]. Consequently, Eq. (184) and Eq. (183) actually give about the same result. In general, the total average mass-transfer rate in the disperser has been evaluated in this model as a function of the following parameters:

$$\bar{N}_T = f(c_i, k, V_v, Q, D, \Phi, a_{32}, \bar{\theta}) \tag{185}$$

These parameters can be determined and predicted, and the theoretical expressions may thus assist in interpretation of mass-transfer data and in prediction of equipment performance. The case of mass transfer without chemical reaction is reported elsewhere (G5).

2. Moving Bubbles with Clean Interfaces under Steady-State Conditions

This model is proposed for steady-state mass transfer without chemical reaction from swarms of moving bubbles with clean interfaces and without interaction between adjacent bubbles.

Using the same principles as in the previous model, Gal-Or and Hoelscher (G5) have used Eq. (143) to evaluate the total rate of mass transfer in the whole vessel, $\bar{N}_T$, by integration over the entire surface of the swarms of bubbles having the distribution of sizes $f^*(a, \bar{a}_v)$ as given by Eq. (17). Combining Eqs. (143), (17), and (178), the result is

$$\bar{N}_T = \frac{24}{3\sqrt{\pi}}\left[\frac{D|\rho_d - \rho_c|g}{2\mu_c + 3\mu_d}\right]^{1/2}(c_0 - c_i)\left(\frac{\Phi V_v}{\bar{a}_v^{\,3}}\right)\left(\frac{\alpha^3}{\pi}\right)^{1/2}\int_0^\infty a^{4.5}e^{-\alpha a^2}\,da$$

$$= \frac{4}{\pi}\left[\frac{D|\rho_d - \rho_c|g}{2\mu_c + 3\mu_d}\right]^{1/2}\left(\frac{\Phi V_v \Gamma(2.75)}{\bar{a}_v^{\,3}\alpha^{5/4}}\right)(c_0 - c_i) \tag{186}$$

or

$$k_L s = 1.04\,\frac{\Phi}{(\bar{a}_v)^{1/2}}\left[\frac{D|\rho_d - \rho_c|g}{2\mu_c + 3\mu_d}\right]^{1/2} \tag{187}$$

On using Eqs. (25) and (22) to eliminate Φ and s, the mass-transfer coefficient is found to be expressed as

$$k_L = 0.371 \left[\frac{D|\rho_d - \rho_c|g}{2\mu_c + 3\mu_d} \right]^{1/2} \sqrt{a_{32}} \tag{188}$$

The comparison of this equation with the experimental data of Calderbank and Moo-Young is shown in Fig. (9).

With k_L evaluated directly from Eq. (143),

$$k_L = 0.379 \left[\frac{D|\rho_d - \rho_c|g}{2\mu_c + 3\mu_d} \right]^{1/2} \sqrt{a} \tag{189}$$

so that if the varying bubble size a is merely replaced by the mean a_{32}, the result is much the same as that derived by taking the bubble-size distribution into account.

This case can also be approached using Kolmogoroff's (K9, H15) theory of local isotropic turbulence to predict the velocity of suspended particles relative to a homogeneous and isotropic turbulent flow. By examining this situation for spherical particles moving with a constant relative velocity, varying randomly in direction, Levich, (L3) has demonstrated that

$$U \doteq \frac{2|\rho_d - \rho_c|\varepsilon_0^{3/4}a^2}{3\sqrt{3}\rho_c v_c^{5/4}} \tag{190}$$

where ε_0 is the intensity of turbulence or power per unit mass and v_c is the kinematic viscosity of the continuous phase. Equation (190) is valid for a particle whose size is substantially less than the scale of an inner turbulent zone λ_0 and whose relative velocity is smaller than the velocity U_λ of turbulent eddies of scale λ. Therefore, the Reynolds number $Ua/v < U_\lambda \lambda_0/v \simeq 1$. Equation (190) means that the particle motion is quasisteady, as if it were moving in some other steady acceleration field—for example, in a gravitational field. Substituting Eq. (190) into Eq. (142) and integrating as in Eq. (186) will give the same dependency on a as in Eqs. (187) and (188). The final result is G5, G9):

$$k_L = 0.287 \left[\frac{D|\rho_d - \rho_c|}{(\mu_d + \mu_c)v_c^{1/4}} \right]^{1/2} \varepsilon_0^{3/8} \sqrt{a} \tag{191}$$

For this case, however, k_L is expressed in terms of ε_0.

Calderbank and Moo-Young (C4) found that mass- and heat-transfer coefficients are largely unaffected by the mechanical power dissipated in the system. According to Eq. (197), this may be the case when the product $\varepsilon^{3/8}\sqrt{a}$ remains constant.

K. Effect of Holdup on Convective Mass Transfer

Some of the principles of "free surface model" proposed by Happel (H10, P7, P8) for studying the rate of sedimentation of solid particles may be adopted for studying heat and mass transfer in gas–liquid dispersions (R9).

For the mass-transfer coefficient of a bubble in a group of bubbles, Ruckenstein (R9) assumes that

$$N_{Sh} = f\left(N_{Re}, N_{Sc}, \frac{U_i}{U}, \Phi\right) \tag{192}$$

where U_i is the velocity with respect to the center of the bubble, in the liquid phase at the interface for $\theta = \pi/2$. The distribution of velocities and concentrations in the immediate vicinity of one bubble is influenced by its neighbors.

For a cellular model of the type given in Section IV,I but with a "free surface" as proposed by Happel for solid particles, Ruckenstein established the form of the function f in two cases:

(a) A group of bubbles for which $N_{Re} \ll 1$, moving through a continuous phase containing surface-active agents.

(b) A group of bubbles for which $N_{Re} \gg 1$, moving through a pure continuous phase.

The model considered is of mass transfer for a liquid in motion between two concentric spheres of radii b and a, the latter being the radius of the bubble:

$$b = a\Phi^{-1/3} = a/\Lambda \tag{193}$$

The concentration at a is considered to be quasiconstant and practically equal to the average concentration in the respective region.

Case (a): $N_{Re} \ll 1$ with Presence of Surface-Active Agents. If there is a sufficiency large quantity of surfactants in the continuous phase, then the interface behaves like a solid surface (see Section III). From Eq. (65) it follows that as a decreases, the quantity of surfactants necessary to make the interface behave like a solid surface becomes smaller. Hence, for $r = a$, V_r, and V_θ, the components of the velocity in the liquid phase, are given by

$$V_r = U\cos\theta, \quad V_\theta = -U\sin\theta \tag{194}$$

where U is the velocity of the bubble.

The boundary of radius b behaves like a free surface. Hence, for $r = b$, the shear stress

$$\tau_{r\theta} = \mu_c \left(\frac{\partial V_\theta}{\partial r} - \frac{V_\theta}{r} \right) = 0 \tag{195}$$

and

$$V_r = 0, \quad \left(\frac{\partial V_r}{\partial \theta} \right)_{r=b} = 0 \tag{196}$$

Assuming

$$(a + y)^n \simeq a^n[1 + n(y/a)] \tag{197}$$

where

$$y = r - a \tag{198}$$

Ruckenstein obtained for the velocity components in the continuous phase with respect to the center of the bubble

$$v_\theta = U\psi(\Lambda)(y/a) \sin\theta \tag{199}$$

$$v_r = -U\psi(\Lambda)(y/a)^2 \cos\theta \tag{200}$$

where

$$\Psi(\Lambda) = \frac{3(1 - \Lambda^5)}{2 - 3\Lambda + 3\Lambda^5 - 2\Lambda^6} \tag{201}$$

For a single solid sphere, $\psi \to 3/2$ when $\Phi \to 0$.

If the thickness of the diffusion boundary layer is smaller than $b - a$ (and also smaller than a), one may consider that the diffusion takes place from the sphere to an infinite liquid. It should be emphasized here that the thickness of the diffusion boundary layer is usually about 10% of the thickness of the hydrodynamic boundary layer (L3). Hence this condition imposes no contradiction to the requirements of the free surface model and Eq. (195). Thus, for $N_{Sc} \gg 1$ (in liquids, N_{Sc} is of the order of 10^3), an equation of the type of Eq. (138) holds, i.e.:

$$v_r \frac{\partial c}{\partial y} + \frac{v_\theta}{a} \frac{\partial c}{\partial \theta} = D \frac{\partial^2 c}{\partial y^2} \tag{202}$$

By solving this equation with the above-mentioned velocity components and, in addition, the boundary conditions:

$$\begin{matrix} c = c_i & \text{for} & r = a \\ c = c_0 & \text{for} & r \to \infty \end{matrix} \tag{203}$$

Ruckenstein demonstrated that

$$N_{Sh} = 0.868 N_{Re}^{1/3} N_{Sc}^{1/3} \{\Psi(\Lambda)\}^{1/3} \tag{204}$$

For very small values of Reynold's numbers (negligible convective diffusion), elementary calculations lead to

$$N_{Sh} = 2/(1 - \Phi^{1/3}) \tag{205}$$

Consequently, for intermediate conditions, Ruckenstein proposed the following approximation:

$$N_{Sh} = (2/1 - \Phi^{1/3}) + 0.868 N_{Re}^{1/3} N_{Sc}^{1/3} \{\Psi(\Lambda)\}^{1/3} \tag{206}$$

Using Happel's "free surface model," which predicts (H10)*

$$U = U_s \frac{3 - \tfrac{9}{2}\Lambda + \tfrac{9}{2}\Lambda^5 - 3\Lambda^6}{3 + 2\Lambda^5} \equiv U_s \Psi^*(\Lambda) \tag{207}$$

Ruckenstein obtained

$$N_{Sh} = (N_{Sh})_0 (\tfrac{2}{3}\Psi\Psi^*)^{1/3} \equiv \mathscr{F}(N_{Sh})_0 \tag{208}$$

where $(N_{Sh})_0$ represents the Sherwood numbers for a single solid particle and the values of $\mathscr{F}$ are given in the accompanying tabulation.

Φ:	0.0	0.2	0.3	0.4	0.5	0.6	0.7	0.8
$\mathscr{F}$:	1.0	0.96	0.91	0.88	0.82	0.77	0.68	0.55

Using Stokes' equation, Eq. (208) gives

$$k_L N_{Sc}^{2/3} = 0.38 \mathscr{F} \left(\frac{(\rho_c - \rho_d)\mu_c g}{\rho_c^2} \right)^{1/3} \tag{209}$$

For values of Φ within the range 0.0–0.7, the product $0.38\mathscr{F}$ varies between 0.26 and 0.38, which is close to the value 0.31 reported in the experimental results of Calderbank and Moo-Young (C4) as given in Eq. (10).

Case (b): $N_{Re} \gg 1$ with Pure Continuous Phase. For this case, the tangential component of the continuous-phase velocity (with respect to the center of the bubble) varies slowly with y. A penetration type of mechanism can be assumed, i.e.,

$$U \, \partial c / \partial z \equiv \partial c / \partial t = D \, \partial^2 c / \partial y^2 \tag{210}$$

where t is the time measured from the moment in which an element of liquid comes into contact with the bubble and z is the vertical distance measured from the front stagnation point.

---

* Eq. (207) can be derived also as a special case of a more general solution by Waslo and Gal-Or (W1) which gives the terminal velocity of an ensemble of drops or bubbles in the presence of surfactants (see Sections IIF and IIIB for more details).

The solution of Eq. (210), with the initial and boundary conditions

$$\begin{array}{lll} c = c_0 & \text{for } t = 0 & \text{and } 0 < y < (b-a) \\ c = c_0 & \text{for } t > 0 & \text{and } y = b - a \\ c = c_i & \text{for } t > 0 & \text{and } y = 0 \end{array} \quad (211)$$

is

$$\frac{c - c_0}{c_i - c_0} = \sum_{n=1}^{\infty} \left\{ \operatorname{erfc} \frac{2n(b-a) + y}{2(Dt)^{1/2}} - \operatorname{erfc} \frac{2(n+1)(b-a) - y}{2(Dt)^{1/2}} \right\} \quad (212)$$

Ruckenstein shows (R9) that the duration t_0 of the contact between the element of fluid and the bubble may be estimated for a group of bubbles as $2a/U$ [compare Eq. (83)].

Consequently,

$$k_L = \frac{-D_0 \int_0^{t_0} (\partial c/\partial y)_{y=0} \, dt}{t_0 (c_i - c_0)}$$

$$\simeq \left(1.13 + 4i \operatorname{erfc} \frac{b-a}{(Dt)^{1/2}} \right) \left(\frac{D}{t_0} \right)^{1/2} \quad (213)$$

If

$$(b-a)/(Dt)^{1/2} \geqslant 2$$

this gives

$$k_L \simeq 1.13(D/t_0)^{1/2} \simeq 1.13(DU/2a)^{1/2} \quad (214)$$

If Eq. (38) is substituted for U, the result is

$$k_L = 0.266 \left[\frac{D\rho_c g}{\mu_c} \right]^{1/2} \sqrt{a} \quad (215)$$

which is similar to Eq. (145) derived in Section IV,G and is in general agreement with the experimental data of Calderbank and Moo-Young (C4).

L. COUPLED HEAT TRANSFER AND MULTICOMPONENT MASS TRANSFER, WITH RESIDENCE-TIME AND BUBBLE-SIZE DISTRIBUTIONS

1. *General Formulation*

In this section, a general formulation will be given for the effect of bubble residence-time and bubble-size distributions on simultaneous and thermodynamically coupled heat- and mass-transfer in a multicomponent gas–liquid dispersion consisting of a large number of spherical bubbles. Here one can assume that the bubble population has a normalized size distribution $f(a)$ so that

$$\int_0^\infty f(a)\,da = 1 \qquad (216)$$

where $f(a)\,da$ represents a fraction in the size range $a \pm \tfrac{1}{2}\,da$. Apparently the particular form of $f(a)$ depends on the physical parameters and on the operating conditions of the dispersed system under consideration (see Section II, C).

The independent transport fluxes of simultaneous (coupled) heat- and mass-transfer in an n-component gas–liquid dispersion consisting of two phases ($\alpha = 1, 2$) can be written as (G3)

$$\mathbf{J}_{Di}^\alpha = -\left\{\sum_{j=1}^{n-1} \Psi_{ij}^\alpha \operatorname{grad} c_j^\alpha + \Psi_{iT}^\alpha \operatorname{grad} T^\alpha\right\} \qquad (217)$$

$$i = 1, 2, \ldots, n-1$$

$$\mathbf{J}_H^\alpha = -\left\{\sum_{j=1}^{n-1} \Psi_{Tj}^\alpha \operatorname{grad} c_j^\alpha + \Psi_{TT}^\alpha \operatorname{grad} T^\alpha\right\} \qquad (218)$$

where the superscript refers to each homogeneous phase α and the subscripts D and H refer to diffusional and heat flux, respectively. The nature of Ψ_{ij}^α can be illustrated in the special case of a binary mixture ($n = 2$). Here, Eq. (217) and Eq. (218) take the more familiar form

$$\mathbf{J}_{D1}^\alpha = -\Psi_{11}^\alpha \operatorname{grad} c_1^\alpha - \Psi_{1T}^\alpha \operatorname{grad} T^\alpha \qquad (219)$$

$$\mathbf{J}_H^\alpha = -\Psi_{T1}^\alpha \operatorname{grad} c_1^\alpha - \Psi_{TT}^\alpha \operatorname{grad} T^\alpha \qquad (220)$$

where

$$\Psi_{11} = \frac{L_{11}^\alpha (\partial \mu_1/\partial c_1)_{p,\,T,\,c_2}}{c_2 T^\alpha} \qquad (221)$$

$$\Psi_{1T}^\alpha = L_{1T}^\alpha / (T^\alpha)^2 \qquad (222)$$

$$\Psi_{T1}^\alpha = \frac{L_{T1}^\alpha (\partial \mu_1/\partial c_1)_{p,\,T,\,c_2}^\alpha}{c_2^\alpha T^\alpha} \qquad (223)$$

$$\Psi_{TT}^\alpha = L_{TT}^\alpha / (T^\alpha)^2 \qquad (224)$$

$$c_j^\alpha = \frac{\rho_j^\alpha}{\rho^\alpha} \quad \left(\sum_{j=1}^n c_j^\alpha = 1\right) \qquad (225)$$

with (221) the diffusion coefficient (222) and (223) the coupling coefficients, (224) the heat conductivity, and (225) the mass fraction coefficient; the L_{ij} are the coefficients of the linear phenomenological laws expressed in terms of gradients of the chemical potentials μ_j^α and of the temperature T^α (G3). To proceed, one may now assume that the volumetric heat capacity $\rho^\alpha c_p^\alpha$ and the coefficients Ψ_{ij}^α are constant for the range of the operating conditions prevailing in the system considered.

The conservation principle for mass and energy in the absence of external fields and internal sources or sinks is expressed as

$$Dc_i^\alpha/Dt = -\text{div}(\mathbf{J}_{Di}^\alpha/\rho^\alpha) \tag{226}$$

$$DT^\alpha/Dt = -\text{div}(\mathbf{J}_H^\alpha/\rho^\alpha c_p^\alpha) \tag{227}$$

where

$$\frac{D}{Dt} = \frac{\partial}{\partial t} + \mathbf{v} \cdot \text{grad}$$

is the substantial time derivative.

If one defines

$$\xi^\alpha = (c_1^\alpha, c_2^\alpha, \ldots, c_{n-1}^\alpha, T_n^\alpha)^T \tag{228}$$

as the transposed n-dimensional vector with components $c_i, c_2, \ldots, c_{n-1}, T_n$ and introduces the matrix

$$\Omega^\alpha = \begin{vmatrix} \Psi_{11}^\alpha/\rho^\alpha & \Psi_{12}^\alpha/\rho^\alpha & \cdots & \Psi_{1,n-1}^\alpha/\rho^\alpha & \Psi_{1T}^\alpha/\rho^\alpha \\ \Psi_{21}^\alpha/\rho^\alpha & \Psi_{22}^\alpha/\rho^\alpha & \cdots & \Psi_{2,n-1}^\alpha/\rho^\alpha & \Psi_{2T}^\alpha/\rho^\alpha \\ \vdots & \vdots & \cdots & \vdots & \vdots \\ \Psi_{n-1,1}^\alpha/\rho^\alpha & \Psi_{n-1,2}^\alpha/\rho^\alpha & \cdots & \Psi_{n-1,n-1}^\alpha/\rho^\alpha & \Psi_{n-1,T}^\alpha/\rho^\alpha \\ \Psi_{T1}^\alpha/\rho^\alpha c_p^\alpha & \Psi_{T2}^\alpha/\rho^\alpha c_p^\alpha & \cdots & \Psi_{T,n-1}^\alpha/\rho^\alpha c_p^\alpha & \Psi_{TT}^\alpha/\rho^\alpha c_p^\alpha \end{vmatrix} \tag{229}$$

Eqs (226) and (227) can be put in the compact matrix form

$$D\xi^\alpha/Dt = \Omega \nabla^2 \xi^\alpha \tag{230}$$

In general, Ω is a nonsymmetrical matrix whose components off the main diagonal are the coupling coefficients among the various fluxes involved.

The equations represented by (230) reduce to the familiar equations for the special ideal case of unsteady-state mass transfer without coupling in a binary system if we let

$$\xi^\alpha = c_1^\alpha, \quad n = 2 \tag{231}$$

whereupon Eq (230) becomes (compare with Eq. (71))

$$\frac{\partial c_1^\alpha}{\partial t} + \mathbf{v} \cdot \text{grad } c_1^\alpha = D^\alpha \nabla^2 c_1^\alpha \tag{232}$$

Equation (230) reduces to the familiar equations for the simplified case of heat transfer (not coupled with mass transfer) if we let

$$\bar{\xi}^\alpha = T^\alpha \qquad (233)$$

whereupon the equation becomes

$$\frac{\partial T^\alpha}{\partial t} + \mathbf{v} \cdot \mathrm{grad}\, T^\alpha = \frac{k^\alpha}{\rho^\alpha c_p^{\,\alpha}} \nabla^2 T^\alpha \qquad (234)$$

In most cases, however, heat transfer and mass transfer occur simultaneously, and the coupled equation (230) thus takes into account the most general case of the coupling effects between the various fluxes involved. To solve Eq (230) with the appropriate initial and boundary conditions one can decouple the equation by making the transformation (G3)

$$\bar{\zeta}^\alpha = M^\alpha \bar{\xi}^\alpha \qquad (235)$$

The decoupled equations are then

$$\frac{D \bar{\zeta}^\alpha}{Dt} = \Lambda^\alpha \nabla^2 \bar{\zeta}^\alpha \qquad (236)$$

where

$$\Lambda^\alpha = [M^\alpha]^{-1} \Omega^\alpha M^\alpha \qquad (237)$$

is a diagonal matrix with elements

$$\Lambda^\alpha = \mathrm{diag}(\lambda_1^{\,\alpha}, \lambda_2^{\,\alpha}, \ldots, \lambda_{n-1}^{\,\alpha}, \lambda_n^{\,\alpha}) \qquad (238)$$

In the case of a binary mixture, the eigenvalues of (238) were shown by Padmanabhan and Gal-Or (P1) to be given by

$$\lambda_{1,2}^\alpha = \frac{1}{2}\left\{ \frac{\Psi_{11}^\alpha}{\rho^\alpha} + \frac{\Psi_{TT}^\alpha}{\rho^\alpha c_p^{\,\alpha}} \pm \left(\left[\frac{\Psi_{11}^\alpha}{\rho^\alpha} - \frac{\Psi_{TT}^\alpha}{\rho^\alpha c_p^{\,\alpha}} \right]^2 + \frac{4 \Psi_{1T}^\alpha \Psi_{T1}^\alpha}{(\rho^\alpha)^2 c_p^{\,\alpha}} \right)^{1/2} \right\} \qquad (239)$$

It follows from the second law of thermodynamics, the Onsager relation $L_{ij} = L_{ji}$, the thermodynamic stability condition (G3)

$$\left(\frac{\partial \mu_1}{\partial c_1} \right)_{p, T, c_2} \geqslant 0 \qquad (240)$$

and the previous definitions that the $\lambda_{1,2}^\alpha$ have real values.

To analyse the equations represented by (236), Gal-Or (G3) has introduced an integral transformation with a suitable kernel $K^\alpha(\bar{\tau}^\alpha, t)$ defined by

$$\langle \bar{\xi}^\alpha(\bar{\tau}^\alpha) \rangle = \int_0^\infty K^\alpha(\bar{\tau}^\alpha, t) \bar{\xi}^\alpha(t)\, dt \qquad (241)$$

so that $\langle \bar{\xi}^\alpha(\bar{\tau}^\alpha) \rangle$ is a time average of $\bar{\xi}^\alpha(t)$ over $K^\alpha(\bar{\tau}^\alpha, t)$ The integral operator (241) can be applied to the decoupled Eq (236) to yield ordinary differential equations. These can then be solved with the appropriate initial and boundary conditions to give the average concentrations and the temperature, as well as average fluxes. The advantage of this general approach is that the kernel $K^\alpha(\bar{\tau}^\alpha, t)$ can be chosen to be the residence-time distribution function for the bubbles as well as for the continuous phase, so that $\langle \bar{\xi}^\alpha(\bar{\tau}^\alpha) \rangle$ becomes the expected value of $\bar{\xi}^\alpha(t)$, and $\bar{\tau}^\alpha$ can be interpreted as a parameter of the residence-time distribution in phase α.

For swarms of spherical bubbles, the $\bar{\zeta}^\alpha$ field may be expected to be approximately spherically symmetric when the origin of coordinates is fixed on the center of mass of a typical particle. Therefore, by using spherical coordinates and the initial condition:

$$t = 0, \quad r \neq a, \quad \bar{\zeta}^\alpha = \bar{\zeta}_0^\alpha = M^\alpha \bar{\xi}_0^\alpha \tag{242}$$

the integral operators (241) can be applied to Eqs (236) to yield the ordinary decoupled differential equations

$$\lambda_i^\alpha \nabla^2 \langle \bar{\zeta}_i^\alpha \rangle = (\mathbf{v} \cdot \nabla)\langle \bar{\zeta}_i^\alpha \rangle + \int_0^\infty K^\alpha(\bar{\tau}^\alpha, t) \frac{\partial \bar{\zeta}_i^\alpha}{\partial t} dt \tag{243}$$

independent of the time variable. Using appropriate boundary conditions (which should include the average distance between adjacent particles, equilibrium relationships between interfacial concentrations and temperatures, as well as the equality of mass- and heat-transfer fluxes on both sides of the interface), (243) can be solved in principle to yield $\langle \bar{\zeta}_i^\alpha \rangle$ as a function of the space coordinates.

In a gas–liquid dispersed system, the total interfacial transfer rate $\overline{W}(\xi)$ from a population of bubbles in the size range $a \pm \frac{1}{2} da$ is then given by

$$d\overline{W}(\xi) = 4\pi N a^2 \langle \mathbf{J}(a) \rangle f(a) \, da \tag{244}$$

where

$$\langle \mathbf{J}(a) \rangle = \langle \mathbf{J}^\alpha(a) \rangle = -M^\alpha \Lambda^\alpha \left(\frac{\partial \langle \bar{\zeta}^\alpha \rangle}{\partial r} \right)_{r=a}$$

$$= -\Omega^\alpha \left(\frac{\partial \langle \bar{\xi}^\alpha \rangle}{\partial r} \right)_{r=a} \tag{245}$$

and

$$N = 3\Phi V_v / 4\pi (\bar{a}_v)^3 \tag{246}$$

is the total number of bubbles in the dispersion whose volume is V_v, Φ is the volumetric dispersed-phase holdup fraction, and $\bar{a}_v$ is the mean volume radius of the bubble population defined by [compare with Eq. (18)]

$$\bar{a}_v = \left[\int_0^\infty a^3 f(a) \, da \right]^{1/3} \tag{247}$$

Here, **J** (a, t) is the instantaneous interfacial flux expressed as a state vector whose components are the diffusional and heat fluxes. Therefore, the expected value $\langle \mathbf{J}(a) \rangle$ takes into account the variations in residence time among the entire bubble population.

Summing up the contribution from all size ranges of the population gives the total average heat or mass transfer $\overline{W}(\xi)$ in the dispersed system, i.e.,

$$\overline{W}(\xi) = \frac{3\Phi V_v}{\bar{a}_v^3} \int_0^\infty a^2 f(a) \int_0^\infty K^\alpha(\bar{\tau}^\alpha, t) \mathbf{J}^\alpha(a, t) \, dt \, da$$

$$= -\mathbf{M}^\alpha \Lambda^\alpha \left\{ \frac{3\Phi V_v}{\bar{a}_v^3} \int_0^\infty a^2 f(a) \left(\frac{\partial \langle \bar{\zeta}^\alpha \rangle}{\partial r} \right)_{r=a} da \right\} \quad (248)$$

Equation (248) is a generalization of the binary uncoupled case described by Eq. (179) with (178) and (180). $\overline{W}(\xi)$ is a state vector whose components are the total-average diffusional and heat rates in the disperser.
Hence, the quantity

$$I_{\text{ensemble}}^\alpha = -\Omega^\alpha \frac{3\Phi V_v}{\bar{a}_v^3} \int_0^\infty \int_0^\infty K^\alpha(\bar{\tau}^\alpha, t) a^2 f(a) \{\cdots\} \, dt \, da \quad (249)$$

can be viewed as a general integral operator for the population which yields the total average interfacial transfer rate when operating on local instantaneous gradients at the interface. This general formulation can be extended to cases involving simultaneous heat and mass transfer in reacting systems, simultaneous mass, heat, and electric charge transfer, etc. (T2).

2. Age Distributions

In the case of transfer in a well-mixed dispersed system, the kernel $K^\alpha(\bar{\tau}^\alpha, t)$ in Eq. (249) can be chosen as

$$K^\alpha(\bar{\tau}^\alpha, t) = [\bar{\tau}^\alpha]^{-1} \exp(-t/\bar{\tau}^\alpha) \quad (250)$$

where $\bar{\tau}^\alpha$ is the holding time of either phase. Frequently, $\bar{\tau}^\alpha$ is found to be the same for both phases when there is no obstruction of the flow of either phase in the exit (P1). Hence, for this case the holding time is also the average residence time.

The integral transformation (241) with kernel (250) is seen to be accomplished by taking the Laplace transforms of Eqs. (236) with e^{-st}, where $s = 1/\bar{\tau}^\alpha$, and dividing the transformed quantity by $\bar{\tau}^\alpha$. Hence, the expected value of $\langle \mathbf{J}(a) \rangle$ is simply given by

$$\langle \mathbf{J}(a) \rangle = -\frac{1}{\bar{\tau}^\alpha} \Omega \left(\frac{\partial [\bar{\xi}^\alpha(s)]_{s=1/\bar{\tau}^\alpha}}{\partial r} \right)_{r=a} \quad (251)$$

Consequently, the tedious and usually impossible evaluation of the inverse transform of the transformed partial differential equations (236) becomes an unnecessary procedure, and a significant mathematical simplification results.

For any general case of residence-time distribution in any real system one can choose $K^\alpha(\bar{\tau}^\alpha, t)$ so that

$$K^\alpha(\bar{\tau}^\alpha, t) = [\bar{\tau}^\alpha]^{-1} \exp\left[-\eta^\alpha\left(\frac{t - e^\alpha}{\bar{\tau}^\alpha}\right)\right] \qquad (252)$$

where $\exp[-\eta^\alpha(t - e^\alpha)/\bar{\tau}^\alpha]$ is the general residence-time distribution proposed by Wolf and Resnick (W5) for real systems. In the Wolf and Resnick distribution, η is a measure of the efficiency of mixing and e is a measure of the phase shift in the system. For a well-mixed vessel, $\eta = 1$ and $e = 0$, whereas for plug flow, η tends to infinity. Assuming that $\bar{\tau}^\alpha$, η^α, and e^α are known experimentally for a given dispersed system, one can now reexpress the integral operator I^α_ensemble in terms of a modified Laplace transform divided by a characteristic time parameter of the system $\bar{\tau}_*^\alpha$, so that (G3)

$$K^\alpha(\bar{\tau}^\alpha, \eta^\alpha, e^\alpha, t) = [\bar{\tau}_*^\alpha]^{-1} \exp(-s^\alpha t) \qquad (253)$$

where

$$s^\alpha = \frac{\eta^\alpha}{\bar{\tau}^\alpha} = \frac{1}{\hat{\tau}^\alpha} \qquad (254)$$

Here $\hat{\tau}$ is a modified average residence time defined by

$$\hat{\tau}^\alpha = \frac{\int_0^\infty t \exp(-s^\alpha t)\, dt/\bar{\tau}_*^\alpha}{\int_0^\infty \exp(-s^\alpha t)\, dt/\bar{\tau}_*^\alpha} = \frac{\bar{\tau}^\alpha}{\eta^\alpha} \qquad (255)$$

and a characteristic time parameter of the system:

$$\bar{\tau}_*^\alpha = \bar{\tau}^\alpha \exp[-\eta^\alpha e^\alpha/\bar{\tau}^\alpha] \qquad (256)$$

To obtain the expected values of the fluxes one must normalize the Wolf-Resnick distribution function. Normalization gives

$$e^\alpha = \bar{\tau}^\alpha \frac{\ln \eta^\alpha}{\eta^\alpha} \qquad (256a)$$

Hence use of the normalized distribution function allows a reduction of the number of independent parameters in the kernel (253).

Thus, using the modified kernel, the set of transport equations given by (236) can be solved for any real system, and again the inverse Laplace transformation becomes an unnecessary procedure. This removes a considerable mathematical difficulty to the derivation of analytical solutions for multi-component mass and heat transfer in any real dispersed system.

3. Moment Equations

The moment equations of the size distribution should be used to characterize bubble populations by evaluating such quantities as cumulative number density, cumulative interfacial area, cumulative volume, interrelationships among the various mean sizes of the population, and the effects of size distribution on the various transfer fluxes involved. If one now assumes that the particle-size distribution depends on only one internal coordinate a, the typical size of a population of spherical particles, the analytical solution is considerably simplified. One can define the nth moment μ_n of the particle-size distribution by

$$\mu_n(a, t) = \int a^n f(a, t)\, da, \quad n = 1, 2, \ldots \tag{257}$$

where the integration extends over all possible values of a.

For most purposes, the integration can be extended from 0 to ∞ without introducing any significant error. Therefore, the following quantities can be defined for the dispersed phase:

Cumulative number density, $\% = 100 \int_0^a f(a, t)\, da / \mu_0(a, t)$ (258)

Cumulative interfacial area, $\% = 100 \int_0^a a^2 f(a, t)\, da / \mu_2(a, t)$ (259)

Cumulative volume, $\% = 100 \int_0^a a^3 f(a, t)\, da / \mu_3(a, t)$ (260)

where $f(a, t)$ is a normalized distribution function.

In addition, the following mean sizes can be defined:

Surface mean radius (Sauter mean):

$$\bar{a}_{32} = \frac{\sum_i n_i a_i^3}{\sum_i n_i a_i^2} \equiv \frac{\mu_3(a, t)}{\mu_2(a, t)} \tag{261}$$

Mean volume radius:

$$\bar{a}_v = \left[\frac{\sum_i n_i a_i^3}{\sum_i n_i} \right]^{1/3} \equiv \left[\frac{\mu_3(a, t)}{1} \right]^{1/3} \tag{262}$$

Volume mean radius:

$$\bar{a}_{43} = \frac{\sum_i n_i a_i^4}{\sum_i n_i a_i^3} \equiv \frac{\mu_4(a, t)}{\mu_3(a, t)} \tag{263}$$

Mean surface radius:

$$\bar{a}_2 = \left(\frac{\sum_i n_i a_i^2}{\sum_i n_i}\right)^{1/2} \equiv \left[\frac{\mu_2(a, t)}{1}\right]^{1/2} \tag{264}$$

Interrelationships among these mean sizes can be evaluated when $f(a, t)$ is known (G3).

The dispersed-phase holdup fraction can now be expressed as

$$\Phi = \tfrac{4}{3}\pi N_v \mu_3(a, t) \tag{265}$$

where N_v is the total number of particles per unit volume of the particulate system (i.e., $N_v = N/V_v$ where N is given by Eq. (246)).

4. *The Spherical Cell Model*

The proposed technique will be used here to illustrate the case of interfacial heat and multicomponent mass transfer in a perfectly mixed gas–liquid disperser. Since in this case the holding time is also the average residence time, the gas and liquid phases spend the same time on the average. If $\bar{\tau}^c = \bar{\tau}^d = \bar{\tau}$, then for small values of $\bar{\tau}$, the local residence times t^c and t^d of adjacent elements of the continuous and dispersed phases are nearly of the same order of magnitude, and hence these two elements remain in the disperser for nearly equal times. One may conclude from this that the local relative velocity between them is negligibly small, at least for small average residence times. Gal-Or and Walatka (G9) have recently shown that this is justified especially in dispersions of high Φ values and relatively small bubbles in actual practice where surfactants are present. Under this domain, Eqs. (66), (68), (69) show that as the bubble size decreases, the quantity of surfactants necessary to make a bubble behave like a solid particle becomes smaller. Under these circumstances $(\mu_d + \gamma) \to \infty$ and Eq. (69) reduces to

$$U_{\text{ensemble}} = U_{\text{Stokes}} \left(\frac{3 - \tfrac{9}{2}\Phi^{1/3} + \tfrac{9}{2}\Phi^{5/3} - 3\Phi^2}{3 + 2\Phi^{5/3}}\right) \tag{266}$$

For example, for equal volumes of gas and liquid ($\Phi = 0.5$), Eq. (266) predicts that the Stokes velocity (which is already very small for relatively fine dispersions) should be reduced further by a factor of 38 due to hindering effects of its neighbor bubbles in the ensemble. Hence in the domain of high Φ values and relatively fine dispersions, one can assume that the particles are completely entrained by the continuous-phase eddies, resulting in a negligible convective transfer, although this does not preclude the existence of finite relative velocities between the eddies themselves.

Using the kernel (250) in (243) with the initial condition (242) and integrating by parts, one obtains (P1)

$$\bar{\tau}^\alpha \lambda_i \nabla^2 \langle \bar{\zeta}^\alpha \rangle = \langle \bar{\zeta}^\alpha \rangle - \bar{\zeta}_0^\alpha \tag{267}$$

where

$$\langle \bar{\zeta}^\alpha \rangle = \mathbf{M}^\alpha \langle \bar{\xi}^\alpha \rangle \tag{268}$$

and

$$\bar{\zeta}_0^\alpha = \mathbf{M}^\alpha \bar{\xi}_0^\alpha \tag{269}$$

In spherical coordinates (267) becomes

$$\frac{d^2}{dr^2}(r\langle \bar{\zeta}^\alpha \rangle) - [\boldsymbol{\mu}^\alpha]^2 (r\langle \bar{\zeta}^\alpha \rangle) = -[\boldsymbol{\mu}^\alpha]^2 r \bar{\zeta}_0^\alpha \tag{270}$$

where

$$\boldsymbol{\mu} = \text{diag}(\mu_1^\alpha, \mu_2^\alpha, \ldots, \mu_n^\alpha) \tag{271}$$

and

$$\mu_i = (\bar{\tau}^\alpha \lambda_i^\alpha)^{-1/2} \tag{272}$$

In the absence of convective effect, the profiles of $\langle \bar{\xi}^\alpha \rangle$ between any two adjacent bubbles exhibits an extremum value midway between the bubbles. Therefore, there exists around each bubble a surface on which $\partial \bar{\xi}^\alpha / \partial r = \partial \langle \bar{\xi}^\alpha \rangle / \partial r = 0$, and hence the fluxes are zero. Using the cell model [Eqs. (158) or (172)] one obtains the following boundary conditions: For $t \geq 0$

$$r = 0, \quad \bar{\zeta}^D \text{ is finite} \tag{273}$$

$$r = b = a\Phi^{-1/3}, \quad \partial \bar{\zeta}^c / \partial r = 0 \tag{274}$$

$$r = a, \quad \bar{\zeta}^c = \mathbf{U} \bar{\zeta}^D \tag{275}$$

$$r = a, \quad \Omega^c \, \partial \langle \bar{\xi}^c \rangle / \partial r = \Omega^D \, \partial \langle \bar{\xi}^D \rangle / \partial r \tag{276}$$

Condition (273) is the requirement that at the center of the bubble the concentrations and the temperature must be finite, and condition (274) follows from the condition that the net average flux is zero on the surface $r = b$ which encloses each bubble. Condition (275) refers to the interfacial concentrations and the temperature on both phases, which are related through known equilibrium partition coefficients m_i. Hence

$$\bar{\xi}^c = \mathbf{m} \bar{\xi}^D \tag{277}$$

where

$$\mathbf{m} = \text{diag}(m_1, m_2, \ldots, m_{n-1}, 1) \tag{278}$$

Combining (277) with (268) yields

$$\mathbf{U} = [\mathbf{M}^c] \mathbf{m} [\mathbf{M}^D]^{-1} \tag{279}$$

which completes the definition of boundary condition (275).

Finally, the boundary condition (276) refers to the continuity of the flux $\mathbf{J}^\alpha = -\Omega^\alpha(\partial \bar{\xi}^\alpha/\partial r)$ at the interface.

The ordinary differential equation (270) can now be solved by employing the boundary conditions (273)–(276) to generate the expected value of $\bar{\xi}^\alpha(r)$ (P1):

$$\langle \bar{\xi}^D(r) \rangle = \bar{\xi}_0^D + (a/r)[\mathbf{M}^D]^{-1}\mathbf{T}^D(r)\{[\mathbf{Q}^c]^{-1}\mathbf{Q}^D \\ - [\mathbf{T}^c]^{-1}\mathbf{U}\mathbf{T}^D\}^{-1}[\mathbf{T}^c]^{-1}\mathbf{M}^c(m\bar{\xi}_0^D - \bar{\xi}_0^c) \quad (280)$$

$$\langle \bar{\xi}^c(r) \rangle = \bar{\xi}_0^c + (a/r)[\mathbf{M}^c]^{-1}\mathbf{T}^c(r)\{\mathbf{I}_n \\ - [\mathbf{T}^c]^{-1}\mathbf{U}\mathbf{T}^D[\mathbf{Q}^D]^{-1}\mathbf{Q}^c\}^{-1}[\mathbf{T}^c]^{-1}\mathbf{M}^c(m\bar{\xi}_0^D - \bar{\xi}_0^c) \quad (281)$$

where:

$$\mathbf{T}^\alpha(r) = \mathrm{diag}(T_1^\alpha(r), T_2^\alpha(r), \ldots, T_n^\alpha(r)) \quad (282)$$

$$T_i^c(r) = \mu_i^c b \cosh\{\mu_i^c(r-b)\} + \sinh\{\mu_i^c(r-b)\} \quad (283)$$

$$T_i^D(r) = \sinh(\mu_i^D r) \quad (284)$$

$$\mathbf{R}^\alpha = \mathrm{diag}(R_1^\alpha, R_2^\alpha, \ldots, R_n^\alpha) \quad (285)$$

$$R_i^C = (\mu_i^C ab - 1)\sinh\{\mu_i^C(a-b)\} + \mu_i^C(a-b)\cosh\{\mu_i^C(a-b)\} \quad (286)$$

$$R_i^D = \mu_i^D a \cosh(\mu_i^D a) - \sinh(\mu_i^D a) \quad (287)$$

$$\mathbf{Q}^\alpha = [\mathbf{M}^\alpha]^{-1} \Lambda^\alpha \mathbf{R}^\alpha \quad (288)$$

and where $\mathbf{I}_n$ is the n-dimensional identity matrix.

The expected value of the interfacial flux is

$$\langle \mathbf{J}(a) \rangle = -\Omega \left. \frac{\partial \langle \bar{\xi}^\alpha \rangle}{\partial r} \right|_{r=a} \quad (289)$$

where $\partial \langle \bar{\xi}^\alpha \rangle / \partial r$ is found from (280) or (281). Hence, using (289) one obtains

$$\langle \mathbf{J}(a) \rangle = -\frac{1}{a}\{\mathbf{Y}^C(a) - m\mathbf{Y}^D(a)\}^{-1}(m\bar{\xi}_0^D - \bar{\xi}_0^C) \quad (290)$$

where

$$\mathbf{Y}^\alpha(a) = [\mathbf{M}^\alpha]^{-1}\{\mathbf{T}^\alpha(a)[\mathbf{R}^\alpha(a)]^{-1}\Lambda^\alpha\}[\mathbf{M}^\alpha] \quad (291)$$

The total interfacial transfer rate in the whole dispersed system, $\overline{W}(\xi)$, is found by summing the interfacial transfer rate over the entire population of the bubbles, taking into account the variation in their sizes. Thus, one obtains

$$\overline{W}(\xi) = \frac{3\Phi V_v}{\bar{a}_v{}^3} \int_0^\infty a^2 f(a) \langle \mathbf{J}(a) \rangle \, da \qquad (292)$$

where V_v is the total volume of the dispersed system.

Substituting the expression for the expected interfacial flux (290) in (292), one finally obtains the required general solution:

$$\overline{W}(\xi) = \frac{3\Phi V_v}{\bar{a}_v{}^3} \int_0^\infty \{\mathbf{Y}^C(a) - \mathbf{m}\mathbf{Y}^D(a)\}^{-1}(\bar{\xi}_0{}^C - \mathbf{m}\bar{\xi}_0{}^D) f(a) \, da \qquad (293)$$

This result can be useful for design purposes when the diffusivities, partition coefficients, feed-stream conditions, dispersed-system volume, gas-phase holdup (or average residence time), and the size distribution are known. When the size distribution is not known, but the Sauter-mean radius of the population is known, (293) can be approximated by

$$\overline{W}(\xi) \approx \frac{3\Phi V_v}{a_{32}} \{\mathbf{Y}^C(a_{32}) - \mathbf{m}\mathbf{Y}(a_{32})\}^{-1}(\bar{\xi}_0{}^C - \mathbf{m}\bar{\xi}_0{}^D) \qquad (294)$$

where $\overline{W}(\xi)$ represents the n-dimensional column vector of the total interfacial transfer rate of each component ($i = 1, 2, \ldots, n - 1$) and the heat flow.

The above equations can be used to predict the output $\bar{\xi}_E{}^\alpha$ in the effluent from a continuous-flow gas–liquid contractor. Let the flow rate of phase α be F^α. Then, a macroscopic balance at a steady state leads to the equation:

$$\overline{W}(\xi) = F^\alpha \mathbf{S}^\alpha (\bar{\xi}_E{}^\alpha - \bar{\xi}_0{}^\alpha) \qquad (295)$$

where

$$\mathbf{S}^\alpha = \text{diag}(1, 1, \ldots, 1, \rho^\alpha c_p{}^\alpha) \qquad (296)$$

is a partitioned matrix containing the identity matrix $\mathbf{I}_{n-1}$ of dimension $(n - 1)$ and the volumetric heat capacity $\rho^\alpha c_p{}^\alpha$ on the diagonal.

Equations (293) and (295) can be combined to give (P1)

$$\bar{\xi}_E{}^\alpha = \bar{\xi}_0{}^\alpha + \frac{3\Phi V_v}{F^\alpha \bar{a}_v{}^3} [\mathbf{S}^\alpha]^{-1} \int_0^\infty \{\mathbf{Y}^C(a) - \mathbf{m}\mathbf{Y}^D(a)\}^{-1}(\bar{\xi}_0{}^C - \mathbf{m}\bar{\xi}_0{}^D) a f(a) \, da$$

$$(297)$$

Making use of the fact that in perfect mixing

$$\bar{\tau} = \Phi V_v / F^D = (1 - \Phi) V_v / F^C \tag{298}$$

the output in each phase can be obtained from (297):

$$\bar{\xi}_E^C = \left\{ I_n + \left(\frac{\Phi}{1-\Phi}\right)[S^C]^{-1}P\right\}\bar{\xi}_0^C - \left\{\frac{\Phi}{1-\Phi}[S^C]^{-1}Pm\right\}\bar{\xi}_0^D \tag{299}$$

$$\bar{\xi}_E^D = \{I_n - [S^D]^{-1}Pm\}\bar{\xi}_0^D + \{[S^D]^{-1}P\}\bar{\xi}_0^C \tag{300}$$

where

$$P = \frac{3\bar{\tau}}{\bar{a}_v^3} \int_0^\infty \{Y^C(a) - mY^D(a)\}^{-1} af(a)\, da \tag{301}$$

The integral P can rarely be evaluated analytically, and the use of a digital computer is required.

Equations (299) and (300) depict the input-output relationships for the concentrations and the temperature in each phase for a given continuous steady-flow dispersed system. Therefore, (299) and (300) can be used in predicting the input-output relationships for a multistage multicomponent gas-liquid system with several continuous stirred vessels in series.

Although the above treatment is based on the assumption of perfect mixing, it was found experimentally that deviations from this ideal situation can be taken into account by introducing the additional parameters of Wolf and Resnick (W5) into the kernel (253).

To evaluate the effects of operating variables on the capacity of the gas-liquid disperser one can make the following substitutions:

(a) substitution of Eq. (17) in the general equations for $W(\xi)$ or $\bar{\xi}_E^\alpha$.

(b) substitution of relationships between the average size and operating variables [such as (16)] into Eq. (17) and the appropriate equation for $W(\xi)$ or $\bar{\xi}_E^\alpha$.

M. Additional Mathematical Studies

Some of the additional mathematical models mentioned below were derived for the characterization of swarms of bubbles whereas others were derived for the specific case of a single bubble or the general case of two-phase contact. Most models for the bubble-liquid contact are limited to the case of a single bubble, and consequently their direct applicability to gas-liquid dispersions is very restricted.

General equations of momentum and energy balance for dispersed two-phase flow were derived by Van Deemter and Van Der Laan (V2) by integration over a volume containing a large number of elements of the dispersed phase. A complete system of solutions of linearized Navier-Stokes equations and its application to boundary problems of flow around a sphere have been presented by Kaufman (K3). Mathematical solutions were developed by Miyauchi and Vermeulen (M8, M9) for general and specific cases of logitudinal dispersion accompanying mass or heat transfer.

Analysis of the effect of particle-size and residence-time distributions on mass (or heat) transfer with an irreversible first-order source in a two-component dispersed system has been reported by Padmanabhan and Gal-Or (P1). Two differential equations (one for each phase) were solved simultanelusly to evaluate the total interfacial transfer rates from the bubble populations. The effect of reversible first-order chemical reaction on multi-component mass transfer in dispersions with residence-time and size distributions has recently been evaluated by Tavlarides and Gal-Or (T2). Jameson and Davidson (J1, J2) have developed a theory to describe the motion of a small single bubble in a vertically oscillating liquid. Their theoretical predictions are in reasonable agreement with experimental data. A theoretical analysis of breakage and coalescence in a dispersed-phase system has recently been reported by Valentas *et al.* (V1). It relates the breakage and coalescence of droplets in a dispersed system to the steady-state size distribution as a function of some operating variables. Haas and Johnson (H1) have developed a model for studying the laminar flow between foam bubbles which shows good agreement with experimental results.

Marchello and Toor (M2) proposed a mixing model for transfer near a boundary which assumes that localized mixing occurs rather than gross displacement of the fluid elements. This model can be said to be a modified penetration-type model. Kishinevsky (K6–K8) assumed a surface-renewal mechanism with eddy diffusion rather than molecular diffusion controlling the transfer at the interface.

Van De Vusse (V4) described the kinetics of simultaneous mass transfer and chemical reaction in gas–liquid systems and derived expressions for the overall reaction rate for the stationary and unsteady-state conditions. He used both the film and the penetration theories and also considered the case of a zero-order reaction. Numerical solutions for gas absorption accompanied by an irreversible chemical reaction of general order according to the film and penetration theories were given by Brian *et al.* (B14–B18) and Hikita and Asai (H14). Lightfoot (L7) presented a method for simplifying the solution of unsteady diffusion problems involving an irreversible first-order reaction. His method permits writing the concentration profiles for reactive system in terms of those for nonreactive systems that have the same geometry and similar boundary conditions.

Beek and Kramers (B6) studied the case of interphase mass transfer where the interfacial area expands or contracts as the transfer proceeds. As examples where the interfacial area may change, they mention mass transfer between a growing bubble and its surroundings and the absorption of gas in liquid in the form of expanding or contracting jets or films.

In recent years an increasing amount of basic and theoretical attention has been given to mass transfer between phases, especially to cases involving drops and bubbles. Much of the recent work has been reviewed (B10, B19, D7, H5, H6, H21, K4, S8, S10, S12) and a list of pertinent references is given in footnote 1. Many of these researches attempt to elucidate the phenomena and micromechanisms affecting transfer to or from a single bubble or drop. It is clear that the more complicated system of agitated gas–liquid dispersions in which swarms of bubbles and complicated liquid-flow patterns are involved will not be completely characterized until a complete understanding has been reached in a form amenable to mathematical treatment for the effects of surface-active agents, holdup, coalescence, breakup etc.

Nomenclature

Any consistent set of units may be used. For the few dimensional equations, the appropriate units are given in the text.

a	Bubble radius	C	Constant in Eq. (28)
a_{32}	Surface mean radius [Eq. (22)]	d_{32}	Surface mean diameter [Eq. (12)]
a_{cr}	Radius at which breakup begins [Eq. (26)]	D	Diffusivity
		D_s^s	Surface diffusion coefficient of surfactants
a_v	Mean volume radius [Eq. (18)]	D_T	Vessel diameter
Δa	Range of bubble-size variation [Eq. (27)]	e	Surface viscosity coefficient [Eq. (46)]
A	Total bubble surface area in the dispersion	$f^*(a, \bar{a}_v)$	Bubble-size distribution [Eq. (17)]
A_T	Vessel cross-sectional area	$\mathscr{F}$	Defined in Eq. (208)
b	Radius of spherical shell [Eq. (158)]	g	Local acceleration of gravity
		G	Elastic modulus
$B_1, B_2,$ B_3, B_4	Constants	G_v	Dimensionless geometrical factor $= L^2 W / D_T^2 H$
c	Concentration	H	Liquid height without gas
c_0	Average gas concentration in fresh liquid	Hu	Volume of gas holdup in dispersion
c_i	Equilibrium concentration at interface	J	Ratio of rate of absorption to rate of physical absorption to tank containing no dissolved gas
c_L	Dissolved gas concentration at distance L	J_n	Normal flux of surfactants into the interface
c_1	c/c_i [Eq. (121)]		
$\bar{C}_A(x, p)$	The Laplace Transform of the concentration of reactant A	k	First-order reaction-rate constant

[1] References pertaining to theoretical approaches to transfer between phases: A1, B2, B5, B8, B13, C3, C6, C7, C10, D1–D5, F3, F4, G14, H5, H6, H17, J3, M1, O1, R,1 R6, R10, S7, T6, V3, V4, V6.

k_1', k'_{-1}	Forward and backward reaction-rate constants, respectively [Eq. (109)]	r	Radius in spherical coordinates
		R	Gas constant
k'	Defined in Eq. (149)	R^*	Molar rate of consumption per unit volume by chemical reaction
k''	Second-order reaction-rate constant	s	Specific bubble-surface area
K	Overall mass-transfer coefficient per unit area	S	Distance from the leading edge of the bubble along its surface
K_p	Drag coefficient	S_1	Defined in Eq. (121)
k_L	Mass-transfer coefficient, liquid film	t	Time
		t'	Defined in Eq. (121)
Ks	Volumetric mass-transfer coefficient	t_0	Time required for the liquid element to move completely around the bubble [Eq. (83)]
l	Characteristic length [Eq. (121)]	T	Absolute temperature
L	Impeller diameter; also characteristic distance from the interface where the concentration remains constant at c_L	U	Magnitude of bubble velocity relative to the liquid
		U_s	Velocity given by Stokes equation
L_i	Impeller blade length	$\mathbf{v}$	Velocity vector
N	Impeller rotational speed; also number of bubbles [Eq. (246)].	v_L	Defined in Eq. (157)
		v_t	Tangential liquid velocity at the surface
N^*	Ratio of absorption rate in presence of chemical reaction to rate of physical absorption when tank contains no dissolved gas	V	Liquid volume in the vessel
		Vs	Superficial gas velocity based on empty cross-sectional area
N_A	Instantaneous mass-transfer rate per unit bubble-surface area	V_v	Total volume of the dispersion in the disperser
N_A^*	Local rate of mass-transfer per unit bubble-surface area	W	Impeller blade width
		x	Parallel vector distance to the bubble interface
$\bar{N}_A$	Average mass-transfer rate per unit bubble-surface area	y	Normal distance from the bubble interface
N_b	Number of bubbles in the vessel at any instant at constant operating conditions	y_1	Defined in Eq. (121)
		Y	Dispersed liquid depth
N_v	Number of bubbles per unit volume of dispersion [Eq. (24)]	**GREEK LETTERS**	
N_v^*	Defined in Eq. (134)	α	Defined in Eq. (21)
N_T	Total instantaneous mass-transfer rate	α^*	Defined in Eq. (102)
		α_1	Defined in Eq. (121)
$\bar{N}_T$	Total average mass-transfer rate in the disperser	$\beta_0 = k\bar{\theta}$	
		γ	Retardation coefficient of surfactants
Q	Volumetric flow rate of dispersed phase	Γ	Surface concentration of surfactants [Eq. (56)]; gamma function
Q^*	Flux of absorbed surfactants		
p	Rate of surface renewal by eddies		
P	Power input to impeller	δ	Thickness of hypothetical stagnant liquid film
P^*	Flux of desorbed surfactants		
P_0	Power input to impeller, no-gas condition	ε_0	Intensity of turbulence or power per unit mass
P_v	Power input to impeller per volume of gas-free liquid	$\bar{\eta}$	Defined in Eq. (177)

θ	Cone angle in spherical coordinates	SUPERSCRIPTS	
$\bar{\theta}_d$	Average residence-time per unit height of dispersion	α	Refers to the phase (c or d).
		c	Refers to the continuous phase
$\bar{\theta}_h$	Average residence-time per unit height of liquid on gas-free basis	d	Refers to the dispersed phase
		SUBSCRIPTS	
λ_n	Roots of Eq. (175)	θ	Refers to the θ component in spherical coordinates
$\Lambda = \Phi^{1/3}$			
μ	Viscosity	i	Refers to component i
ν	Kinematic viscosity	j	Refers to subreactor j; also to component j
ξ	Defined in Eq. (167)		
$\bar{\xi}$	Defined in Eq. (177)	r	Refers to the r component in spherical coordinates
ρ	Density		
σ	Surface tension	DIMENSIONLESS NUMBERS	
$\bar{\tau}$	Average residence time		
τ_0	Defined in Eq. (86)	N_{As}	Astarita number, $\rho_c ga/G$, Eq. (45)
Υ	Defined in Eq. (177)		
ϕ	Polar angle in spherical coordinates	N_{Le}	Levich number, $\mu_c U/\rho_c ga^2$
		N_{Fr}	Froude number, LN^2/g
Φ	Fractional gas holdup	N_{Pe}	Peclet number, $N_{Re} N_{Sc}$
ψ	Defined in Eq. (82)	N_{Re}	Reynolds number, $L^2 NS/\mu$ or aUS/μ
ψ_0	Defined in Eq. (87)		
Ψ'	Defined in Eq. (201)	N_{Sc}	Schmidt number, $\mu/\rho D$
Ψ^*	Defined in Eq. (207)	N_{Sh}	Sherwood number, KL/D
Ω	Defined in Eq. (177)	N_{We}	Weber number, $L^3 N^2 \rho/\sigma$

References

A1. Aiba, S., *A.I.Ch.E. J.* **4**, 485 (1958); *Chem. Eng. (Tokyo)* **15**, 354 (1951); **20**, 290, 298, 571 (1956); *Preprints, 15th Anniv. Congr., Soc. Chem. Engrs. Japan, November 1961*, p. 291.
A2. Anderson, T. T., *A.I.Ch.E. J.* **10**, 776 (1964).
A3. Astarita, G., *Symp. Mechanics Viscoelastic Fluids*, 58th Ann. Meeting, A.I.Ch.E. Philadelphia, December 1965.
A4. Astarita, G., and Apuzzo, G., *A.I.Ch.E. J.* **11**, 815 (1965).
A5. Astarita, G., and Marrucci, G., *Ind. Eng. Chem. Fundamentals* **2**, 4 (1963).
B1. Backstrom, H. S., *J. Am. Chem. Soc.* **49**, 1460 (1929); *Trans. Faraday Soc.* **24**, 601, 706 (1928).
B2. Baird, M. H. J., and Davidson, J. F., *Chem. Eng. Sci.* **17**, 87 (1962).
B3. Bankoff, S. G., *A.I.Ch.E. J.* **10**, 776 (1964).
B4. Bayens, C., Ph.D. Thesis, The Johns Hopkins Univ., Baltimore, Maryland, 1967.
B5. Beek, W. J., and Bakker, C. A. P., *Appl. Sic. Res.* **10A**, 241 (1961).
B6. Beek, W. J., and Kramers, H., *Chem. Eng. Sci.* **16**, 909 (1962).
B7. Bentwich, M., Private communication, 1964.
B8. Bentwich, M., Szwarcbaum, G., and Sideman, S. Report to the Israel Nat. Council for Res. and Develop., No. 3 (August 1963).
B9. Bird, R. B., Stewart, W. E., and Lightfoot, E. N., "Transport Phenomea," Sect. 17.4-5. Wiley, New York, 1960.
B10. Bischoff, K. B., and Himmelblau, D. M., *Ind. Eng. Chem.* **57**, 54 (1965).
B11. Blakebrough, N., Hamer, G., and Walker, A. W., *The Chem. Eng.* A71 (1961).

B12. Boussinesq, J., *Compt. Rend.* **156**, 1124 (1913).
B13. Bowman, C. W., Ward, D. M., Johnson, A. I., and Trass, O., *Can. J. Chem. Eng.* **39**, 9 (1961).
B14. Brian, P. L. T., *A.I.Ch.E. J.* **10**, 5 (1964).
B15. Brian, P. L. T., and Beaverstock, M. C., *Chem. Eng. Sci.* **20**, 47 (1965).
B16. Brian, P. L. T., and Bodman, S., *Ind. Eng. Chem. Fundamentals* **3**, 339 (1964).
B17. Brian, P. L. T., Baddour, R. F., and Matiatos, D. C., *A.I.Ch.E. J.* **10**, 727 (1964).
B18. Brian, P. L. T., Hurley, J. F., and Hasseltine, E. H., *A.I.Ch.E. J.* **7**, 226 (1961).
B19. Brodkey, R. S., "The Phenomena of Fluid Motions," Addison-Wesley, Reading, Massachusetts, 1967.
C1. Calderbank, P. H., *Trans. Inst. Chem. Engrs.* (*London*), **36**, 443 (1958).
C2. Calderbank, P. H., *Trans. Inst. Chem. Engrs.* (*London*) **37**, 173 (1959).
C3. Calderbank, P. H., and Korchinski, I. J. O., *Chem. Eng. Sci.* **6**, 65 (1956).
C4. Calderbank, P. H., and Moo-Young, M. B., *Chem. Eng. Sci.* **16**, 39 (1961).
C5. Callow, D. S., Gillet, W. A., and Pirt, S. J., *Chem. Ind.* (*London*) 418 (1957).
C6. Carslaw, H. S., and Jaeger, J. C., "Conduction of Heat in Solids," Oxford Univ. Press, London and New York, 1959.
C7. Chiang, S. H., and Roor, H. L., *A.I.Ch.E. J.* **5**, 165 (1959).
C8. Clark, M. W., and Vermeulen, T., Power requirements for mixing of liquid-gas systems, LRL Rept. UCRL-10996, Berkeley, California, August 1963.
C9. Cooper, C. M., Fernstrom, G. A., and Miller, S. A., *Ind. Eng. Chem.* **36**, 504 (1944).
C10. Crank, J., "The Mathematics of Diffusion," Oxford Univ. Press, London and New York, 1956.
D1. Danckwerts, P. V., *Trans. Faraday Soc.* **46**, 300 (1950).
D2. Danckwerts, P. V., *Trans. Faraday Soc.* **46**, 701 (1950).
D3. Danckwerts, P. V., *Trans. Faraday Soc.* **47**, 1014 (1951).
D4. Danckwerts, P. V., *Ind. Eng. Chem.* **43**, 1460 (1951).
D5. Danckwerts, P. V., *Ind. Chemist* **30**, 102 (1954).
D6. Danckwerts, P. V., Kennedy, A. M., and Roberts, D., *Chem. Eng. Sci.* **18**, 63 (1963).
D7. Davies, J. T., Mass transfer and interfacial phenomena, *Advan. Chem. Eng.* **4**, 3 (1963).
D8. Davies, R., and Taylor, G. I., *Proc. Roy. Soc.* (*London*) **A200**, 375 (1950).
D9. Davies, J. T., Kilner, A. A., and Tatcliff, G. A., *Chem. Eng. Sci.* **19**, 583 (1964); Davies J.T., and Mayers, G. R. A., *Ibid.* **16**, 55 (1961).
D10. Deindoerfer, F. H., and Humphrey, A., *Ind. Eng. Chem.* **53**, 755 (1961).
D11. Dobbins, W. E., "Biological Treatment of Sewage and Industrial Wastes," Vol. I, p. 141. Reinhold, New York, 1955.
D12. Dukler, A. E., and Wicks, M., "Modern Chemical Engineering," Vol. 1. Reinhold, New York, 1963.
E1. Elsworth, B., Williams, V., and Harris-Smith, B., *J. Appl. Chem.* **7**, 261 (1957).
E2. Emmery, R. E. and Pigford, R. L., *A.I.Ch.E. J.* **8**, 171 (1962).
F1. Fair, J. R., Lambright, A. J., and Andersen, J. W., *Ind. Eng. Chem. Process Design Develop.* **1**, 33 (1962).
F2. Foust, H. C., Mack, D. E., and Rushton, J. H., *Ind. Eng. Chem.* **36**, 517 (1944).
F3. Friedland, W. C., Peterson, M. H., and Sylvester, J. C., *Ind. Eng. Chem.* **48**, 2180 (1956).
F4. Friedlander, S. K., *A.I.Ch.E. J.* **3**, 43 (1957).
F5. Friedlander, H. N., and Resnick, W., *Advan. Petrol. Chem. Refining* **1**, 527–570 (1958).
F6. Friedman, A. M., and Lightfoot, E. N., *Ind. Eng. Chem.* **49**, 1227 (1957).

F7. Fuller, E. C., and Crist, R. H., *J. Am. Chem. Soc.* **63**, 1644 (1941).
G1. Gaden, E. L., *Sci. Rept. Ist. Super. Sanita* **1**, 161 (1961).
G2. Gal-Or, B., D.Sc. thesis, Technion–Israel Inst. Technol., Haifa, Israel, 1964.
G3. Gal-Or, B., *Intern. J. Heat Mass Transfer*. In press.
G4. Gal-Or, B., *A.I.Ch.E. J.* **12**, 604 (1966).
G5. Gal-Or, B., and Hoelscher, H. E., *A.I.Ch.E. J.* **12**, 499 (1966).
G6. Gal-Or, B., and Resnick, W., *Chem. Eng. Sci.* **19**, 653 (1964).
G7. Gal-Or, B., and Resnick, W., *A.I.Ch.E. J.* **11**, 740 (1965).
G8. Gal-Or, B., and Resnick, W., *Ind. Eng. Chem. Process Design Develop.* **5**, 15 (1966).
G9. Gal-Or, B., and Walatka, V. V., *A.I.Ch.E. J.* **13**, 650 (1967).
G10. Gal-Or, B., J. P. Hauck, and H. E. Hoelscher, *Intern. J. Heat Mass Transfer*, **10**, 1559 (1967).
G11. Garner, F. H., *Trans. Inst. Chem. Eng. (London)* **28**, 88 (1950).
G12. Garner, F. H., and Hale, A. B., *Chem. Eng. Sci.* **4**, 157 (1955).
G13. Garner, F. H., and Lane, T. T., *Trans. Inst. Chem. Eng. (London)* **37**, 162 (1959).
G14. Garner, F. H., Foord, A., and Tayeban, M., *J. Appl. Chem.* **2**, 315 (1959).
G15. Gill, W. N., *Chem. Eng.* **69**, 198 (1962).
G16. Gill, W. N., and Nunge, R. J., *Chem. Eng. Sci.* **17**, 683 (1962).
G17. Griffith, R. M., *Chem. Eng. Sci.* **12**, 198 (1960).
H1. Haas, P. A., and H. F. Johnson, *Ind. Eng. Chem. Fundamentals* **6**, 225 (1967).
H2. Hadamard, J. S., *Compt. Rend.* **152**, 1735 (1911).
H3. Hamer, G., and Blakebrough, N. *J. Appl. Chem.* **13**, 517 (1963).
H4. Hamielec, A. E., McMaster Univ., Hamilton, Ontario, Canada, Personal communication, 1965.
H5. Hamielec, A. E., Hoffman, T. W., and Ross, L. L., *58th A.I.Ch.E. Meeting, Philadelphia, December 1965*; Hamielec, A. E., Johnson, A. I., and Houghton, W. T. *ibid*.
H6. Hamielec, A. E., Storey, S. H., and Whitehead, J. M., *Can. J. Chem. Eng.* **41**, 246 (1963).
H7. Hammerton, D., and Garner, R. H., *Trans. Inst. Chem. Eng. (London)* **32**, 518 (1954).
H8. Hanhart, J., Kramers, H., and Westerterp, K. R., *Chem. Eng. Sci.* **18**, 503 (1963).
H9. Hanratty, T. J., *Am. Inst. Chem. Eng.* **2**, 359 (1956).
H10. Happel, J., *A.I.Ch.E. J.* **4**, 197 (1958); **5**, 174 (1959).
H11. Harriott, P., *Can. J. Chem. Eng.* **40**, 60 (1962).
H12. Heuss, J. M., King, C. T., and Wilke, C. R., *A.I.Ch.E. J.* **11**, 866 (1965).
H13. Higbie, R., *Trans. Am. Inst. Chem. Eng.* **31**, 365 (1935).
H14. Hikita, H., and Asai, S., *Intern. Chem. Eng.* **4**, 332 (1964).
H15. Hinze, J. O., "Turbulence," McGraw-Hill, New York, 1953.
H16. Hinze, J. O., *A.I.Ch.E. J.* **1**, 283 (1955).
H17. Howarth, W. J., *Chem. Eng. Sci.* **18**, 47 (1963); **19**, 33 (1964).
H18. Huang, C. J., and Kuo, C. H., *A.I.Ch.E. J.* **9**, 161 (1963).
H19. Huang, C. J., and Kuo, C. H., *A.I.Ch.E. J.* **11**, 901 (1965).
H20. Hulburt, H. M., and Katz, S., *Chem. Eng. Sci.* **19**, 555 (1964).
H21. Hyman, D., *Advan. Chem. Eng.* **2**, 120 (1962).
H22. Hyman, D., and Van Der Bogaerde, J. M., *Ind. Eng. Chem.* **52**, 751 (1960).
J1. Jameson, G. J., *Chem. Eng. Sci.* **21**, 35 (1966).
J2. Jameson, G. J., and J. F. Davidson, *Chem. Eng. Sci.* **21**, 29 (1966).
J3. Johnson, A. I., and Hamielec, A. E., *A.I.Ch.E. J.* **6**, 145 (1960).
J4. Johnson, A. I., Hamielec, A. E., and Houghton, W. T., *58th A.I.Ch.E. Meeting, Philadelphia, December 1965*. See also *A.I.Ch.E.J.* **13**, 379 (1967).

J5. Johnson, D. L., Saito, H., Polejes, J. D., and Hougen, O. A., *A.I.Ch.E. J.* **3**, 411 (1957).
J6. Jorrisen, W. P., *Z. Phys. Chem.* **22**, 54 (1897).
K1. Karow, E. O., Bartholomew, W. H., and Stat, M. R., *J. Agr. Food, Chem.* 302 (1953).
K2. Karwat, H., *Chem. Ingr.-Tech.* **31**, 588 (1959).
K3. Kaufman, R. N., *Intern. Chem. Eng.* **5**, 8 (1965).
K4. Kintner, R. C., *Advan. Chem. Eng.* **4**, 52 (1963).
K5. Kintner, R. C., Horton, T. J., Graumann, R. E., and Amberkar, S., *Can. J. Chem. Eng.* **39**, 235 (1961).
K6. Kishinesvky, M. Kh., *J. Appl. Chem. U.S.S.R. (English transl.)* **24**, 593 (1951).
K7. Kishinevsky, M. Kh., *J. Appl. Chem. U.S.S.R. (English transl.)* **27**, 415 (1954).
K8. Kishinevsky, M. Kh., *J. Appl. Chem. U.S.S.R. (English transl.)* **28**, 881 (1955).
K9. Kolmogoroff, A. N., *Compt. Rend. U.R.S.S.* **30**, 301 (1941); **31**, 538 (1941); **32**, 16 (1941); *Dokl. Akad. Nauk S.S.S.R.* **66**, 825 (1949).
L1. Lamb, H., "Hydrodynamics." Dover, New York, 1945.
L2. Levenspiel, O., and Bischoff, K. B., *Advan. Chem. Eng.* **4**, 95 (1963).
L3. Levich, V. G., "Physiochemical Hydrodynamics." Prentice-Hall, Englewood Cliffs, New Jersey, 1962.
L4. Levich, V. G., *Intern. Chem. Eng.* **2**, 78 (1962).
L5. Lightfoot, E. N., *A.I.Ch.E. J.* **4**, 499 (1958).
L6. Lightfoot, E. N., *A.I.Ch.E. J.* **8**, (1962).
L7. Lightfoot, E. N., *A.I.Ch.E. J.* **10**, 278 (1964).
L8. Lindblad, N. R., *J. Colloid Sci.* **19**, 729 (1964).
L9. Lindland, K. P., and Terjesen, S. G., *Chem. Eng. Sci.* **6**, 265 (1956).
L10. Lochiel, A. C., *Can. J. Chem. Eng.* **43**, 40 (1965).
M1. Marangozis, J., and Johnson, A. I., *Can. J. Chem. Eng.* **39**, 152 (1961).
M2. Marchello, J. M., and Toor, H., *Ind. Eng. Chem. Fundamentals* **2**, 8 (1963).
M3. Marrucci, G., *Ind. Eng. Chem. Fundamentals* **4**, 224 (1965).
M4. Marrucci, G., Univ. of Napoli, Italy, Personal communication, 1966.
M5. Maxon, W. D., and Johnson, M. J., *Ind. Eng. Chem.* **45**, 2554 (1953).
M6. Metzner, A. B., and Taylor, J. S., *A.I.Ch. E. J.* **6**, 109 (1960).
M7. Michel, B. J., and Miller, S. A., *A.I.Ch.E. J.* **8**, 262 (1962).
M8. Miyauchi, T., and Vermeulen, T., *Ind. Eng. Chem. Fundamentals* **2**, 113 (1963).
M9. Miyauchi, T., and Vermeulen, T., *Ind. Eng. Chem. Fundamentals* **2**, 304 (1963).
M10. Moore, D. W., *J. Fluid Mech.* **6**, 113 (1959); **16**, 161 (1963); **27**, 737 (1965).
M11. Morris, R. M., Chem. Eng. Dept., Univ. of Natal, Durban, South Africa, Personal communication, 1966.
N1. Newman, J., *Chem. Eng. Sci.* **22**, 83 (1967).
N2. Nieuwenhuizen, J. K., *Chem. Eng. Sci.* **19**, 367 (1964).
N3. Nysing, R. A. T. O., Hendrikz, R. H., and Kramers, H., *Chem. Eng. Sci.* **10**, 88 (1958).
O1. Olander, D. R., *Chem. Eng. Sci.* **18**, 123 (1963); **19**, 275 (1964).
O2. Oldshue, J. Y., *Proc. Bioeng. Symp., Rose Polytechnic Inst., Terre Haute, Indiana, May, 1953*.
O3. Oldshue, J. Y., *Ind. Eng. Chem.* **48**, 2194 (1956).
O4. Ottmers, D. M., and Rase, H. F., *Ind. Eng. Chem. Fundamentals* **3**, 106 (1964).
O5. Oyama, Y., and Endoh, K., *Chem. Eng. (Tokyo)* **19**, 2 (1955).
P1. Padmanabhan, L., and Gal-Or, B., *A.I.Ch.E.J.* (1968). In press.
P2. Padmanabhan, L., and Gal-Or, B., *Chem. Eng. Sci.* In press.
P3. Parker, H. H., *Chem. Eng.* **71**, 165 (1964).

P4. Pavlushenko, I. S., Braginskii, L. N., and Brylou, V. N., *J. Appl. Chem. U.S.S.R.* (*English transl.*) **34**, 773 (1961).
P5. Perlmutter, D. D., *Chem. Eng. Sci.* **16**, 287 (1961).
P6. Perry, J. H., "Chemical Engineers' Handbook," 4th ed., pp. 18–78, McGraw-Hill, New York, 1963.
P7. Pfeffer, R., *Ind. Eng. Chem. Fundamentals* **3**, 380 (1964).
P8. Pfeffer, R., and Happel, J., *A.I.Ch.E. J.* **10**, 605 (1964).
P9. Phillips, D. H., and Johnson, M. J., *Ind. Eng. Chem.* **51**, 83 (1959).
P10. Polejes, J. D., Ph.D. thesis, Univ. of Wisconsin, 1959.
P11. Preen, B. V., Ph.D. thesis, Univ. of Durham, South Africa, 1961.
R1. Ratcliff, G. A., and Reid, K. J., *Trans. Inst. Chem. Engrs.* (*London*) **39**, 423 (1961).
R2. Reinders, W., and Dingemans, P., *Rec. Trav. Chim.* **53**, 231 (1934).
R3. Richards, J. W., *Brit. Chem. Eng.* **8**, 153 (1963).
R4. Richards, G. M., Ratcliff, G. A., and Danckwerts, P. V., *Chem. Eng. Sci.* **19**, 325 (1964).
R5. Roberts, D., and Danckwerts, P. V., *Chem. Eng. Sci.* **17**, 961 (1962).
R6. Robinson, R. G., *Ind. Eng. Chem. Fundamentals* **2**, 191 (1964).
R7. Rodger, W. A., Trice, V. G., and Rushton, J. H., *Chem. Eng. Progr.* **52**, 515 (1956).
R8. Rowe, P. N., and Partridge, B. A., *Chem. Eng. Sci.* **19**, 81 (1964).
R9. Ruckenstein, E., *Chem. Eng. Sci.* **15**, 131 (1964).
R10. Ruckenstein, E., and Berbente, C., *Chem. Eng. Sci.* **19**, 329 (1964).
R11. Rushton, J. H., *Chem. Eng. Progr.* **50**, 587 (1954).
R12. Rushton, J. H., Gallagher, J. B., and Oldshue, J. Y., *Chem. Eng. Progr.* **52**, 319 (1956).
R13. Rybczynski, W., *Bull. Intern. Acad. Sci., Cracovie* (A) **40** (1911).
S1. Sachs, J. P., and Rushton, J. H., *Chem. Eng. Progr.* **50**, 597 (1954).
S2. Saito, H., Sakai, T., and Sakata, O., *26th Ann. Meeting Soc. Chem. Engrs., Japan, 1961*.
S3. Savic, P., Natl. Res. Lab. Rept. (Canada), MT-22 (1953).
S4. Schechter, R. S., and Farley, R. W., *Brit. Chem. Eng.* **8**, 37 (1963).
S5. Schultz, J. S., and Gaden, E. L., *Ind. Eng. Chem.* **48**, 2209 (1956).
S6. Scriven, L. E., and Pigford, R. L., *A.I.Ch.E. J.* **4**, 439 (1958).
S7. Shinnar, R., and Church, J. M., *Ind. Eng. Chem.* **52**, 253 (1960).
S8. Sideman, S., and Shabtai, H., *Can. J. Chem. Eng.* **42**, 107 (1964).
S9. Snyder, J. R., Hagerty, P. F., and Molstad, M. C., *Ind. Eng. Chem.* **49**, 689 (1957).
S10. Soo, S. L., *Ind. Eng. Chem. Fundamentals* **4**, 426 (1965). See also "Fluid Dynamics of Multiphase Systems" by S. L. Soo, Blaisdell, London, 1967.
S11. Steel, B., and Maxon, W. D., *Ind. Eng. Chem.* **53**, 739 (1961).
S12. Sweeny, R. R., Davis, R. S., and Hendrix, C. D., *Ind. Eng. Chem.* **57**, 72 (1965).
T1. Tadaki, T., and Maeda, S., *Kagaku Ku* **28**, 270 (1964) [English ed. **2**, 195 (1964)].
T2. Tavlarides, L. L., and Gal-Or, B., *Chem. Eng. Dept., University of Pittsburgh* (1968). To be published.
T3. Thorsen, G., and Terjesen, S. G., *Chem. Eng. Sci.* **17**, 137 (1962).
T4. Timson, W. J., and Dunn, C. G., *Ind. Eng. Chem.* **52**, 799 (1960).
T5. Toor, H. L., and Marchello, J. M., *A.I.Ch.E. J.* **4**, 97 (1958).
T6. Trambouze, P., Trambouze, M. T., and Piret, E. L., *A.I.Ch.E. J.* **7**, 138 (1961).
V1. Valentas, K. J., Bilous, O., and Amundson, N. R., *Ind. Eng. Chem. Fundamentals* **5**, 271, 533 (1966).
V2. Van Deemter, J. J., and Van Der Laan, E. T., *Appl. Sci. Res.* **10A**, 102 (1961).
V3. Van De Vusse, J. G., *Chem. Eng. Sci.* **10**, 229 (1959).
V4. Van De Vusse, J. G., *Chem. Eng. Sci.* **16**, 21 (1961).

V5. Vassilatos, G., Trass, O., and Johnson, A. I., *Can. J. Chem. Eng.* **40**, 210 (1962).
V6. Vermeulen, T., *Ind. Eng. Chem.* **45**, 1664 (1953).
V7. Vermeulen, T., Preprints, *25th Anniv. Congr., Soc. Chem. Engrs., Japan, November 1961* p. 300.
V8. Vermeulen, T., Williams, G. M., and Langlois, G. E., *Chem. Eng. Progr.* **51**, 85-F (1955).
W1. Waslo, S., and Gal-Or, B., *Chem. Eng. Sci.* In press.
W2. Westerterp, K. R., D.Sc. thesis, Technische Hogeschool, Delft, The Netherlands, 1962.
W3. Westerterp, K. R., *Chem. Eng. Sci.* **18**, 157 (1963).
W4. Westerterp, K. R., Dierendonck Van, L. L., and Kraa, J. A., *Chem. Eng. Sci.* **18**, 157 (1963).
W5. Wolf, D., and Resnick, W., *Ind. Eng. Chem. Fundamentals* **2**, 287 (1963).
Y1. Yagi, S., and Inoue, H., *Chem. Eng. Sci.* **17**, 411 (1962).
Y2. Yoshida, F., and Akita, K., *A.I.Ch.E. J.* **11**, 9 (1965).
Y3. Yoshida, F., and Miura, Y., *Ind. Eng. Chem. Process Design Develop.* **2**, 263 (1963).
Y4. Yoshida, F., Ikeda, A., Imakawa, S., and Miura, Y., *Ind. Eng. Chem.* **52**, 435 (1960).

AUTHOR INDEX

Numbers in parentheses are reference numbers and indicate that an author's work is referred to although his name is not cited in the text. Numbers in *italics* show the page on which the complete reference is listed.

A

Ackerman, F. J., 181 (R4), *206*
Ackerman, G. H., 115(T3), *136*
Ackermann, P., 77(K7, K8, K9, K12), *135*
Adams, G. K., 31, 37, *66*
Adams, J. M., 231, *289*
Adler, S. B., 180, *203*
Adlington, D., 125, 126, 127, 129, *133*
Adnams, D. J., 216(A1), *287*
Adorni, N., 272, *287*
Aiba, S., 316(A1), *390*
Akiba, R., 57, *66*
Akita, K., 113, 115, *137*, 314(Y2), *395*
Aladyev, I. T., 218, 226(A4), 233(A4), 234(A4), *287*
Alberda, G., 96, *135*
Alessandrini, A., 217, *287*
Allan, D. S., 21(B4), 22(B4), 23(B4), *67*
Allison, J. B., 78(R1), *136*
Altman, D., 11, 12(A4), 20, *66*
Amberkar, S., 316(K5), *393*
Amundson, N. R., 387(VI), *394*
Andersen, J. W., 314(F1), 319(F1), *391*
Anderson, F. A., 57, 62, *66*
Anderson, R., 7(B14), 13(B8), 16, 17(A10), 18(A6, A7, A8, A9), 19(B8), 20(A8), 23(B8), 24(B8, B14), 25(B8), 26(B14), 28(J3), 46, 47(A6), 66, 67, 68, *69*
Anderson, T. T., 311, *390*
Anderson, W. H., 38(I1), *68*
Angelin, L., 154(S4), *206*
Angelus, T. A., 56, 66, *69*

Apuzzo, G., 317(A4), 318, *390*
Asai, S., 387, *392*
Astarita, G., 317(A4), 318, 319, *390*
Auble, C. M., 63(A12), *67*
Aungst, W. P., 56(W1), *69*
Azbel, D. S., 115, *133*
Azizyan, A. G., 115, *133*

B

Babcock, B. D., 104, *133*
Backstrom, H. S., *390*
Baddour, R. F., 387(B17), *391*
Badger, W., 209(S16), *293*
Baer, A. D., 11, 12, 13, 53(R5), 67, *69*
Bagley, R., 224, *287*
Bailey, A. E., 75(B2), *133*
Baird, M. H. I., 112, *133*, *390*
Baker, E., 209(S16), *293*
Bakker, C. A. P., *390*
Balder, J. R., 195(B1), 196, 197, *203*
Bankoff, S. G., 126, *133*, 208(B7), *287*, 311, *390*
Barker, J. J., 122, *133*
Barnett, P. G., 209, 210, 231, 232, 239, 240, 241, 242, 244(B6), 267, 269, 270, 271, 273, 274, 275, 276, 277, 279, 280, 282, 283, 284, *287*
Barrère, I. M., 43, 45, *67*
Barte, D. R., 208(B7), *287*
Bartholomew, W. H., 300(K1), 321(K1), *393*
Bartley, C. E., 31(H12), 32(H12), 33(H12), 35(H12), *68*

397

Basset, J., 169, *203*
Bastress, E. K., 21, 22, 23, *67*
Batch, J. M., 213(M5), 218(M5), 260(H3), 263(H2), 264(B8, W1), *281, 289, 291, 293*
Bayens, C., 309(B4), 310(B4), *390*
Beachell, H. C., 40, *67*
Beattie, J. A., 144, *203*
Beaverstock, M. C., 387(B15), *391*
Becker, K. M., 209, 218, 227, 231, 251, 260, 272, *288*
Beckstead, M., 57, 62, *67*
Beek, W. J., 387, *390*
Beenakker, J. J. M., 144(R6), *206*
Benedict, M., 171, *203*
Bennett, A. W., 221, 222, *288*
Benson, F. R., 76(B6), *133*
Benson, P. R., 147, 148(P8), 149(P8), *205*
Benson, H. E., 77(B7, C12), *133*
Bentwich, M., *390*
Berbente, C., *394*
Berenson, P. J., 210, *288*
Bertoletti, S., 217(A5), 238, 239, 251, 272(A2), 279, *287, 288*
Beyer, R. B., 13, 19, 23, 24, 25(B8), *67*
Biancone, F., 279, *288*
Bibby, R., 117(C7), *133*
Biderman, R., 264(M11), *291*
Bienstock, D., 77(B7), *133*
Bilous, O., 387(VI), *394*
Bircumshaw, L. L., 18(B9), 36, *67*
Bird, J. F., 53(M5), *68*
Bird, R. B., 145(H6), 153(H6), *204*, 335(B9), *390*
Birdseye, D. E., 208(B21), *288*
Birthler, R., 75(K1), *134*
Bischoff, K. B., 314(L2), 388(B10), *390, 393*
Black, C., 147, *203*
Blakebrough, N., 316(B11), 318(H3), 323, *390 392*
Bliss, H., 123, *135*
Bode, M., 209(B13), 272(B13), *288*
Bodman, S., 387(B16), *391*
Borchers, E., 115(K13), 118(K14, K15), *135, 136*
Boussinesq, J., 319, *391*
Bowman, C. W., 112, *133, 134, 391*
Bowring, R. W., 260, *288*
Boyden, J. E., 224(S6), *292*
Bradley, H. H., Jr., 10(P8), 11(P8), 12(P9), 15(P8), 17(P8), 20(P8), *69*

Braginskii, L. N., 304(P4), 325(P4), *394*
Branlick, W. J., 113, 115, *133*
Brian, P. L. T., 387, *391*
Brodkey, R. S., 388(B19), *391*
Brown, R. S., 7(B14), 16(A10), 17(A10), 18(A6, A7, A8, A9), 20(A8), 24, 26(B14), 28(J3), 29, 46, 47(A6), 55, 64, *66, 67, 68*
Brown, S. A., 63(A12), *67*
Brown, W. B., 152(B6), *203*
Brož, Z., 102, *134*
Brusie, J. P., 76(B10), *133*
Brylou, V. N., 304(P4), 325(P4), *394*
Bundy, R. D., 208(G1), 232, *288, 289*
Byrd, C. R., 75(C8), *133*

C

Calderbank, P. H., 83, 112, 117, 119, 121(C1, C2), 122, *133, 135*, 305(C1, C2), 306, 308, 313(C2), 326, 348(C2), 349, 358, 362, 371(C4), 373, 374, *391*
Callow, D. S., *391*
Carl, R., 236(M1), *291*
Carlson, L. W., 21, 22, 23, 59, *67*
Carpenter, C., 52(C2), *67*
Carr, J. B., 209, *288*
Carslaw, H. S., *391*
Carver, J. R., 275(S18), *293*
Casterline, J. E., 230, *288*
Cathro, K. J., 209, *288*
Chao, K. C., 165(C1, P13), 173(P13), 176, *204, 205*
Chapperlear, P. S., 149(L2), 152(L2), 172(L2), *205*
Chelmer, H., 260(T4), *293*
Chervenak, M. C., 75(C8), *133*
Chiang, S. H., *391*
Chueh, P. L., 147, 151(C3b, P8a), 163, 165(C3), 176, 178(C3, C4), 180, *204, 205*
Church, J. M., 311, *394*
Ciepluch, C. C., 58(P5), 59(P5), *67, 69*
Clerici, G. C., 250, *288*
Clusius, K., 142, *204*
Coates, R. L., 53(R5), *69*
Colahan, W. J., 208(B7), *287*
Collier, J. G., 223, 232(M15), *288, 292*
Companile, A., 279(B20), *288*
Connolly, J. F., 146(C8), 170, 190(C8), *204*
Cook, W. H., 279, *289*
Coppock, P. D., 111, *133*

AUTHOR INDEX

Corcoran, W. J., 13(B8), 19(B8), 23(B8), 24(B8), 25(B8), *67*
Core, T. C., 208(C7), *289*
Corr, H., 76(C10), *133*
Cose, D. A., 28(J3), *68*
Costello, C. P., 231, *289*
Coulson, J. M., 209(C9), 211, *289*
Cova, D. R., 109, *133*
Cowan, P. L., 13(M4), 14(M4), 18(M4), *68*
Crandall, G. S., 123(D3), *133*
Crawford, B. L., Jr., 32, *69*
Crist, R. H., *392*
Clark, M. W., 313, 319(C8), 323, 324, *391*
Cooper, C. M., 300, 303, 304(C9), 313, 315, 359, 360, *391*
Crank, J., *391*
Crowe, C. T., 49, *67*
Crowell, J. H., 77(C12), *133*
Crump, J. E., 55, 57(P10), *67*, *69*
Curl, R. F., 152(C9), *204*
Curl, R. F., Jr., 147, 148, 152(P3), *205*
Currin, H. B., 209, 210(T6), *293*
Curtiss, C. F., 145(H6), 153(H6), *204*

D

Danckwerts, P. V., 89, *133*, 301(R4, R5), 302(R4, R5), 343(D4), *391*, *394*
Datta, R. L., 111, *133*
Davenport, A. J., 190(D1, D2), *204*
Davidson, J. F., 112, 127, *133*, *136*, 387, *390*, *392*
Davis, H. S., 123, *133*
Davies, J. T., 328, 332(D7), 333, 362(D9, D11), 388(D7), *391*
Davies, R., *391*
Davis, R. S., 388(S12), *394*
De Bortoli, R. A., 209, 233, 240, 258, 259, *289*
Dehority, G. L., 10(P8), 11(P8), 15(P8), 17(P8), 20(P8), *69*
Deindoerfer, F. H., 298(D10), *391*
de Kraa, J. A., 121(W5), 122(W5), *137*
De Maria, F., 92, *133*
De Soto, S., 24, *67*
De Swaan Arons, J., 190, 192, *204*
Dew, J. E., 236(M1), *291*
De Waal, K. J. A., 92, 100, *133*
Diepen, G. A. M., 151, 190, 192, *204*
Dimmock, T. H., 208(D3), *289*
Dingemans, P., *394*

Dobbins, W. E., 339, *391*
Dodds, W. S., 91, *133*
Dodé, C. R., 169, *203*
Dodge, B. F., 154(D6), 167(D7), 175(D6), 190, *204*, *205*
Dodonov, L. D., 218(A3), *287*
Doroshchuk, V. E., 226(A4), 228, 223(A4), 234(A4), *287*, *289*
Dressler, R. G., 77(K2), *134*
Dukler, A. E., 319(D12), *391*
Dunlop, R., 49, *67*
Dunn, C. G., 328(T4), *394*
Dunn, W. E., 93, 98, *134*
Durant, W. S., 233, *289*

E

Ebeling, R. W., 16(A10), 17(A10), *66*
Eckenfelder, W. W., Jr., 120, 121, *134*
Eckert, C. A., 151, 161, 163, 175(E1), *204*, *205*
Edmister, W. C., 165(P13), 173(P13), 180, *205*, *206*
Edwards, P. A., 260, 266, *289*
Ehrlich, P., 190(E3), *204*
Eisel, J. L., 57, *67*
Elgin, J. C., 95, *134*, 200, *204*
Ellion, M. E., 210, *289*
Elliott, D. F., 208(S13), 213(S13), 214(S13), 215(S13), 247(S13), 248(S13), 277(S14), 280(S13), *293*
Elsworth, B., 304(E1), *391*
Emmery, R. E., 303, *391*
Endoh, K., 323, *393*
Engel, F. C., 209, *293*
Engelhardt, F., 77(K12), *135*
Ericsson, E. O., 76(E3), *134*
Erikson, O., 209(B13), 272(B13), *288*
Esikov, V. I., 277(S5), 278(S5), *292*
Evans, F., 83(C6), 119(C6), *133*

F

Fair, J. R., 113(B9), 115(B9), *133*, 314(F1), 319, *391*
Faktorovitc,h, L. E., 228(S17), 229(S17), 245(S17) 246(S17), *293*
Farber, E. A., 213, 236, 237, *289*
Farkas, E. J., 85, 117, 120, *134*, *136*
Farley, R., 77(F3), 83(C6), 114, 119(C6), 120, *133*, *134*
Farley, R. W., 328(S4), *394*

Fauske, H., 209(I1), *290*
Feigelman, S., 75(C8), *133*
Fernstrom, G. A., 300(C9), 303(C9), 304(C9), 313(C9), 315(C9), 359(C9), 360(C9), *391*
Field, J. H., 77(B7, C12), *133*
Firstenberg, H., 209(G2), *289*
Fishman, N., 13, 20, *67*
Fitzsimmons, D. E., 260(H3), 263(H2), 264(W1), *289, 293*
Fleming, R., 12(P9), *69*
Foley, D. J., 213(M5), 218(M5), *291*
Foord, A., *392*
Foust, H. C., 122, *134*, 312, 313(F2), 314, *391*
Franck, E. U., 151(F1, F2), 154(F3), *204*
Frank-Kamentskii, D. A., 9, *67*
Frazer, J. H., 9, *67*
Freeman, P. I., 190(D1, R9), *204, 206*
Frid, F. P., 228, *289*
Friedland, W. C., *391*
Friedlander, H. N., 296(F5), *391*
Friedlander, S. K., *391*
Friedly, J. C., 54, *67*
Friedman, A. M., 304(F6), 316(F6), *391*
Friedman, H. A., 24(D1), *67*
Friedman, R., 37, 38(F5), *67*
Friend, L., 180(A1), *203*
Fuller, E. C., *392*
Fullman, C. H., 24, 25(F6), *67*

G

Gaden, E. L., 319, *392, 394*
Gaden, E. L., Jr., 120, *134*
Galimi, G., 279(B20), *288*
Gall, D., 77(H1), 19(H1), *134*
Gallagher, J. B., 304(R12), 316(R12), 321(R12), *394*
Gal-Or, B., 122, *134*, 296(G3, G5, G6, W1), 297(G1, G6), 298(G5, G6, G8), 299(G5, G6, G9), 303, 308(G3), 309(G4), 310, 311(G5), 313(G10), 315, 316, 317(G7, G9), 321, 322, 323, 325, 329, 330, 331, 332(G5, G9), 334, 348, 349, 350(G7, G9), 353, 355(G6), 356, 358(G2), 359(G9), 360, 361, 362, 363, 364(G5, G7), 368, 369, 370(G5, G9), 373, 375 (G3), 376(G3), 377, 379,(P2, T2), 380(G3), 382(G3), 383(P2), 387, *392, 393, 394, 395*
Gambil, W. R., 208(G1), *289*

Gambill, W. R., 232(B23), *288*
Gamson, B. W., 149(L2), 152(L2), 172(L2), *205*
Garner, F. H., 112, *134*, 328, 362, *392*
Garner, R. H., 327, *392*
Garriba, S., 250(C4), *288*
Gaspari, G. P., 217 (A5, B19), 238(B19), 239(B19), 251(B18), 279(B17), *287, 288*
Geckler, R. D., 41, *67*
Gill, W. N., 344, 346, *392*
Gillet, W. A., *391*
Gilliland, E. R., 126, *136*
Ginnell, R., 32, *69*
Glaser, M. B., 98, 99, *134*
Goffi, M., 279(B20), *288*
Goldmann, K., 209(F3), *289*
Gonikberg, M. G., 190(G1), *204*
Gorring, R. L., 105, *134*
Govier, G. W., 114, *134*
Grace, T. M., 209, *289*
Graham, E. B., 190(E3), *204*
Grassie, N., 40, *67*
Grassmann, P., 111, *134*
Graumann, R. E., 316(K5), *393*
Grayson, H. G., 175(G2), *204*
Green, L., Jr., 41(S1), 51, *68, 69*
Green, S. J., 209(D2), 233, 240(D2), 258(D2), 259(D2), *289*
Griffith, P., 210, *289*
Griffith, R. M., 112, *134, 392*
Günther, K., 114, *136*
Guggenheim, E. A., 153(G3), *204*
Guha, D. K., 109(R3), *136*
Gunn, R. D., 145(P9), 147(P9), 151(P9), 181(R4), *2045, 206*
Guyer, A., 111, *134*

H

Haarer, E., 76(C10), *133*
Haas, P. A., 387, *392*
Hadamard, J. S., 318(H2), *392*
Haga, I., 227)B14), 231)B14), *288*
Hagerty, P. F., *394*
Hale, A. B., 328, 362(G12), *392*
Hall, C. C., 77(H1), 119, *134*
Hall, K. P., 43(S7), 48(S7), *69*
Halvorson, H. O., 79(H2), *134*
Hamer, G., 316(B11), 318(H3), 323, *390, 392*

AUTHOR INDEX

Hamielec, A. E., 350(J4), 351, 353(H4, J4), 388(H5, H6), *392*
Hammer, H., 83(K18), 84(K16), 86(K16), 108(K18), 119(K18), 120(K16), *135*
Hammerton, D., 112, *134*, 327, *392*
Hanhart, J., 122(K21), *135*, 314, 315, *392*
Hanratty, T. J., 347, *392*
Hansen, R. H., 39(H1), *68*
Hansson, P. T., 227(B14), 231(B14), *288*
Happel, J., 371(P8), 373(H10), *392*, *394*
Harriott, P., 122, *134*, *392*
Harrison, D., 97, *134*
Harris-Smith, B., 304(E1), *391*
Hart, R. W., 53, *68*
Hasseltine, E. H., 387(B18), *391*
Hauck, J. P., 313(G10), *392*
Heath, G. A., 36, *68*
Hector, A., 77(P2), *136*
Heidmann, M. F., 55(P6), *69*
Hellwig, L. R., 75(C8, H6), *133*, *134*
Hench, J. E., 264(H1), *289*
Hendrikz, R. H., 301(N3), *393*
Hendrix, C. D., 388(S12), *394*
Henne, H. J., 83(K18), 108(K18), 119(K18), *135*
Hennico, A., 198, *204*
Herborg, G., 209(B13), 272(B13), *288*
Herington, E. F. G., 180, 184, *204*
Hermance, C. E., 14, *68*
Hernborg, G., 260(B11), *288*
Hesson, G. M., 260, 263, 264(B8, W1), *287*, *289*, *293*
Heuss, J. M., *392*
Hewitt, G. F., 209, 211(H4), 218, 219, 221(B15), 222(B15), 225, 235, *288*, *289*, *290*
Hicks, B. L., 9, *67*, *68*
Higbie, R., 337(H13), *392*
Highet, J., 86, 120, *136*
Hightower, J. D., 37, 48, *68*
Hikita, H., 387, *392*
Hildebrand, J. H., 153(H4), 175, 201(H4), *204*
Himmelblau, D. M., 388(B10), *390*
Hines, W. S., *290*
Hinze, J. O., 311, 370(H15), *392*
Hirschfelder, J. O., 145(H6), 153(H6), *204*
Hirst, L. L., 77(K2), *134*
Hixon, A. W., 120, *134*
Hoelscher, H. E., 296(G5), 298(G5), 299(G5), 309, 310, 311(G5), 313(G10), 332(G5), 348, 362, 364(G5), 368, 369, 370(G5), *392*
Hoffman, H. W., 208(H8), *290*
Hoffman, T. W., 351(H5), 388(H5), *392*
Hofmann, H., 97, 176, *134*
Hoog, H., 75(H9), *134*
Hoogendoorn, C. J., 99, 105, 107, *134*
Hornberger, P., 76(C10), *133*
Horne, W. A., 75(H11), *134*
Hort, E. V., 76(B10), *133*
Horton, M. D., 54, 55, 57(E1), *67*, *68*
Horton, T. J., 316(K5), *393*
Hougen, O. A., 104(B1), 121(J4), 123(J4), *133*, *134*, 303(J5), 304(J5), *393*
Houghton, G., 111(H13), 112, 115, *134*, *135*
Houghton, W. T., 350(J4), 353(J4), 388(H5), *392*
Howard, C. L., 232, 272, *290*
Howarth, W. J., *392*
Huang, C. J., 341, 342(H19), 343, 347, 362(H18, H19), *392*
Huggett, C., 31, 32(H12), 33(H12), 35(H12), *68*
Huggins, C. M., 152(P3), *205*
Hughmark, G. A., 115, *134*
Hulburt, H. M., *392*
Hultgren, G. O., 149(P21, 152(P2), 153, 174(P2), *205*
Humphrey, A., 298(D10), *391*
Hurley, J. F., 387(B18), *391*
Hyman, D., 304(H22), 314, 322, 388(H21), *392*

I

Ibiricu, M. M., 10(P8), 11(P8), 15(P8), 17(P8), 20(P8), *69*
Ikeda, A., 304(Y4), 306(Y4), 308(Y4), 313(Y4), 315(Y4), 316(Y4), 322(Y4), *395*
Ilinskaya, 169, *205*
Imakawa, S., 304(Y4), 306(Y4), 308(Y4), 313(Y4), 315(Y4), 316(Y4), 322(Y4), *395*
Inoue, H., 300, 301(Y1), *395*
Irwin, O. R., 38, *68*
Isbin, H. S., 209, *289*, *290*
Ivashkevich, A. A., 210, *290*
Ivey, H. J., 231, *290*

J

Jacket, H. S., 231(J1), *290*
Jackson, R., 114, *134*

Jacobson, M., 181(R4), *206*
Jaeger, J. C., *391*
Jahnberg, S., 227(B14), 231(B14), *288*
Jameson, G. J., 387, *392*
Janssen, E., 209, 260, 264(J3), 267(J3), 272, *290*
Jaroudi, R., 63(J1), 64, *68*
Jeffrey, D. W., 101(L2), *135*
Jens, W. H., 210, 233, *290*
Jensen, G. E., 28, 29, 30(J2), 64, *67*, *68*
Jepson, G., 83(C6), 119(C6), *133*
Jesser, B. W., 95, *134*
Jiji, L. M., 222, 223, *290*
Joffe, J., 152(J1), *204*
Johnson, A. I., 112, 122, *133*, *134*, *135*, 301(V5), 302(V5), 316(V5), 319(V5), 350, 353, 388(H5), *391*, *392*, *393*, *395*
Johnson, D. L., 121(J4), 123, *134*, 303(J5), 304(J5), *393*
Johnson, H. F., 387, *392*
Johnson, M. T., 304(M5), *393*, *394*
Johnson, W. E., 33, 35(J4), 38, *68*
Jones, S. J. R., 122, *133*
Jorrisen, W. P., *393*

K

Kaiser, J. R., 78(Z1), 104(Z1), 108(Z1), *137*
Kakar, A. S., 126(V6), 129(V6), *137*
Kakarala, C. R., 275(S18), *293*
Kakihara, M., 113(Y2), *137*
Kandalic, G. A., 146(C8), 190, *204*
Károlyi, J., 75(K1), *134*
Karow, E. O., 300, 321, *393*
Karwat, H., 304(K2), *393*
Kasarnovsky, J. S., 167, *205*
Kastens, M. L., 77(K2), *134*
Kato, Y., 109, 110, 114, 123, *134*
Katz, D. L., 105, *134*
Katz, S., *392*
Kaufman, R. N., 387(K3), *393*
Kay, W. B., 153, *204*
Kays, W. H., 22(K1), *68*
Kearsey, H. A., 209(H5), 211(H4), 218(H4), 219(H4), 221(B15), 222(B15), 224, 225(H4), 235(H6), *288*, *289*, *290*
Keeler, R. N., 145(P10), *205*
Keeys, R. K. F., 221(B15), 222(B15), *288*
Kennedy, A. M., 301(D6), 302(D6), *391*
Kennel, W. E., 236(M1), *291*

Kervinen, J. A., 209, 264(J3), 267(J3), 272, *290*
Kihara, T., 146(K2, K3), *204*
Kilner, A. A., 328(D9), 332(D9), 362(D9), *391*
King, C. T., *392*
Kintner, R. C., 316(K5), 327(K4), 333, 388(K4), *393*
Kirby, G. J., 209, 222, 245, 246, 247, *290*
Kishinesvky, M. Kh., 387, *393*
Kister, A. T., 145(R3), 180, *205*, *206*
Kölbel, H., 77(K5, K7, K8, K9, K12), 83, 84, 85, 86, 108, 112, 115, 116, 118, 119, 120, *134*, *135*
Kolár, V., 102, *134*
Kolmogoroff, A. N., 370, *393*
Konkov, A. S., 224, *290*
Korchinski, I. J. O., *391*
Kraa, J. A., 300(W4), 301(W4), 302(W4), 308(W4), *395*
Krakoviak, A. I., 208(H8), *290*
Kramers, H., 96, 122, *135*, 301(N3), 314(H8), 315(H8), 387, *390*, *392*, *393*
Kreiter, M. R., 213(M5), 218(M5), *291*
Krichevsky, I. R., 167, 169, *205*
Krönig, W., 74(K22), *135*
Kuniguta, E., 97, *136*
Kuo, C. H., 341, 342(H19), 343, 347, 362(H18, H19), *392*
Kuss, E., 154(K7), *205*
Kwong, J. N. S., 149(R2), *205*

L

Lacey, P. M. C., 209(H5), 211(H4), 218(H4, H5), 219(H4, H5), 221(B15), 222(B15), 225(H4), 235(H6), *288*, *289*, *290*
Lake, A., 51, *68*
Lamb, H., 318, *393*
Lambright, A. J., 314(F1), 319(F1), *391*
Landau, J. V., 190(L1), *205*
Landers, L. C., 58, *68*
Lane, A. D., 223(C5), *288*
Lane, M., 97(H5), *134*
Lane, T. T., 362(G13), *392*
Langemann, H., 85(K17), 112, 115(K13), 116(K17), *135*
Langlois, G. E., 305(V8), 308(V8), 327(V8), *395*
Lanzo, C. D., 227(L10), *291*
Lapidus, L., 96, 97, *135*, *136*

Larkins, R. P., 101, *135*
Lau, S., 181(R4), *206*
Lee, 125, 127, *135*
Lee, D. H., 215, 216, 217, 231, 232, 234, 237, 238, 251(L2), 252(L2), 256, 272, 275, 277, 279, 280, *290*, *291*
Lee, D. M., 230, *288*
Le Goff, P., 101, *136*
Leland, T. W., 149(L2), 152(L2), 153(R5), 172, *205*, *206*
Leonard, J. H., 112, *135*
Lenoir, J. M., 50, 51(L2), *68*
Leppert, G., 210, *290*
Lerner, B. J., 113(B9), 115(B9), *133*
Lesage, J., 272(A2), *287*
Le Tourneau, B. W., 209(D2), 233(D2), 240(D2), 258(D2), 259(D2), *289*
Levenspiel, O., 93, 101, *136*, 314, (L2), *393*
Levich, V. G., 311, 317, 319(L3), 329, 330, 332, 362(L3, L4), 370, *393*
Levy, S., 209, 224(S6), 268(L8), *291*, *292*
Lewis, G. N., 144, 152(L4), *205*
Li, P.-S., 112, *135*
Lichtenstein, I., 98, *134*
Lightfoot, E. N., 304(F6), 316(F6), 335(B9), 336, 338, 347, 348, 349, 350, 387, *390*, *391*, *393*
Lindblad, N. R., 329, *393*
Lindland, K. P., 332, *393*
Lindroos, A. E., 190, *205*
Lippman, D., 152(P3), *205*
Lips, J., 99, 105, 107, *134*
Litt, M., 99, *134*
Littman, H., 123, *135*
Lo Bianco, L., 209(F3), *289*
Lochiel, A. C., 112, *133*, *135*, 333, 362(L10), *393*
Lombardi, C., 209(G2), 217(A5, B19), 238(B19), 239(B19), 251(B18), 272(A2), 279(B17), *287*, *288*, *289*
Long, R. A. K., 209, *292*
Longwell, P. A., 52(C2), *67*
Lottes, P. A., 233, *290*
Lowdermilk, W. H., 227, 231, *291*
Lurie, H., 208(R3), *292*
Lyckman, E. W., 161, 163, *205*
Lyons, M. F., 224(S6), *292*

M

Maas, R., 118(K19), *135*

McAdams, W. H., 236, *291*
McAfee, J., 75(H11), *134*
McAlevy, R. F., 13, 14, 18, *68*
Macbeth, R. V., 209, 244(M3, M4), 246(M2), 248(M3), 249, 251(T1), 252, 259(M3), 260, 264(M4), 266, 267, (M4, M10), *291*, *293*
McCabe, E. A., 260(T4), *293*
McCarter, R. J., 91(D6), *133*
McCarthy, J. L., 76(E3), *134*
McClure, F. T., 53, *68*
McDonald, A. J., 63(J1), 64(J1), *68*
Macdonald, I. P. L., 232(M15), *292*
McEwen, L. H., 213, 218, *291*
McGie, M. R., 55, *68*
McGlashan, M. L., 147, *205*
Mack, D. E., 122(F4), *134*, 312(F2), 313(F2), 314(F2), *391*
McKinney, A. W., 268(L8), *291*
MacLaren, R. O., 13(B8), 19(B8), 23(B8), 24(B8), 25(B8), *67*
McLean, A. M., 111(H12, H13), 115(H13), *134*
McNelly, M. J., 209(C9), 211, *289*
Madorsky, S. L., 39(M1), *68*
Maeda, S., 112, 117, *136*, 319(T1), *394*
Maennig, H.-G., 83, 119, *135*
Maenning, H.-G., 83(K18), 108(K18), *135*
Majer, R. J., 36, *568*
Manchanda, K. D., 129, *135*
Mann, C. A., *136*
Marangozis, J., 122, *135*, *393*
Marchello, J. M., 339, 340, 362(T5), 387, *393*, *394*
Marklund, T., 51, *68*
Marrucci, G., 296(M3), 317(M3), 318(M4), 350, *390*, *393*
Martins, J., 118(K14), *135*
Masnovi, R., 233, *289*
Massimilla, L., 76(V7), 124, 130, *135*, *137*
Majuri, N., 124(M5), *135*
Mathisen, R. P., 227(B14), 231(B14), *288*
Matiatos, D. C., 387(B17), *391*
Matzner, B., 236, 255, 260(M10), 261, 264(M6, M9, M11), 266, *291*
Maxon, W. D., 304(M5), 323, 324(S11), *393*, *394*
Mayers, G. R. A., 362(D11), *391*
Meiklejohn, G. T., 111, *133*
Meisl, U., 84(K16), 86(K16), 120(K16), *135*

AUTHOR INDEX

Mejdell, G. T., 104(B1), *133*
Metzner, A. B., 316(M6), *393*
Michel, B. J., 323, *393*
Mickley, H. S., *68*
Milioti, S., 249, *291*
Miller, C. L., 24, *68*
Miller, P., *205*
Miller, S. A. 300(C9), 303(C9), 304(C9), 313(C9), 315(C9), 323, 259(C9), 360(C9), *391, 393*
Mills, N. M., 31(H12), 32(H12), 33(H12), 35(H12), *68*
Minden, C. S., 236(M1), *291*
Miropolski, Z. L., 226(A4), 228(S17), 229(S17), 233(A4), 234(A4), 245(S17), 246(S17), 287, *291, 293*
Mirshak, S., 233, *289*
Miura, Y., 301(Y3), 302, 304(Y4), 306(Y4), 307, 308(Y4), 313(Y4), 315(Y5), 316(Y4), 322(Y4), *395*
Miyauchi, T., *135*, 387, *393*
Modnikova, V. V., 224, *290*
Moeck, E. O., 232, 260, *291, 292*
Molstad, M. C., 110, *136, 394*
Moore, D. W., 317(M10), *393*
Moo-Young, M. B., 371(C7), 121, *133*, 306, 308(C4), 326, 348, 349, 358, 362, 271(C4), 373, 374, *391*
Morash, R. T., 63(M8), *68*
Morey, G. N., 151(M4), *205*
Morozov, V. G., 208(S10), 213(S10), 214(S10), *292*
Morris, R. M., 120, *135*, 318(M11), *393*
Most, W. J., 45, *68*
Mostinski, I. L., 228(M13, S17), 229(S17), 245(S17), 246(S17), *291, 293*
Moulton, R. W., 112(L5), *135*
Müller, K., 118(K15, K19), *135*
Muirbrook, N. K., 178(C4), 180, 181, *204, 205*
Mullins, J. C., 142, *205*
Mullis, B. G., 21, *68*
Murli, P. S., 126(V6), 129(V6), *137*
Murphy, P., 51(Z1), *69*
Muzzy, R. J., 55(B13), *67*
Myers, A. L., 146(P11), 147(P11), *205*
Myers, C. O., 149(M7), *205*
Myers, J. E., 102, 103, *137*
Myrat, C. D., 190(M8), *205*

N

Nachbar, W., 33, 35(J4), 38, 41, 48, *68, 69*
Napier, D. H., 111(D2), *133*
Nemphos, S. P., 40, *67*
Newitt, D. M., 111(D2), *133*
Newman, B. H., 18(B9), 36, *67*
Newman, J., 329, *393*
Newton, R. H., 167(D7), *204*
Nichols, P., 11, 12(A4), 20, *66*
Nicklin, D. J., 115, *135*
Nielsen, F. B., 24, 25(F6), 63(N2), 64, *67, 69*
Nieuwenhuizen, J. K., 316(N2), *393*
Noel, M. B., 208(N2), *292*
Norman, W. S., 91, *135*
Notely, T. N., 40, *69*
Nugent, R. G., 37(F5), 38(F5), *67*
Nukiyama, S., 212, *292*
Nunge, R. J., 346, *392*
Nysing, R. A. T. O., 301(N3), *393*

O

Obertelli, J. D., 215, 216, 217, 232, 237, 238, 260, 266, 277, 279, 280, *289, 291*
O'Connell, J. P., 147, 157(O1), 170(O1), 197, *205*
Odiozo, R. C., 78(Z1), 104(Z1), 108(Z1), *137*
Østergaard, K., 124, 126, 127, 128, *135, 136*
Okada, K., 95, *136*
Olander, D. R., *393*
Oldshue, J. Y., 121, *135*, 304(O3, R12), 316(R12), 321(R12), *393, 394*
Orentlicher, M., 170(O3), *205*
Otake, T., 95, 97, *136*
Othmer, D. F., 79(Z2), *137*
Ottmers, D. M., 319(O4), *393*
Oxborn, J. R., 51(Z1), *69*
Oyama, Y., 323, *393*

P

Pace, R. D., 223(C5), *288*
Padmanabhan, L., 377, 379(P1, P2), 383 (P1, P2), 387, *393*
Parker, D. B. V., 40, *69*
Parker, H. H., *393*
Parr, R. G., 32, *69*
Partridge, B. A., 311, *394*
Pasveer, A., 111, *136*

AUTHOR INDEX

Pavlushenko, I. S., 304(P4), 325, *394*
Pearson, D. A., 76(E3), *134*
Penner, S. S., 41(S1), 45, 48, *69*
Perlmutter, D. D., *394*
Perroud, P., 208(P1), *292*
Perry, J. H., 320(P6), *394*
Persson, P., 209, *288*
Peterlongo, G., 217(B19), 238(B19), 239(B19), 251(B18), 272(A2), *287, 288*
Petersen, D. E., 152(P3), *205*
Peterson, M. H., *391*
Pfaff, S., 53(S6), *69*
Pfeffer, R., 371(P7, P8), *394*
Pfister, X., 111, *134*
Phillips, D. H., *394*
Phillips, T. R., 36, *67*
Pichler, H., 77(P2), *136*
Piconnell, P., 236(M1), *291*
Pigford, R. L., 91(S6), *136*, 180(A1), *203*, 303, 328, *391, 394*
Piret, E. L., *136, 394*
Pirt, S. J., *391*
Pitzer, K. S., 147, 148, 149(P2), 152(C9, P2, P3), 153, *204, 205*
Platz, J., 85(K17), 116(K17), *135*
Polejes, J. D., 121(J4), 123(J4), *134, 136*, 303(J5), 304(J5, P10), *393, 394*
Poll, A., 83(C6), 119(C6), *133*
Polomik, E. E., 268(L8), *291, 292*
Polyakov, V. K., 277(S5), 278(S5), *292*
Poppe, G., 187, *205*
Potter, D. J. B., 147, *205*
Povinelli, L. A., 55(P4), 58, 59(P5), *69*
Powling, J., 49, *69*
Prausnitz, J. M., 144(P5), 145(P5, P9, P10), 146(P11, S3), 147(P9, P11), 148(P8), 149(P8), 151(C3b, E2, P8a, P9), 156(P7), 157(O1), 159(P6, P7), 161(L6), 163(L7), 165(C3, P13), 170(O1, O3),, 173(P13), 174(P2), 175(E1), 176, 178(C3, C4), 180(C4), 181(M5), 195(B1), 196(B1), 197(B1), *203, 204, 205, 206*
Preiser, S., 209(F3), *289*
Preen, B. V., 305(P11), *394*
Pressburg, B. S., 115, *134*
Price, E. W., 10(P8), 11, 12, 15, 17(P8), 20, 37, 48, 54, 55, 57(E1), *67, 68, 69*
Price, R. H., 123, *136*
Prigogine, I., 172(P15), 193(P15), 195, *205*
Prost, C., 101, *136*

Pulling, D. J., 209(H5), 211(H4), 218(H4, H5), 219(H4, H5), 225(H4), 235(H6), *289, 290*

Q

Quinn, E. P., *292*

R

Rabinowitz, G., 209(F3), *289*
Raff, R. A. V., 78(R1), *136*
Randall, M., 144(L3), 152(L4), *205*
Randles, J., 209, *292*
Ras, M. N., 109(R3), *136*
Rase, H. F., 319(O4), *393*
Ratcliff, G. A., 301(R4), 302(R4), *394*
Ray, D. J., 77(F3), 114, 120, *134*
Rebiere, J., 208(P1), *292*
Redlich, O., 145(R3), 149(R2), 180, 181(R4), *205, 206*
Reid, K. J., *394*
Reid, R. C., 153(R5), *206*
Reinders, W., *394*
Renon, H., 151(E2), *204*
Resnick, W., 122, *134*, 296(F5, G6), 298(G6, G8), 299(G6), 313, 314(W5), 315, 316, 317(G7), 323, 348, 350(G7), 353, 355(G6), 356, 358(G6), 359(G8), 360, 361, 364(G7), 380, 386, *391, 392, 395*
Reuss, J., 144(R6), *206*
Reynolds, J. M., 210, *292*
Rice, D. W., 55, 57(P10), *68, 69*
Rice, O. K., 32, 40, *69*
Richards, G. M., 301(R4), 302(R4), *394*
Richards, J. W., 320(R3), 321, *394*
Richardson, D. L., 21(B4), 22(B4), 23(B4), *67*
Richardson, M. J., 145(R10), 147(R10), 151(R10), *206*
Ritchie, P. D., 111(H12, H13), 115(H13), *134*
Roarty, J. D., 231(J1), *290*
Roberts, D., 301(D6, R5), 302(D6, R5), *391, 394*
Robillard, G., 50, 51(L2), *68*
Robinson, J. M., 208(R3), *292*
Robinson, R. G., *394*
Rodger, W. A., 327, *394*
Roor, H. L., *391*
Roozeboom, H. W., 187, *206*
Rosen, G., 33, *69*

AUTHOR INDEX

Ross, L. D., 99, 104, *136*
Ross, L. L., 351(H5), 388(H5), *392*
Ross, R. C., *68*
Rosselli, G. M., 180(A1), *203*
Rowe, P. N., 311, *394*
Rowlinson, J. S., 145(R8, R10), 147(R8, R10), 151(R10), 190(B4, D1, D2, M8, R8, R9), *204, 205, 206*
Roy, N. K., 109, *136*
Rubin, L. C., 171(B4), *203*
Ruckenstein, E., 112, *136*, 296(R9), 347(R9), 362(R9), 371, 374, *394*
Rumbel, K. E., 37(F5), 38(F5), *67*
Rushton, G. H., *394*
Rushton, J. H., 122(F4), *134*, 304(R12), 312(F2), 313(F2), 314(F2), 316(R12), 320(R11), 321, 327(R7), *391, 394*
Rutsch, W., 111(V8), *137*
Ryan, N. W., 11, 12, 13(B2), 40, 53, *67, 69*
Rybczynski, W., 318(R13), *394*

S

Sachs, P., 209, *292, 394*
Saito, H., 121(J4), 123(J4), *134*, 301(S2), 303(J5), 304(J5), *393, 394*
Sakai, T., 301(S2), *394*
Sakata, O., 301(S2), *394*
Sala, R., 250(C4), *288*
Salt, D. L., 13(B2), *67*
Salt, K. J., 215, 216(A1), *287, 292*
Salzman, P. K., 38(I1), *68*
Sater, V. E., 93, 101, *136*
Sato, K., 208(C7), *289*
Satterfield, C. N., 78(S2), 122, *136*
Savic, P., 328(S3), *394*
Sawochka, S. G., *292*
Schechter, R. S., 328(S4), *394*
Scheffer, F. E. C., 151, *204*
Schiesser, W. E., 97, *136*
Schiewetz, D. B., 123, *136*
Schmidt, B. K., 78(Z1), 104(Z1), 108(Z1), *137*
Schmidt, K. R., 224, *292*
Schneider, G., 187, 188, 189, 190(S1), *206*
Schoenemann, K., 76(S4), 85(S4), 96, 103, 106, *136*
Schultz, J. S., *394*
Schultz, R., 41, *69*
Schuman, S. C., 75(H6), *134*

Scorah, R. L., 213, *289*
Scott, R. L., 153(H4), 172, 175, 193(S2), 201(H4), *204, 206*
Scriven, L. E., 328, *394*
Scurlock, A. C., 37(F5), 38(F5), *67*
Seader, G. D., 165(C1), 173, 176, *204*
Seader, J. D., 21, 22, 23, 59, *67*
Sehgal, R., 57, 62(S2), *69*
Shabtai, H., 388(S8), *394*
Shair, F. H., *205*
Shannon, L. J., 18(A7–A9), 20(A8), 46(A9), *66, 69*
Sherwood, A. E., 146(S3), *206*
Sherwood, T. K., 78(S2), 85, 91(S6), 120, 122, *136*
Shingu, H., 77(S7), *136*
Shinnar, R., 14(H4, H5), *68*, 311, *394*
Shitsman, M. E., 228(M13, S17), 229(S17), 245(S17), 246(S17), *291, 293*
Shulman, H. L., 95, 110, *136*
Sideman, S., 388(S8), *390, 394*
Siegel, B. L., 227(L10), *291*
Silvestri, M., 209, 217(B19), 229(S4), 230, 238(B19), 239(B19), 251(B18), *288, 292*
Siemes, W., 85, 114, 115, 117, 118(K19), *135, 136*
Signorini, P., 124(M5), *135*
Singh, S., 209(I1), *290*
Slesser, C. G. M., 86, 120, *136*
Slyngstad, C. E., 75(H6), *134*
Smirnov, N. I., 115, *133*
Smith, J. M., 46, *69*
Smith, S. L., 77(H1), 119(H1), *134*
Smolin, V. N., 277, 278, *292*
Snyder, J. R., *394*
Soldaini, G., 217(A5, B19), 238(B19), 239(B19), 272(A2), *287, 288*
Solimando, A., 124(M6), *135*
Sollami, B. J., 91(D6), *133*
Soo, S. L., 311, 388(S10), *394*
Sorgato, I., 154(S4), *206*
Sorlie, T., 224(S6), *292*
Sortland, L. D., *206*
Spalding, D. B., 33, *69*
Spitzner, H., 75(K1), *134*
Squillace, E., 124(M6), *135*
Squyers, A. C., *68*
Staniforth, R., 209, 218, 219, 220, 221, 244, 245, 285, *292*
Stat, M. R., 300(K1), 321(K1), *393*

Stavrovski, A. A., 228(M13, S17), 229(S17), 245(S17), 246(S17), *291*, *293*
Steel, B., 323, 324(S11), *394*
Steinlc, M. E., 55(B13), *67*
Stemerding, 105, 106, *136*
Sterman, L. S., 208(S9, S10), 213(S9, S10), 214, *292*
Stevens, G. F., 208, 209(S8), 213, 214, 215, 218(S8), 219, 220, 221(S8), 244, 245, 247, 248, 277, 280, 285, *292*, *293*
Steward, W. E., *68*
Stewart, N. C., 129, *136*
Stewart, P. S. B., 127, *136*
Stewart, W. E., 335(B9), *390*
Stiushin, N. G., 208(S9, S10), 213(S9, 10), 214(S10), *292*
Stock, B. J., 208(S13), 210, *293*
Storch, H. H., 77(B7, C12), *133*
Storey, S. H., 388(H6), *392*
Strand, C. P., 115(T3), *136*
Strand, L., 57, 62(S2), *69*
Strand, L. D., 57(A5), 62(A5), *66*
Stranski, J. N., 111(V8), *137*
Strehlow, R. A., 57(A5), 62(A5), *66*
Strittmater, R., 53, 56(W1), *69*
Stroebe, G., 209, *293*
Stutzman, L. F., 91(D6), *133*
Styrikovitch, M. A., 226(A4), 228, 229, 233(A4), 234(A4), 245, 246, *287*, *293*
Subbotin, B. I., 231, *293*
Summerfield, M., 13, 14(H4, H5, M3, M4), 18(M3, M4), 43, 45(M9), 48, *68*, *69*
Sutherland, G. S., 43(S7), 48(S7), *69*
Swan, C. L., 268(L8), *291*
Sweeny, R. R., 388(S12), *394*
Swenson, H. S., 275(S18), *293*
Sylvester, J. C., *391*
Szwarcbaum, G., *390*

T

Taback, H. J., 43(S7), 48(S7), *69*
Tacconi, F. A., 251(B18), *288*
Tadaki, T., 112, 117, *136*, 319(T1), *394*
Tait, R. W. F., 209, *288*
Tanno, M., 57, *66*
Tatcliff, G. A., 328(D9), 332(D9), 362(D9), *391*
Tavlarides, L. L., 379(T2), 387, *394*
Tayeban, M., *392*
Taylor, G. I., *391*

Taylor, J. S., 316(M6), *393*
Temkin, M., 191, *206*
Terjesen, S. G., 328, 332, *393*, *394*
Theisen, P. I., 127, 128, *136*
Thompson, B., 209, 244, 251(T1), 252, *293*
Thompson, E., 125, 126, 127, 129, *133*
Thompson, G. T., 16(A10), 17(A10), *66*
Thompson, R. E., 180, *206*
Thompson, G., 123(D3), *133*
Thorp, A. G., 210(T6), *293*
Thorsen, G., 328, *394*
Timmermans, J., 187, 188, *206*
Timson, W. J., 328(T4), *394*
Tippets, F. E., 209, *293*
Tödheide, K., 154(F3), *204*
Tong, L. S., 209, 210, 251, 260, *293*
Toor, H. L., 339, 340, 362(T5), *393*, *394*
Towell, G. D., 115, *136*
Tozzi, A., 250(C4), *288*
Trambouze, M. T., *394*
Trambouze, P., *394*
Trass, O., 301(V5), 302(V5), 316(V5), 319(V5), *391*, *395*
Treybal, R. E., 122, *133*
Trice, V. G., 327(R7), *394*
Troy, M., 209(D2), 233(D2), 240(D2), 258(D2), 259(D2), *289*
Tsuchiya, Y., 113(Y2), *137*
Turner, R., 127, *136*
Turnquist, C. E., 145(R3), *206*

U

Udalov, V. S., 218(A3), *287*
Ullrich, C. F., 95(S9), *136*
Urban, W., 75(U1), *136*

V

Valentas, K. J., 387, *394*
Vance, W. H., 112(L5), *135*
Van Deemter, J. J., 75(V1), *136*, 386, *394*
Van der Bogaerde, J. M., 304(H22), *392*
Van der Laan, E. T., 100, 107, *136*, 386, *394*
Vanderwater, R., 209(I1), *290*
Van de Vusse, J. G., 123, *136*, 387, *394*
Van Dierendonck, L. L., 121(W5), 122(W5), *137*, 300(W4), 301(W4), 302(W4), 305(W4), 306(W4), 308(W4), *395*
Van Driesen, R. P., 75(C8, H6, V4), 129, *133*, *134*, *136*
Van Gasselt, M. L. G., 208(V1), *293*

Van Mameren, A. C., 92, 100, *133*
Van Meel, D. A., 208(V1), *293*
Van Ness, H. C., 179(V1), *206*
Vassilatos, G., 301(V5), 302, 316(V5), 319(V5), *395*
Vermeulen, T., 93(D7), 98(D7), *134, 135,* 198, *204,* 305(V8), 308, 313, 319(C8), 323, 324, 327, 387, *391, 393, 395*
Verschoor, H., 114, *136*
Viswanathan, S., 126, 129, *137*
Volpicelli, G., 76(V7), 130, *137*
Von Bogdandy, L., 111, *137*
Von Elbe, G., 58, *69*

W

Walatka, V. V., 298(G9), 299(G9), 317(G9), 332(G9), 334(G9), 349, 350(G9), 370(G9), 382, *392*
Walker, A. W., 316(B11), *390*
Wall, T., Jr., *136*
Walne, D. J., 97(H5), *134*
Wansborough, R. W., 232(B23), *288*
Ward, D. M., *391*
Waslo, S., 296(W1), 329, 330, 331, 334(W1), 373, *395*
Watermeier, L. A., 53(56), 56, *69*
Waters, E. D., 264, *293*
Webb, G. B., 171(B4), *203*
Webb, M. J., 43(S7), 48(S7), *69*
Weber, H. H., 97, 105, 106, *137*
Weckermann, F. J., 272(A2), *287*
Weekman, V. W., Jr., 102, 103, *137*
Weigand, K., 142, *204*
Weil, L., 208(P1), *292*
Weiland, W. F., 231, *291*
Weinstock, J. J., 200, *204*
Weir, N. A., 40, *67*
Weiss, A., 209(D2), 233(D2), 258(D2), 259(D2), *289*
Weiss, E. B., 95, *134*
Weiss, W., 85, 117, *136*
Wells, N., 95(S9), *136*
Wenograd, J., 45(M9), *68*
West, F. B., 112(L5), *135*
Westerterp, K. R., 121, 122(K21), *135, 137,* 300, 301(W2, W4), 302, 305(W2, W4), 306(W4), 308, 314(H8), 315(H8), 320(W3), *392, 395*
Westphal, W. R., 63(A12), *67*
White, R. R., 92, 101(L2), *133, 135*
Whitehead, J. M., 388(H6), *392*
Wicks, M., 319(D12), *391*
Wiebe, R., 168, *206*
Wikhammer, G. A., 232(M15), *292*
Wilke, C. R., 93(D7), 98(D7), *134, 392*
Williams, F. A., 43(B3), 45(B3), 54, *67, 69*
Williams, G. M., 305(V8), 308(V8), 327(V8), *395*
Williams, V., 304(E1), *391*
Winnick, J., 187, *206*
Winter, E. E., 223(C5), *288*
Wintle, C. A., 215, 216(A1), *287, 292*
Wirrick, T. K., 7(B14), 24(B14), 26(B14), *67*
Wolf, D., 314(W5), 380, 386, *395*
Wolk, R., 75(C8), *133*
Wood, R. W., 209(S8), 213(S13), 214(S13), 218(S8), 221(S8), 247(13), 248(S13), 277(S14), 280(S13), 285, *292, 293*
Word, T. T., 93(D7), 98(D7), *134*
Wormald, C. J., 147, *205*

Y

Yagi, S., 300, 301(Y1), *395*
Yoshida, F., 113, 115, *137,* 301(Y3), 302, 304(Y4), 306, 307, 308, 313, 314(Y2), 315(Y5), 316(Y4), 322(Y4), *395*
Yoshitome, H., 113, *137*
Yount, R. A., 56, *69*

Z

Zabor, R. C., 78(Z1), 104, 108, *137*
Zalai, A., 75(K1), *134*
Zandbergen, P., 193(Z1), 194, *206,*
Zavattarelli, R., 217(A5, B19), 238(B19), 239(B19), 272(A2), 279(B17), *287, 288*
Zenkevitch, B. A., 210, 231, *293*
Zenz, F. A., 79(Z2), *137*
Zerbe, J. E., 231(J1), *290*
Ziegler, W. T., 142, *205*
Zucrow, M. J., 51, *69*
Zudkevitch, D., *204*

SUBJECT INDEX

A

Absorption
  film
    model, 335–336
    penetration model, 339–344
  gas, 346–347
  gaseous reactants, 82
  liquid, agitated, 335–339
  volumetric, 359–360
Acetone-carbon disulfide system, 187
Age distributions, 379–381
Aluminum, 49–50, 57
  depressurization and, 58
Amagat's law, 144–145
Ammonia, 302
Ammonium perchlorate, 36–38
  burned and quenched, 48–49
  intermediate species, 47
  self-deflagration, 37–38
Argon
  fugacity, 143
  pressure, 142

B

Barnett local-conditions hypothesis, 242–245, 267–270, 275–278
Boiling
  forced convection, 213–214
  pool, 212–213, 231
Bubble(s)
  column slurry
    holdup, axial dispersion, 114–117
    mass transfer, 109–114
    operation, 108–120
    reactors, 80
  flow operation, 80, 104–108
    holdup, axial dispersion, 105–108
    mass transfer, 104
  size
    average, 307–308
    distribution, 308–312
    in swarm, 361, 378, 386
    steady-state, 369–371
    unsteady-state, 362–369
  velocities, 316–319
    estimates, 316–317
    experiments, 317–319
Burning
  erosive, 50–51
  rate, 5
    catalysts, 36
    curves, 33
    pressure, 34
Burnout
  annular
    flow regime, 218–221
    test sections, 267–273
  detection, 214–218
  equivalent-diameter hypothesis, 273–274
  flow patterns, 222–223
  forced-convection boiling, 207–293
  general formula, 244, 267
  heat-flux-controlled systems, 212–214
  inlet subcooling, 235–238
  local-conditions concept, 241–246
  low-velocity regime, 246–249
  mass velocity, 253–256
  modeling, 280–287
  nonuniform heating, 274–280
  post, 223–225
  pressure, 257–258
  rectangular channels, 258–259
  rod bundles, 260–267
  in round tubes, 251–258
  system parameters, 222–235
    contractions, 238–241
    gravity and, 231–232
    liquid purity, 232–233
    wall thickness, 233–235
  temperature-controlled systems, 210–212
  tube diameter, 256
  water
    data, 249–273
    flux for, 226
n-Butane, 177
Butynediol synthesis, 76

409

C

Calcium acid sulfite, 76
Carbon dioxide
  absorption, 301
  thermodynamic consistency, 183
CCl_2F_2, 208
Cell model, 382–386
Chamber filling, 29
$CHFCl_2$, 208
Church's equations, 176–177
Combustion
  instability, 52–57
    bulk-coupled, 56–57
    pressure-coupled, 52–55
    velocity-coupled, 55–56
  steady-state, 29–51
    prediction, 30
    pressure plateaus, 34
    propellants, 31–50
  termination, 57–64
    depressurization, 58–62
    fluid-injection, 63–64
    L*, 62–63
Contactor, gas-liquid, 322

D

Danckwert's surface age distribution, 343
Desorption, 82
Diffusion
  with chemical reaction, 350–353
  steady-state convective, 347–350
Dispersions, gas-liquid, 295–395
  approaches, 299–300
  efficiency
    contacting, 300–303
    operating, 296–299

E

Enthalpy, 143
Equivalent-diameter hypothesis, 273–275
Ethylene
  hydrogenation, 83–84
  polymerization, 78

F

Film
  model, 335–336
  penetration
    model, 336–344
    theory, 362
  theory, 340
Fisher-Tropsch
  process, 76–77
  reaction, 119
  synthesis, 114
Fizz reaction, 32
Flame
  propagation, 24–29
  thickness, 44–45
Flow patterns, 316–319
Fluidization, 80–81
  gas-liquid, 123–130
    holdup, 126–128
    mass transfer, 124–126
    residence time, 126–128
  liquid-solid, 126
Fluid-phase equilibria, 139–206
  coefficient, 170–171
    activity, 155–160
  concentrated solutions, 170–184
  equations of state, 171–172
  fugacity, 141–160
  gas in liquid, 166–170
  in liquid mixtures, 172–173
  pressure
    effect, 160–166
    high, 139–206
  thermodynamics of, 139–206
Freon, 244, 246
  boiling, 214
  burnout curves, 248
Fugacity, 196
  argon, 143
  gas mixtures
    equations of state, 145–152
    Lewis rule, 144–145
    pseudocritical hypothesis, 152–154
  liquid mixtures, 154–160
    activity coefficients, 155–158
    constant pressure, 158–160
    standard states, 155–158

SUBJECT INDEX

G

Gas,
   critical velocities, 109
   phase ignition theory, 13–15
   residence time, 314
   solid reactions, 18, 26, 76–78
Gas-gas equilibria, 190–194
Gas-liquid
   dispersions
      design, 319–327
      impellers, 319–322
      models, 333–338
      power, 322–324
      scale-up, 324–327
      surface-active agents, 327–333
      transfer rate, 334
   distribution, 386–388
   at high pressure, 166–170
   particle operations, 71–137
      fixed-bed, 79–80
      isothermal, 83–90
      model, 88
      process(es), 73–76
      steps, 81–82
      suspended-bed, 80–81, 89
      theory, 81–90
      transport phenomena, 90–130
Gibbs-Duhem equation, 158–160, 179–180
Gibbs energy, 176, 184, 191, 195, 197
Glycerol, 302, 349
Granular diffusion flame theory, 43–50
Gravity, 231–232

H

Heat(ing)
   flux, 20
   nonuniform, 274–280
   transfer, 102–103, 108, 118–119, 123, 128–129, 361–371
   igniters, 21–23
Helium-xenon system, 192–194
Henry's law, 84–85, 166, 170, 201
Hydrogenation
   α-methylstyrene, 86, 104, 120
   carboxylic acids, 76
   cyclohexene, 86
   ethylene, 83–84, 119
   liquid petroleum, 73–75
   unsaturated fats, 75

I

Igniter(s)
   design, 21–24
   pyrotechnic, 23–24
   rocket exhaust, 21–23
      aft end, 23, 28
      heat end, 21–23
Ignition
   approaches
      kinetic, 9
      thermodynamic, 9
   chamber-pressure transient, 7–8
   delay time, 20, 24–25
   hot-wire, 11–12
   hypergolic, 24
   pressure transients, 29–30
   theory
      gas-phase, 13–15
      heterogeneous, 18–20
      hypergolic, 15–18
      thermal, 9–13
   time, 10–11
Impeller, 319
   vaned-disk, 322–323
Inlet
   hard vs soft, 226–231
   subcooling, 236–238
   throttling, 228–229

J

JPN, 12

K

Kirby local-conditions hypothesis, 245–246
Krichevsky-Ilinskaya equation, 169–170
Krichevsky-Kasarnovsky equation, 166–169

L

L* termination, 62–63
Lewis fugacity rule, 144–145
Liquid-liquid equilibria
   in binary systems, 184–190
   in ternary systems, 194–203
      phase diagram, 196–202

SUBJECT INDEX

M

Margules parameters, 200–202
  phase behavior and, 202
Mass transfer, 303–307
  bubble
    gas, moving, 344–345
    size distribution, 361–371
  chemical reaction
    irreversible, 341
    reversible, 341–344
  coupled, 374–386
  gas-liquid dispersions, 326
  holdup effect, 371–374
  multicomponent, 374–386
  residence time, 353–361
Methylethylketone-water system, 188
4-Methylpiperdine, 189
2-Methylpyridene, 189
Miscibility, 194–203
  in gases, 190
  pressure, and, 185–190
Molar volumes
  liquid, 174
  partial, 160–166
    dilute solutions, 161–162
    saturated liquids, 162–166
Moment equations, 381–382

N

Navier-Stokes equations, 318, 386–387
Nitrocellulose, 31
Nitroglycerine, 31–32
Normalization
  binary solutions, 156–157
  multicomponent solutions, 157–158
Nusselt number, 118

O

Olefins
  hydration, 79
  oxidation, 77–78
Onsager relation, 377

P

PBAA/AN, 17
PBAN, 64

Peclet number, 352
  trickle operation, 92–93
    gas phase, 94
    liquid phase, 103
Penetration theory, 340
Polymers, 38–40
n-Propane, 178
Propellants, solid
  combustion, 4–50
    flameless, 45
    granular diffusion flame theory, 43–50
    sandwich model, 41–43
  composite, 3, 35–50
    modified double-base, 3
  double-base, 2, 31–35
  energy equation, 24
  erosive burning, 51
  ignition, 6–29
  metallized, 49
  rocket motors, 3–6

R

Raschig rings, 97–98, 102, 106–107
Reaction kinetics, 104, 108, 119–120, 123, 129–130
Reynolds number, 118, 318, 323, 348, 352–353, 370
Redlich-Kwong equation, 181
Rybczynski-Hadamard formula, 318, 332, 348

S

Scaling
  factor, 284–286
  law, 280–284
Scatchard-Hildebrand equation, 173–175
Separation-process, 203
Solubility parameters, 174
Sorbital, 76
State, equation of
  empirical
    binary constants, 151
    high densities, 149–152
  gaseous mixtures, 191
  Redlich-Kwong, 163, 173
    constants, 164
  vapor, 171–172
  virial, 145–149

Stirred-slurry
  operation, 120–123
    holdup, axial dispersion, 122–123
    mass transfer, 120–122
  reactors, 80
Subcooling, 236–238
  inlet, 261
Subreactors, 363
Sulfite-oxidation, 300–301
Summerfield, combustion equation, 44–45
Surface-active agents, 327–333
  experiment, 327–329
  theory, 329–333

T

Thrust, 4
Trickle-flow, 79
  fixed-bed, 90–104
  holdup and axial dispersion
    gas phase, 92–94
    liquid phase, 94–102

mass transfer
  gas-liquid, 91
  liquid-solid, 91
Total-power hypothesis, 279–280
Triphenylmethane sulfur-system, 189

V

van der Laan's equation, 100, 107
  dilated, 176–179
van der Waals constants, 191–192
Vapor-liquid equilibria
  activity coefficients, 173–179
  constants, 179
  equation of state, 171–172
  high pressure, 170–184
  thermodynamic consistency, 179–184

W

Water-glycerol systems, 349–350
Wolf-Resnick distribution function, 380